Sustainable Engineering Forum

AA002847

Held at the 2024 AIChE Annual Meeting

San Diego, California, USA
27-31 October 2024

ISBN: 979-8-3313-1665-5

Printed from e-media with permission by:

Curran Associates, Inc.
57 Morehouse Lane
Red Hook, NY 12571

CURRAN ASSOCIATES INC. proceedings.com

AIChE
120 Wall Street, FL 23
New York, NY 10005-4020

Phone: (800) 242-4363
Fax: (203) 775-5177

www.aiche.org

Additional copies of this publication are available from:

Curran Associates, Inc.
57 Morehouse Lane
Red Hook, NY 12571 USA
Phone: 845-758-0400
Fax: 845-758-2633
Email: curran@proceedings.com
Web: www.proceedings.com

TABLE OF CONTENTS

ADVANCES IN LIFE CYCLE ASSESSMENT

96a A Tale of Two Mines: Insights to the Similarities and Differences of Surface and Underground Critical Mineral Mines Life Cycle Analyses 1
Jenna Trost, Jennifer Dunn

96b Prospective Impact Analysis Combining Integrated Assessment Modeling and Life Cycle Assessment for Alternative Industrial Heat Sources 3
Tapajyoti Ghosh, Patrick Lamers

96c Evaluating the Manufacturing Emissions and End-of-Life Impact of Paperboard Cartons Vs. PET Bottles 5
Nur H. W. Firdaus, Julia Shamshina, Jerrad Legako, Kalavathy Rajan

96d Microbially Produced Proteins to Reduce Global Pressure on the Planetary Boundaries 6
Samir Meramo, Eleonora Passuto, Sumesh Sukumara

96e Enhancing Direct Air Capture and Utilization (DACCU) Pathways through Life Cycle Assessment, Cost Analysis, and Optimization 7
Nick Badger, Mustafa Erguvan, Shahriar Amini

96f Advancing the Economic and Environmental Sustainability of Rare Earth Element Recovery from Phosphogypsum Using a Selective Bio-Based Adsorptive Separation 8
Adam Smerigan, Rui Shi

96g Marine Scrubbers Vs. Low-Sulfur Fuels: A Comprehensive *Well-to-Wake* Life Cycle Assessment Supported by Measurements Aboard an Ocean-Going Vessel 9
Patricia Stathatou, Ievgenii Petrunia, Torsten Barenthin, George Gotsis, Paul Jeffrey, Christopher Fee, Scott Bergeron, Marios Tsezos, Michael Triantafyllou, Neil Gershenfeld

96h Integrated Environmental, Economic, and Social Lifecycle Assessment: A Novel 'Hub and Spoke' Methodological Framework 11
Alex Newman

MACHINE LEARNING AND ANALYTICS FOR SUSTAINABILITY

229g A Machine Learning Approach for Examining Flushing Operations in Packaging Plants 13
Robert Hesketh, Barnabas Gao, Swapana S. Jerpoth, Emmanuel Aboagye, C. Stewart Slater, Kirti Yenkie

229b Deep Learning-Enabled Modeling Framework to Evaluate the Financial Feasibility of Hydrothermal Liquefaction Waste Valorization Systems 14
Ali Ahmad, Yalin Li

229c Sustain-GPT: A Large Language Model for Creating a Structural Database from Unistructural Text Resources to Develop Life Cycle Inventories 17
Avan Kumar, Amrita Sen, Bhavik Bakshi

229d Machine Learning Enabled Life Cycle Assessment for Early-Stage Sustainable Process Design 21
Kirti Yenkie, John Pazik, Jared Longo, Matthew Conway, Milo Barkow, Emmanuel Aboagye, Robert Hesketh

229e Utilizing Multi-Criteria Decision Analysis for Holistic Evaluation and Selection of Wastewater Nitrogen Recovery Technologies 23
Chayse Lavallais, Jennifer Dunn

229f Air Quality Monitoring with Thingspeak 24
Aycan Hacioglu, Ramnarayan Krishnamurthy

CLIMATE CHANGE AND ENGINEERING SUSTAINABILITY- I

498a Climate Change, Extreme Events, and Their Effects on Aboveground Storage Tanks 25
Raymond Smith

498b Climate Disaster or Not? Using Science, Philosophy, and Psychology to Comprehend the Future of Humankind on a Warming Planet 26
Jonathan Stickel

498c Deriving Techno-Socio-Economic Policies for Global Sustainability 27
Siddhi Manjarekar, Urmila Diwekar

498d Chemical Engineering in Cement Decarbonization 28
Joshua W. Allen, James Bresson, Jason L. Bronkema, Janny Cai, Eric Liu, Cody Finke, Hugo Leandri, Shadi Saberi

498f Optimization of CO2 Pipeline Networks for Climate Change Mitigation 29
Aditya Nathan, Christian Lastoskie

498g Enhancing CO_2 Sweep and Storage Efficiency: Development and Evaluation of a Novel Hydrogel for CO_2 EOR and Storage Projects 31
Maryam S. Paroushi, Xu Y. Tian, Baojun Bai

DESIGN FOR A CIRCULAR ECONOMY

46a An Optimization-Based Approach for Developing a Circular Economy of Swine Wastewater 38
Ahteshamul Haq, Madison Kratzer, Prathap Parameswaran, Vikas Khanna

46b From Fossil Fuel to Green Powertrains for Vehicles: A Circular Economy Assessment 39
Kenneth Martinez

46c Guiding the Transition to a Sustainable and Resilient Circular Economy: An Eco-Inspired Quantitative Framework 41
Farshid Nazemi, Brian D. Fath, Bhavik Bakshi

46d Potential to Recover Copper from Electronic Waste and Implications for the Circular Economy 44
Andrea Hicks

46e Enabling Circular Aquaculture with a Biofilm-Based Waste-to-Feed Biotechnology 46
Jin Wang, Q. Peter He, Reza Khorasani

46f Plastic Waste Prevention: Environmental Impact Assessment of Advanced Circular Solutions as a Substitute for Single-Use Virgin Plastic Materials 47
Annamaria Vujanović, Jan Puhar, Martin Leban, Anja Pajntar

46g Material, and Circular Economy Challenges in Advancing Automotive Plastics Recycling through Molecular Recycling Technologies 48
Mahshid Mokhtarnejad, Gary Hawkins, Hendrik Mainka

46h Carbon Fiber-Reinforced Polymer (CFRP) in Automobiles: Achieving Net-Zero Emission and Circular Economy 49
Reena Sharma, Aniket Mali, Bhavik Bakshi

SUSTAINABLE CHEMISTRY AND ENGINEERING- I

354a Dozn™2.1 - A Quantitative Green Chemistry Evaluator for a Sustainable Future 51
Ettigounder Ponnusamy

354b A Data Engineering Framework for Operationalizing the Alternative Assessment of Chemicals and Chemical Processes 52
John D. Chea, Michael Gonzalez

354c Climate Policies Can Substantially Reduce the Environmental Footprint of Green Chemicals 53
Abhinandan Nabera, Antonio J. M. Fernandez, Ioan-Robert Istrate, Javier Pérez-Ramírez, Gonzalo Guillen-Gosalbez

354d Life Cycle Assessment for Polyamide 6: Production, Recycling, Disposal 55
Caroline Ganzer, Ann-Joelle Minor, Lisciele Soto, Liisa Rihko-Struckmann, Kai Sundmacher

354e Heat and Mass Transfer Analysis of Temperature-Gradient Thermolysis Reactors for Plastic Upcycling 58
Ruiyang Ji, Nuwayo E. Munyaneza, Carlos Posada, Guoliang Liu, Rui Qiao

354f Discovering Chemical Reaction Pathways and Processes for Net-Zero Greenhouse Gas Emissions 59
Sunghoon Kim, Bhavik Bakshi

PROCESS DESIGN: INNOVATION FOR SUSTAINABILITY

100a Framework and Results for Roadmapping to a Global Net-Zero Chemical Economy in an Uncertain Climate Future 61
Amrita Sen, Vyom Thakker, George Stephanopoulos, Bhavik Bakshi

100b Nylon 6 Depolymerization: Reactor Design, Catalysts, Kinetics, Process Modeling, and Data Analytics 64
Adam McNeeley, Niket Sharma, Y. Liu

100c Comparative Study of Eco-Friendly Bioleaching Solvents for Sustainable Recovery of Critical Materials and Minerals from Spent Batteries 67
Haoran Wu, Meltem Urgun-Demirtas, Stina Daniel

100d Facile Remanufacturing and Re-Lithiation for Spent Lithium-Ion Battery Electrodes 68
Kasim Adewuyi

100f Chemical Recycling of Polycaprolactone Via Reactive Melt Processing 70
Anubhav Sarmah, Aristotle Zervoudakis, Michaela Pfau-Cloud, Marc A. Hillmyer, Christopher Ellison

100g D o E - Assisted Development of Sustainable Mesoporous Silica Synthesis for Tailored Drug Delivery Applications. 71
Benjamin Roe, Carlos Brambila, Solomon F. Brown, Robert Milton, Siddharth Patwardhan

RESILIENT AND SUSTAINABLE SUPPLY CHAINS AND PRODUCT SYSTEMS

160a Towards Resilient Pharmaceutical Supply Chains: Integration of Manufacturing Uncertainty........ 72
Miriam Sarkis, Nilay Shah, Maria M. Papathanasiou

160b Enhancing Resilience of the Energy-Water Nexus to Climate Change: A Holistic Framework and Dynamic Modeling Approach........ 74
Hesan Elfaki, Dhabia Al-Mohannadi

160c Cost-Effective Strategies Towards Net-Zero Plastic Packaging Based on a Resilient and Sustainable Circular Economy 75
Farshid Nazemi, Amrita Sen, Fabio Sporchia, George Stephanopoulos, Federico Pulselli, Bhavik Bakshi

160d Sustainable Supply Chain Design Under Uncertainty within the "Safe and Just Space": Modeling Framework and a Case Study of Li-Ion Battery........ 78
Ying Xue, Bhavik Bakshi

160e Towards Circular Food Supply Chains: Integrated Design of Food Packaging and Waste Management Infrastructures........ 80
Paola A. M. Briones, Aurora Del Carmen Munguía-López, Victor M. Zavala, Styliani Avraamidou

160f Assessing the Implications of Recycling Targets and Scale-up of Post-Consumer Plastics Recycling Infrastructure: A Reverse Supply Chain Optimization Approach........ 82
Tapajyoti Ghosh

160g Trojan Horse Strategy for Sustainable PET Plastics: A Comparative Evaluation of Techno-Economic and Lifecycle Impacts 83
Vedanti Sharma, Dhananjay Dileep, Mark M. Wright, Eric Cochran

SUSTAINABLE APPROACHES TO EMERGING CONTAMINANTS

231a Effect of Ambient Condition and Terrain Structure on the Dispersion of Accidentally Released Toxic Chemicals........ 84
S. Moumita Manjur, Sidney Lin

231b Enhanced Removal of Pharmaceutical Compounds from Aqueous Solution Using Deep Eutectic Solvents -Impregnated Graphene Oxide........ 92
Anjali Goyal, Tarek Lemaoui, Enas Nashef, Shadi Hasan, Hassan Arafat

231c Exploring Hydrothermal Carbonization of Exotic Fruit Hydrochars for Acesulfame-K and Carbamazepine Removal from Water........ 94
María F. Lecaro, Yuliana P. Gallardo, Andrés Lagos, Darío Niebieskikwiat, Luis M. Prócel, Marco León, Andrea C. Landázuri

231d Managing PFAS-Contaminated Adsorbents with a Life-Cycle Perspective 102
Margaret O'Connell, Petros Stamoulis, William Dichtel, Rosa Gwinn, Jennifer Dunn

231e Breakthrough Extraction of Aqueous Perfluorooctanoic Acid Using Natural Deep Eutectic Solvents for Sustainable Water Remediation 103
Sana Eid, Tarek Lemaoui, Ahmad Darwish, Maguy A. Jaoude, Fawzi Banat, Shadi Hasan, Enas Nashef

231f Understanding the Fate and Transport of Perfluorooctanesulfonic Acid (PFOS) and Microplastics in Aquatic Environment 104
Jasneet Pala, Gabriel Barbosa, Christoffer Turner, Milad Esfahani

231g Bioengineering Under Siege: The Unseen Threat of Co-Existing PFAS and Microplastics on Ecosystem Health 105
Mileisha Velazquez, Katherine Deliz-Quiñones

SUSTAINABLE END-OF-LIFE MANAGEMENT OF MATERIALS AND WASTE

663a Sustainability Assessment in Additive Manufacturing End-of-Life Material Management Practices 106
John D. Chea, Gerardo Ruiz-Mercado, Raymond Smith, David E. Meyer, Michael Gonzalez, William Barrett

663b Using Laser Irradiation to Separate Polymer Adhesives from Silicon in End-of-Life PV Panels for Recycling Purpose 107
Touhid Bin Anwar, Christopher J. Bardeen

663c Electrochemical Recovery of Rare Earth Elements from Neodymium-Iron-Boron Based Magnets Coupled with Iron Recovery 108
Alamgir M. Haque, Christian E. Alvarez-Pugliese, Gerardine Botte

663d Polymer-Degrading Potential of a Diverse Set of Environmentally Ubiquitous Microorganisms Using High-Throughput Biodegradation Testing 109
Alex Zappi, Katharina Fransen, Gabrielle Godbille-Cardona, Natalie Mamrol, Kristala Prather, Bradley Olsen

663e Carbon Sequestration Potential of Stabilized Hydrochar from Cattle Manure 110
Cordel Bever, Chuck Coronella

663g End-of-Life Options for Materials in Wind Turbine Blades 111
Caroline Cameron

DECARBONIZATION BY BIOMASS: TECHNOLOGY, POLICY, AND SUSTAINABILITY

430a Sustainable Pathways to Acrylic Acid - The Impact of Biomass Source on the Catalytic Dehydration of Lactic Acid to Acrylic Acid 112
Christopher Nicholas

430b Directed Mineralization of Waste Carbon to Develop Negative Emission Cements 113
David Kenney, Andrew Charlebois, Nima Rahbar, Shuai Wang, Michael Timko, Andrew R. Teixeira

430c CO_2 Capturing Technology through Lignin-Based Polymers 114
Hoyong Chung

430d Large-Scale Decarbonization through Synergetic Utilization of Forest Biomass Resources 115
Yuan Yao

430e Multiscale Analysis for the Substitution of Natural Gas Boilers Using Biomass 116
Roberto C. García, Guillermo G. Iglesias, Mariano Martin

430f Effect of Zeolite on Hydrothermal Carbonization of Loblolly Pine and Its Impact on Carbon Capture 118
Swarna Saha, Toufiq Reza

430g Biobased Films Using Ecuadorian Agro-industrial Residues - In the Framework of a Circular Bioeconomy, Circular Engineering, and Sustainability 119
Briggitte L. Intriago, Lourdes M. Orejuela-Escobar, Ana G. A. Sosa, Inés H. Guerra, Arleth G. Brito, Andrea C. Landázuri, Alfredo Valarezo, Edgar J. Guerrón-Orejuela

CO2 CAPTURE FROM INDUSTRIAL POINT SOURCES

97a Techno-Economic Analysis of Liquid Absorbent-Based Carbon Capture for Achieving Net-Zero Natural Gas Power Generation 125
Nicholas Cross, Mengyao Yuan, Wenqin Li

97b Comparative and Predictive Analyses of Different Adsorption Processes for Carbon Capture from Cement Plants Flue Gas Streams 126
Abdullah Al Moinee, Fateme Rezaei

97c Deploying an *in Situ* Raman Methodology for Online, Live Solvent Composition Analysis for Process Monitoring of CO_2 Capture 127
Dushyant Barpaga, Abhishek Kumar, Po K. Tse, Amanda Lines, Samuel Bryan, David Heldebrant

97d Pilot Testing of FrostCC™, a Novel Cryogenic Carbon-Capture Technology, at the National Carbon Capture Center 129
Jonathan Stickel, Jeffrey Young, Widad Yossri, Michael McNelis, Miles Abarr

97e Thin-Film Composite Membranes Based on Poly(1,3 dioxolane) and Uio-66-NH_2 for CO_2/N_2 Separation 130
Gengyi Zhang, Narjes Esmaeili, Fathy M. E. Attia, Haiqing Lin

97f Development of All-Polymer Sorbents for High Capacity CO_2 Capture 131
Ankana Roy, Hannah Holmes, Lisa S. Baugh, David C. Calabro, Johannes Leisen, Yi Ren, Simon C. Weston, Ryan Lively, M. G. Finn

97g CFD Modeling of High-Flux Plate-and-Frame Membrane Modules for Carbon Capture at Industrial and Power Generation Plants. 132
Cheick Dosso, Hector Pedrozo, Thien Tren, Lingxiang Zhu, Victor A. Kusuma, David Hopkinson, Lorenz Biegler, Grigorios Panagakos

97h Corrosion Analysis of Non-Aqueous Solvent for Carbon Capture 134
Katy Jessop, Paul Mobley, Jak Tanthana, Vijay Gupta, Marty Lail

CONCENTRATED SOLAR POWER GENERATION AND CHEMICAL PROCESSING I

499a Particles as Heat Transfer Medium to Drive Endothermal Chemical Reactions By Solar Energy 136
Martin Roeb

499b Experimental Evaluation of a Continuous Oxidation Reactor for High-Temperature Discharge of Thermochemical Energy Storage Particles 139
Juvenal Ortiz-Ulloa, Owen Ramsey, Lucas Freiberg, Fuqiong Lei, Michael Hayes, Philipp Schimmels, Like Li, Andre Benard, Joerg Petrasch, Kelvin Randhir, Nesrin Ozalp, James F. Klausner, Nick AuYeung

499c Experimental Evaluation of an Indirectly-Irradiated Packed-Bed Solar Thermochemical Reactor for Calcination–Carbonation Chemical Looping 140
Lifeng Li, Alireza Rahbari, Mahdiar Taheri, Roelof Pottas, Bo Wang, Morteza Hangi, Leanne Matthews, Lindsey Yue, José Zapata, Peter Kreider, Alicia Bayon, Chi-Hwa Wang, Terrence W. Simon, Joe Coventry, Wojciech Lipiński

499d Counter-Current Regenerative Redox Chemical Looping for Syngas Production: Packed Bed Reactor Design, Evaluation, and Feasibility Study 141
Zachary Hart, Alon Lidor

499e Numerical Heat Transfer Simulation and Optimization of Indirectly Heated Porous Foams in Thermochemical Hydrogen Generation 144
Durgesh Ranjan, Elizabeth Gager, Caroline Hill, Juan C. Nino, Jonathan Scheffe

499f A Photothermal Reactor with a Selective Solar Absorber to Produce Liquid Fuels Via Ethylene Oligomerization 146
Aisulu Aitbekova, Magel Su, Matthew Salazar, Harry Atwater

499g Decarbonizing Ethylene Production Using Concentrating Solar Power 147
John Fawcett, Abhaya K. Datye, Andrew T. De La Riva, Christopher Riley, H. Evan Bush, Erik Spoerke, Dresean Abeyta, Benjamin Harvey

DESIGN, ANALYSIS, AND OPTIMIZATION OF SUSTAINABLE ENERGY SYSTEMS AND SUPPLY CHAINS I

98a An Integrative Framework for Optimal Decarbonization Strategies in the Chemicals Sector 148
Sheikh M. Ahmed, Joan Cordiner

98b Design of a Supply Chain for the Production of Biodegradable Plastics Based on the Revalorization of Sewage Sludge 149
Jose E. Roldán-San Antonio, Mariano Martin

98c Transition Toward a Decarbonized Steel Industry: A Supply Chain Analysis for the United States 151
Yeonseo Jeong, Qi Zhang

98d Supply Chain Model for Bioenergy with Carbon Capture and Storage (BECCS) in Iowa. 152
Mrinal M. Redij, Mark M. Wright, Lisa Schulte

98e The Energy and Materials Transition Nexus– Multiscale Modeling and Optimization 153
Betsie M. Flores, Marco De Sousa, Rahul Kakodkar, Shayan Niknezhad, Efstratios Pistikopoulos

98f Strategic Supply Chain Optimization for Bio-Jet Fuel Production in the Brazilian Sugarcane Industry 154
Madelynn Watson, Aline V. Da Silva, Pedro Gerber Machado, Celma O. Ribeiro, Claudio A. O. Do Nascimento, Alexander Dowling

98g State-Task Network Representation of Early-Stage Process for Identification of Best Processing Route – in Application of Reactive Carbon Capture 156
Wonsuk Chung, Ung Lee

98h Integration of Temporal Factors into Spatially Explicit Optimization of Second Generation Biofuels Supply Chains 158
Phuc Tran, Eric O'Neill, Christos Maravelias

SUSTAINABLE APPROACHES FOR CHEMICAL PRODUCTION

662a Techno-Environmental Analysis of Olefins Production from Syngas Via Sorption-Enhanced Synthesis 161
Aurora Cavaliere, Massimiliano Materazzi, Paola Lettieri, Andrea Paulillo, Nilay Shah, Semra Bakkaloglu

662c Automatic High Throughput Reaction Pathway Identification for More Sustainable Chemicals Production 162
Dion Jakobs, Lucas F. Dos Santos, Gonzalo Guillen-Gosalbez

662d A Computational Toolbox Supporting the Development of Safe and Sustainable by Design Chemicals and Materials 165
Dimosthenis Sarigiannis, Fotini Nikiforou, Achilleas Karakoltzidis, Anna Agalliadou, Tomas Rydberg, Maja Halling, Chiara L. Battistelli, Emilio Benfenati, Cecilia Bossa, Evert Bouman, Émilien Bourgé, Milena Brouwer-Milovanovic, Annabel Hill, Eleni Iacovidou, Ivo Iavicoli, Tomi Kanerva, Therese Kärnman, Veruscka Leso, Jenny Lindén, Magnus Lofstedt, Bernd Nowack, Araceli S. Jiménez, Susanne Resch, Gianluca Selvestrel, Kirsi Siivola, Anezka Sharma, Vrishali Subramanian, Rosella T. Leggieri, Martijn Van Bodegraven, Joanke Van Dijk, Jaco Westra, Ziye Zheng, Antonis Gypakis, Spyros Karakitsios

662e Optimal Solvent Selection Using Computer-Aided Molecular Design (CAMD) to Obtain Octacosanol from Sugarcane Rind 167
Nistala V. Subrahmanyam, Sharad Bhartiya, Vinay A. Juvekar, Urmila Diwekar

662g Navigating the Route of Polyester Textiles to Net-Zero Emissions 171
Aniket Mali, Bhavik Bakshi

ENGINEERING GEOLOGIC CARBON DIOXIDE STORAGE SYSTEMS I

289a Application of Modified Meshgraphnets for Subsurface Prediction during CO_2 Sequestration 173
Paul Holcomb, Alex Sun, Chung Shih, Guoxiang Liu, Hema Siriwardane

289b Monitoring Carbon Storage Sites Using Electromagnetics: Recovering a Baseline Conductivity Model at the Kemper Carbonsafe Site 175
Colton Kohnke, Yaoguo Li, Richard Hammack

289c Permeability and Porosity of Tuscaloosa Mudstone at CO_2 Injection Site 177
Vyacheslav Romanov, Yee Soong, Igor Haljasmaa

289d Insight into the CO2 Pipeline Transportation Cost: New Cost Calculation Model with Regional Variation and CO2 Pipeline Network. 178
Kwang H. Baek, Sheik Tanveer, Pingping Sun, Amgad Elgowainy

289e Unveiling the CO_2 Storage Resources of the St. Peter Sandstone and Everton Formation in the Illinois Basin 180
Mansour Khosravi, Zohreh Askari, Roland Okwen

289f Offshore Stored CO2 Residence Time: Accurate Assessment with Geochemical, Seismic, and Modelling Integration ... 182
Richard Coffin, Richard Coffin, Maria Vasilyeva, Hussain Abdulla

289g Research and Development of Carbon Capture and Storage in the Illinois Basin ... 183
Carl Carman

289h Novel Carbon Storage Monitoring Methods ... 185
Trevor Richards, John Hunt, Cesar Barajas-Olalde, Justin Kovacevich, Kyle McBride, Bethany Kurz

FUEL AND ENERGY DECARBONIZATION I

701a A Multi-Objective Decision-Making Framework for Renewable Energy Transportation ... 187
Halil Iseri, Funda Iseri, Eleftherios Iakovou, Efstratios Pistikopoulos

701b Decarbonizing Natural Gas Sector: Natural Gas/Hydrogen Blends and Synthetic Gas Production Pathways ... 188
Kyuha Lee, Vincenzo Cappello, Pingping Sun, Amgad Elgowainy, Kwang H. Baek, Pallavi Bobba

701c A Technoeconomic Analysis and Life Cycle Assessment for Long Distance Hydrogen Transportation Via Hydrogen Carriers ... 189
Gasim Ibrahim, Bosong Lin, Haoxiang Lai, Guiyan Zang

701d Modeling and Analysis of Electrofuel Production Using CO_2 and Nuclear Energy Via Methanol Intermediate ... 190
Hernan E. Delgado, Vincenzo Cappello, Pingping Sun, Clarence Ng, Kwang H. Baek, Amgad Elgowainy

701e Techno-Economic Analysis and Life Cycle Assessment of Production, Storage and Transportation of Hydrogen Carriers ... 191
Maanasa Bhat, Gasim Ibrahim, Sili Deng, Guiyan Zang

701f CO_2-to-Sustainable Synthetic Fuel Process: Scenario-Based Techno-Economic and Carbon Footprint Analysis ... 192
Jinsu Kim, Yanhui Yuan, Ben McCool, Matthew Realff

701g The Primary Aqueous Aluminum-Copper Battery ... 193
Sirshendu Guha

NOVEL APPROACHES TO CO2 UTILIZATION I

353b On the Good Practices of Electrochemical Reactive CO_2 Capture ... 199
Avishek Banerjee, Chudi Yue, Jounghwan Choi, Carlos Morales-Guio

353c Free-Standing Porous Solid Electrolyte Layer for Efficient CO_2 Electrolysis ... 200
Sarah Adaryan, Rafael Verduzco, Haotian Wang

353d What's inside the Black Box? Understanding Enzymatic Conversion of CO_2 ... 201
Jenny Hancock, Siddharth Patwardhan

353e Systems Level Prediction of Best Product Conversion from CO_2 for the Novel Fast Growing Cyanobacterium *Synechococcus Elongatus* PCC 11801 ... 202
Lokesh V. Babu, Pramod Wangikar

NOVEL ENERGY STORAGE AND SYSTEM INTEGRATION

230a Efficient Thermal Storage: Unveiling a 2D Computationally Efficient Borehole Thermal Energy Storage (BTES) Model for Renewable Energy Integration.. 203
Mohamed Bahr, Piyush K. Kumawat, Blake Billings, Palash Panja, Kody Powell

230b Thermodynamic Characterization of a Novel Compressed-Air Energy Storage Technology: Flooded-Mineshaft Compressed Air Energy Storage (FM-CAES) 204
Neil Stacey, Diane Hildebrandt, Peter Stacey, James Fox, David Glasser

230c Optimal Scheduling of Pumped Thermal Energy Storage Plants for Multi-Vector Energy Provision.. 205
Alp Albay, Mehmet Mercangöz

230d Optimization of Liquid Air Energy Storage Systems in Future US Electricity Markets 207
Shaylin Cetegen, Paul I. Barton

230e Techno-Economic Analysis of Integrating Different Energy Storage Models in Highly Renewable Energy Systems.. 209
Ali Khosravani, Kody Powell

230f Optimal Design and Integration of a Green Hydrogen Production, Compression and Storage System for Solar Photovoltaic Power Smoothing.. 210
Yifan Wang, Paul Brooker, James Fenton

230g Durational Testing of Lithium-Ion Battery Cells Using Grid-Informed Duty Cycles 211
Jaesung Kim, Dongwei Zhao, Audun Botterud, Fikile Brushett

EARLY-CAREER RESEARCHERS IN SUSTAINABLE ENERGY

431a Integrating Carbon Capture Technologies in Hard-to-Abate Industries: A Techno-Economic Analysis... 212
Bosong Lin, Haoxiang Lai, Lingyan Deng, Guiyan Zang

431b Synergistic Conversion of Captured CO2 and Green H2 to Olefins and SAF By Microwave Catalytic Processing 213
Yujuan Chen, Brandon Robinson, Jianli Hu

431c Comparative Techno-Economic and Life Cycle Assessment of Reactive CO_2 Capture to Synthetic Natural Gas.. 215
Alvina Aui, Hannah M. Goldstein, Wenqin Li, Nathan C. Ellebracht, Simon H. Pang

431d Accelerated Deep Learning-Based Data Assimilation and CO2 Injection Optimization at the Illinois Basin-Decatur Carbon Sequestration Project 216
Masahiro Nagao, Takuto Sakai, Chin H. Chan, Akhil Datta-Gupta

431e Evaluation of Thermophotovoltaics as a Decarbonization and Clean Energy Storage Technology: Techno-Economic Analysis and Uncertainty Assessment 219
Manish Mosalpuri, Mark M. Wright, Fatima Toor

431f Investigating the Design of Global-Scale Power-to-Methanol Production Systems: Large- or Small-Scale Chemical Plants? 221
Tibor Svitnic, Kai Sundmacher

BIOLOGICAL CONVERSIONS AND PROCESSES FOR RENEWABLE FEEDSTOCKS

351a Experimental Studies on Enzymatic Ring-Opening Polymerization for Sustainable Polyester Synthesis 223
Victor Sharma, Peter V. Bonnesen, Yangyang Wang, Jong K. Keum, John F. Cahill, Yue Yuan

351b Alternative Feedstock Utilization by the *Rhodococcus Jostii* Strain PET for Biomanufacturing Beyond Earth 224
Hakyung Lee, Eunseo Lee, Yuxin Tian, Richa Guleria, Alexandra Smith, Luke Roberson, Mark Blenner, Daniel H. Yeh, Yinjie Tang, Jinjin Diao, Tae S. Moon

351c Enhancing Biological CO_2 Conversion to Ethanol by *Clostridium Ragsdalei*: Insights from Batch and Continuous Operations 225
Rahul Thunuguntla, Hasan Atiyeh, Mari S. Chinn, Ralph S. Tanner

383g Separation of 2,3-Butanediol from Fermentation Broth Via Cyclic and SMB Adsorption over Nano-MFI Zeolites 226
Jianpei Lao, Qiang Fu, Marco Avendano, Yadong Chiang, Matthew Realff, Sankar Nair

351e Continuous Production of Natural Blue Phycocyanin in Pilot-Scale Bioreactor 227
Mohammad R. Rahman, M. Tabish Noori, Klaus Hellgardt

CHEMICAL AND CATALYTIC CONVERSIONS AND PROCESSES FOR RENEWABLE FEEDSTOCKS

429a Kinetic Modeling of the Epoxidation of High Oleic Palm Oil Using Amberlite IR-120 As Catalyst 228
Steven Solarte, Alvaro Orjuela, Laura R. Conde

429b The Effects of Solvent on the Mechanisms and Kinetics of Levulinic Acid Production from Biomass Using a Polymeric Acid Catalyst 230
Zhexi Zhu, Ranil Wickramasinghe, Xianghong Qian

429c Catalytic Pyrolysis of Lignocellulosic Biomasses: A Sugar Cane Bagasse and Salicornia Study 231
Jeimy J. B. Martinez, Sandro González-Arias, Ricardo García-Morales, Octavio Elizalde-Solis, Edgar Ramirez-Jimenez

429d Thermodynamic Evaluation of Mixed Metal Redox Carriers for Biomass Utilization Using Chemical Looping 241
Sudeshna Gun, Tanay Jawdekar, Sonu Kumar, Falguni Akulwar, Liang-Shih Fan

FEEDSTOCK LOGISTICS FOR BIOREFINERIES

661b Understanding the Effect of Particle Characteristics on Flowability of Biomass Slurries 242
Glenn Matamoros, Sampath Gunukula, Clayton Wheeler, C. Luke Williams

661d Characterization of Sulfur Profile and Fate Study of Loblolly Pine Using Pyrolysis-GC/MS-FPD 243
Gary Lopez, Jian Shi

661e How Do Feedstock Variabilities Impact the Convertibility of Lignocellulosic Biomass? A Data-Oriented Case Study on Corn Stover in the United States 244
Xihui Kang, Chang Dou, Ling Ding, Yining Zeng, Bryon Donohoe, Kenneth Sale, Ning Sun

661c Air Classification of Forestry Residues for Fast Pyrolysis 245
Nepu Saha, Jordan Klinger, Tiasha Bhattacharjee, Yidong Xia, Vicki Thompson, Oluwafemi Oyedeji, James E. Parks, Mehrdad Shahnam, Yupeng Xu

LIFE CYCLE ANALYSIS OF BIO-BASED FUELS, ENERGY, AND CHEMICALS

99g Agile Systems Analysis Framework to Assess the Sustainability Implications of Cellulosic Biorefineries Using Consolidated Bioprocessing 246
Kay Glass, Muhammad H. Wadgama, Rui Shi

99b Feasibility of Photovoltaic Energy: Life Cycle Assessment, Circular Economy, and Criticality Indicators 247
Natalia Cano, Laura Franco-García, Rebecca Saive

99c Life Cycle Sustainability Assessment (LCA) of Nature-Inspired Bio-Syngas Technologies for Olefin Synthesis 249
Semra Bakkaloglu, Aurora Cavaliere, Paola Lettieri, Andrea Paulillo, Massimiliano Materazzi, Nilay Shah

99g Biogas Utilization in Refuse Power Plants (BURP) 250
Ahmed Essam

99e Carbon Footprint Assessment of Production of Ethylene from U.S. Corn Ethanol 251
Kirti Richa, Pahola T. Benavides, Ulises G. Alvarez, Jennifer Port, Troy Hawkins

99f Soil Carbon: An Overlook Component of Biofuel Life Cycle Assessment 253
Yuan Yao, Kai Lan, Bingquan Zhang

PRODUCTION OF LOW-CARBON FUELS FORM BIOMASS: DESIGN, SIMULATION, AND ECONOMIC ANALYSIS II

561a Pulp and Paper Sludge Valorization: Techno- Economic and Life Cycle Assessment for Enhanced Methane Production 254
Erfan Hosseini, Selen Cremaschi, Zhihua Jiang

561b Integration of Biogas Model (ADM1) for Economic and Environmental Analysis of RNG Production 257
Katherine Wild, Juan C. Dominguez, Lisa Schulte, Mark M. Wright

561c Environmental Sustainability of Biofuels for Maritime Decarbonization on a Planetary Scale 258
Eric Tan

561e Life Cycle Assessment of Functionalized Ultraporous Superactivated Hydrochars for Carbon Capture 259
Swarna Saha, Toufiq Reza

561f Process Design and Economics for Production of Sustainable Aviation Fuel (SAF) from Biomass Derived Intermediate 260
Kevin Dyke, Devin Fan, Michael A. Lucci, Grant Zeszutek, Dhruv Patel, Muhammad H. Wadgama, Rui Shi

REACTOR ENGINEERING FOR BIOMASS FEEDSTOCKS I

730a From Waste to Sweetness: Catalytic Strategies and Reaction Engineering for Upcycling Greek Yogurt Acid Whey 261
Wenjia Wang, Owen J. Dziedzic, Claire Lesnjak, Xiaolei Shi, Jarryd Featherman, James Miller, Scott A. Rankin, George W. Huber

730b Ru/CNT Catalyst for Supercritical Water Gasification of Biomass 262
Yukihiko Matsumura, Mizuki Kodama, Mohammed A. M. Ali, Takashi Noguchi, Mengli Zhang, Shuhei Inoue

730c Overall Transesterification Rate of Palm Oil with Methanol Catalyzed by Acid in Oil-Methanol Two Phase: A Novel Two-Phase Overall Reaction Model 263
Yoshimitsu Uemura, Yuki Mitsutake, Fon Y. Han, Thanh H. Trinh, Katsuki Kusakabe

730d Characterization of Torrefied Woody Biomass 264
Yoshinobu Yokogawa, Hajime Kida, Kenichiro Tanoue

SUSTAINABLE DESIGN AND MODELING OF BIOREFINERIES I

161a Techno-Economic Analysis and Life Cycle Assessment of Non-Thermal Plasma Co-Upcycling of Waste Plastics and CO_2 to High-Value Chemicals 265
Samirah Gnangbe, Harish Radhakrishnan, Xianglan Bai, Mark M. Wright

161b Techno Economic Analysis and Life Cycle Assessment for Synthetic Spider Silk Production 267
Melika Tajipour, Bojing Jiang, Pallavi Dubey, Fuzhong Zhang, Marcus Foston, Mark M. Wright

161c Techno-Economic Analysis of a Modular Bioprocess System for Producing Polyhydroxyalkanoates (PHAs) from Food Waste Via Heterogeneous Fermentation. 268
M. S. T. Sharifa Khatun, Arna Ganguly, Zhiwu Wang, Mark M. Wright

161d Design and Analysis of a Direct Pathway Process for the Enantioselective Production of Biomass-Derived (*R*)-γ-Valerolactone 269
Seungwon Lee, Jaehyun Cha, Byeongchan Ahn, Joungho Park, Kyoungseon Min, Wangyun Won

161e Life Cycle Assessment and Sensitivity Analysis of a Plant-Based Meat Alternative Containing Heme Protein 272
Kay Glass, Hariteja Nandimandalam, Rui Shi

TRANSLATING BENCH-SCALE RESEARCH INTO COMMERCIAL MANUFACTURING PRACTICE

731a Multiple Criteria Productivity Analysis Applied to Perovskite Solar Devices from Lab to Factory Production 273
Nathan Dodd, Rachael Rothman

731b Integrated Systems for Continuous CO_2 Separations from Industrial Flue Gas Sources. 274
Rohan Sartape, Vamsi V. Gande, Meenesh Singh

731c Critical Technoeconomic Appraisal of Product-Specific Sequential and Integrated CO2 Capture and Electrochemical Utilization 275
Ahmed Badreldin, Ying Li

731d Effective CO_2 Utilization Using Coke Oven in Iron Making Process for Sustainability 276
Gihun Kwon, Chang H. Rhee, Younghun Kim, Sangwook Joo, Beom-Sik Kim, Hai W. Park

731e Proposal for an Upcycling Process to Commercialize All Components of By-Product Oil in Edible Oil Refinery 278
Naomi Shibasaki-Kitakawa, Kousuke Hiromori, Makiko Kato

731f Continuous Active Pharmaceutical Ingredient (API) Manufacturing from Biomass-Derived Building Blocks: An Integrated Process to Enable Sustainable Manufacturing 280
Yuma Miyai, Hung L. Lee, Aibolat Koishybay, Hubert Hellwig, Loic Bovy, Tianjie Zhu, Rodolfo Romañach, Muhammad N. Alam, Cornelis Vlaar, Jean-Christophe Monbaliu, Allan Myerson, Torsten Stelzer

731g Evaluating Glycerol Carbonate Production from CO_2 and Glycerol: An Integrated Process Analysis 281
Octavio Del-Mazo-Alvarado, Carlos Prieto, Antonio Sánchez, Cesar Ramírez, Adrián Bonilla-Petriciolet, Mariano Martin

POSTER SESSION: SUSTAINABILITY SCIENCE AND ENGINEERING, BIOREFINERIES, AND ENERGY

573a Techno-Economic Analysis and Life Cycle Assessment of Hydroxylamine Eco-Manufacturing Via Wastewater Electrochemical Reduction 283
Manish Mosalpuri, Wenzhen Li, Mark M. Wright

573c Sustainable Resources for the Synthesis of Aromatic Feedstock 285
Marvin Wittek, Andreas Widjaja, Guido Waidmann, Alfons Drochner, Bastian Etzold

573d Using Electrochemical CO_2 Reduction to Recover Lithium Carbonate from Spent Lithium-Ion Battery Compounds 287
Shelby Atherton

573e Techno-Economic Analysis of Itaconic Acid Production Using a Biorefinery Approach 288
Nkechi M. Nwaogaraku, Mark M. Wright

573f Optimizing Building Envelope Design for Sustainable Energy Performance 291
Sahar Fereidoni, Leila Fereidooni, Alibakhsh Kasaeian

573g Redefining Lignin Valorization by Integrating Chemical Depolymerization and Microbial Funneling to Produce Specialty Bioproducts 292
Sarada Sripada, Samuel Scholz, Anna Qualls, Maximiliano A. Garcia, Bumkyu Kim, Timothy J. Donohue, Daniel R. Noguera, Steven D. Karlen, Canan Sener

573h Techno-Economic Analysis of Green Hydrogen Production in South Korea and Saudi Arabia, Using Monte Carlo Simulation Based on Meteorological Data 293
Yewon Kim, Sanghun Lee

573i Advancing Green Ammonia Synthesis in Dynamic Offshore Wind Power-to-X Systems 294
Philipp Rentschler, Niklas Brinker, Christoph Klahn, Roland Dittmeyer

573j A Case Study of Valorizing Landfill Gas to Renewable Energy in an Industrial Zone 297
Jian Fang, Helen Lou, Andrew W. Kolp, Daniel Chen

573k Assessing the Sustainability of Recovering Rare Earth Elements from a Toxic Wastewater Slurry: A Combined Life Cycle Assessment and Technoeconomic Analysis Study 298
Adam Smerigan, Rui Shi

573l Advancing Automotive Plastics Recycling through Molecular Recycling Technologies: Innovations for a Circular Economy and Sustainability 299
Mahshid Mokhtarnejad, Gary Hawkins, Hendrik Mainka

573m Exploring the Effect of Facility Construction in Biofuel Life Cycle Analyses through Greenhouse Gas Emissions 300
Huy Nguyen, Rui Shi

573n Electrically Enhanced Thermochemical CO_2 Splitting for Fuel Production 302
Jayni Hashimoto, Jordan Monroe, Alicia Bayon, Alonzo Mendez, Olivia Tamburro, Christopher L. Muhich

573o Tribological Properties of Bio-Based Lubricants Obtained from Castor Oil Fatty Acids and Isopentanol 303
Marliete M. Neta, Igor Figueredo, Paulo R. C. F. R. Filho, Celio Cavalcante, Francisco M. T. De Luna

573p Studying Precipitation Rates Under Semicontinuous Operation of a CO_2 Mineralization Reactor Utilizing Flue Gas and Produced Water 304
Quinn Bennett

573q A Review of Existing Data and ASTM E681 Testing for Flammability Limits of Hydrofluorocarbons and Their Mixtures: Towards a Predictive Model 305
Kevin Turner

573r Excess Absorption of Type II Porous Liquids for CO_2/H_2 Separation 306
Lu Lu, Ryan Lively

573s Perovskite Oxides as New Family of Tunable CO_2 Sorbents 307
Seyedamin Razavi, Vahid Rahmanian, Mahe Rukh, Runxia Cai, Mohammadreza Kosari, Saad A. Khan, Fanxing Li

573t Assessing and Unlocking the Potential for Carbon Mineralization in Peridotite Rock for Carbon Dioxide Removal 308
Soyoung Choi, Ah-Hyung A. Park

573u Surface Wettability and Pore Size Effect on CO_2 Diffusivity in Water-Saturated Kaolinite Nanopores 309
Minjunshi Xie, Zhehui Jin

573v Hydrogen Storage in Silica-Kaolinite Shale and Its Impact on the Integrity: Molecular Modeling and Experimental Investigation 310
Alankaa Al-Harbi, Mohammed J. Almarri, Safwat Abdel-Azeim, Mohammed Ali Saleh, Ibnelwaleed Hussein

573w Optimizing the Reverse Supply Chain Network for Photovoltaic Panels Under Uncertainty 311
Funda Iseri, Halil Iseri, Natasha Chrisandina, Shivam Vedant, Eleftherios Iakovou, Efstratios Pistikopoulos

573x Directing Climate Mitigation throughout Energy Transformation: Feasibility of Synthetic Methane Integrated with Direct Air Capture 312
Changgwon Choe, Hankwon Lim

573y Optimal Compressor Operational Policies for LNG Storage Tanks Under Unloading Mode Operation: A Numerical Approach 313
Suraj P. Singh, Rajagopalan Srinivasan, Iftekar Karimi

573z Understanding Microbial Processes of Algal-Fixed Carbon Using a Co-Culture Porous Microplate 315
Hyungseok Kim, Vanessa L. Brisson, John R. Casey, Jeffrey A. Kimbrel, Kristina A. Rolison, Christopher A. Vaiana, Peter K. Weber, Rhona K. Stuart, Xavier Mayali, Cullen R. Buie

573ab Fabricating an Economical and Sustainable Fermentation Medium to Produce Drop-in-Fuel and Value-Added Bioproducts from Alternative Feedstocks 316
Shraddha Maitra, Bruce S. Dien, Kristen Eilts, Nurzhan Kuanyshev, Yoel Cortes-Pena, Jeremy Guest, Yong-Su Jin, Vijay Singh

573ad Systematic Analysis on the Energy Transition Pathways for Sustainable Natural Gas Production 319
Meire E. G. R. Domingos, Daniel F. Orrego, François Maréchal

573ae LATP-incorporated Cellulose Membrane for Selective Lithium Extraction from Salt Lake Brine under an Electrochemical System 322
Jiwoo Lee, Dongju Seo, Shik R. Kong, Gyudae Sim, Youngjune Park

729f Optimal Design of the Bio-Jet Fuel Supply Chain Under Uncertainty 323
Shunmei Wu, Congqin Ge, Zhihong Yuan

SUSTAINABLE ENGINEERING FORUM PLENARY/AWARD SESSION (INVITED TALKS)

290a Atmospheric Methane Removal: A National Academies Research Agenda 324
Christopher W. Jones

290f Carbon Utilization Markets, Infrastructure, Research, and Development: A National Academies Assessment 325
Ah-Hyung A. Park

290c Convergent Research and Innovation at the NYS Center for Plastics Recycling Research and Innovation 326
Amit Goyal

290d Systems Engineering for Renewable Energy: Novel Approaches for New Problems 327
Christos Maravelias

290e Chemical-Free UV Unit That Degrades PFAS in Landfill Leachate Using Non-Toxicboron Nitride 328
Abraham Nelson

ENGINEERING GEOLOGIC CARBON DIOXIDE STORAGE SYSTEMS II

352a Overview of DOE-FECM's Carbon Storage Infrastructure Projects 329
William Aljoe, Darin Damiani

352b Regional Resource Assessment for CO_2 Storage Via Mineralization Process in New Mexico and Surrounding Areas 331
Jiyue Wu, Sai Wang

352c CO_2 Storage Site Selection: A Case Study from Oklahoma, US 338
Priyank Jaiswal, Aaditya Khanal

352d Implications for Developing Multiple CCS Projects in Same Geological Setting – US Gulf Coast Case Study 341
Seyyed Hosseini, Zhichang Wang

352e Can Fault Activation Cause Long-Term Caprock Integrity Issues? Findings from a Self-Sealing Field Experiment 342
Jens Birkholzer, Yves Guglielmi, Christophe Nussbaum, Frederic Cappa, Tanner Shadoan, Jonathan Ajo-Franklin, Florian Soom, G. W. Lanyon, Paul Cook, Chet Hopp, Verónica R. Tribaldos, Michelle Robertson, Todd Wood, Senecio Schefer

352f Olivine Dissolution in CO_2 Saturated Water Under Reservoir Conditions 344
Qiaoyun Chen, Martin Trusler

352g A Quantitative Approach to Assessing CO_2 Permanence for California Air Resources Board (CARB) CCS Protocol 346
Zahra Finnigan, Stacy Kouba, Travis C. McGrath, Mark C. Ciardelli, Nicholas Azzolina

352h Impact of Impurities on CO_2 Transport in Saline Aquifers 348
Prince Otabir, Aaditya Khanal

ENGINEERING GEOLOGIC CARBON DIOXIDE STORAGE SYSTEMS III

432a Managing a Gigaton Carbon Sequestration Future: A Framework for Prediction of Fault Activation Caused by Basin-Scale CO_2 Storage 349
Jens Birkholzer, Stanislav Glubokovskikh, Yves Guglielmi, Abdullah Cihan, Jonny Rutqvist, Meng Cao, Preston Jordan, Matthew Reagan

432b Three-Dimensional Pore Networks in Miocene Stevens Sandstone, California: Implications for CO_2 Injection and Trapping 352
Liaosha Song

432c Progress and Challenges in Developing CCS Clusters in the Midwest and Northeast Quadrant of United States 354
Neeraj Gupta, Mark Kelley, Joel Sminchak, Jared Hawkins, Joy Frank-Collins, Matthew Young, Christopher Korose, Dawn Deel

432d Applicable Insights of MICP-Derived Capillary Pressure on CO_2 Injection for Wyoming Carbonsafe Project: When Does Capillary Pressure Affect the Pressure Front the Most? 356
Ying Yu, Tao Bai, John Jiao, Jonathan McLaughlin, Scott Quillinan

432e Leveraging 3D Seismic to Understand Induced Seismicity Risk Associated with CO_2 Injection: An Example from the One Earth Energy Sequestration Project in Gibson City, Illinois 359
Kendall Taft, Sherilyn Williams-Stroud

432f Assessment of CO_2 Storage and Mineralization in Hawaiian Basalt: Subsurface Characterization and Modeling Results 361
Donald De Paolo, Nicole Lautze, Donald Thomas, Bhavna Arora, Ziqiu Xue, Saeko Mito, Shuo Zhang, Pramod Bhuvankar, Xunfeng Lu, John Christensen

432g Water Thin Films on Kaolinite Gibbsite and Edge Surfaces and Their Effects on Surface Wettability in Relation to Geological Carbon Sequestration 362
Minjunshi Xie, Zhehui Jin

432h Key Challenges of CO2 Transport Via Pipelines 363
Isaac Gamwo, Deepak Tapriyal, Ward A. Burgess

DESIGN, ANALYSIS, AND OPTIMIZATION OF SUSTAINABLE ENERGY SYSTEMS AND SUPPLY CHAINS II

228a Optimal Design of Hydrogen-Blended Natural Gas Pipeline Network Considering Separation System 364
Shiya Gu, Yachao Dong, Jian Du

228b Operational, Economic, and Environmental Analysis of Hybrid Renewable Energy Systems 365
Ali Khosravani, Elaheh Safaei, Matthew De Haan, Blake Billings, Michael Reynolds, Kerry Kelly, Kody Powell

228c The Influence of Technical, Economic, and Site-Specific Factors on Optimal Ammonia Production from Renewables 367
Alexander Klimek, Tibor Svitnic, Caroline Ganzer, Kai Sundmacher

228d Drivers of Blue and Green Hydrogen Deployment in Net-Zero Emissions Scenarios 369
Jun W. Law, Dharik Mallapragada

228e A System-Level Decision-Making Analysis of Liquid Organic Hydrogen Carrier (LOHC) Processes: A Case Study of Korea-Australia 371
Byeongchan Ahn, Wangyun Won

228f Impact of Technology Development and Climate Characteristics on the Joint Expansion Planning of Green Hydrogen and Renewable Energy 373
Mingyu Kim, Yechan Choi, Seongmin Heo

228g Ammonia as an Energy Carrier: Cost and Emission in the Global Ammonia Supply Chain 375
Woojae Shin, Bosong Lin, Haoxiang Lai, Daniel R. Cohn, Leslie Bromberg, Guiyan Zang

228h Assessing the Key Factors in the Transition to a Hydrogen-Based Economy through Agent-Based Simulation 376
Venkataramanan Swaminathan, Rajagopalan Srinivasan

SUSTAINABLE ENERGY - AREA PLENARY

162a How Carbon Capture and Storage Can Fail and Why It Won't 378
D. Nathan Meehan

162b Tuning Process Parameters for Enhanced Low CO_2 Concentration Capture at a 0.7 MWe Post-Combustion Pilot Plant 379
Reynolds Frimpong, Heather Nikolic, Kunlei Liu

162c Electricity Planning for Ecosystems and Society: A Spatial-Temporal Approach Considering Cost, Societal Benefit and Environmental Justice 380
Xinyu Zhen, Bhavik Bakshi

162d Carbon Accounting for Sustainable Supply Chains 383
Nadin Moustafa, Nilay Shah

162e A Sequential Decision-Making Framework for Long-Term Energy Transition Planning Under Uncertainty 386
Molly McDonald, Christos Maravelias

162f CO2 Conversion to Syngas Coupled with a Membrane Separation: Multi-Scale Modeling and Simulation. 388
Seckin Karagoz

CONCENTRATED SOLAR POWER GENERATION AND CHEMICAL PROCESSING II

559a Inverse Design of Perovskite Materials through the Crossfit CEF and Bayesian Data Selection Algorithms for Solar Thermochemical Hydrogen Production 389
Steven Wilson, Matthew Park, Ellen B. Stechel, Christopher Muhich

559b Mixed Metal Oxides for Enhanced Solar Fuel Production Via Thermochemical Redox Cycling 390
Asim Riaz, Christopher I. Bodger, Jingjing Chen, Adrian Lowe, Takuya Tsuzuki, Wojciech Lipiński

559c An Efficient and Scalable Approach to Solar Thermochemical Fuel Production Via an Iron Aluminate-Based, Pressure-Swing Redox Cycle 391
Kent J. Warren, Liam Taylor, Dragan Mejic, Robert Anderson, Justin Tran, Deepak Dileepkumar, Alan W. Weimer

559d Electric Field Enhanced Thermochemical CO_2 Splitting 392
Jayni Hashimoto, Jordan Monroe, Alicia Bayon, Alonzo Mendez, Olivia Tamburro, Christopher L. Muhich

559e Kinetics and Techno-Economic Analysis of the Boudouard Reaction System Applied to Long Distance Thermal Energy Transport 393
Riley Choquette, Benjamin Phillips, Abigail Joseph, Annette Von Jouanne, Alexandre Yokochi

559f Ammonia-Based Thermochemical Energy Storage: Design and Optimization of Heat Recovery Cycle Using Conductive Ammonia-Selective Absorbents 394
Mahdi Malmali, Tanveer Mahtab, Sufian Humayun, Shahriar Rahman, Amir Safikhani, Alon McCormick

559g Hydrogenation and Hydrodeoxygenation of Oxygenates and Nitroarenes at Ambient Conditions Via a Two-Step Redox Cycle 395
Chongyan Ruan, Kunran Yang, Ryota Akutsu, Junchen Liu, H. Henry Lamb, William Martin, Caitlin Beckett, Benjamin Lessin, Fanxing Li

FUEL AND ENERGY DECARBONIZATION II

729a >90% Purity Syngas Generation for Direct Reduction of Iron Ore by Integration of Biomass-Based Chemical Looping with Carbon Dioxide Capture 397
Ishani K. Kudva, Shekhar Shinde, Tanay Jawdekar, Sudeshna Gun, Rushikesh Joshi, Ashin Sunny, Sonu Kumar, Liang-Shih Fan

729b Advancing Low-Carbon Hydrogen Production: Innovations and Challenges 399
Teerawit Prasomsri, Junchen Liu, Ashten Molley, Brooke Ames

729c Life Cycle Analysis of NO_x Emissions from Combustion of Hydrogen Blends in Natural Gas 400
Xiaoyue Pi, Vincenzo Cappello, Pradeep Vyawahare, Pingping Sun, Amgad Elgowainy

729d Real-Time Emission-Driven Modeling and Optimization Framework Under Uncertainty for Sustainable Hydrogen Production 401
Marco De Sousa, Saatvi Suresh, Betsie S. M. M. Flores, Rahul Kakodkar, Harsh B. Shah, Dustin Kenefake, Iosif Pappas, Xiao Fu, C. Doga Demirhan, Mete Mutlu, Shayan Niknezhad, Efstratios Pistikopoulos

729e Screening Cobalt-Based Single Atom Alloy Catalysts for Low-Temperature Fischer-Tropsch Synthesis 402
John El Berch, Rashad Ahmadov, Marc Porosoff, Giannis Mpourmpakis

NOVEL APPROACHES TO CO2 UTILIZATION II

433a Life Cycle Analysis and Techno-Economic Analysis Tools to Evaluate Carbon Conversion Technologies 404
Sheikh Moni, Michelle Krynock, Samuel Henry, Gregory Hackett

433b Analysis of a Novel Porous Bi-Functional Polymer Catalyst for CO_2 Conversion to Formic Acid: Techno-Economic Analysis and Life Cycle Assessment 406
Ikenna Okeke, Canan Karakaya, Michelle Kidder, Aye Meyer

433d High-Temperature CO_2 Electroreduction into Carbon Nanostructures in Molten Salt Electrolyte for Carbon-Negative MSCO2RR 407
Andrew Wong

433e Two-Dimensional Steady State Numerical Simulation of Electrochemical CO_2 Conversion Using a Zero-Gap Electrolyzer 408
Ahmad Ijaz, SeyedSepehr Mostafayi, Mohammad Asadi, Javad Abbasian, Hamid Arastoopour

433f Effect of Sorbent Incorporation to a $ZnZrO_2$ Catalyst in Its CO_2 Capture Capacity and Catalytic Activity 409
Laura Proano, Katlo Galefete, Guanhe Rim, Christopher Jones

433g Oxide-Encapsulated Electrocataysts for Selective and Stable CO_2 Upgrading from Reactive Carbon Capture Solutions 411
Zhexi Lin, Nathaniel Blake, Zhirui He, Xueqi Pang, Reza Mirshekari, Oyinkansola Romiluyi, Yoon J. Son, Suryansh Kabra, Daniel Esposito

NOVEL APPROACHES TO CO2 UTILIZATION III

500a Synergistic Effect between Plasma and Catalyst on Dry Reforming of Methane 412
Dong-Wook Kim, Jinwoo Kim, Jaehee Chang, Jun K. Kim, Jong-Min Lee, Hyuk J. Kwon

500b Autothermal Sabatier Reactor for Biogas Conversion to Renewable Natural Gas: Design, Experimental Proof-of-Concept, and Parametric Study 414
Yichen Zhuang, David Simakov

500c Towards Self-Regenerative Sorption-Enhanced Catalysts for Integrated CO_2 Capture and Dry Reforming of Methane 416
Kandis L. Gilliard-Abdulaziz, Seongbin Jo

500d Material and Process Optimization of Reactive Carbon Capture to Methanol Using Cu-ZnO-Al_2O_3 Dual Function Materials 417
Chae Jeong-Potter, Martha A. Arellano-Treviño, Wilson McNeary, Alexander Hill, Daniel Ruddy, Anh To

500e Hydrophobic h-BN Supported Cu/ZnO Catalysts for CO_2 Hydrogenation to Methanol 418
Jui-Chi Hu, Jeffrey Wu, Hsi-Yen Hsu, Yen-Chih Chen

500f Process Modeling and Design for Methanol Synthesis Using Membrane Reactor and Vacuum Pressure Swing Adsorption 420
Akihide Wakita, Tomoyuki Yajima, Yuta Nishikawa, Kohei Yoshikawa, Nobuyuki Shigaki, Yoshiaki Kawajiri

500g Dehydration Membrane Reactor for the Production of Valuable Chemicals from CO_2 and H_2 422
Weiwei Xu

NOVEL APPROACHES TO CO2 UTILIZATION IV

560a Experimental and Computational Efficiency Analysis of Perovskites-Driven Solar Thermochemical CO2 Conversion Cycle 423
Rahul Bhosale, Sayma Akhter

560b Integrated Technology for Cost-Effective CO_2 Capture and Formic Acid Production: Modeling, Optimization, and Economic Analysis 424
Maojian Wang, Jinliang Ma, Miguel A. Zamarripa, Benjamin P. Omell

560c Tri-Reforming of CH_4 over Non-Noble (Cu, Ni) Metal Catalyst Supported on Basic Oxide (MgO, ZrO_2) for the Production of Syn-Gas with the H_2/CO Ratio of 1.5-2 426
Akansha Pandey, Prakash Biswas

560d Low Intensity Pathways for Carbon Mineralization: Utilizing Mafic Resources in Arizona 427
Adam Sjolund, Thomas Sasser, Olivia A. Wrenn, Lisa A. Thompson, Klaus S. Lackner, Jennifer Wade

560e Transition Metal-Free Dual Function Materials for Highly Selective CO_2 Capture and Conversion 428
Anh To, Alexander J. Hill, Wilson McNeary, Chae Jeong-Potter, Martha A. Arellano-Treviño, Daniel Ruddy

560g Direct Carbonylation of Glycerol with CO_2 to Organic Carbonates Catalyzed by Cerium Oxide/Zinc Oxide 429
Wei-Hong Hong, Bing-Hung Chen

CLIMATE CHANGE AND ENGINEERING SUSTAINABILITY- II

558b Electrification as a Pathway to Green Living: Economic and Environmental Outcomes in U.S. Multifamily Housing 430
Matthew Williams, Chris Parker, Amanda Dillon, Blake Billings, Kody Powell

558c Direct Methane Capture Via Aerobic Methanotrophs 431
Waaseyaaban Landgren, Robert Handler, David Shonnard, Mary E. Lidstrom

558d A Novel "Dry" Biofilm Reactor for Atmospheric Methane Removal 432
Reza Khorasani, Q. Peter He, Jin Wang

558e Eco-Efficient Processing Clusters: Blockchain-Enabled Peer-to-Peer Resource and Emission Trading Framework 434
Manar Oqbi, Dhabia Al-Mohannadi

558f A Systems Approach for Renewable Fuels and Innovations for the Shipping Industry 435
Nefeli Roufogali, Konstantinos Pyrgakis, Lazaros Karaoglanoglou, Antonios Kokossis

558g A 1-2-3 of Performing Industrial Life Cycle Assessments 440
Seshasayee Mahadevan, Deboleena Chakraborty, Mahesh Sawant, Chad English, Paul Doll

SUSTAINABLE CHEMISTRY AND ENGINEERING- II

434b Novel Fruit Hydrochars: Eco-Friendly Solutions for Waste Management Challenges in Electronics and Agriculture Applications 441
Erick Mendizábal, María F. Lecaro, Camily Rivadeneira, Yuliana P. Gallardo, Mathilde Reffet, Juan F. Posso, Pierina Mirabá, Victoria Alomía, Marco León, Darío Niebieskikwiat, Luis M. Prócel, Andrea C. Landázuri

434d Enhancing Dispersion Stability of Alpha-1,3-Glucan Using Adipic Acid Derivatization: A Novel Approach for Industrial Applications 450
Nigel Twi-Yeboah, Alan Allgeier, Khanh Le Vinh, Christian Lenges, Jure Zlopasa, Natnael Behabtu

434e Acid-Catalysed Reflux-Free Silanisation at Room Temperature: Technical and Environmental Benefits... 451
Deniz Etit, Tsaone Gosiamemang, Ethan Errington, Miao Guo, Jerry Heng

434f Prospective NIPU and BioPU Models: A Push for Polyurethanes Sourced from Biomass..................... 455
Jennifer B. Dunn, Ulises R. Gracida-Alvarez, Troy Hawkins, Yasheen Jadidi, Chloe Liang

PRODUCTION OF LOW-CARBON FUELS FORM BIOMASS: DESIGN, SIMULATION, AND ECONOMIC ANALYSIS I

501b Comparative Study of Sustainable Aviation Fuel Produced from Wastewater Grown Algae Vs. Farmed Algae 457
Abhishek Kumar, Peter Valdez

501c Environmental-Techno-Economic Analyses of Biodiesel Production from Animal Fat Waste Using Hydrothermal Process 458
Pyeong-Gon Jung, Semie Kim, Young-Il Lim

501d Techno-Economic Analysis with Uncertainties for Alternative Fuels Enabling Near and Mid-Term U.S. Rail Sector Decarbonization 459
Yuan Jiang, Shuyun Li

501e Implications of Integrating Carbon Capture and Storage Technology into Sustainable Aviation Fuel (SAF) Production 460
Alberto Ruiz, Mariano Martin, Katie Chong, Röder Mirjam, Thornley Patricia

501f Techno-Economic and Life Cycle Analysis of $BioH_2$ Production from Waste Streams through Process Integration 463
Arna Ganguly, Xinyu Liu, Hernan Delgado, Lili Sun, Pingping Sun, Amgad Elgowainy

SUSTAINABLE DESIGN AND MODELING OF BIOREFINERIES II

232b Charting Development Pathways for Precision Fermentation through Agile System Design and Analyses... 464
Sarang Bhagwat, Christopher V. Rao, Huimin Zhao, Vijay Singh, Jeremy Guest

232d Evaluation of Alternative Processes for Grapeseed Oil Production from a Techno-Economic and Environmental Perspective 466
Rosa M. Fonseca-Pérez, Alberto Ruiz, César Ramírez-Márquez, Adrián Bonilla-Petriciolet, Mariano Martin

232c Integrated Techno-Economic and Life Cycle Analysis of Reductive Catalytic Fractionation and Base-Catalyzed Depolymerization Hybrid Processing for Lignin Valorization........ 468
Zahra Ebrahimpourboura, Aditya Ponukumati, Rhiannon Carr, Yu Gao, Zeyu Shang, Anirudh Krishnamurthy, Tae S. Moon, Marcus Foston, Mark M. Wright

232f Techno-Economic and Life Cycle Analysis of Conversion of Natural Gas and Biomass to Hydrogen and Performance Carbons........ 469
Farhan M. Haidary, Phuong N. Thi, Eric Cochran, Jean-Philippe Tessonnier, Steven Crossley, Mark M. Wright

Author Index

96a A Tale of Two Mines: Insights to the Similarities and Differences of Surface and Underground Critical Mineral Mines Life Cycle Analyses

Jenna Trost, Jennifer Dunn

Problem

Decarbonization technologies are potential solutions to reducing greenhouse gas emissions and mitigating climate change. However, the decarbonization transition will undoubtedly be mineral-intensive with critical minerals like cobalt, lithium, and nickel serving as the backbone of many decarbonization technologies. Mining is the primary acquisition method for critical minerals but wields many environmental and social effects – both positive (i.e. job opportunities and boosting the local economy[1]) and negative (i.e. land degradation[2], reduced water quality[3], and community displacement[2]). Life cycle analysis (LCA) is a tool that can assess effects and sustainability. Newer interest in critical minerals has sparked LCA research on critical minerals acquisition, but there are limited studies. Most of studies only investigated mine operations – such as extraction and chemical processing – and often excluded closure and reclamation. To the best of our knowledge, there are no studies that consider the whole life cycle of the mine – encompassing the first ground break to reclamation and mine closure.

Additionally, the location of most studies mirrored top critical mineral-producing countries, like Australia, Chile, and China. Few studies investigated U.S. mines. Moreau et al. (2021) compared manual versus automated operations for six different underground copper mines, two of were in the U.S. They found automation reduced global warming potential, acidification potential, and human toxicity potential up to 20%. Twin Metals Minnesota (TMM), a mining company in Minnesota, completed a prospective LCA of their proposed underground copper-nickel mine. Their LCA, like many others, included basic operations like mining, beneficiation, and electricity burdens but excluded land clearing, water treatment, and reclamation. Notably, they assumed electricity will come from wind farms in Iowa. They calculated the global warming potential (GWP) per kilogram of copper and nickel in concentrate. TMM found that explosive emulsions and quicklime were the largest GWP contributors.

Methods

We conduct LCAs on the whole mine life cycle of two proposed copper-nickel mines in Minnesota – which would mark the first non-ferrous mines in the state. One mine, PolyMet, is surface and the other, Tamarack, is underground. We employ two function units: one ton of mineral product equivalent and one year of mining operation. We define the system boundary as cradle-to-grave to encompass the entire life cycle of the mine, from first ground break to mine reclamation and closure. We use foreground system processing and site information from legal documents as an appropriate, open data source. We use Greenhouse Gases, Regulated Emissions, and Energy Use in Technologies (GREET) Model5, and literature for background system inventory data. We include uncertainty and scenario analyses to improve the robustness and validity of our LCAs. We calculate energy consumption, water use, and greenhouse gas emissions.

Results

The results will illuminate interesting dynamics between surface and underground mining LCAs and provide insights into the associated environmental costs. For example, PolyMet may require more land clearing and result in a larger carbon stock loss than Tamarack. Conversely, Tamarack may require more fuel emissions to dig and haul rock and ore out miles beneath the surface. This work will showcase how two different mine types with different life expectancies and operations can be compared using LCA. Their geographic proximities – a mere 90 miles apart – are an advantage to truly understanding the burdens, similarities, and differences of surface and underground mining. We will implement the same background data for many processes. For example, electricity from the Minnesota grid mix will source both projects. Reducing data variances limits additional uncertainties and allows dynamics to shine.

Implications

These results will help guide decision-making on what type of mine to *implement* for a given ore deposit to minimize environmental costs and will expand the limited LCA research of critical mines in the United States. These results can guide the development of a critical mineral LCA framework to standardize comparison between mines regardless of type, mineral product, and geographic location. These results expand, which can help guide policy and decision-making. Also, the two mines are close to Indigenous reservations so future work could encompass including Indigenous perspectives into critical mineral LCAs to broaden the definition of sustainability.

96b Prospective Impact Analysis Combining Integrated Assessment Modeling and Life Cycle Assessment for Alternative Industrial Heat Sources.

Tapajyoti Ghosh, Patrick Lamers

Background/Objectives:

The U.S. government's commitment to achieving a net-zero greenhouse gas (GHG) emissions economy by 2050 aligns seamlessly with the global climate mitigation objectives outlined in the Paris Agreement. The overarching goal is to limit the average temperature increase to 1.5 °C or less by 2100, compared to preindustrial levels. To realize this ambitious domestic mid-century target, there is an imperative need for the rapid adoption of energy-efficient technologies and the decarbonization of pivotal sectors such as power and transportation. This involves embracing electrification, fuel switching, and the expansion of variable renewable energy sources and storage technologies. Furthermore, an increased emphasis on electrification is crucial for both the buildings and industrial sectors.

While the power and transportation sectors, contributing 29% and 25% to total U.S. national GHG emissions, have extensively outlined and modeled their decarbonization strategies, meeting the 2035 and 2050 targets demands a concerted effort due to the substantial scale of the power sector and the heterogeneous nature of the transport sector. The industrial sector, responsible for 23% of total U.S. GHG emissions, poses a unique challenge due to its activities that are challenging to electrify. Addressing these activities requires the exploration of technologies that are either less understood or have not yet been scaled. Finding innovative solutions in this realm is crucial for achieving the broader climate goals.

Approach/Activities:

Emerging technologies call for the application of forward-looking life cycle assessment (LCA) methodologies. These methodologies enable the comprehensive consideration of technology scaling and process enhancements, including learning by doing. Additionally, the future system context in which these technologies are envisioned to operate holds equal significance in many instances. Background scenarios systematically generated by integrated assessment models (IAMs) proficiently incorporate the evolving dynamics of the energy-economy-land-climate system. These IAM scenarios are harmonized across socioeconomic and climate change mitigation pathways, enhancing the comparability of prospective LCAs utilizing various IAMs.

In this context, we present an open-source framework for prospective LCA, known as the Life-cycle Assessment Integration into Scalable Open-source Numerical models (LiAISON). LiAISON facilitates the examination of non-linear relationships between technology foreground and the future energy system background, encompassing a spectrum of midpoint and resource-use metrics.

Results/Lessons Learned:

The LiAISON framework has been used to assess the environmental impact of various hydrogen production methods in the United States. This involved combining data from two integrated assessment models, IMAGE and GCAM, to enable life cycle assessment (LCA) across different scenarios. The framework is currently applied to a case study involving industrial heat supply in the US. Several technologies and strategies play a role in decarbonizing industrial heat. Electrification, where traditional fossil fuel-based heating systems are replaced with electric alternatives powered by renewable energy, is one approach. Another method involves the use of green hydrogen produced through electrolysis using renewable energy sources. Additionally, improving energy efficiency in industrial processes helps reduce overall heat demand, contributing to the decarbonization goal. The study not only analyzes LCA results with temporal and geospatial details for two technologies but also aims to create a foundational framework that can be extended to incorporate other scenarios generated by integrated assessment models and U.S. open-source life cycle inventory databases.

96c Evaluating the Manufacturing Emissions and End-of-Life Impact of Paperboard Cartons Vs. PET Bottles

Nur Hendri Wahyu Firdaus, Julia Shamshina, Jerrad Legako, Kalavathy Rajan

Liquid packaging board (LPB) could be an eco-friendly alternative for single use plastics, such as polyethylene terephthalate (PET) bottles, and play a critical role in reducing petrochemical use as well as mitigating micro- and nano-plastic pollution. However, comprehensive life cycle assessment is required to prove the environmental benefits of LPB over PET bottles. The life cycle assessment (LCA) conducted in this study, at the manufacturing stage as well as at three different end-of-life scenarios, i.e., recycling, incineration, and landfilling, evaluates 11 environmental impact categories for LPB and PET containers. For the same container volume and number of packaging units, PET bottles caused 18% higher negative environmental impacts at the manufacturing stage. Similarly, at the end-of-life, PET bottles caused an average of 68% higher negative environmental impact if it were to be landfilled or recycled. These results underscore the pressing need for more sustainable packaging solutions amid the growing concern for ecosystem health.

96d Microbially Produced Proteins to Reduce Global Pressure on the Planetary Boundaries

Samir Meramo, Eleonora Passuto, Sumesh Sukumara

There is increased awareness about the global environmental sustainability challenges modern society faces including climate change, the need for more sustainable agriculture, and economic growth. The Planetary Boundary (PB) framework was introduced to communicate the safe operating space for humanity to thrive towards sustainable development while keeping Earth's systems stable at the Holocene-like state. However, six out of nine planetary boundaries have been transgressed, where agriculture and food systems are major impact drivers in categories like climate change, freshwater change, or biogeochemical flows. In a food chain context, meat is the major contributor to global environmental impacts. Therefore, an absolute environmental sustainability assessment is presented here in line with the PB framework and the Life Cycle Assessment (PB-LCA) of biologically produced meat alternatives to show the environmental hotspots and assess replacement scenarios based on global consumption trends.

The PB-LCA is built on calculating environmental impact categories pertaining to the Earth System Processes described in the PB framework. As the LCA and PB categories do not match either in their control variables or in the impact level in the cause-effect chain, the Environmental Footprint (EF) v3.1 method, the life cycle impact assessment method used for which absolute thresholds based on the Planetary Boundaries are available. The above enables the harmonization of LCA calculations and the definition of safe operating space in the PB framework. Relevant impact categories in the EF v3.1 were selected, which show more relation to the PB, including climate change, water deprivation, marine and freshwater eutrophication. Land system change is measured using land occupation as a proxy, and biodiversity loss is calculated based on the loss in mean species abundance as a proxy for functional diversity boundary. After impact characterization, the absolute sustainability ratios are calculated to determine the system's performance in absolute terms. This step requires allocating the share of the safe operating space for the system.

The share of the safe operating space is downscaled based on equal per capita principles, and allocation was defined based on economic and grandfathering approaches. The LCA results of the microbially produced meat show that increasing the energy efficiency of the production process could lead to reducing the impact below the absolute threshold for climate change or land use change. Reduction of all impact categories was observed with promising emissions avoided in climate change, land use, and biodiversity loss. The above is attributed to lower resources and land transformed for red meat consumption, resulting in less soil degradation for pasture, grazing, and crops for animal feed, along with less fertilizers and lower energy consumption, even for the scenarios with renewable energy.

96e Enhancing Direct Air Capture and Utilization (DACCU) Pathways through Life Cycle Assessment, Cost Analysis, and Optimization

Nick Badger, Mustafa Erguvan, Shahriar Amini

The urgency of climate change calls for innovative carbon management strategies. Direct Air Capture and Utilization (DACCU) emerges as a crucial solution for reducing atmospheric CO_2 levels while supplying valuable feedstocks for various industrial applications. This study focuses on optimizing key DACCU pathways: the production of formic acid, Fischer-Tropsch Synthesis products, liquid CO_2, and methanol. These pathways are evaluated for their potential to contribute to carbon sequestration and utilization efforts.

Life Cycle Assessment (LCA) methodology is employed to assess the environmental impacts associated with each DACCU pathway. The analysis covers energy requirements, greenhouse gas emissions, and potential environmental tradeoffs, offering a holistic perspective on the sustainability of each process. Additionally, detailed costing information is presented to better understand the economic feasibility and market competitiveness of these technologies.

Mixed-integer linear programming model is used to investigate future pathways for direct air capture and CO_2utilization with a focus on costs and carbon reduction. The system is optimized on the whole supply chain costs and on a number of environmental objectives using the U.S. as an example.

The outcomes of this study provide valuable insights for a wide range of stakeholders, including industry leaders and policymakers. It emphasizes the importance of advanced DACCU pathways in addressing climate change, supported by thorough LCA results, cost analyses, and optimization strategies. This comprehensive approach brings to the forefront the need for more sustainable and economically viable carbon capture and utilization practices.

96f Advancing the Economic and Environmental Sustainability of Rare Earth Element Recovery from Phosphogypsum Using a Selective Bio-Based Adsorptive Separation

Adam Smerigan, Rui Shi

Transitioning to green energy technologies requires more sustainable and secure rare earth elements (REE) production. The current production of rare earth oxides (REOs) is completed by an energy and chemically intensive process (beneficiation, leaching, separation by solvent extraction, and refinement) from the mining of REE ores. Investigations into a more sustainable supply of REEs from secondary sources, such as toxic phosphogypsum (PG) waste, is vital to securing the REE supply chain. PG is a waste byproduct of fertilizer production produced at hundreds of millions of tonnes per year. With a minimum REE content of ~0.02 wt% REEs, PG has the potential to meet the annual REE consumption within the United States (~9,000 tons REE/year). Additionally, PG is stored indefinitely in 'stacks' which are vulnerable to the release of toxic and radioactive waste into the environment. The extraction of REEs from this waste may make remediation feasible. However, it is unclear how to recover the dilute REEs from PG waste (conventional solvent extraction is inefficient and has high environmental impact). In this work, we propose a treatment train (technological readiness level < 3) for the recovery of REEs from PG which includes a bio-inspired adsorptive separation to generate a stream of pure REEs. We assessed its financial viability and life cycle environmental impacts via life cycle assessment (LCA), techno-economic analysis (TEA), and identify targeted improvement opportunities through global uncertainty/sensitivity analysis and scenario analysis. We determined that this system can be profitable (net present value of $200 million, internal rate of return of 17%, and minimum selling price (MSP) of $48/kg REO) and shows reduced environmental impact in several ReCiPe 2016 impact categories (land use, ecotoxicity, eutrophication, ionizing radiation, material resource depletion, and human toxicity) when compared to conventional REO production and PG stacking. However, the system underperforms in other impact categories (climate change, terrestrial acidification, ozone depletion, photochemical oxidant formation, particulate matter formation, and fossil resource depletion). The endpoint analysis shows that the system outperforms conventional REO production in ecosystem quality (93.0% of the impact) and resource depletion (96.2% of the impact) but underperforms in human health (213% of the impact). The life cycle environmental impacts and financial viability are primarily driven by chemical consumption in the leaching, concentration, and wastewater treatment process sections ($15/kg REO). In addition to high chemical costs, the large capital cost of the selective adsorption resin ($18/kg REO) limits profitability. Scenario analysis shows that the system is profitable at capacities larger than 100,000 kg/hr PG for PG with REE content above 0.5 wt%. The most dilute PG sources (0.02-0.1 wt% REE) are inaccessible using the current process scheme (limited by the cost of acid and subsequent neutralization) requiring further examination of new process schemes and improvements in technological performance. Overall, this study evaluates the sustainability of a first-of-its-kind REE recovery process from PG and uses these results to provide clear direction for advancing sustainable REE recovery from secondary sources.

96g Marine Scrubbers Vs. Low-Sulfur Fuels: A Comprehensive *Well-to-Wake* Life Cycle Assessment Supported By Measurements Aboard an Ocean-Going Vessel

Patricia Stathatou, Ievgenii Petrunia, Torsten Barenthin, George Gotsis, Paul Jeffrey, Christopher Fee, Scott Bergeron, Marios Tsezos, Michael Triantafyllou, Neil Gershenfeld

Maritime transport serves as the backbone of the global economy, with ocean-going vessels transporting more than 80% of the total volume of internationally traded goods[1]. The shipping sector currently contributes around 3% of the world's annual greenhouse gas (GHG) emissions, alongside sulfur dioxide (SO_2), nitrogen oxides (NO_x), and particulate matter (PM) emissions, which can cause air pollution and climate change, while elevating ozone levels impacting human health[2]. Efforts to mitigate environmental pollution from maritime transport have been prioritized recently, leading to the implementation of stricter regulations. The International Maritime Organization (IMO), among other measures, established the Emission Control Areas (ECAs), where NO_x and SO_x emissions are more rigorously controlled, and set upper limits for the sulfur content of marine fuels under the Annex VI of the International Convention for the Prevention of Marine Pollution from Ships (MARPOL). Since January 1, 2020, this maximum allowable sulfur content has been reduced to 0.1% m/m inside ECAs and 0.5% m/m outside ECAs, aiming to further limit the global SO_x emissions[3,4].

To abide by these regulations, ocean-going bulk carrier vessels, which have been traditionally burning heavy fuel oil (HFO) with higher sulfur contents (1-3%), have three options: i) switch to low sulfur fossil fuels, such as marine gas oil (MGO) and very low-sulfur fuel oil (VLSFO); ii) burn biofuels, such as biodiesel, which have significantly low sulfur contents; iii) implement exhaust gas cleaning systems or scrubbers to remove SO_x emissions while continuing burning HFO. The first two options are considerably more expensive, while the current availability of biofuels is very low to accommodate demand. Due to the economic feasibility of installing marine scrubbers that would allow for continuing burning less expensive HFO, the third option is the most widely used, with over 4,000 vessels being fitted with scrubbers in 2023 compared to less than 800 in 2018[5]. The most prevalent scrubber type is the wet, open-loop one, which uses sea water to neutralize SO_x in the exhaust gases, discharging the generated washwater from this process to the sea without treatment.

While the widespread adoption of marine scrubbers is a positive step towards SO_x emission reduction, growing concerns are being raised regarding the release of washwater into the sea and its potential impact on marine life and biogeochemical processes. The acidic nature of washwater, combined with the presence of various organic and inorganic contaminants, such as heavy metals, Polycyclic Aromatic Hydrocarbons (PAHs), and eutrophication agents, may pose potential threats to marine ecosystems. Yet, there is no conclusive scientific evidence demonstrating significant harm to aquatic life from scrubber washwater. Additionally, comprehensive comparisons accounting for the environmental impacts of HFO/scrubber production and operation, versus the production and combustion of low sulfur fuels are lacking. Therefore, we conducted a first-of-its-kind, holistic, *Well-to-Wake* life cycle assessment (LCA) comparing climate change, air pollution, acidification, eutrophication and ecotoxicity impacts associated with the production of a marine scrubber as well as the production, bunkering and combustion of HFO, MGO and VLSFO under similar operating conditions, informed by real-time, air and washwater emission data collected aboard an ocean-going bulk carrier vessel.

In this talk, the results of this comprehensive, measurement-informed LCA study will be presented. Scrubber production emissions were found to be negligible throughout its lifetime compared to fuel emissions. Scrubber/HFO had significantly lower *Well-to-Tank* impacts than MGO and VLSFO across all the considered LCA impact categories. For the *Tank-to-Wake* assessment, CO_2, CO, NO_x, SO_2 and $PM_{2.5}$ emission data were collected for all three fuels aboard the Hedwig Oldendorff vessel, following the ISO 8178 standard[6]. In parallel, during scrubber operation, seawater and washwater samples were collected and analyzed for over 60 quality parameters. Fuel and lubricant samples were also collected to close the mass balance. The *Tank-to-Wake* results demonstrated notable reductions in almost all air-related impact categories for scrubber/HFO compared to low-sulfur fuels, with lower emissions observed, except for PM formation. Specifically, HFO released on average approximately 10% less g $PM_{2.5\text{-}Eq}$/MJ after the scrubber compared to VLSFO, but 26% more compared to MGO. With regards to washwater emissions, all the analyzed parameters were significantly below the IMO thresholds. The *Tank-to-Wake* marine eutrophication and ecotoxity LCA impact indicators were higher for scrubber/HFO due to lack of washwater generation in case of low-sulfur fuels. However, the measured, net concentrations of all organic and inorganic pollutants released through washwater into the sea were below the US EPA Aquatic life criteria for toxic chemicals, suggesting no likely adverse impacts on receiving water bodies and ecosystems. Overall, the presented study will challenge the notion that *'end-of-pipe'* solutions outperform *'start-of-pipe'* ones, emphasizing the need for holistic, measurement-informed LCA approaches. The adoption of such *cradle-to-grave* approaches by policy makers, supported by robust data while accounting for a variety of impacts and not just focusing on particular emissions or parts of the considered supply chains, would enable the effective assessment of different fuel systems, pollution abatement and decarbonization technologies avoiding perverse incentives and expediting maritime decarbonization. The implications of these findings for the scientific community, the shipping industry, and policy makers will be also discussed, paving the way for adopting the LCA rationale in other transportation sectors.

References

1. UNCTAD, Review of Maritime Transport: Towards a Green and Just Transition. 2023.
2. IMO 4th GHG Study
3. IMO: International Maritime Organization, Prevention of Air Pollution from Ships, https://www.imo.org/en/OurWork/Environment/Pages/Air-Pollution.aspx, accessed August 9, 2021.
4. https://www.imo.org/en/MediaCentre/HotTopics/Pages/Cutting-GHG-emissions.aspx
5. https://www.statista.com/statistics/1099286/number-of-scrubbers-in-vessels/
6. ISO: International Organization for Standardization INTERNATIONAL STANDARD ISO 8178-2: Reciprocating Internal Combustion Engines: Exhaust Emission Measurement. Part 2: Measurement of Gaseous and Particulate Exhaust Emissions under Field Conditions , Geneva, Switzerland, 2008.

96h Integrated Environmental, Economic, and Social Lifecycle Assessment: A Novel 'Hub and Spoke' Methodological Framework

Alex Newman

Sustainability is best viewed as being underpinned by three constituent impact areas, or 'strands': environment, economics, and society. Despite this, methodology development in life cycle assessment initially focussed purely on environmental aspects, predominantly utilising the ISO 14040 standards, while neglecting economic and societal aspects. A fully harmonised and integrated assessment framework remains elusive, preventing a detailed and holistic assessment of impact burden shifting between the environmental, economic, and social strands.

In recognition of this, the work presented herein outlines a novel 'hub and spoke' framework for the holistic life cycle assessment of chemical value chains. Its modular structure maximises utility in the context of chemical value chains, with hubs representing core process under an organisations control, and spokes representing competing feedstock generation routes and suppliers. In many application cases a significant number of possible hub and spoke permutations will be present. Through the incorporation of a novel aggregation procedure these are assessed simultaneously, removing the need for manual evaluation of each possibility. As with traditional methodologies, aspects such as system boundary specification, indicator selection, and allocation procedures are left at the practitioner's discretion.

The framework's underpinning methodology is ISO 14040 aligned across all three strands, building upon a strong methodological backbone in which a majority of practitioners are already well versed. Furthermore, it is compatible with broadly utilised characterisation methods such as ReCiPe, TRACI, and CML. Additionally, parallel integration of multi criteria decision making (MCDM), in the form of hybrid AHP-TOPSIS, weights and aggregates the assessed impact indicator results to generate a single overall 'sustainability score'. Consequently, value chain recommendations are tailored to the commissioning organisations priorities. These overall value chain scores for each permutation are presented in parallel to the traditional objective indicator results, framing the subjective MCDM derived value chain scores and providing much valued reporting transparency.

Methodological validation was carried out through a proof-of-concept study, examining value chain performance across 19 environmental, economic and social impact indicators. The study focused on the production of soda ash in India via the Hou process, considering 35 alternative routes for feedstock production. This verified significant efficacy, reducing practitioner workloads by up to 68% and simultaneously assessing 14,580 value chain permutations. Recommendations from the model were checked against the parallel objective result, confirming that AHP-TOPSIS can be used within lifecycle assessment to screen large decision spaces without compromising scientific validity. Furthermore, it has been demonstrated that the developed lifecycle inventory structure facilitates a 'plug and play' philosophy in which seamless re-use in subsequent assessments is possible.

Implications of this work are far reaching, benefiting stakeholders throughout value chains, beyond improvements in environmental performance. Organisations can better inform strategic decision making around sustainable development, minimising risk of unintended secondary economic or social effects associated with planned projects or initiatives. Meanwhile, consumers benefit from the inclusion of both economic and social aspects. The increased cost of goods, often associated with more environmentally sustainable processes, are minimised, while safeguarding against impacts on quality and quantity of life.

In summary, the holistic life cycle assessment methodology presented in this work offers a robust and versatile tool for the identification of sustainable value chains. Two major advancements presented include the integration of AHP-TOPSIS and the modular approach to life cycle inventory structure.

229g A Machine Learning Approach for Examining Flushing Operations in Packaging Plants

Robert Hesketh, Barnabas Gao, Swapana S. Jerpoth, Emmanuel Aboagye, C. Stewart Slater, Kirti Yenkie

A typical oil manufacturing industry formulates over a thousand unique oil products by blending different base oils with numerous groups of additives. Each of these products is packaged in a container at the industry and it is not practical to have a separate production line for each product. So, the system of blending vessels, ancillary equipment, and associated pipeline network must be reused multiple times for numerous batch productions. Each product has a given set of product quality requirements and the blending of two oils must be minimized. In general, the use of external agents to clean the line between filling operations is not used since this introduces additional contaminants to the oil. So, the equipment is cleaned between filling operations by using the upcoming product batch to clean or flush the filling equipment and associated pipelines. This results in the formation of commingled/mixed oil that is regarded as a low-value product since it doesn't meet product specifications. This project reports on our work in identifying optimization opportunities in the filling operations. To achieve desired product specifications, attain quality targets, and maximize the productivity of assets, it is of paramount importance to enhance process control, minimize human errors and improve the resource management footprint of these industries. To this end, we present a machine learning approach to learn from the data of existing operations and strategically optimize the flushing operations.

229b Deep Learning-Enabled Modeling Framework to Evaluate the Financial Feasibility of Hydrothermal Liquefaction Waste Valorization Systems

Ali Ahmad, Yalin Li

Organically rich solid waste valorization processes have the potential to zero-waste disposal and minimize greenhouse gas emission by eliminating landfill of waste as well as creating high-value-added products from it. Hydrothermal liquefaction (HTL) is a feasible process that converts biomass into an energy-dense biocrude oil that can be upgraded and refined to make liquid fuels, platform chemicals, as well as materials (e.g., bio-based asphalt binder) from various feedstocks. However, the tradeoff between deployment and scale-up options for viable feedstocks, as well as uncertainty around biocrude yield due to feedstock variability makes it difficult to determine the ideal configurations and operating parameters HTL valorization systems. To solve this problem, we incorporate a deep learning (DL) regression model that forecasts biocrude yield based on the biochemical composition of feedstock profiles. The results show the high predictability of the model with a mean squared error (MSE) < 0.02 for validation data. When integrated into techno-economic analysis (TEA), the model can be leveraged to deliver insights for cost-competitive new waste valorization pathways or upgrading of existing facilities with lower environmental impacts.

Introduction

Before scaling up or commercializing novel technologies, TEA is often carried out to evaluate the financial viability of the system and identify cost drivers for optimization. A minimum selling price (MSP) of product—the lowest price to achieve a net present value of zero—is commonly used to determine the process economic feasibility (Luo, et al., 2021) . As product yield typically has a significant impact on the process's profitability, wet lab experiments are needed to provide accurate data for TEA. Depending on the variation of potential feedstocks and operating conditions, these experiments can be costly and time-consuming, especially when multiple system configurations exist.

With the advent of artificial intelligence (AI), data-driven models have been increasingly used in the modeling of complex reaction systems where the use of mechanistically based models is challenging (e.g., HTL valorization systems). To this end, machine learning (ML) approaches have been used to predict biocrude yield based on the characteristics of feedstocks during HTL (Cheng, et al., 2022) (Omidkar, Alagumalai, Li, & Song, 2024) (Cheng & Luo, 2022) , but these models have not been further used to provide insight on the financial feasibility of HTL systems. In this work, a regression-supervised deep learning (DL) model was developed to predict the yield of the biocrude from various HTL feedstocks, and the developed model was integrated into TEA to evaluate the economic feasibility of an HTL valorization system for the production of asphalt binder with co-production of biofuel additives and fertilizers.

DL Model Development

Literature data was leveraged to construct a preliminary dataset of biochemical composition (lipid, protein, cellulose, hemicellulose, carbohydrate, lignin, and ash) containing various feedstocks and associated HTL biocrude yield (709 data points from 190 publications). The data set used here had a small data size and uneven data distribution in a wide range (0–100) of response variable. All data preparation and model development were performed using Python libraries, including *scikit-learn* for data pre-processing, dataset splitting, normalization, and missing value handling (Abraham, et al., 2014) and *Tensorflow* for model development (Si, Tarnoczi, Wiens, & Du, 2019) . All data were unified on a dry weight basis and normalized to a uniform scale prior to use. Unreported values were treated with a "replacement with mean" strategy. Pearson and spearman correlation coefficients between the biochemical composition features were then evaluated to eliminate features with strong linear correlations (cutoff = 0.7). *KFold* (value=3) was then used to split the dataset into 3 groups for cross-validation.

A sequential DL model with 4 layers was developed. Among the four dense layers, three (64 units) were equipped with ReLU activation and one (1 unit) with linear activation function. Finally, Adam Optimizer was used to optimize the model with a learning rate of 0.01 that follows stochastic gradient descent method (Kingma P. & Ba, 2015) . Several other supervised neural network algorithms, including multi-layer perceptron and support vector regression, were also evaluated, but the DL model with sequential layers offered better performance across model metrics and were used in subsequent analyses. Model output shows a generic trend to pick up rapid changes in biocrude yield due to variation in feedstock biochemical composition (Figure A). Notably, the assembled dataset had model compounds with 100% lipid, protein, or carbohydrate contents with nearly 100% biocrude yield, which was captured by the model, showing the robustness of the DL model. The model was further evaluated using a validation split of 0.33 (i.e., the validation dataset consists of 33% of the data). The general trend of both training and validation losses was similar across epochs. Although a larger difference was observed around 50 epochs (Figure B), presumably due to overfitting from the "replacement-by-mean" strategy used to offset missing value problems, both training and validation loss values remained extremely low (<0.02 across the epochs), revealing the robustness of the model in predicting biocrude yields.

TEA

The predicted yield of biocrude can then be utilized to evaluate MSP of a hydrothermal system for biobinder production (Figure C). Specifically, after gravity separation and skimming, the collected biocrude and char products were fractioned by distillation, with the light and medium fraction used as biofuel additives and the heavy fraction as biobinder. Aqueous products were treated by three potential approaches (sand filtration, electrochemical oxidation, or fungi treatment), after which the collected liquid could be used as liquid fertilizer to supplement crop nutrient (particularly N) requirements. TEA was then performed using discounted cash flow rate of return (Snowden-Swan, et al., 2017) using the predictive outputs from the DL model for feedstocks of varying biochemical compositions. Overall, this work demonstrates the utility of this integrated modeling framework to provide insights on system economic performance for feasibility analysis and process optimization.

References

Abraham, A., Pedregosa, F., Eickenberg, M., Gervais, P., Mueller, A., Kossaifi, J., . . . Varoquaux, G. (2014). Machine learning for neuroimaging with scikit-learn. *Frontiers in Neuroinformatics*, 1-10.

Cheng, F., & Luo, H. (2022). Evaluating the minimum fuel selling price of algae-derived biofuel from hydrothermal liquefaction. *Bioresource Technology Reports*, 100901.

Cheng, F., Belden, E. R., Li, W., Shahabuddin, M., Paffenroth, R. C., & Timko, M. T. (2022). Accuracy of predictions made by machine learned models for biocrude yields obtained from hydrothermal liquefaction of organic wastes. *Chemical Engineering Journal*.

Kingma P., D., & Ba, J. (2015). Adam: A Method for Stochastic Optimization. *3rd International Conference for Learning Representations*. San Diego: ICLR .

Luo, H., Cheng, F., Yu, B., Hu, L., Zhang, J., Qu, X., . . . Luo, Z. (2021). Full-scale municipal sludge pyrolysis in China: Design fundamentals, environmental and economic assessments, and future perspectives. *Science of The Total Environment*, 795.

Omidkar, A., Alagumalai, A., Li, Z., & Song, H. (2024). Machine learning assisted techno-economic and life cycle assessment of organic solid waste upgrading under natural gas. *Applied Energy*.

Si, M., Tarnoczi, T. J., Wiens, B. M., & Du, K. (2019). Development of Predictive Emissions Monitoring System Using Open Source Machine Learning Library – Keras: A Case Study on a Cogeneration Unit. *IEEE Access*, 113463 - 113475.

Snowden-Swan, L. J., Zhu, Y., Bearden, M. D., Seiple, T. E., Jones, S. B., Schmidt, A. J., . . . Fox, S. (2017). *Conceptual Biorefinery Design and Research Targeted for 2022: Hydrothermal Liquefaction*

Processing of Wet Waste to Fuels (No. PNNL-27186). Richland, WA (United States): Pacific Northwest National Lab. (PNNL).

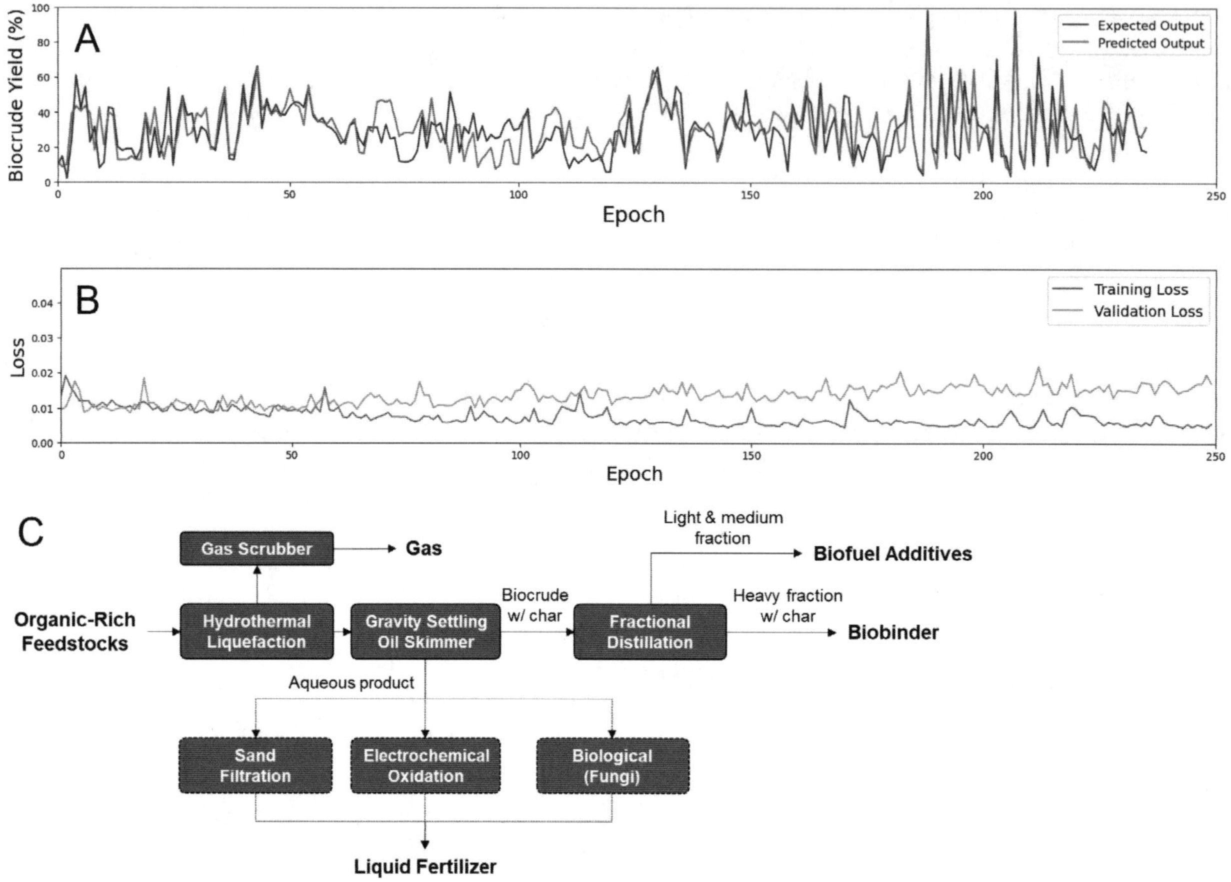

229c Sustain-Gpt: A Large Language Model for Creating a Structural Database from Unistructural Text Resources to Develop Life Cycle Inventories

Avan Kumar, Amrita Sen, Bhavik Bakshi

Technological advancement, aimed at improving the quality of life for humankind and equitable access to resources for all, often has an adverse impact on nature. The systematic accounting of such impact in the design and deployment of technologies is, however, a recent development [1]. Life-cycle assessment (LCA) is the methodology most commonly used to assess the overall environmental impact of a technology or product across all steps in its value chain. LCA thus serves as a tool to compare the impacts of an innovation over conventional alternatives, as well as between multiple innovations [2]. One of the most crucial and tedious steps in LCA is collecting datasets for subsequent analysis [3]. These datasets are a combination of background and foreground process data, and thus necessitate domain knowledge of the process being studied as well as access to life-cycle inventories (LCI). LCIs vary in the granularity of data included, depending on the ubiquity of the process, as well as the spatial information covered. Some sources of inventories include EcoInvent, USLCI, GaBi etc. Additionally, the scientific literature on LCA studies boasts a rich source of inventory data, which may or may not be covered in background datasets. The source of these inventories and, hence, the quality of the data vary across articles, thus introducing uncertainty.

These unstructured datasets constitute ~80% of all available life-cycle inventories and exist in the form of text, tables, and images. Systematic mining can potentially exploit the size of these datasets and their hidden knowledge. Natural language processing (NLP), a subfield of machine learning (ML), is a suitable candidate for this application. The breakthrough in the field of NLPs was the introduction of transformer architecture in 2017 [4]. This was followed up by the introduction of two contributor models: the Bidirectional Encoder Representation of Transformer (BERT) [5] and the Generative Pre-trained Transformer (GPT) [6]. Recently, BERT, a pre-trained model trained on a large corpus with a hyperparameter count range of 110M to 340M, was further fine-tuned on a custom dataset using a transfer learning approach. Since then, BERT has been applied in such niche scientific fields as SciBERT [7] , BioBERT [8], PatentBERT [9] , and Recycle-BERT [10]. Furthermore, the GPT model has also been utilized in various applications such as BioGPT[11], GPT-GNN[12], and versions of GPTs [13], [14] . A real-life application of GPT-3 is ChatGPT, which works on questions and answers and text generation tasks. The transformer-based models are called Large Language Models (LLMs) if their hyperparameter counts are in billions. In 2023, the largest-sized LLM was launched by Meta, Llama-2, with hyperparameter counts ranging from 7B to 175B [15].

In this work, we are leveraging the capabilities of the Llama-2 model and proposing Sustain-GPT, a large language model that would be pre-trained and fine-tuned on research articles. It will help us retrieve quantitative and qualitative information from chemical process-based literature. The extracted knowledge can be utilized in various problem-solving approaches by creating databases such as custom LCI databases, reaction databases of chemical processes with uncertainty in values. As a preliminary result, we are showcasing the results from Sustain-GPT in the case of LCI subset creation for the methanol synthesis process. We are utilizing the full-text research articles and abstracts extracted from the Elsevier database using an API key and institute token with the simple, specific keyword 'life cycle assessment for methanol synthesis .' The total number of full text articles downloaded is ~5k. Figure 1(a) shows the trend of number of reliable publications in the last 24 years. We pre-trained the Llama-2 model on our custom corpus, which is the full text of research articles. Further, we fine-tuned our custom pre-trained model for a question-and-answer task. Additionally, it was encapsulated in the form of a chatbot framework using the open-source Python library Gradio. To fine-tune the model, we generated a total of ~500 pairs of questions and answers related to the field of methanol LCA. The learning curve is shown in Figure 1(b); it reflects how loss values are decreasing with steps, and the proximity of the lines indicates the perfect fitting nature of the model. The model tested and retrieved quantitative information related to LCI creation for the functional unit of 1 kg of methanol by asking a set of questions to the Sustain-GPT model. The responses are compared with the standard LCI database and tabulated in Table 1. The results are close to standard data. Thus, the preliminary results are promising.

Sustain-GPT's scope covers quantitative knowledge extraction for industrial processes used to synthesize various chemicals. The variables of interest include product yield, process costs, reagent use, etc. towards populating the input and output flows for an inventory. To do this, we will retrain our proposed model on reliable literature for various chemical synthesis processes. The overarching LCI data retrieval system will consist of super prompts. These super prompts, essentially a combination of sub-prompts, progressively populate the various fields of an LCI. Thus, the creation of an LCI only requires a single super prompt from the user. Additionally, this will allow the compilation of foreground LCI for hitherto uninvestigated processes. Finally, the sourcing of data from multiple publications will yield a range of values for the variables and a sense of associated uncertainty. This will allow users the freedom to populate the input and output flows for their LCA study.

References

[1] R. G. Hunt and W. E. Franklin, "LCA - How it Came about - Personal Reflections on the Origin and the Development of LCA in the USA," International Journal of Life Cycle Assessment, vol. 1, no. 1, pp. 4–7, 1996, doi: 10.1007/BF02978624.

[2] W. Heijungs, 'Environmental life cycle assessment of products-Guide', *Technical Report of Centre of Environ. Sci.*, 1992.

[3] R. Heijungs and S. Suh, The computational structure of life cycle assessment. 2002. Accessed: Apr. 05, 2024. [Online]

[4] A. Vaswani et al., "Attention Is All You Need."

[5] J. Devlin, M.-W. Chang, K. Lee, K. T. Google, and A. I. Language, "BERT: Pre-training of Deep Bidirectional Transformers for Language Understanding." [Online]. Available: https://github.com/tensorflow/tensor2tensor

[6] A. R. Openai, K. N. Openai, T. S. Openai, and I. S. Openai, "Improving Language Understanding by Generative Pre-Training." [Online]. Available: https://gluebenchmark.com/leaderboard

[7] I. Beltagy, K. Lo, and A. Cohan, "SciBERT: A Pretrained Language Model for Scientific Text," EMNLP-IJCNLP 2019 - 2019 Conference on Empirical Methods in Natural Language Processing and 9th International Joint Conference on Natural Language Processing, Proceedings of the Conference, pp. 3615–3620, Mar. 2019, doi: 10.18653/v1/d19-1371.

[8] J. Lee et al., "BioBERT: a pre-trained biomedical language representation model for biomedical text mining," academic.oup.com, Accessed: Apr. 04, 2024. [Online]. Available: https://academic.oup.com/bioinformatics/article-abstract/36/4/1234/5566506

[9] J.-S. Lee and J. Hsiang, "PatentBERT: Patent Classification with Fine-Tuning a pre-trained BERT Model," May 2019, Accessed: Apr. 04, 2024. [Online]. Available: http://arxiv.org/abs/1906.02124

[10] A. Kumar, B. R. Bakshi, M. Ramteke, and H. Kodamana, "Recycle-BERT: Extracting Knowledge about Plastic Waste Recycling by Natural Language Processing," ACS Sustain Chem Eng, vol. 11, no.

32, pp. 12123–12134, Aug. 2023, doi: 10.1021/acssuschemeng.3c03162.

[11] R. Luo et al., "BioGPT: generative pre-trained transformer for biomedical text generation and mining," Brief Bioinform, vol. 23, no. 6, pp. 1–11, Nov. 2022, doi: 10.1093/BIB/BBAC409.

[12] Z. Hu, Y. Dong, K. Wang, K. W. Chang, and Y. Sun, "GPT-GNN: Generative Pre-Training of Graph Neural Networks," Proceedings of the ACM SIGKDD International Conference on Knowledge Discovery and Data Mining, pp. 1857–1867, Aug. 2020, doi: 10.1145/3394486.3403237.

[13] A. Radford, J. Wu, R. Child, D. Luan, D. Amodei, and I. Sutskever, "Language Models are Unsupervised Multitask Learners", Accessed: Apr. 04, 2024. [Online]. Available: https://github.com/codelucas/newspaper

[14] T. B. Brown et al., "Language Models are Few-Shot Learners," Adv Neural Inf Process Syst, vol. 2020-December, May 2020, Accessed: Apr. 04, 2024. [Online]. Available: https://arxiv.org/abs/2005.14165v4

[15] H. Touvron et al., "Llama 2: Open Foundation and Fine-Tuned Chat Models," Jul. 2023, Accessed: Apr. 04, 2024. [Online]. Available: https://arxiv.org/abs/2307.09288v2

[16] P. Levi, J. C.-E. science & technology 2018, "Mapping global flows of chemicals: from fossil fuel feedstocks to chemical products," ACS Publications

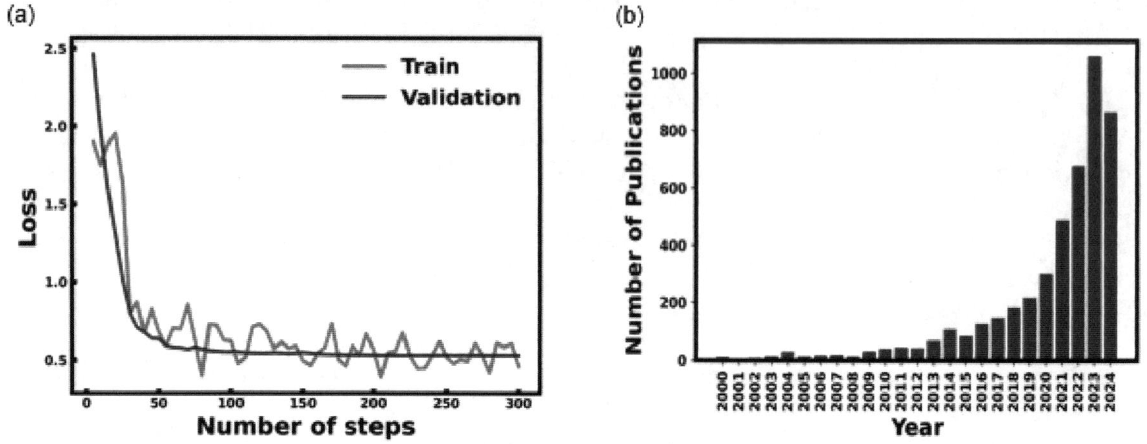

Figure 1: (a) This bar plot shows the overall trend of the number of annually published journal articles for the last 24 years, and (b) the variation of loss values vs. the number of steps in the form of a learning curve for fine-tuning step.

Table 1: The performance evaluation for our proposed model, Sustain-GPT, with a standard LCI database of methanol synthesis.

Variable name	*Outcome of Sustain-GPT*	*LCI data (own calculations)[16]*
H_2 required	1.3 kg	1.25 kg
CO_2	1.35 kg	1.375 kg
CO	1.06 kg	0.875 kg
Energy	10.3 GJ	6.9 GJ
H_2O	1 kg	0.56 kg
Carbon footprint	1.1 kg CO2e	0.6 - 2.3 kg CO2e

229d Machine Learning Enabled Life Cycle Assessment for Early-Stage Sustainable Process Design

Kirti Yenkie, John Pazik, Jared Longo, Matthew Conway, Milo Barkow, Emmanuel Aboagye, Robert Hesketh

Holistic assessment during early-stage process design is important for developing environmentally sustainable processes. The complexity in assessing sustainability arises from the multifaceted nature of the design problem, where the initial stages are open to interpretation and lack sufficient information, making it challenging to evaluate sustainability. As the design progresses, the potential for enhancing sustainability is further constrained by preliminary decisions, requiring the consideration of interdisciplinary factors including environmental impacts, health and safety, and social considerations (Argoti et al., 2019). Furthermore, impact assessment requires collecting and analyzing a wide range of data, including physicochemical properties, molecular structure, toxicity, and fate of these chemicals in the environment (Finnveden et al., 2009). Machine Learning (ML) algorithms are proving to be a promising solution to address these challenges, as they can predict the environmental impacts of novel chemicals with high accuracy even with limited availability of data (Karka et al., 2019). Thus, early-stage ML assessments enable informed decisions for sustainable process design.

The proposed study uses ML to predict the Midpoint and Endpoint categories of the Life Cycle Inventory (LCI) for chemicals. The midpoint LCIs comprise directly observable and measurable emissions such as ozone layer depletion, PM10 concentration, aquatic eutrophication, respiratory effects, non-renewable energy use, etc., whereas endpoint LCIs are the final consequences or damage that includes a combination of midpoint categories that contribute towards the same endpoint LCI indicator. Endpoint LCIs are comprised of four summary indicators including human health, ecosystem quality, climate change, and resource utilization. Thus, depending on the data availability, we can either predict midpoint or endpoint LCIs and utilize them in conducting the entire Life Cycle Assessment (LCA) for a chemical in its manufacturing phase (cradle-to-gate), use phase (gate-to-gate) and end-of-life (gate-to-grave/gate-to-cradle) phases.

Thus, our work presents a novel approach to conducting the complete LCA for a given chemical and its associated processes in the early design stages. Many ML algorithms were tested, and their results were compared, for example, eXtreme gradient boosting (XGBoost), artificial neural networks (ANN), and advanced Transformer Models such as Graph Neural Networks (GNNs). The LCI data for known chemicals were acquired from the Ecoinvent database that is used in widely accepted LCA software such as SimaPro and GaBi. The feature set comprised over 200 molecular descriptors and 23 thermodynamic properties for each chemical. A stepwise feature selection criterion was used to reduce the number of features and hyperparameter tuning was employed in the final ML model. The dataset was split into training, testing, and validation sets, and the model performance was evaluated on the test set using the R-squared, Root-Mean-Squared-Error (RMSE), and loss functions.

These ML-predicted LCIs were used to conduct complete LCAs for case studies and the results followed similar trends as observed in published work. This provides an accuracy check on the ML predictions, and the anticipation is that these ML models will enable LCI predictions for novel chemicals, thus, allowing early-stage LCA of the entire process.

References:

Argoti, A., Orjuela, A., Narváez, P.C., 2019. Challenges and opportunities in assessing sustainability during chemical process design. Curr. Opin. Chem. Eng. 26, 96–103. https://doi.org/10.1016/j.coche.2019.09.003

Finnveden, G., Hauschild, M.Z., Ekvall, T., Guinée, J., Heijungs, R., Hellweg, S., Koehler, A., Pennington, D., Suh, S., 2009. Recent developments in Life Cycle Assessment. J. Environ. Manage. 91, 1–21. https://doi.org/10.1016/j.jenvman.2009.06.018

Karka, P., Papadokonstantakis, S., Kokossis, A., 2019. Predictive LCA - a systems approach to integrate LCA decisions ahead of design, in: Computer Aided Chemical Engineering. Elsevier, pp. 97–102. https://doi.org/10.1016/B978-0-12-818634-3.50017-5

229e Utilizing Multi-Criteria Decision Analysis for Holistic Evaluation and Selection of Wastewater Nitrogen Recovery Technologies

Chayse Lavallais, Jennifer Dunn

Municipal wastewater contains high concentrations of ammonia that if left untreated can lead to a variety of environmental problems such as eutrophication and the production of nitrous oxide (N_2O). Traditionally, wastewater treatment plants (WWTPs) have utilized nitrogen removal technologies such as nitrification-denitrification to produce diatomic nitrogen that is released into the atmosphere. However, nitrification-denitrification is an energy-intensive process that does not produce a usable product. As the emphasis on developing a nitrogen circular economy continues to grow, there is interest in WWTPs implementing nitrogen recovery technologies into their infrastructure. Although multiple technologies that produce different products are being developed, there is no technology that is the industrial standard within WWTPs. As these technologies continue to be developed, and new technologies become available, it will be important for a WWTP to identify which technology should be integrated into their facility. Therefore, a holistic and consistent approach is needed to guide the decision-making process.

This study develops that framework to evaluate 9 different nitrogen recovery technologies to rank which technologies would have the biggest impact on nitrogen circularity. These technologies are evaluated against five indicators: technical, environmental, economic, social, and circularity. Life cycle assessments, techno economic analysis, and process modeling are used to assign values to each of the indicators for the different technologies. For each of those indicators, objective weighting methods such Standard Entropy and CRiteria Importance Through Intercriteria Correlation (CRITIC) are used to identify the importance of each category in the evaluation. Then multi-criteria decision analysis (MCDA) methods such as Analytic Hierarchy Process (AHP) and Technique for Order Preference by Similarity to Ideal Solution (TOPSIS) are used to compare and rank the different technologies against each other. Ultimately, this framework can be used by existing WWTP facilities looking to integrate nitrogen recovery into their facility to identify the technology that best fits within their system to improve the nitrogen circular economy.

229f Air Quality Monitoring with Thingspeak

Aycan Hacioglu, Ramnarayan Krishnamurthy

Numerous environmental scientists work with air quality data collected from field sensors, employing various computational tools for preprocessing, visualization, and modeling. In this talk, we will illustrate our use of ThingSpeak, an Internet of Things (IoT) platform with advanced MATLAB analytics, to obtain, analyze, and visualize CO2 data. Additionally, we will elaborate on how we employed the platform to trigger actions, such as activating a fan for ventilation when critical concentrations are reached. MATLAB provides numerous options for analyzing the data, including curve fitting, machine learning, and statistical methods. Since ThingSpeak aggregates the data in the cloud, it facilitates easy access and sharing. ThingSpeak proves useful in academic research, classroom settings, and industrial applications and we will share some success stories from the field. Moreover, as MATLAB is widely used in engineering and science, it promotes seamless collaboration among environmental engineers and other engineering disciplines in working with environmental data.

498a Climate Change, Extreme Events, and Their Effects on Aboveground Storage Tanks

Raymond Smith

Environmental managers, modelers, and decision makers need to understand the systems under their control, concern, and authority. Aboveground storage tanks (ASTs) are often overlooked in evaluations of environmental systems, but their emissions can be substantial and the vulnerabilities associated with them significant [1]. In particular, climate change and extreme events can have highly non-linear effects on ASTs, for instance, when an increase in extreme event intensity leads to qualitatively different behavior. This presentation will describe vulnerabilities and how models show emissions increasing for higher ambient temperatures and wind speeds. One result is that emissions of aromatics increase even more than VOCs. The results also describe how lack of maintenance leads to dramatically increased emissions. Among the applications for this information are risk assessments, life cycle assessments, and environmental product declarations. With improved estimates of emissions from the storage of materials such as fuels, one can more accurately compare the effects of systems that include these stored materials.

[1] Raymond L. Smith, Joe Terriquez, Eben Thoma, Michael A. Gonzalez, Derek Johnson, Hans Buenning, Fran Kremer, Jacob D. Carpenter, and Nigel N. Clark, "Climate Change, Extreme Events, and Their Potential Effects on Aboveground Storage Tanks," *EM Magazine*, 24–29, September (2023).

The views expressed in this presentation are those of the author and do not necessarily represent the views or policies of the U.S. Environmental Protection Agency.

498b Climate Disaster or Not? Using Science, Philosophy, and Psychology to Comprehend the Future of Humankind on a Warming Planet

Jonathan Stickel

Anthropogenic climate change poses a serious threat to humanity and Earth as we know it. For several decades, the accepted approach for mitigating climate change has been focused on reducing greenhouse gas emissions. What we need to do, climate activists claim, is to avoid using energy where we can, use existing energy more efficiently, and to convert energy sources from fossil fuels to renewables. Although these conservation and technical solutions are available, for a cost, they are not being implemented nearly fast enough. Considering the potentially massive magnitude and scope of climate change, assuredly altering human life and the entire Earth in profound ways, it is worth taking a step back and looking at climate change with a broader lens.

Drawing on knowledge from science, philosophy, and psychology, this presentation examines the likely outcomes of climate change, what it means if we fail to mitigate it, and why so little progress is being made. By accepting the limitations of human behavior and society, mitigation measures are discussed that better align with humans' evolutionary psychology and our existing institutions. Engineering solutions are emphasized, specifically carbon capture and storage, carbon dioxide removal, and geoengineering. While these solutions are no panacea and come with their own challenges—technical, economical, and philosophical—they will continue humanity's long history of mastering the environment to enable progress.

498c Deriving Techno-Socio-Economic Policies for Global Sustainability

Siddhi Manjarekar, Urmila Diwekar

Sustainable development presents an ongoing global challenge, necessitating innovative approaches to address environmental concerns while promoting economic prosperity. This research builds upon the Generalized Global Sustainability Model (GGSM), which provides a comprehensive framework for analyzing ecological and economic dynamics. Drawing from previous work, we propose an extension to the GGSM by integrating water pricing as a key policy variable. Water stress, a critical aspect of sustainability, is addressed through this addition, allowing for a more nuanced examination of resource management policies.

Utilizing an optimal control framework, the research defines global sustainability across ecological, economic, and social dimensions, focusing on minimizing variance in Fisher Information, a key sustainability indicator. The study finds that optimizing for Fisher Information effectively addresses multiple sustainability objectives, simplifying the multi-dimensional problem framework. Seven techno-socioeconomic policy decision profiles that can make the planet sustainable for the foreseeable future are obtained using this framework.

This research contributes to the ongoing discourse on sustainable development by emphasizing the pivotal role of water pricing in shaping environmental policies and economic decisions. By extending the GGSM to include water pricing, we not only enhance the model's utility but also broaden its applicability for policymakers and researchers. Our findings are not just theoretical but aim to inform evidence-based policymaking and promote holistic approaches to sustainability in a rapidly changing world.

498d Chemical Engineering in Cement Decarbonization

Joshua W. Allen, James Bresson, Jason L. Bronkema, Janny Cai, Eric Liu, Cody Finke, Hugo Leandri, Shadi Saberi

Cement is the binding agent in concrete, the most widely used material on Earth after water. Cement production accounts for 7.5% of global CO2 emissions due to (1) the use of CO_2-bearing limestone ($CaCO_3$) as a feed rock and (2) high process energy requirements. Brimstone has developed a breakthrough, deeply decarbonized process to co-produce ordinary portland cement (OPC) and supplementary cementitious material (SCM) by using carbon-free calcium silicate rocks instead of limestone. Brimstone is the first company to make OPC that meets the ASTM C150 standard without using limestone as a feed rock. The Brimstone process involves leaching of calcium silicate rocks instead of CO_2-bearing limestone, purification of the resulting brine to selectively recover a calcium-containing product, and thermal treatment to create OPC and regenerate the leaching agent. Brimstone is currently scaling its industrial process to deliver deeply decarbonized industry-standard cement and SCM at competitive market prices without the need for carbon capture technologies.

Chemical engineering has a critical role in the successful scale-up of the Brimstone process. This is illustrated through the development of the leaching reactor. Development of a suitable compositional model for the feed rock enabled accurate capture of the leaching reaction exotherm. Kinetic models allowed for translation of batch laboratory data to a large-scale, continuous commercial reactor. Thermodynamic models of electrolytes were tuned to accurately represent the solubility of the various metal salts in the post-leach brine. Process development unifies this information into a process model that establishes the foundation for scale-up to pilot, demonstration, and commercial plants.

498f Optimization of CO2 Pipeline Networks for Climate Change Mitigation

Aditya Nathan, Christian Lastoskie

Over the last century, the United States has invested in an extensive network of wastewater collection systems to remove pathogens, dissolved organic matter, and industrial effluents from municipal wastewater streams. The investment secured a thousand-fold decrease in waterborne diseases and dramatically reduced fouling of surface waters from eutrophication. A comparable, nation-spanning investment is now needed to meet ambitious goals for arresting global climate change: namely, investment in *gaseous* pollutant collection systems to transport CO_2 recovered from major industrial sector greenhouse gas (GHG) emitters. Stabilizing the climate requires far more than electrifying automobile fleets and building wind and solar farms. Gigaton scale CO_2 capture, utilization, and storage (CCUS) is needed to cost-effectively achieve deeply decarbonized systems, e.g. 80% GHG emission reductions by 2050[1], and negative emissions systems, crucial to meet 1.5 degrees Celsius warming limits. Existing pipeline infrastructure is inadequate in scale and location to transport captured CO_2 from carbon-emitting heavy industries to geologic sinks and CO_2-utilizing manufacturers[2]. Deployments of CO_2 capture and pipeline facilities are challenged by high capital costs, weak policy incentives for capturing CO_2, patchwork intra-state CO_2 pipeline laws, and an absence of federal authority over interstate CO_2 pipeline siting. Overcoming cost and pipeline siting regulatory constraints could significantly accelerate CCUS deployment.

Large industrial CO_2 sources by sector include refineries (188 Mt CO_2/yr), pulp and paper (143), iron and steel (83), petrochemicals (80), and cement (68). The CO_2 content in the exhaust gases from these sources also varies widely, with high fractions in ammonia, ethanol, and petrochemical plants (30-99.9+%) and medium to low concentrations from cement (14-33%), iron and steel (20-27%), and pulp and paper (8%) manufacturing[3]. The work requirement and the corresponding cost to separate CO_2 from the exhaust at purity suitable for pipeline transport varies in inverse proportion to the CO_2 content of the exhaust, ranging up to \$70/ton CO_2 for capture from low concentration sources (e.g., natural gas power plants) using current state-of-art amine solvent capture technologies.

A sequential strategy is thus indicated for scaling industrial sector CCUS, with inexpensive near-term capture from industrial facilities that emit exhaust with high CO_2 content, followed by the addition of CO_2 from sources with progressively lower exhaust CO_2 concentrations, allowing time for the maturation of technology alternatives to amine scrubbing that may enable lower-cost CO_2 separation from dilute gas streams. Tandem to the sourcing considerations is the sale value of the captured CO_2 at its utilization site (or, in the case of carbon sequestration, the additional cost of CO_2 injection into a storage reservoir), as well as tariffs assessed to pipeline users for compression and transport of captured CO_2 from its industrial source to its intended endpoint. Presently the scale of CO_2 utilization is only 1% of total emissions, mostly for EOR operations in west Texas using CO_2 that is pipelined from natural underground reservoirs in New Mexico. Additional EOR operations, as well as expanded CO_2 utilization in other industries, are thought to be viable if CO_2 from anthropogenic sources is available for purchase at the typical oil-linked price of 40% of the per-barrel oil price per ton of CO_2. Investment in CO_2-dedicated pipeline infrastructure, financed in part or in whole by the government, can thus accelerate CCUS adoption to scales sufficient to alter the planet's climate change trajectory by removing the "chicken-or-egg" barrier of linking large-scale CO_2 emission sources to CO_2 utilization or storage sinks.

In this presentation, we report on a methodology that was developed to optimize the sequential construction of CO_2 capture facilities and pipeline networks, to convey CO_2 from hard-to-decarbonize industrial processes to sites for CO_2 utilization as a feedstock. Capture costs and available CO_2 flow volumes from regional industrial emission sources in the Atlantic, Pacific, Midwest, and Southeastern U.S. were compiled from EPA datasets. Placement of trunk pipeline segments for aggregated CO_2 flows, within existing rights of way (e.g., corridors for natural gas pipelines, electricity transmission lines, interstate highways) was then determined using ArcGIS, and pipe diameters, lengths, and costs to build and operate pipeline networks connecting CO_2 sources to the trunk lines were calculated. Financial analysis was next conducted to determine the economic viability of pipeline networks of varying spatial extent and installation timetables under financing scenarios with partial and full government backing, taking into account pipeline tariffs, oil-linked CO_2 sale prices, availability of Section 45Q carbon sequestration credits, equity target internal rate of return, and technology advancements leading to lower future CO_2 capture costs.

A crucial systems interdependency of the CO_2 pipeline network is that the shared cost per unit flow in the pipeline increases as the flow volume decreases. The cost of CCUS abatement for one pipeline user is therefore dependent on the costs incurred by all other users on the same network, and whether or not those other prospective CO_2 "sellers" elect to use the pipeline system to bring their own CO_2 to market. The pipeline network is thus an inherently dynamic system, such that the financial analysis of the sized networks must be carried out iteratively until an economically viable network design is attained.

A case study is presented for the construction and operation of dedicated CO_2 pipeline network for the transport and delivery of captured carbon dioxide from industrial emitters along the Eastern seaboard of the United States to a terminal in Louisiana for enhanced oil recovery operations. Pipeline funding scenarios that involve full commercial financing, full government financing, and private-public partnership were investigated with the use of the Department of Energy's FE/NETL CO_2 Transport Cost Model. The range of capacity additions that are possible, beyond a baseline of 68 Mt CO_2/yr for full government financing of the pipeline, were explored for different technology advancement scenarios leading to a decrease in future onsite CO_2 capture costs.

1. Sustainable Development Solutions Network and the Institute for Sustainable Development and International Relations. Pathways to deep decarbonization. Deep Decarbonization Pathways Project (2015).
2. National Energy Technology Laboratory and the Great Plains Institute. Siting and Regulating Carbon Capture, Utilization and Storage Infrastructure. U.S. Department of Energy Workshop Report (2017).
3. P.C. Psarras, S. Comello, P. Bains, P. Charoensawadpong, S. Reichelstein, and J. Wilcox. Carbon Capture and Utilization in the Industrial Sector. *Environmental Science and Technology* (2017) **51**, 11440-11449.

Enhancing CO_2 Sweep and Storage Efficiency: Development and Evaluation of a Novel Hydrogel for CO_2 EOR and Storage Projects

Maryam Sharifi Paroushi, Missouri university of science & technology, Rolla, MO.
Tian, Xu yang, Missouri university of science & technology, Rolla, MO
Baojun Bai, Geosciences and Geological and Petroleum Engineering, Missouri University of Science and Technology, Rolla, MO

Abstract

In addressing climate change, carbon capture and sequestration (CCS) is crucial for mitigating emissions from major stationary sources worldwide. The integration of CCS with EOR presents promising prospects, enabling the simultaneous mobilization of oil reserves and sequestration of CO_2 underground. Hydrogel technology offers reliable solutions to key challenges in underground CO_2 storage, such as controlling CO_2 conformance within reservoirs and preventing CO_2 leakage through cap rock integrity maintenance. By strategically utilizing hydrogel, CCS-EOR projects can enhance storage efficiency and comply with environmental standards, significantly contributing to global climate change mitigation efforts. However, traditional polymer gels are not stable at CO_2 acid conditions. To solve the problem, we have developed a novel polymer gel that can be utilized as a plugging agent for either improving CO_2 sweep efficiency by controlling the preferential fluid flow problem or used as a blocking agent to prevent CO_2 leakage through wellbore or caprocks. This study extends the application of polymer gel systems to the field of CCS, addressing the unique challenges posed by CO_2 injection and storage in underground reservoirs. First, we screen polymers that are capable of resisting CO_2 at reservoir conditions. Then, the polymer concentration optimization was conducted through bottle tests and core flooding experiments to determine a critical concentration range conducive to stable gel formation in porous media. Subsequently, formula optimization was pursued by assessing the effects of varying polymer concentrations, crosslinker concentrations, and salinity levels, considering the specific conditions encountered in CCS reservoirs. Rheology studies compared static bottle tests with dynamic viscosity measurements under different salinity conditions to simulate the flow state in porous media. Furthermore, the CO_2-resistant properties of the polymer gel were evaluated, encompassing stability assessments and studies on plugging efficiency under simulated CO_2 injection conditions. Through comparison, it was determined that gel modified by polymer AN 125 was the most CO_2-resistant polymer due to the lowest initial viscosity, a lower syneresis degree, and a higher residual resistance factor. Overall, this research provides valuable insights into developing and applying polymer gel systems for sealing CO_2 leakage and enhancing the security and efficiency of CCS operations.

Keywords: EOR, Sulfonate Polymer crosslinking gel, Sweep Efficiency Improvement, Storage Efficiency Improvement, Leakage Control.

Introduction

CO_2-enhanced oil recovery (EOR) projects face several challenges impacting both their efficiency and environmental compliance. A primary issue is CO_2's high mobility compared to crude oil, which often causes it to bypass oil-rich areas and reach production wells prematurely. This early breakthrough reduces oil recovery efficiency, as CO_2 flows through the easiest paths, leaving substantial portions of oil untouched. Achieving miscibility between CO_2 and crude oil, essential for lowering oil viscosity and enhancing extraction, is another challenge, as it depends on specific pressure and temperature conditions that can be difficult to establish in some reservoirs. Environmental concerns, particularly the risk of CO_2 leakage in EOR and Carbon Capture and Storage (CCS) applications, require strict containment and monitoring practices, which add to operational complexity and costs. Reservoir heterogeneity variation in rock characteristics like porosity and permeability further complicates CO_2 distribution, as it often flows through higher-permeability zones, leaving other areas unswept. To address these issues, EOR projects use techniques like polymer gels and other mobility control agents to achieve a more uniform CO_2 distribution, improving overall recovery across varied reservoir conditions [1], [2], [3], [4].

Polymer gels enhance CO_2-EOR by functioning as blocking agents to reduce CO_2 mobility in high-permeability zones. When injected, they create a more uniform flow profile, redirecting CO_2 to sweep through previously bypassed oil-rich areas, thus improving recovery efficiency. Additionally, polymer gels help achieve better CO_2 conformance in reservoirs with heterogeneous rock properties. By selectively targeting high-permeability pathways, they reduce permeability and improve CO_2 distribution, leading to more complete oil recovery across diverse rock structures [5]. In addition to improving mobility and sweep efficiency, polymer gels address environmental concerns in CO_2-EOR combined with CCS. CO_2 leakage is a major risk in these projects, and polymer gels contribute to containment by sealing fractures and high-permeability channels that could otherwise act as escape pathways for CO_2. This capability not only enhances oil recovery but also ensures that CO_2 remains stored within the reservoir, aligning with regulatory standards and environmental safety goals [6], [7]. Studies have shown that polymers like hydrolyzed polyacrylamide (HPAM), especially when cross-linked with agents like chromium, can make them suitable for a variety of challenging reservoir environments [5]. However, despite their advantages, polymer gels face several challenges that may limit their effectiveness in CO_2-EOR applications [8], [9].

One major concern is the instability of traditional polymer gels in the acidic conditions common in many reservoir environments. Acidic conditions can lead to hydrolysis and degradation of polymer chains, especially in polyacrylamide-based gels, weakening the gel's structure and diminishing its ability to block high-permeability pathways effectively. This degradation not only reduces gel performance but also necessitates frequent re-injection of fresh formulations, increasing both operational costs and complexity [3],[10].

To address these challenges, ongoing research is focused on advancing polymer gel systems to enhance their stability, efficiency, and reliability for CO_2-EOR and CCS applications, ensuring their effectiveness in a wide range of reservoir conditions, focused on screening CO_2-resistant polymers and optimizing their concentration for stable gel formation in porous media. It involved testing polymer concentrations, crosslinker levels, and salinity conditions alongside rheology studies to simulate flow in porous media. The CO_2 resistance, stability, and plugging efficiency of the polymer gel were evaluated under simulated CO_2 injection conditions.

Experimental Materials and Procedures

First, polymers screened that can be capable of resisting CO_2 at reservoir conditions. Then, the polymer concentration optimization was conducted through bottle tests. Table 1 shows the different classifications and codes of each step [11]. The experimental conditions included the Sydansk Code grade C, which was considered for gelation time. Gel preparation was conducted at a temperature of 45 °C, and all experiments were performed with a salinity of 27,500 ppm. The salinity composition consisted of 2.50 wt% NaCl, 0.12 wt% $MgCl_2$, 0.06 wt% $CaCl_2$, and 0.015 wt% KCl. Additionally, all experiments were conducted with 5000 ppm of polymer at 45 °C.

Table 1: Classification and Codes for Each Step (Sydansk Code [11])

Code	Shape
A	No detectable gel formed
B	Highly flowing gel
C	Flowing gel
D	Moderately flowing gel
E	Barely flowing gel
F	Highly deformable nonflowing gel
G	Moderately deformable nonflowing gel
H	Slightly deformable nonflowing gel
I	Rigid gel
J	Ringing rigid gel

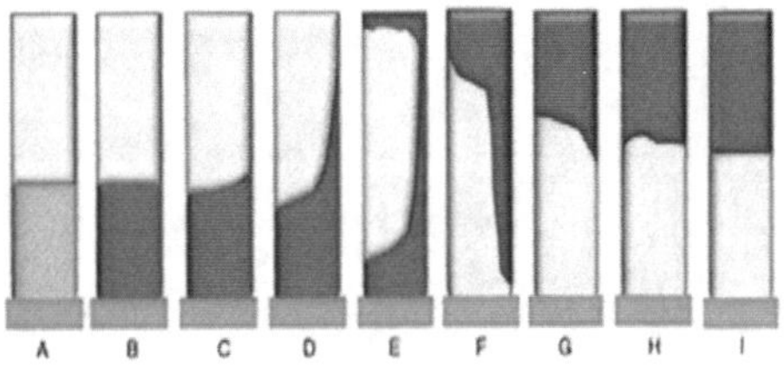

Results and Discussion

Bottle Test Evaluation

Optimizing the polymer concentration through bottle tests and core flooding experiments was necessary to determine the critical concentration range that promotes stable gel formation within porous media. Figure 1 illustrates the classification of bottle test results based on gel stability and quality across various polymer-gel codes. At low concentrations of polymer or crosslinker, no gel forms, and the solution remains liquid. As concentration increases to 5000ppm, a weak gel begins to appear, though it lacks the strength needed for effective reservoir application. In the moderate concentration range, the gel becomes more cohesive, exhibiting better integrity, but may still fall short of stability under reservoir conditions. The optimal concentration is identified where the gel is strong and stable, making it suitable for use in porous media. Beyond this optimal range, at very high concentrations (more than 10000 ppm), the gel becomes excessively stiff, which can hinder its flow and distribution within the reservoir. This classification aids in pinpointing the concentration range conducive to stable gel formation under reservoir conditions.

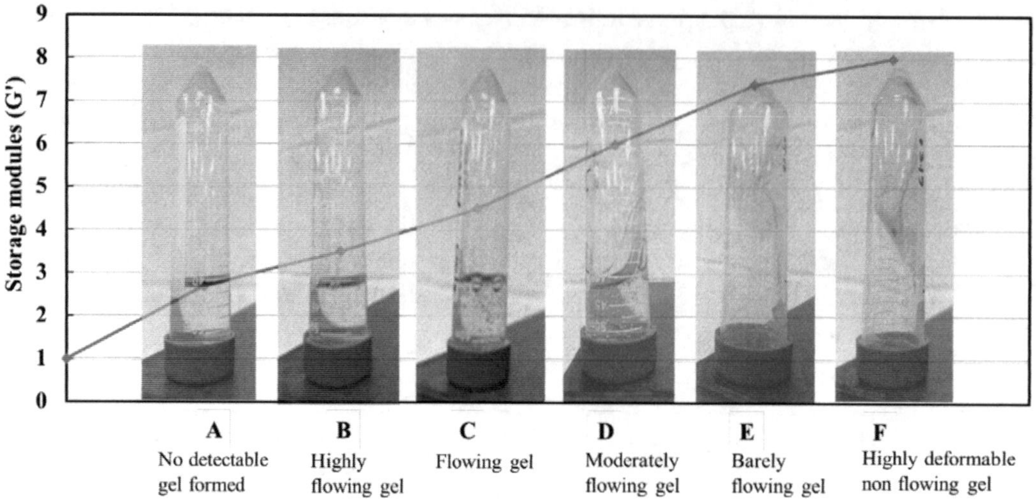

Figure 1. Bottle Test Classification for Polymer Gel Stability

Initial screening focused on evaluating each polymer's CO_2 resistance, stability, and compatibility under reservoir conditions. By comparing various polymers, we aimed to determine which would best withstand CO_2 exposure while maintaining effective gelation and stability. Our findings revealed that Polymer AN125 was identified as the most CO_2-resistant polymer in our studies due to its lower syneresis degree, higher residual resistance factor, and strong pH stability, indicated by its minimal viscosity reduction under acidic conditions. These characteristics make AN125 an ideal candidate for CO_2-EOR applications, as it maintains its structural integrity and performance in the challenging environments typical of reservoir conditions.

However, despite its resilience, gelation challenges arise with AN125, as achieving the desired gelation time is critical for ensuring effective blocking and conformance control in reservoirs. To address this, we experimented with various crosslinkers (Ni-Acetate·$4H_2O$, $NiCl_2 \cdot 6H_2O$, $CoCl_2 \cdot 6H_2O$, $ZrCl_4$, ZrAc, $AlCl_3$, $CrCl_3$, CrAc) to optimize the gelation time, aiming to produce a stable gel that can withstand CO_2 injection and flow dynamics in porous media and ended up with these two crosslinkers, $CrCl_3$ and CrAc, as the most suitable options.

Crosslinker Concentration Effects on Gelation Time and gel-Strength

In our optimization process, we investigated several factors that influence gelation time, including crosslinker concentrations, polymer concentration, polymer molecular weight, temperature, salinity, and ion strength. Each of these factors plays a crucial role in determining the rate and stability of gel formation. Crosslinker concentration affects the network density of the gel, while polymer concentration and molecular weight influence gel strength and viscosity. Temperature can accelerate gelation, whereas salinity and ion strength impact gel structure and stability due to interactions with the polymer and crosslinker. By understanding and controlling these variables, could fine-tune AN125-based gels to achieve optimal performance in CO_2-EOR applications, as illustrated in Figure 2 (a), which presents the effects of crosslinker concentrations on gelation time for $CrCl_3$. Gelation time and gel strength (G') plotted against polymer crosslinker concentration. This figure illustrates how varying concentrations of $CrCl_3$ influence the time required for the gel to form, and Figure 2 (b) displays the effects of crosslinker concentrations on gelation time for CrAc and provides insights into the gelation kinetics when using CrAc as a crosslinker.

As seen in Figure 2 (a & b), the relationship between concentration and gelation time shows a decrease in gelation time with increasing crosslinker concentration for both $CrCl_3$ and CrAc. The impact of $CrCl_3$ concentrations on gel strength indicates that while higher concentrations enhance gel strength, they do not reach the levels observed with CrAc. When comparing the performance of CrAc and $CrCl_3$, it is evident that CrAc is the superior crosslinker, exhibiting higher gel strength. While $CrCl_3$ demonstrates a shorter gelation time, which can be beneficial in applications near the wellbore where rapid gel formation is essential, CrAc consistently outperforms $CrCl_3$ in terms of overall stability and strength in most scenarios. This makes CrAc a more reliable choice for long-term applications in CO_2 injection and enhanced oil recovery processes.

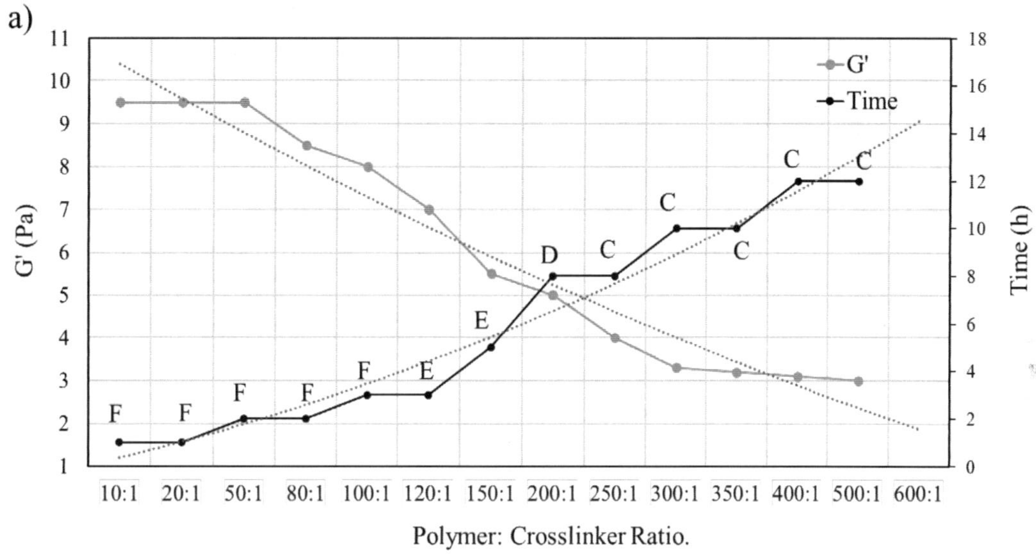

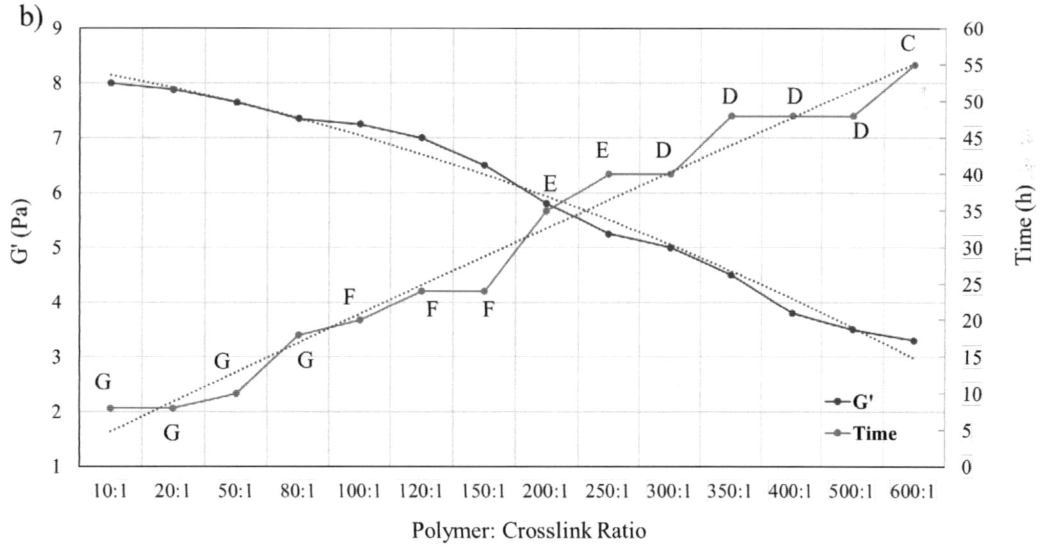

Figure 2. Effects of crosslinker concentrations on gelation time. (a) $CrCl_3$. (b) CrAc.

CO_2 Resistance Evaluation of the Gel

The CO_2 resistance test was conducted under controlled conditions of 1150 psi pressure and 45 °C temperature for 24 hours to evaluate the stability of the gel. The initial weight of the sample was measured at 17.27 g (M_1), which decreased to 16.64 g (M_2) after exposure, resulting in a weight loss of approximately 3%. This slight weight reduction suggests that the gel exhibits a reasonable level of resistance to CO_2, indicating its potential suitability for applications in CO_2 injection processes. The relatively low weight loss implies that the gel maintains its structural integrity under high-pressure conditions, which is crucial for preventing CO_2 leakage in enhanced oil recovery scenarios. Further investigations may be necessary to optimize the gel formulation and assess its long-term performance under varying operational conditions.

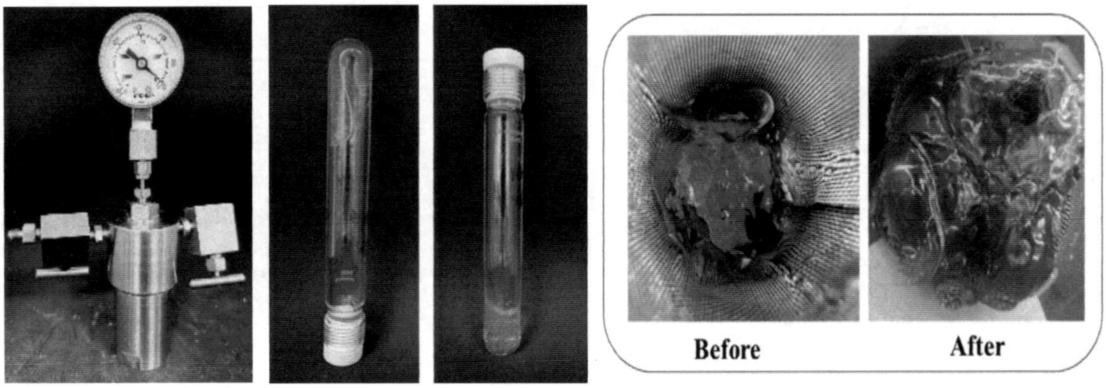

Figure 3. CO_2 resistance test equipment and gel before and after CO_2 exposure

Conclusion

The polymer gel developed for CO_2 applications exhibits excellent CO_2 resistance and stability, particularly under supercritical CO_2 conditions, making it a promising candidate for enhanced oil recovery processes. As highlighted in the introduction, the ability to effectively manage CO_2 emissions is important for mitigating climate change, and the use of polymer gels in CO_2 injection not only enhances oil recovery but also contributes to carbon capture and storage efforts. The gelation process of this polymer gel is both controllable and adaptable, allowing for optimization to meet the specific requirements of diverse field environments. The results from pH testing demonstrate that the gel maintains its stability across a broad pH range of 3 to 9, indicating its versatility for various subsurface conditions. Furthermore, the CO_2 resistance test yielded encouraging results, with a minimal weight loss of only 3% observed, in conjunction with high gel strength (G'), which underscores the gel's durability under pressure and its potential effectiveness in preventing CO_2 leakage. Additionally, the gel has exhibited hydrothermal stability at 100 °C for over two months, with ongoing observations to assess its long-term performance. These findings support the assertion made in the abstract that this polymer gel not only enhances oil recovery efficiency but also plays a critical role in CO_2 sequestration initiatives. Overall, the results demonstrate the gel's potential to contribute significantly to sustainable energy solutions by optimizing resource extraction while minimizing environmental impacts.

References

[1] O. Massarweh and A. S. Abushaikha, "A review of recent developments in CO2 mobility control in enhanced oil recovery," Sep. 01, 2022, *KeAi Communications Co.* doi: 10.1016/j.petlm.2021.05.002.

[2] M. Al-Shargabi, S. Davoodi, D. A. Wood, V. S. Rukavishnikov, and K. M. Minaev, "Carbon Dioxide Applications for Enhanced Oil Recovery Assisted by Nanoparticles: Recent Developments," Mar. 29, 2022, *American Chemical Society*. doi: 10.1021/acsomega.1c07123.

[3] H. Z. Jiang *et al.*, "Performance and enhanced oil recovery efficiency of an acid-resistant polymer microspheres of anti-CO2 channeling in low-permeability reservoirs," *Pet Sci*, vol. 21, no. 4, pp. 2420–2432, Aug. 2024, doi: 10.1016/j.petsci.2024.02.002.

[4] W. Pu, D. Du, H. Fan, B. Chen, C.-D. Yuan, and M. A. Varfolomeev, "CO2-responsive preformed gel particles with interpenetrating networks for controlling CO2 breakthrough in tight reservoirs," *Colloids Surf A Physicochem Eng Asp*, vol. 613, p. 126065, 2021, doi: https://doi.org/10.1016/j.colsurfa.2020.126065.

[5] Y. Zhao, X. Wen, Z. Rui, and Y. Ding, "A novel dual-network CO2-responsive particle gel for mitigating CO2 channeling and leakage in hydrocarbon recovery and carbon storage," *Chemical Engineering Journal*, vol. 498, p. 155187, 2024, doi: https://doi.org/10.1016/j.cej.2024.155187.

[6] X. Sun, B. Bai, Y. Long, and Z. Wang, "A comprehensive review of hydrogel performance under CO2 conditions for conformance control," *J Pet Sci Eng*, vol. 185, Feb. 2020, doi: 10.1016/j.petrol.2019.106662.

[7] Z. Liu, P. Wei, Y. Qi, X. Huang, and Y. Xie, "High stretchable and self-healing nanocellulose-poly(acrylic acid) composite hydrogels for sustainable CO2 shutoff," *Carbohydr Polym*, vol. 311, p. 120759, 2023, doi: https://doi.org/10.1016/j.carbpol.2023.120759.

[8] Y. Xiong *et al.*, "A Research Progress of CO2-Responsive Plugging Channeling Gels," *Energy Engineering*, vol. 121, no. 7, pp. 1759–1780, 2024, doi: https://doi.org/10.32604/ee.2024.048536.

[9] X. Sun, B. Bai, Y. Long, and Z. Wang, "A comprehensive review of hydrogel performance under CO2 conditions for conformance control," *J Pet Sci Eng*, vol. 185, Feb. 2020, doi: 10.1016/j.petrol.2019.106662.

[10] J.-N. Deng *et al.*, "CO2-responsive preformed particle gels with high strength for CO2 conformance control in heterogeneous reservoirs," *Fuel*, vol. 379, p. 133040, 2025, doi: https://doi.org/10.1016/j.fuel.2024.133040.

[11] Z. Ali Lashari, W. Kang, H. Yang, H. Zhang, and B. Sarsenbekuly, "Macro-rheology and micro-rheological study of composite polymer gel at high salinity and acidic conditions for CO2 shut off treatment in harsh reservoirs for improving oil recovery," in *Society of Petroleum Engineers - SPE/PAPG Pakistan Section Annual Technical Symposium and Exhibition 2019, PATS 2019*, Society of Petroleum Engineers, 2019. doi: 10.2118/201175-MS.

46a An Optimization-Based Approach for Developing a Circular Economy of Swine Wastewater

Ahteshamul Haq, Madison Kratzer, Prathap Parameswaran, Vikas Khanna

Swine farming generates substantial amounts of wastewater, a trend that will continue with increasing protein demand resulting from increase in global population. Swine wastewater contains high levels of organic matter, nitrogen, and phosphorus making it particularly attractive for resource recovery. Emerging advanced anaerobic biotechnologies hold promise for recovering energy, nutrients, and other value-added chemicals from high strength wastewater streams. We have previously developed a physics-based model of an integrated anaerobic membrane bioreactor (AnMBR) treatment train. AnMBR couples anaerobic digestion with a membrane unit to decouple solids retention time from hydraulic retention time thereby enabling high carbon conversion. The treatment train also consists of nitrogen and phosphorus recovery via adsorption and coagulation/flocculation, respectively. Detailed life cycle assessment (LCA) and techno-economic assessment (TEA) of this treatment train highlighted the strong effect of plant capacity on economic feasibility and greenhouse gas (GHG) emissions. Available swine wastewater is also geographically unevenly distributed because of wide variation in capacity of swine farms across the US. A holistic systems-level approach is needed to evaluate the best strategy for swine wastewater management taking into account location and capacity of swine farms, transportation costs, treatment costs, GHG abatement potential, and value of recovered products.

This work presents a comprehensive optimization-based framework for evaluating environmentally and economically feasible strategies for swine wastewater management with a specific focus on AnMBR technology. The system is modeled as a mixed integer non-linear programming (MINLP) problem. Net profit quantified via TEA is used to represent the economic objective and life cycle GHG emissions evaluated via process-based LCA is used to represent the environmental objective. Specifically, we determine the optimal network of wastewater treatment plants for swine farms. Multiobjective optimization is also conducted to evaluate the tradeoffs between economic and environmental objectives and the resulting impact on optimal wastewater treatment network. Preliminary findings indicate that not every county needs a dedicated swine wastewater treatment plant; instead, pipeline-based routing to fewer centralized plants could be more favorable. Detailed sensitivity analysis is performed around key variables such as wastewater composition, available wastewater volume, transportation costs, to quantify their impact on optimal management options. The implications of our findings for developing circular economy solutions for animal wastewater streams will be described.

46b From Fossil Fuel to Green Powertrains for Vehicles: A Circular Economy Assessment

Kenneth Martinez

The transportation sector is dominated by internal combustion engine vehicles (ICEVs) using gasoline, and therefore significantly contributes to greenhouse gas emissions. Meanwhile, battery electric vehicles (BEVs), the primary consumers of lithium, are expected to drive the global demand for lithium, leading to environmental and social issues. This scenario underscores the need for a shift towards more sustainable transportation options, such as fuel cell electric vehicles (FCEVs) and ICEVs running on sustainable fuels. A comprehensive framework is essential for evaluating conventional and emerging fuels and powertrain technologies, facilitating informed policy making for sustainable transportation infrastructure development. While existing studies often focus on a limited range of vehicle types and fail to incorporate multiple impact dimensions into a unified assessment, recent research employing a circular economy (CE) perspective by Baratsas et al, 2022, shows promise in providing a holistic comparison of different technologies [1,2]. This work broadens the scope of the previously developed Circular Economy (CE) assessment tool, known as MICRON [2], to include the evaluation of vehicle power train technologies. The CE assessment tool is aligned with five circular economy goals: i) minimizing waste and pollutants, ii) decreasing the utilization of natural resources, iii) boosting the proportion of renewable resources, iv) lowering emissions, and v) increasing the durability of products. This quantitative evaluation is based on the General Reporting Initiative (GRI) standards and Life Cycle Assessment (LCA) databases. The CE impact scores are calculated through a range of quantitative indicators associated with the five circular economy goals. An overall CE index is obtained by averaging each of these category-specific scores. The overall CE index ranges from 0 to 1, where 0 represents linear operations and 1, circular operations. In this work, MICRON was modified to better address the assessment of vehicle power trains, by including vehicle-specific indicators and metrics, including waste generated per km, water and energy consumed per km, and share of renewable material used to manufacture the vehicle's powertrain. A levelized approach is implemented to account for both the operation and manufacturing of the vehicle, considering its entire lifetime.

The updated MICRON tool is used to evaluate various vehicle powertrains, including hydrogen ICEVs, hydrogen FCEVs, methanol/gasoline ICEVs, BEVs, and gasoline ICEVs, focusing on the current and most promising fuel production pathways for each fuel. The nickel cobalt manganese (NCM) and Li-ion batteries are also considered for BEVs. Our initial circular assessments focused exclusively on the fuel production pathways and have revealed hydrogen from biogas as the most circular fuel, attributed to its low emissions and significant renewable energy share, as shown in Figure 1. For methanol, production using hydrogen from wind-powered water electrolysis and biogas stands out for its low emissions. Similarly, the Haber-Bosch process for ammonia production, powered by wind energy electrolysis, ranks highest in circularity due to efficient waste recycling and reduced emissions. Renewable gasoline also demonstrates high circularity, benefiting from its recyclability and the use of captured carbon in its production process. These findings emphasize the importance of transitioning to such circular technologies to achieve sustainable reductions in greenhouse gas emissions. The project has further expanded the MICRON framework to encompass the transportation and storage of these fuels, as well as their utilization in vehicles, providing a comprehensive circular metric of the vehicle's entire life cycle. These results can help policymakers and stakeholders make informed decision-making to achieve sustainable transportation systems.

Figure 2. Circular economy indexes for each fuel production pathways [3-10]. 'NG' denotes natural gas, and 'SMR' refers to steam methane reforming. 'WE' indicates water electrolysis. 'HB' stands for the Haber-Bosch process. 'EF' represents ethanol fermentation, a bio-based production route.

References:

1. Ahmed AA, et al. (2023) Sustainability. 15(1):171.
2. Baratsas SG, et al. (2022) Comput. Chem. Eng. 160:107697.
3. Mann M. et al. (2004) National Renewable Energy Lab., US Department of Energy.
4. Hajjaji, N., et al. (2016) International Journal of Hydrogen Energy, 41 (14), pp.6064-6075.
5. Spath P.L, et al. (2001) NREL/TP-570-27637, National Renewable Energy Laboratory
6. Li, J. et al. (2018) Journal of Cleaner Production
7. Matzen, M., et al. (2016) Journal of Cleaner Production
8. Eggemann, L., et al. (2020) Journal of Cleaner Production, Volume 271
9. Zupko, R., (2019) The International Journal of Life Cycle Assessment, Vol. 24, pp. 1793–1804
10. Biçer, Y., et al (2017) Environmental Management, Vol. 59, pp. 842–855, 2017.

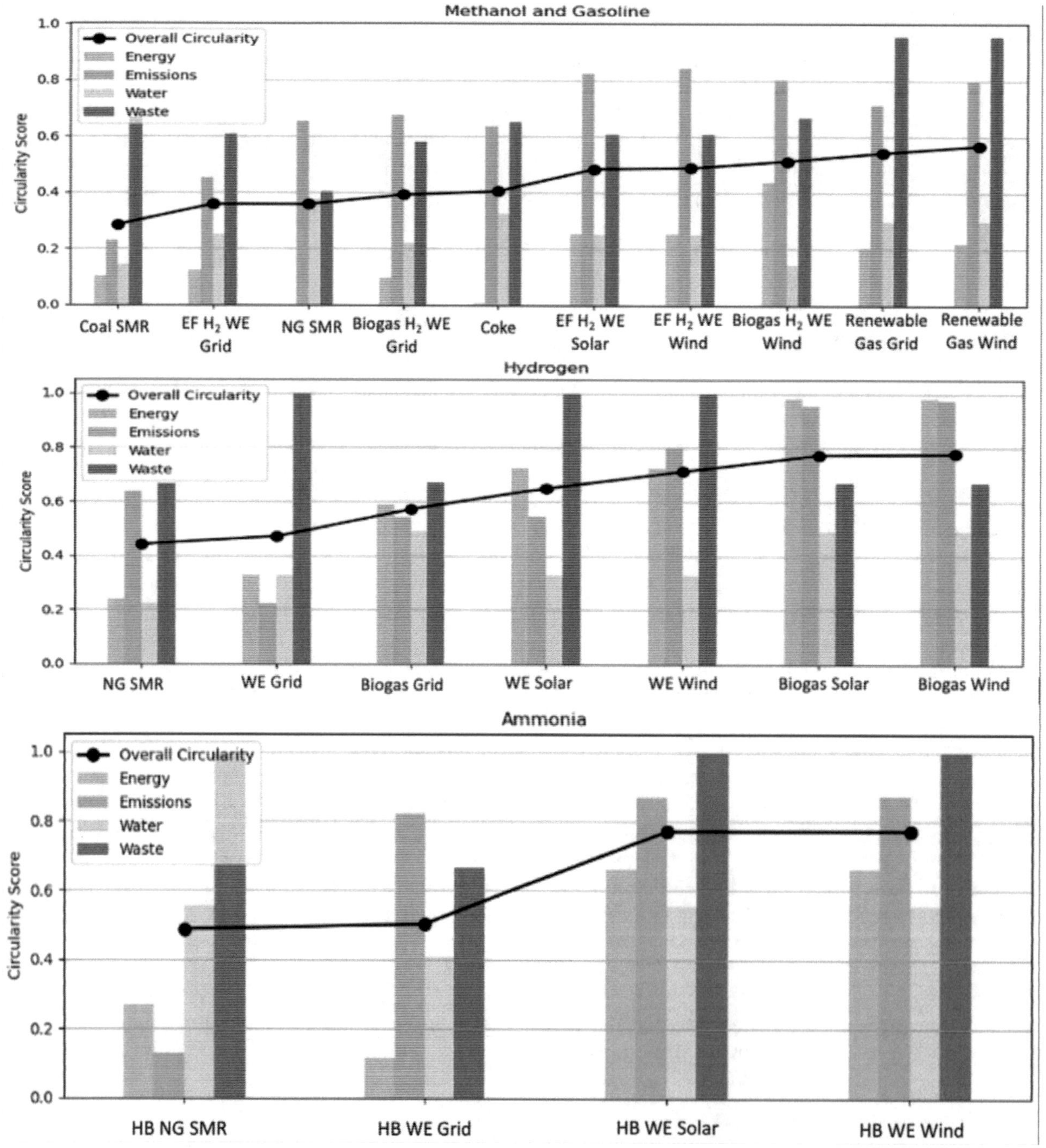

46c Guiding the Transition to a Sustainable and Resilient Circular Economy: An Eco-Inspired Quantitative Framework

Farshid Nazemi, Brian D. Fath, Bhavik Bakshi

Circular Economy (CE) presents an alternative business model to the prevailing business-as-usual economy. Unlike the linear consumption pattern of finite resources without product recovery at the end of their life, CE emphasizes closing the loop for products. This approach has the potential to tackle various environmental issues, including resource depletion, climate change, and biodiversity loss, by promoting the reuse, recycling, and regeneration of materials within the economy.

A considerable body of literature examines the potential of the CE and has developed indices and metrics to assess the level of circularity across different scales, ranging from product-level to industrial parks and the global economy. Product-level metrics, in particular, are extensively studied and are designed to measure the circularity of individual products within an economy. These product-level CE metrics serve as valuable tools for companies and other stakeholders to evaluate the circularity of their products and supply chains. By identifying the degree of circularity and pinpointing bottlenecks, these metrics can inform strategies to enhance the overall circularity of products and supply chains.

While existing metrics primarily emphasize circularity strategies centered around recovery rates based on material and energy flows, they often overlook other crucial aspects of Circular CE strategies. CE strategies extend beyond supply chain and technological interventions aimed at improving recovery rates of materials and products at the end-of-life stage. Changes in consumption patterns and business models also play a significant role in enhancing product circularity and facilitating the transition to a sustainable CE. Unlike strategies focused solely on product recovery, these approaches are often evaluated qualitatively. To our knowledge, there is currently no comprehensive systematic framework available to quantitatively assess the degree of circularity achieved through various CE strategies, including those targeting consumption patterns and business models.

Furthermore, alongside promoting circularity, CE strategies should also strive to foster a sustainable and resilient economy, which ultimately constitutes the overarching objective of CE. Achieving environmental and economic sustainability necessitates access to comprehensive information regarding material, energy, economic, and environmental flows. However, such detailed data is often lacking during the initial stages of designing a CE. This gap in information availability poses a challenge, and closing this gap is essential for enabling informed decision-making and enhancing the effectiveness of CE initiatives at an initial stage of a design.

Focusing solely on enhancing the circularity of a supply chain may inadvertently lead to network designs that lack resilience and flexibility. Therefore, it is crucial to incorporate resilience considerations into the evaluation of CE solutions to assess their robustness in the face of system shocks. A resilient CE should be able to withstand perturbations with minimal damage, highlighting the importance of capturing resilience in CE supply chains. While dynamic engineering resilience metrics have been utilized to assess the impact of perturbations on system survival, these methods require extensive static and dynamic information about the system, which may be lacking at the early design stage [1]. Hence, there is a need for a framework that can evaluate resilience at an early design stage based solely on product flow information. Such a framework would enable the assessment of CE solutions' resilience potential and inform decision-making processes to enhance the robustness of supply chain designs.

In this study, we aim to overcome the above-mentioned limitations of existing CE frameworks and address gaps in the literature by developing an ecologically inspired (eco-inspired) framework rooted in the principles of Ecological Network Analysis (ENA). This framework comprises three sets of metrics. The first focuses on capturing the circularity and resource efficiency achieved through CE strategies. These metrics align with conventional CE metrics, measuring factors such as product recovery rates at the end of life and virgin resource consumption rates. The second set of metrics assesses the sustainability and resilience of a CE system during the early design stages, utilizing physical flow network of products. Importantly, this set of metrics does not require detailed economic, environmental, or dynamic flow data. Lastly, the third set of metrics is designed to evaluate the effectiveness of various CE strategies, ranging from closed- and open-loop recycling to changes in consumption patterns such as product refusal. These metrics serve to guide the improvement of CE designs by informing decisions about adopting or modifying CE strategies.

We have defined the following metrics by tailoring their counterparts in ENA to suit the unique characteristics of product-level CE:

- Detritivory to herbivory ratio (D:H): This metric serves as a conventional circularity metric, reflecting the recovery rate within a product-level CE.
- Resource Consumption Efficiency (RCE): This metric measures the rate of resource consumption required to produce a certain quantity of product.
- Network Intensity (NI): NI quantifies the intensity of a network based on the flow quantities within it. It can be used as an initial tool to compare the environmental and economic intensity of different CE strategies.
- Robustness: Robustness, akin to ecological resilience, assesses the degree of order within a CE network. It gauges the network's ability to withstand shocks in the event of a disturbance.
- Loop Tightness: This metric evaluates the tightness of loops within a CE, favoring closed-loop recycling with tighter loops.
- Mean Circularity Level: Based on a CE hierarchy of different design strategies, this metric assigns scores to various CE strategies, with refusing (reducing consumption) being the most favorable and littering (waste disposal to the environment) the least favorable. It provides a holistic assessment, capturing not only product recovery strategies but also changes in consumption patterns or product design.

This framework was applied to a case study involving multi-layer plastic films utilized in Sheet Molding Compound Production (SMC film). SMCs are commonly employed in the manufacturing of bathtub products. SMC films are used for their carrier and barrier properties but result in a significant amount of waste annually. This waste presents a substantial opportunity for recovery and valorization at the end of their useful life.

In this study, we proposed several circularity strategies to promote the circularization of SMC films. These strategies include:

- Reuse: Mechanical and solvent-based recovery techniques are employed to clean the SMC film, allowing for its reuse in the same production process [2].
- Recycle: A solvent-based method is utilized to extract various polymer layers in the film and recycle the resin to produce new film [3].
- Downcycle: The film waste undergoes pelletization and injection molding processes to create other types of products. This approach represents an open-loop recovery, as the recovered material is not reintegrated into the same product system [2].
- Diverse End-of-Life (EoL): A combination of the previous three end-of-life alternatives is

employed to treat the film waste. This scenario demonstrates the impact of utilizing a diverse set of EoL strategies on the robustness of a CE network.

- Refuse: This scenario considers CE strategies beyond EoL recovery to reduce the demand for SMC films. For example, changes in consumer behavior driven by environmental awareness can decrease SMC film consumption. Similarly, innovations in product design and business models, such as adopting in-mold coating technology [4] or implementing a product-as-a-service model [5] for bathtub products, can reduce the production and consumption of SMC films.

Using the eco-inspired framework, we conduct a comparative analysis and ranking of the circularity strategies based on their circularity, resilience, and sustainability. This framework facilitates the identification of bottlenecks in circularity, resilience, and sustainability, allowing for modifications to the CE strategies to achieve targets in all three areas. The results of the framework reveal that a combination of the refuse and diverse EoL strategies prove to be the most effective in meeting circularity, resilience, and sustainability targets. This scenario was identified as our best case scenario. Furthermore, by comparing the circularity, resilience, and greenhouse gas (GHG) emissions of the best case scenario with the business-as-usual linear economy for SMC films, significant improvements were observed. Specifically, there was a 77% improvement in circularity, a 51% enhancement in resilience, and a substantial 90% reduction in GHG emissions. These results highlight the practical application of the eco-inspired framework in designing circular, sustainable, and resilient solutions for a wide range of products. The framework offers valuable insights for designing CE strategies that achieve environmental, economic, and social sustainability goals.

Reference:

[1] Chatterjee, A., & Layton, A. (2020). Mimicking nature for resilient resource and infrastructure network design. *Reliability Engineering & System Safety, 204*, 107142.

[2] Nazemi, F. et al. (2024). Life cycle and circularity assessment of multilayer films used in Sheet Molding Compound Process (Under Preparation).

[3] Costamagna, M., Massaccesi, B. M., Mazzucco, D., Baricco, M. & Rizzi, P. Environmental assessment

of the recycling process for polyamides - Polyethylene multilayer packaging films. Sustainable Materials

and Technologies 35, e00562 (2023).

[4] Ko, S., Ouyang, X., Straus, E. J., Lee, L. J. & Castro, J. M. Processability study of in-mold coating for sheet molding compound compression molded parts. Polymer Engineering Science 59, 1688–1694(8 Aug. 2019).

[5] Tukker, A. Product services for a resource-efficient and circular economy – a review. Journal of Cleaner Production 97, 76–91 (2015).

46d Potential to Recover Copper from Electronic Waste and Implications for the Circular Economy

Andrea Hicks

Introduction:Copper is by no means a new or exotic material, dreamed up in the laboratories of scientists. It is in fact a very old metal, employed since antiquity. Nevertheless, it is a key metal to enable a decarbonized future and much has been written projecting future copper demands. However, in order to truly understand the future and its environmental and economic implications, it is necessary to first look to the past. This work seeks to present a view of the historical evolution of copper consumption, which can be used to inform the current trajectories of the future in the context of sustainability and resource use.

Copper is crucial to enable the transition to a decarbonized future This is largely due to its role in critical technologies and infrastructure. However, copper production is not without environmental, economic, and societal costs of its own. Typically, the most efficacious manner in which to reduce these burdens is by keeping already mined and refined copper in circulation. However, given the projected copper demands, due to continuing population and economic growth, there will be continued increasing requirements for primary metal production. Beyond that, some uses of copper such as copper containing pesticides are dissipative in nature, meaning that they are unable to cycled in the economy due to the nature and goals of their use. The environmental impact of mining virgin copper ranges from 5.34 to 7.33 kg of CO_2e per kg of refined copper dependent on method, while recovered or secondary copper has an environmental impact of 1.58 kg of CO_2e per kg of refined copper. Due to projected increases in the demand for copper it is estimated that the associated carbon footprint of copper will increase between 100%-200% between 2010 and 2050. However, depending on the quantity of copper, the nature of its use, and the matrix it is embedded in, it may not be environmentally nor economically advantageous to recover the copper. At the same time global copper reserves are currently estimated at 2.1 billion tons of copper, with an estimation of undiscovered resources at 3.5 billion tons. However, not all of this copper is easy to mine and refine. This work seeks to look to the history of copper consumption, largely in the US, in order to generate insight into future copper consumption and its sustainability implications. And to posit the idea of the greater potential of consumer electronic waste as one source of future copper and discuss the conditions under which this might occur.

Methods: Historical copper data is mined from reports by the United States Geological Survey (USGS). It should be noted that the numbers differed or were later revised in some of the reports. And thus, there is some uncertainty around the exact numbers reported. In general, when this occurred the revised reported number was included in the data collected. If this was unclear, then the value presented with more precision was maintained. The real economic cost of copper is calculated from the cost as presented in the USGS reports. In particular using the London Price Average Cents per pound as reported. In some of the earliest price reports, the copper price is simply provided without further reference. Historical series data was used to convert the dollars values to 2022 equivalents or earlier years. The relationship between copper ore grade and the energy needed for the production of usage copper has been modeled previously considering ore grades and gross energy requirements. Where E, the gross energy requirement, is expressed in GJ/metric ton of mined copper as a function of the ore grade (% copper ore). This is applied to the average annual US ore grade obtained from the annual USGS reports. The same methodology was applied for the energy embodied in the copper e-waste. The data was adopted from published studies which characterized the quantity of copper in various e-wastes.

Results: The average primary copper ore produced in the US has decreased over the last century, which has also led to a corresponding increase in the energy required for mining and beneficiation facilities. As primary available ore grade continues to decrease and the energy required to make that ore into useful copper continues to correspondingly increase, the idea arises when the primary copper industry will run into the mineralogical barrier. This barrier as expressed by Skinner, and defined as when the rock would need to be pulled apart chemically in order to gain resources such as copper, instead of mining ore itself. It also introduces the question as to whether copper based waste which was not previously considered economically feasible to mine at different points in industry (such as tailings and e-waste containing copper), will become viable sources of copper feedstock. Essentially, when does the energy and corresponding environmental impacts of copper become costly enough where it is more environmentally efficient to mine old wastes?

Mining spent consumer electronic waste for resources is not a new idea. The figure presents the copper content, expressed in percent mass, of different small post-consumer electronics. And is compared with the energy embodied in the copper from mining and beneficiation when that same grade of ore would be mined from virgin stocks. This is not the energy required to recover the copper from these materials, as that has the potential for significant variability, but simply the minimum amount of energy would need to be spent if virgin ores of this same grade were mined . This is at best a significant simplification, however, it allows for the consideration of energy that would need to be spent to mine comparative virgin ores, and suggests that if the embodied energy cost is less than what is shown in the figure, that there is the potential for significant benefit if copper is recovered from spent electronics. As of 2018 the average copper ore processed in the US had a yield of 0.48%. This is significantly lower than what is available in several categories of spent electronics: batteries (12.98%±10.61%), smartphones (27.83±13.87%), cellphones (36.25%±6.52%) and sometimes printed circuit boards (PCB) (15.86±9.86%). This suggests some potential for considering spent electronics as a source of post-consumer secondary copper, in order to reaching increasing copper demands. Recent work by Zeng et al. found that in China recovering copper from spent urban electronics could be completed at economic costs that we at least comparable to those of virgin mining.

Conclusions: Post-consumer electronic waste is one potential source of copper to meet current demands as many electronics contain a higher mass percentage of copper than what is currently being mined. This suggests the potential for environmentally and economically beneficial copper recovery from waste electronics in order to meet current and future copper demands. Although challenges exist with respect to recovering the waste electronics from consumers and the copper along with other valuable metals. Washing machines are explored as an example for the potential of copper recovery.

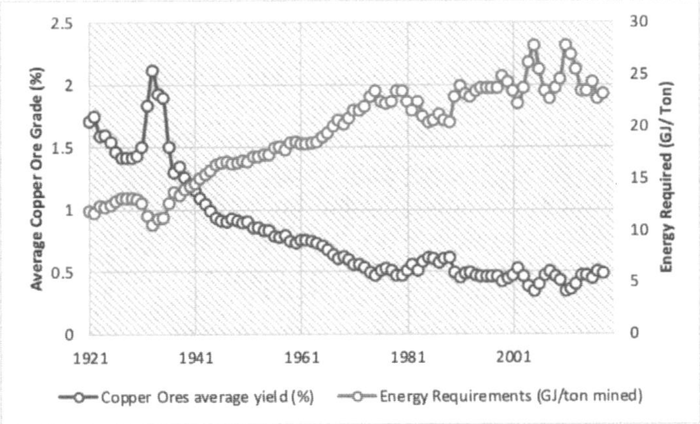

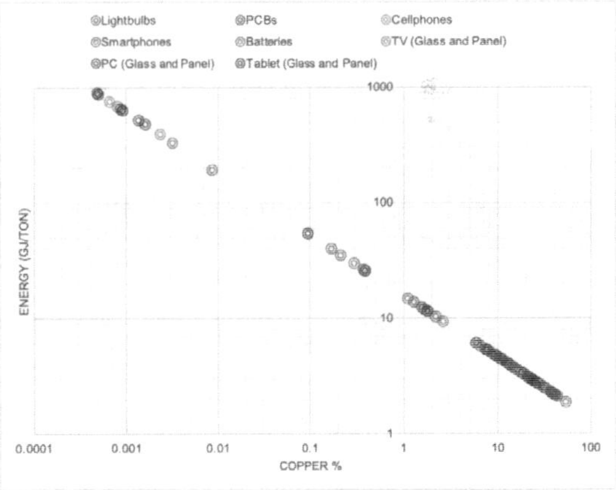

46e Enabling Circular Aquaculture with a Biofilm-Based Waste-to-Feed Biotechnology

Jin Wang, Q. Peter He, Reza Khorasani

In response to the fast-growing global population and increasing demand for animal protein, aquaculture is the fastest-growing animal food production sector worldwide. With growing competition for natural resources (land, water, etc.) among all production sectors, the intensification of aquaculture systems has emerged as a promising pathway for increasing aquaculture production. Accompanying the aquaculture intensification is the increased demand for feed and increased waste generation. On the one hand, aquafeed is recognized as the critical factor limiting the aquaculture industry's growth. On the other hand, aquatic waste is often discharged into the environment without proper treatment, causing eutrophication and threatening the sustainability of local ecosystems. Therefore, sustained intensification of aquaculture production must be coupled with a sufficient supply of sustainable nutrients (feed) and efficient waste management.

To achieve sustained aquaculture production, we propose a a novel, sustainable and profitable waste-to-feed (W2F) biotechnology to enable a circular aquaculture economy. As shown in Figure 1, the proposed W2F biotechnology integrates the commercially proven anaerobic digestion (AD) with a recently patented circulating coculture biofilm photobioreactor (CCBP) to convert aquaculture waste into single cell protein (SCP) as aquafeed supplements, treated clean water and fertilizer.

Although AD has many environmental and technical advantages over aquaponics and bioflocs technology, current AD-based waste-to-value (W2V) technologies have their own limitations. To date, the most prominent AD-based W2V technologies are: (1) combined heat and power (CHP) and (2) biomethane (BM) that utilize AD-generated biogas, and (3) microalgae cultivation utilizing AD-generated digestate (and CO_2) for biofuels. None of these technologies have been proven economically feasible in the US due to several reasons. One is the low value of the products (electricity and natural gas); the other is the limited government incentives (compared to some European countries); finally, it is the high cost in treating contaminants in biogas. We hypothesize that if the AD-produced biogas and nutrient-rich AD effluent can be converted into a higher-value product with low operation cost, it will promote broad installations of AD, especially at small to mid-scale farms, and capture the enormous environmental benefits brought by AD.

In this talk, we will present our recent findings in the optimization of biogas production through co-digestion of fish sludge and food waste. Then we will present our recent progress in utilizing methanotroph-microalgae coculture biofilm to simultaneously convert AD-produced biogas and nutrient-rich AD effluent into microbial biomass, which will serve as the feedstock for SCP, a higher-value aquafeed supplement. We discuss how the proposed CCBP addresses the long-standing challenges of biological methane conversion (mass transfer of gas substrate with small solubility) and microalgae cultivation (mass transfer of gas substrate and light attenuation) by quantitative comparative studies.

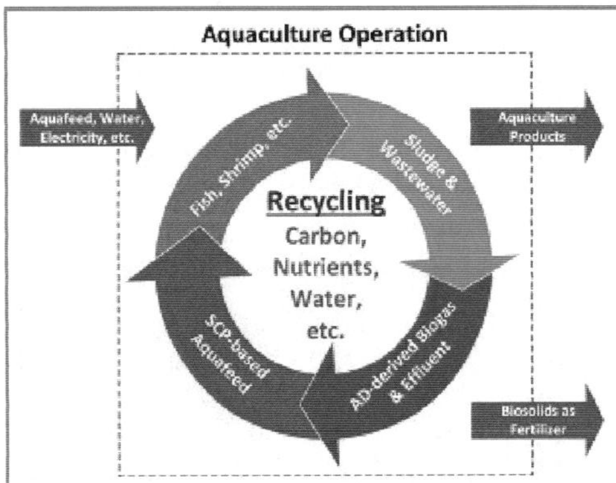

Environmental and economic impact:

- **Significant reduction in GHG emission is achieved through AD**, where aquaculture waste is broken down into biogas, nutrient-rich effluent, and biosolids in a controlled and contained fashion.
- **SCP and treated clean water** are produced from biogas and nutrient-rich AD effluent in CCBP, where a novel coculture is immobilized at a very high concentration on a conveyor belt, which significantly intensifies the conversion.
- **High-quality aquafeed supplements and fertilizers** are derived from SCP and AD-produced biosolids.

Together, the proposed W2F biotechnology will enable a **sustainable circular aquaculture industry**.

Fig. 1 The envisioned circular aquaculture enabled by W2F technologies

46f Plastic Waste Prevention: Environmental Impact Assessment of Advanced Circular Solutions As a Substitute for Single-Use Virgin Plastic Materials

Annamaria Vujanović, Jan Puhar, Martin Leban, Anja Pajntar

Single-use plastic materials pose a significant threat to the environment due to their resource depletion and contributing to plastic pollution. The problem of plastic waste should be addressed not only through elimination and treatment of existing waste, but also with advanced litter prevention measures and the development of circular alternatives. Such measures are being undertaken as part of the Horizon EU Mission project INSPIRE (Innovative Solutions for Plastic Free European Rivers), which is proposing new sustainable alternatives for widely used products and depicting their benefits by comparing them to existing conventional products. The emergence of advanced circular solutions as a plastic prevention measure has shown to be a promising path forward in minimizing waste. However, there is a need for quantification of impacts of the impact such solutions make towards sustainability, where the three pillars of environmental, economic, and social sustainability should all be considered.

One of the key sectors producing single-use plastic waste worldwide is tourism. The spread of plastic pollution in heavily visited areas during the tourist season shows an urgent need for scalable solutions to be widely adopted, which could drive behavioral change and promote sustainable alternatives among consumers. The solution proposed in this study is a plastic-free shampoo formulation, which can substitute conventional shampoo packaged in plastic bottles. The analysis focuses on evaluating the environmental benefits of adopting this alternative on a large scale, considering a case study of a tourist group within a camp using the sustainable shampoo solution.

To assess the environmental implications, the study applies the life cycle assessment (LCA) methodology, considering various stages from raw material extraction to product disposal. Two scenarios are compared: one where the tourists use conventional shampoo packaged in single-use plastic bottles, and another where they use plastic-free shampoo products, produced as natural water-free powder formulations packaged in paper material. The circular solution is a shampoo that does not contain plastic particles or water, ensuring that fewer natural resources are used for its production. The analysis is based on the functional unit of one wash of average length hair for a person in Europe, which requires a 30 mL bottle of liquid shampoo in the conventional case and a 3.5 g package of sustainable shampoo powder in the alternative case. This unit is used to compare the two products based on their material demands for the shampoo as well as its packaging.

The results regarding the case of shampoo use show that adopting plastic-free shampoo formulations significantly reduces environmental burdens compared to conventional shampoo packaged in plastic bottles. The sustainability of the water-free formulations coupled with paper packaging materials minimizes resource depletion and energy use, thereby reducing environmental impacts. The global warming potential, evaluated as one of the impact categories, exemplifies the benefits of the circular solution, showing a reduction from 1,726.1 kg CO_2 eq. to 144.9 kg CO_2 eq., or by 92 % when considering the hair washing needs of the entire tourist camp. As the conventional liquid shampoo requires additional raw materials such as emulsifiers, fillers, foam stabilizers, and moisturizers, the benefits of reducing the need for these materials is clearly shown by the impact category results. The most signifcant reductions in environmental impacts are caused by the reduced demand for electricity consumption during both the formulation mixing and packaging processes. The research therefore points at the importance of transitioning from linear consumption models reliant on single-use plastics to circular solutions and highlights the potential for behavioral shifts among consumers towards more sustainable choices, fostering the principles of a circular economy in everyday life.

46g Material, and Circular Economy Challenges in Advancing Automotive Plastics Recycling through Molecular Recycling Technologies

Mahshid Mokhtarnejad, Gary Hawkins, Hendrik Mainka

The pressing global plastic waste crisis necessitates a transformative shift towards sustainable materials utilization. In this regard, Volkswagen Group of America Inc. (VWGoA) and Eastman Chemical Company are taking the lead in designing and driving circular economy solutions for automotive plastics by employing innovative molecular recycling technologies. This approach facilitates the conversion of traditionally hard-to-recycle plastic waste streams into their molecular components for the production of virgin-grade raw materials. These feedstocks are then used to produce specialty plastics and fibers with quality and performance equivalent to their virgin-grade counterparts. Our work targets a wide range of plastic waste streams including mixed plastics, colored plastic materials, post-consumer and industrial car components, and textiles, with the ultimate goal of confronting recycling challenges while providing environmental sustainability. We have shown that by using this molecular-level chemical recycling process, we can not only offer a high-quality alternative to virgin fossil-based plastics, but also reduce greenhouse gas emissions by 25-50% compared to conventional petroleum-derived automotive plastics. Our team has demonstrated the automotive viability of these sustainable materials by creating multiple demonstration parts on production equipment: an Atlas Cross Sport Spoiler Assembly and Passat R-Line Diffuser. These parts have successfully met all essential mechanical and thermal criteria, including scratch and chemical resistance, adhesion, tensile and flexural modulus as well as Vicat softening and glass temperature tests. This exemplifies the efficacy of the molecular-level chemical recycling process in recycling mixed plastics for automotive application. These components serve as examples of our commitment to sustainability and circular economy principles, paving the way for a cleaner and more resource-efficient automotive industry.

46h Carbon Fiber-Reinforced Polymer (CFRP) in Automobiles: Achieving Net-Zero Emission and Circular Economy

Reena Sharma, Aniket Mali, Bhavik Bakshi

The urgent challenge of climate change highlights the need of sustainable solu- tions, especially in the field of transportation, greenhouse gas emissions, largely from burning fossil fuels like carbon dioxide (CO_2) are a major concern [1]. In the United States, transportation contributes to 97% of the sector's global warming potential [2]. In recent years, regulations were enacted, imposing CO_2 emission performance standards for new passenger cars and vans globally. These standards aim to achieve a reduction of emissions by 2032, as well as incentive mechanisms to promote the adoption of zero and low-emission vehicles worldwide [3]. So there has been a push for stricter regulations on vehicle emissions. To meet these regulations, car manufacturers are increasingly using such materials that make vehicles lighter [4]. Plastics and polymer composites, which are already widely used in car interiors, exteriors and lighting, are now being used in other parts of the vehicles [5]. However, these mate- rials contribute significantly to greenhouse gas emissions and plastic waste, which go against the goal of achieving net-zero emissions [6]. In response, many manufactur- ers are aiming to produce vehicles with net-zero emissions which requires exploration of alternative options for vehicle materials. While there are various approaches to decrease tailpipe emissions, the utilization of alternative lightweight materials or processes that result in lighter end products is often seen as a pivotal solution. As individual materials approach their performance limits, there is growing interest in combining the advantages of different materials into a single application. One of the materials, that will aid in our goal is carbon fiber-reinforced polymer (CFRP) which is recognized for its exceptional lightweight properties which can reduce weight and also hold promise for enhancing vehicle efficiency [7]. However, these composites present significant challenges, from assembly to recyclability at their end of their life cycle, in addition to generally being more expensive than conventional materials such as steel and aluminum. Adding to it, there are uncertainties surrounding the energy consumption of CFRP production and recycling and limited knowledge regarding fuel consumption benefits. In response to these pressing environmental challenges, this work aims to demonstrate a commercially viable engineering solution that mitigates greenhouse gas emissions, aligning with regulatory requirements for improved vehicle fuel efficiency. Central to this approach is the development of a road map that not only aims for net-zero emissions within the automobile industry, but also finds circular and cost-effective solutions.

In this work, we perform a detailed life-cycle assessment (LCA) by utilizing industrially relevant data and a comprehensive system boundary [8]. We find that GHG emissions incorporating CFRP are significantly less than the metal counterparts. We also examine trade-off solutions between costs and emissions using various alternative resins. Our findings indicate that employing formaldehyde resin represents a cost-effective solution, while utilizing PET resin emerges as an effective strategy for reducing emissions. We also perform hotspot analysis which reveals that electricity generated from bituminous coal and energy production are the primary contributors to emissions in our process. To move towards achieving net-zero emission in the future in the automobile industry, we intend to integrate renewable energy sources and carbon capture technologies, all while considering economic factors. We also aim to include emerging technologies and project future scenarios in order to get a road map to net-zero emissions [9]. Taking end-of-life into account, the recovery, recycling, and down cycling of carbon fiber represent integral processes within the sustainable governance of carbon fiber materials, being a matter of significant concern within the automotive industry.

Therefore our study also prioritizes the early integration of various end-of-life recycling scenarios in the product development to optimize sustainability and circularity outcomes. The successful demonstration of our proposed scheme would facilitate the integration of sustainable technologies into automobile industry, thereby reducing CO_2 emissions and plastic waste. By prioritizing circularity and sustainability in materials design and vehicle engineering, our efforts contribute to advancing towards a circular economy and mitigating climate change ensuring a sustainable future.

References:

[1] A. F. Ghoniem, Needs, resources and climate change: Clean and efficient conver- sion technologies, Progress in energy and combustion science 37 (2011) 15–51.

[2] EPA-420-F-18-008, Environmental protection agency, "greenhouse gas emissions from a typical passenger vehicle, (March 2018). URL: https://tinyurl.com/ bwmm8a6s.

[3] EPA-420-F-24-016, Multi-pollutant emissions standards for model years 2027 and later light- duty and medium-duty vehicles: Final rule, (March 2024). URL: https://www.epa.gov/system/files/documents/2024-03/420f24016.pdf.

[4] F. Czerwinski, Current trends in automotive lightweighting strategies and mate- rials, Materials 14 (2021) 6631.

[5] M. A. Fentahun, M. A. Savas, Materials used in automotive manufacture and material selection using ashby charts, Int. J. Mater. Eng 8 (2018) 40–54.

[6] V. V. Rajulwar, T. Shyrokykh, R. Stirling, T. Jarnerud, Y. Korobeinikov, S. Bose, B. Bhattacharya, D. Bhattacharjee, S. Sridhar, Steel, aluminum, and frp- composites: The race to zero carbon emissions, Energies 16 (2023) 6904.

[7] W. Zhang, J. Xu, Advanced lightweight materials for automobiles: A review, Materials & Design 221 (2022) 110994.

[8] V. Thakker, B. R. Bakshi, Designing value chains of plastic and paper carrier bags for a sustainable and circular economy, ACS Sustainable Chemistry & Engineering 9 (2021) 16687–16698.

[9] V. Thakker, B. R. Bakshi, Toward sustainable circular economies: A computa- tional framework for assessment and design, Journal of Cleaner Production 295 (2021) 126353.

354a Dozn™2.1 - a Quantitative Green Chemistry Evaluator for a Sustainable Future

Ettigounder Ponnusamy

MilliporeSigma (The life science business of Merck KGaA, Darmstadt, Germany) developed and launched DOZN™2.0 in 2017, a unique web-based greener alternative scoring matrix, also known as a quantitative green chemistry evaluator. This tool is based on the 12 principles of green chemistry for customers to evaluate their relative greenness of their processes which provide a framework for learning about green chemistry and designing or improving materials, products, processes, and systems. DOZN™2.0 scores products based on metrics for each principle and aggregates the principle scores to derive a final aggregate score. DOZN™2.0 calculates scores based on manufacturing inputs, GHS and SDS data which provide a green score for each substance. DOZN™2.0 is flexible enough to encompass the diverse portfolio of products. The DOZN™2.0 system has also been verified and validated by a third party to ensure best practices are applied. Based on customer feedbacks, we upgraded the tool to DOZN™2.1 and launched in December 2022. Through DOZN™2.1 customers now have access to calculate the green scores of their processes and products. This free, web-based tool provides users with more data so that they are properly equipped to improve their sustainability. DOZN™2.1 keeps data privacy top of mind - allowing customers to score their processes/products in a safe and secure manner.

MilliporeSigma (The life science business of Merck KGaA, Darmstadt, Germany) developed and launched DOZN™2.0 in 2017, a unique web-based greener alternative scoring matrix, also known as a quantitative green chemistry evaluator. This tool is based on the 12 principles of green chemistry for customers to evaluate their relative greenness of their processes which provide a framework for learning about green chemistry and designing or improving materials, products, processes, and systems. DOZN™2.0 scores products based on metrics for each principle and aggregates the principle scores to derive a final aggregate score. DOZN™2.0 calculates scores based on manufacturing inputs, GHS and SDS data which provide a green score for each substance. DOZN™2.0 is flexible enough to encompass the diverse portfolio of products. The DOZN™2.0 system has also been verified and validated by a third party to ensure best practices are applied. Based on customer feedbacks, we upgraded the tool to DOZN™2.1 and launched in December 2022. Through DOZN™2.1 customers now have access to calculate the green scores of their processes and products. This free, web-based tool provides users with more data so that they are properly equipped to improve their sustainability. DOZN™2.1 keeps data privacy top of mind - allowing customers to score their processes/products in a safe and secure manner.

354b A Data Engineering Framework for Operationalizing the Alternative Assessment of Chemicals and Chemical Processes

John D. Chea, Michael Gonzalez

The chemical industry created a strong foundation for many consumer, commercial, and industrial products to directly improve the quality of life for society. In modern times, chemical and chemical process design inherently has some form of sustainability principles instilled to meet the primary goals of crucial stakeholders and product users. However, these processes are rarely analyzed for sustainability from a multi-stakeholder perspective due to the variation of sustainability assessment methods and the low availability and accessibility of usable data, leading to unforeseen problems related to process efficiency, environment, energy, costs, health impacts, and social needs. This research introduces a data engineering framework that extracts and processes the abundance of relevant chemical data and parameters required for alternative assessment (AA) to enhance and direct the sustainability of the overall chemical and chemical process. This framework itemizes and categorizes the critical parameters required by the existing sustainability assessment methods, then automates the multi-stakeholder AA data extraction and conversion process. Using this data engineering framework, stakeholders in the chemical industry can quickly acquire the necessary data to analyze the sustainability implications of potential process investment and improvement opportunities, streamlining the decision-making process.

354c Climate Policies Can Substantially Reduce the Environmental Footprint of Green Chemicals

Abhinandan Nabera, Antonio José Martin Fernandez, Ioan-Robert Istrate, Javier Pérez-Ramírez, Gonzalo Guillen-Gosalbez

The chemical industry poses a formidable challenge towards decarbonization, primarily due to its heavy reliance on fossil fuels as carbon and energy sources, leading to 5.6 Gt CO_2-eq y^{-1}, accounting for 10% of global GHG emissions [1]. Specifically, traditional routes of platform chemicals like ammonia and methanol are catalytic processes, currently heavily relying on natural gas for their production and resulting in significant CO_2 emissions (560 Mt y^{-1}) [2]. About one-third of these emissions are direct, while the remaining are attributed to energy acquisition and the upstream value chain [1]. This challenging dependence on global energy systems and supply chains also presents an opportunity to reduce environmental impacts in platform chemical production through other key sector decarbonization strategies.

In recent years, emerging catalytic technologies have played a crucial role in advancing low-carbon production routes for these chemicals. Greener production routes for ammonia and methanol, based on green hydrogen, have been put forward more recently, with their environmental performance evaluated through life cycle assessments (LCAs) [3–5]. In these LCA studies, a process model is developed utilizing mass and energy balances of the main plant (*i.e.*, foreground system), which is combined with life cycle emissions data based on various technologies across the chemical supply chains (*i.e.*, background system). It is common in LCA studies to assume a fixed background system, *i.e.*, a set of technologies providing inputs to the chemical production plant, reflecting the current state of the technosphere [6–8].

However, the targets set by climate policies will necessitate substantial changes in the economy, including the decarbonization of energy production and transportation. Neglecting these future technological changes when evaluating the environmental impacts of chemicals can result in spurious conclusions. Hence, in this work, we perform a prospective LCA to evaluate the environmental impacts of fossil and green ammonia and methanol pathways until 2050 by considering potential or predicted changes in the power, materials, and transportation sectors.

Therefore, our study investigates the extent to which future economic trends affect the environmental impacts of fossil and green ammonia and methanol production pathways. To enable the assessment of current and future environmental impacts based on expected socioeconomic and technological trends, we utilize future projections of the economy derived with an Integrated Assessment Model (IAM). IAMs like IMAGE comprehensively capture interactions among society, the biosphere, and the climate system [9]. Specifically, in this work, we adopt the 'middle-of-the-road' shared socioeconomic pathway (SSP2) to perform our assessments. In parallel, IMAGE also projects different representative concentration pathways (RCPs), specifically RCP6, RCP2.6, and RCP1.9, presenting a range of radiative forcing values in 2100. RCP6 is a pathway that limits global mean surface temperature to below 3.5 °C, while RCP2.6 and RCP1.9 aim to restrict temperature to 2 °C and 1.5 °C, respectively [10]. Combining these SSPs with RCPs, a range of scenarios is generated entailing specific changes in the economy to perform more accurate environmental assessments.

In terms of climate change impacts, we found that the solar and wind-based ammonia or methanol production routes show a promising trend, with significant reductions anticipated by 2050 (in the range 55–110%), in contrast to fossil-based pathways, which will reduce impacts only marginally (4–11%). Under more ambitious climate policies, such as the 1.5 °C scenario, these impacts are expected to decrease substantially. Specifically, by 2050, both the solar- and wind-based routes are projected to achieve almost a 70% reduction in impacts. Furthermore, a regional analysis reveals that for solar- and wind-based ammonia and methanol production, impact reductions vary significantly by region, emphasizing the importance of location-based dynamics for strategic planning. We believe these results equip policymakers and technology investors with tools to navigate the evolving landscape of the chemical industry, ensuring informed decisions.

References

[1] F. Bauer, J.P. Tilsted, S. Pfister, C. Oberschelp, V. Kulionis, Mapping GHG emissions and prospects for renewable energy in the chemical industry, Current Opinion in Chemical Engineering 39 (2023) 100881. https://doi.org/10.1016/j.coche.2022.100881.

[2] P. Gabrielli, L. Rosa, M. Gazzani, R. Meys, A. Bardow, M. Mazzotti, G. Sansavini, Net-zero emissions chemical industry in a world of limited resources, One Earth 6 (2023) 682–704. https://doi.org/10.1016/j.oneear.2023.05.006.

[3] S.C. D'Angelo, A.J. Martín, S. Cobo, D.F. Ordóñez, G. Guillén-Gosálbez, J. Pérez-Ramírez, Environmental and economic potential of decentralised electrocatalytic ammonia synthesis powered by solar energy, Energy Environ. Sci. 16 (2023) 3314–3330. https://doi.org/10.1039/D2EE02683J.

[4] J.R. Gomez, J. Baca, F. Garzon, Techno-economic analysis and life cycle assessment for electrochemical ammonia production using proton conducting membrane, International Journal of Hydrogen Energy 45 (2020) 721–737. https://doi.org/10.1016/j.ijhydene.2019.10.174.

[5] A. González-Garay, M.S. Frei, A. Al-Qahtani, C. Mondelli, G. Guillén-Gosálbez, J. Pérez-Ramírez, Plant-to-planet analysis of CO_2 -based methanol processes, Energy Environ. Sci. 12 (2019) 3425–3436. https://doi.org/10.1039/C9EE01673B.

[6] J.D. Medrano-García, V. Giulimondi, A. Ceruti, G. Zichittella, J. Pérez-Ramírez, G. Guillén-Gosálbez, Economic and Environmental Competitiveness of Ethane-Based Technologies for Vinyl Chloride Synthesis, ACS Sustainable Chem. Eng. 11 (2023) 13062–13069. https://doi.org/10.1021/acssuschemeng.3c03006.

[7] I. Ioannou, S.C. D'Angelo, A.J. Martín, J. Pérez-Ramírez, G. Guillén-Gosálbez, Hybridization of Fossil- and CO_2-Based Routes for Ethylene Production using Renewable Energy, ChemSusChem 13 (2020) 6370–6380. https://doi.org/10.1002/cssc.202001312.

[8] Á. Galán-Martín, V. Tulus, I. Díaz, C. Pozo, J. Pérez-Ramírez, G. Guillén-Gosálbez, Sustainability footprints of a renewable carbon transition for the petrochemical sector within planetary boundaries, One Earth 4 (2021) 565–583. https://doi.org/10.1016/j.oneear.2021.04.001.

[9] Integrated Assessment of Global Environmental Change with IMAGE 3.0, PBL Netherlands Environmental Assessment Agency, 2014.

[10] R. Sacchi, T. Terlouw, K. Siala, A. Dirnaichner, C. Bauer, B. Cox, C. Mutel, V. Daioglou, G. Luderer, PRospective EnvironMental Impact asSEment (premise): A streamlined approach to producing databases for prospective life cycle assessment using integrated assessment models, Renewable and Sustainable Energy Reviews 160 (2022) 112311. https://doi.org/10.1016/j.rser.2022.112311.

354d Life Cycle Assessment for Polyamide 6: Production, Recycling, Disposal

Caroline Ganzer, Ann-Joelle Minor, Lisciele Soto, Liisa Rihko-Struckmann, Kai Sundmacher

Conventionally, polymers are produced from fossil resources via a myriad of intermediate monomers. At the end of life, they are disposed in landfill or incinerated. The resulting emissions contribute to anthropogenic climate change, and the process is unfavorable with respect to other environmental KPI. Circular economy aims at transforming the existing routes by replacing raw materials with wastes, and closing the cycle of disposal and production. Various steps on the circular economy ladder have been described (Keijer et al., 2019), ranging from reuse and redistribution to recycle and recover. Material can be recycled either through the biosphere or technosphere (Tóth et al., 2022; Meys et al., 2021). Hence, several options exist to supply polymers to the economy sustainably. They include mechanical recycling, chemical recycling, bio-based synthesis, and derivation from captured carbon. Combinations of synthesis, recycling, and end-of-life options differ in cost, carbon and water footprint, land use, as well as overall sustainability. In this work, we perform life cycle assessment (LCA) to compare the various combination for polyamide 6 (nylon 6) with respect to economic and environmental indicators.

For some polymers, mechanical recycling, i.e., melting the polymer, not changing its chemical structure, is possible and cost-effective. However, it requires the waste polymer to be sorted carefully; any impurities can destroy the structure of the polymer. This is a major complication, considering many products contain combinations of polymers. The melting process can also change the properties of the polymer with every cycle, yielding sub-standard product (Coates et al., 2020; Achilias et al., 2014). As a result, chemical recycling, specifically the depolymerization of the polymer to its monomer, has been a research focus (Coates et al., 2020; Achilias et al., 2014). It enables the separation of the monomer and re-polymerization to pure, high-quality polymer. Subjecting the waste polymer to pyrolysis and synthesizing the monomer from pyrolysis oil also constitutes a chemical recycling pathway (Somoza-Tornos et al., 2020). Another option is the recycling of material through the biosphere. When the raw material for polymer production is sustainably sourced biomass, it can be incinerated at the end of life without incurring net emissions to the atmosphere. Alternatively, the cycle can be closed following the incineration by capturing CO_2 and using it as carbon source for the synthesis, a concept referred to as carbon capture and utilization (CCU) (Meys et al., 2021). If the polymer remains sourced from fossil resources, the combustion emissions can be abated by capturing the CO_2 and sequestering it (CCS). CCS could be combined with bio-based synthesis, potentially yielding net negative emissions.

Polyamides (PA) represent 8% of current polymer production at ~7 Mt/a (Rihko-Struckmann, 2022) and are utilized in the automotive industry, for electrics and electronics, consumer goods, healthcare, and construction. They are at present produced from fossil benzene and toluene via multiple monomers. Efforts to collect and recycle PA at scale both by mechanical and chemical processes have been realized commercially by DuPont since the 1990s (Kasserra, 1998) and also practiced by other companies (Coates et al., 2020). Several options for depolymerization as well as research on pyrolysis of PA exist (Deulkar, 2022; Minor et al., 2023). Still, only a small fraction of polyamide is currently recycled. Biological decomposition of PA appears extremely difficult.

In this work, we examine production, recycling, and disposal options for polyamide 6, which is synthesized from the monomer caprolactam. We include as virgin production routes the synthesis from fossil benzene, and synthesis from agricultural waste via hydroxymethylfurfural and adipic acid. Recycling options are mechanical recycling, and chemical recycling back to caprolactam by hydrolysis using phosphoric acid (Minor et al., 2023). In case of recycling via CO_2, the synthesis comprises methanol synthesis, methanol-to-aromatics, and the conventional step from benzene to caprolactam (Rihko-Struckmann, 2022; Deulkar, 2022). Incineration and landfill are considered as end-of-life treatment for the amount of PA-6 which is not recycled.

We gather LCAs from literature wherever available, and perform LCA using ECOINVENT 3.10 based on techno-economic analyses in literature for the individual processes. Then, we compare combinations of pathways and technologies. Each case combines virgin production, recycling option, and end-of-life treatment. For instance, fossil production combined with mechanical recycling, incineration, and CCS, or bio-based production without recycling plus landfill. Thereby we identify optimal routes with respect to cost, emissions, and other LCA indicators including resource depletion and biodiversity. Further, we vary recycling rates to evaluate how the impacts may evolve over time if ambitions for circularity in plastics are realized, and identify the lowest possible environmental impact PA utilization can have. It is of interest to quantify the trade-offs between the environmental benefits and economic drawbacks of recycling options. Recent work on modelling circular economy for plastics suggests similar emissions reductions can be achieved with input of large quantities of either biomass or (low-carbon) electricity (Meys et al., 2021). Hence, the optimal strategy may depend on the national context including natural resources as well as decarbonization and circular economy agenda. Further, estimating the cost of a low-carbon system, and the (additional) cost of a circular system could help inform public policy needed to achieve the transition. Detailed analysis of the distinct routes could also identify potential pitfalls in designing green PA production as well as key technologies and future research needs.

References

Tom Keijer, Vincent Bakker, and J. Chris Slootweg. Circular chemistry to enable a circular economy. *Nature Chemistry*, 11(3):190–195, 2019. doi: 10.1038/s41557-019-0226-9.

András József Tóth, Dániel Fózer, Péter Mizsey, Petar Sabev Varbanov, and Jirí Jaromír Klemeš. Physicochemical methods for process wastewater treatment: powerful tools for circular economy in the chemical industry. *Reviews in Chemical Engineering*, 0(0), 2022. doi: 10.1515/revce-2021-0094.

Raoul Meys, Arne Katelhon, Marvin Bachmann, Benedikt Winter, Christian Zibunas, Sangwon Suh, and André Bardow. Achieving net-zero greenhouse gas emission plastics by a circular carbon economy. *Science*, 374(6563):71–76, 2021.

Geoffrey W Coates and Yutan DYL Getzler. Chemical recycling to monomer for an ideal, circular polymer economy. *Nature Reviews Materials*, 5(7):501–516, 2020.

Dimitris S Achilias, Lefteris Andriotis, Ioannis A. Koutsidis, Dimitra A. Louka, Nikolaos P. Nianias, Panoraia Siafaka, Ioannis Tsagkalias, and Georgia Tsintzou. Recent advances in the chemical recycling of polymers (PP, PS, LDPE, HDPE, PVC, PC, Nylon, PMMA), chapter 1, pages 3–64. *IntechOpen*, 2012.

Liisa Rihko-Struckmann and Kai Sundmacher. Closing the circular cycles for polyamides. Smart Process Systems Engineering Symposium 2022: Chemical and Biological Decomposition Processes, 2022.

H Peter Kasserra. Recycling of polyamide 66 and 6. Science and Technology of Polymers and Advanced Materials: Emerging Technologies and Business Opportunities, pages 629–635, 1998.

Ankush Kishor Deulkar. Material flow analysis of polyamide in Germany. *MSc thesis, Otto von Guericke University Magdeburg*, 2022.

Ann-Joelle Minor, Ruben Goldhahn, Liisa Rihko-Struckmann, and Kai Sundmacher. Opportunities for chemical recycling processes of nylon 6 to caprolactam: Review and techno-economic assessment. *Chemical Engineering Journal*: 145333, 2023.

Ana Somoza-Tornos, Andres Gonzalez-Garay, Carlos Pozo, Moisès Graells, Antonio Espuña, and Gonzalo Guillén-Gosálbez. Realizing the potential high benefits of circular economy in the chemical industry: ethylene monomer recovery via polyethylene pyrolysis. *ACS Sustainable Chemistry & Engineering*, 8(9):3561–3572, 2020.

354e Heat and Mass Transfer Analysis of Temperature-Gradient Thermolysis Reactors for Plastic Upcycling

Ruiyang Ji, Nuwayo Eric Munyaneza, Carlos Posada, Guoliang Liu, Rui Qiao

Driven by the steady increase of global production and consumption of plastics, the pollution caused by waste plastics continues to escalate. There is thus an urgent need for innovative technologies to address waste plastic issues. The recent development of catalyst- and hydrogen-free temperature-gradient thermolysis and upgrading processes (Xu et al., Science, 381, 666, 2023) are promising upcycling technologies that can transform discarded plastics into valuable chemicals. Nevertheless, the effective control of thermolysis to produce hydrocarbons of desirable chain lengths with high value remains a challenge. Fundamentally, such a challenge originates from the fact that the operation and product of temperature-gradient thermolysis is governed by the intricate interplay between a host of thermal, chemical, and physical processes. Here, we develop a mathematical model for temperature-gradient thermolysis of polyethylene by capturing the essential elements of the underlying physicochemical and transport processes. Simulations based on this model and the accompanying experiments revealed the essential role of operation temperatures in affecting the chain length distribution of hydrocarbons produced, and pointed to new opportunities for targeted chain length selection and enhanced reaction rates through the meticulous control of heat and mass transfer in the reactor. The insights gained here will help guide the design of next-generation thermolysis reactors for plastic upcycling, thus supporting the advancement toward large-scale industrial deployment.

354f Discovering Chemical Reaction Pathways and Processes for Net-Zero Greenhouse Gas Emissions

Sunghoon Kim, Bhavik Bakshi

In response to the pressing issue of embodied greenhouse gas (GHG) emissions within the chemical industry, this research underscores the necessity of discovering new pathways including unconsidered chemicals and reactions to innovatively address these environmental challenges [1]. Traditional methods, while producing realistic results, often rely on established pathways and materials, which limits their effectiveness in significantly reducing GHG emissions [2]. Our research seeks to transcend these limitations by identifying and exploring alternative pathways that mitigate GHG emissions and fundamentally transform chemical lifecycle management. Our work diverges by advocating for the exploration and discovery of alternative pathways that not only mitigate GHG emissions but also revolutionize chemical lifecycle management from a foundational chemical perspective [3]. Standard practices in constructing chemical process superstructures tend to neglect the incorporation of novel reactions and chemicals, constrained by the limitations of existing information frameworks.

Our study delves deep into the molecular and reaction level to reconceptualize the production and recycling processes. Employing a double-direction method [4], this approach integrates forward and retrosynthetic analyses to construct an extensive reaction network that covers the entire lifecycle of chemical products, including raw material procurement, manufacturing, usage, and end-of-life stages. The forward analysis collates reaction and chemical data pertinent to both general production and end-of-life pathways. Simultaneously, the retrosynthetic analysis compiles a list of precursor chemicals leading to the target product. This dual analysis facilitates the screening of chemical candidates, enabling the establishment of a circular reaction network that promotes material reuse and minimizes waste. Through retrosynthetic analysis [5], we challenge the conventional boundaries by screening for and integrating innovative chemicals and reactions. This approach not only broadens the superstructure of potential pathways but also uncovers previously unexplored avenues for sustainability.

Using chemical similarity analysis in the development of the chemical reaction network, this strategy not only ensures the identification and integration of chemical candidates that foster material reuse and minimize waste but also addresses the environmental and economic limitations inherent in traditional strategies [6]. To refine this network further, we adopt a hierarchical process design approach. This approach is structured into three distinct levels and scopes, facilitating targeted screening at each scale-up stage—from level 1 (chemistry) to level 2 (reaction), and subsequently from level 2 to level 3 (process). This configuration enables the exclusion of non-viable strategies and the accentuation of promising ones for further development. Moreover, our analysis of Technology Readiness Levels (TRLs) [7] is instrumental in identifying research gaps. By pinpointing these gaps, we set a clear direction for future efforts aimed at realizing processes that are not only cost-effective and low in GHG emissions but also boast a high degree of circularity.

To demonstrate the practical application and validate the effectiveness of our methodology, we initially apply it to a natural gas-based methanol production network. Our approach, employing a double-direction search, we search the tentative chemicals and reactions for methanol production: 453 chemicals and 527 reactions using ASKCOS [8]. Subsequent screening of these chemicals, utilizing Tanimoto similarities, narrows the field to eight chemicals. Integrating this refined chemical set, along with data collated through forward analysis, allows us to construct an extended reaction network from the conventional natural gas based methanol network. Further refinement is achieved by implementing hierarchical process design and screening across three defined scopes and scales. Our findings reveal that the introduction of additional reforming reactions to the conventional methane reforming technology, specifically steam methane reforming, when combined with strategies for recycling CO2 emitted at the end-of-life stage, markedly improves both the economic and environmental performance relative to the prevailing commercialized methanol production networks. This finding sets a foundation for applying our methodology to other chemical products, such as polyethylene terephthalate (PET), potentially offering innovative pathways to significantly reduce embodied GHG emissions.

This research contributes a novel framework to the chemical industry, advocating for a shift towards more environmentally sustainable production methods by focusing on the development of innovative pathways. By addressing the challenge of reducing embodied GHG emissions from a chemical perspective, this study illuminates the path toward a more sustainable and environmentally responsible chemical production paradigm. The novelty of our work lies in its comprehensive approach to rethinking chemical production and lifecycle management through the lens of chemistry.

References

[1] Kula, K., Klemeš, J. J., Fan, Y. V., Varbanov, P. S., Gaurav, G. K., & Jasiński, R. (2023). Environmental footprints and implications of converting GHG species to value-added chemicals: a review. *Reviews in Chemical Engineering*, (0).

[2] Mion, A., Galli, F., Mocellin, P., Guffanti, S., & Pauletto, G. (2022). Electrified methane reforming decarbonises methanol synthesis. *Journal of CO2 Utilization*, *58*, 101911.

[3] Zhang, C., & Lapkin, A. A. (2023). Reinforcement learning optimization of reaction routes on the basis of large, hybrid organic chemistry–synthetic biological, reaction network data. *Reaction Chemistry & Engineering*, *8*(10), 2491-2504.

[4] Xu, Z., & Mahadevan, R. (2022). Efficient Enumeration of Branched Novel Biochemical Pathways Using a Probabilistic Technique. *Industrial & Engineering Chemistry Research*, *61*(25), 8645-8657.

[5] Lopez, L. M., Zhang, Q., Dollar, O., Pfaendtner, J., Shanks, B. H., & Broadbelt, L. J. (2024). Application of automated network generation for retrosynthetic planning of potential corrosion inhibitors. *Molecular Systems Design & Engineering*.

[6] Weber, J. M., Guo, Z., Zhang, C., Schweidtmann, A. M., & Lapkin, A. A. (2021). Chemical data intelligence for sustainable chemistry. *Chemical Society Reviews*, *50*(21), 12013-12036.

[7] Roh, K., Bardow, A., Bongartz, D., Burre, J., Chung, W., Deutz, S., ... & Mitsos, A. (2020). Early-stage evaluation of emerging CO 2 utilization technologies at low technology readiness levels. *Green chemistry*, *22*(12), 3842-3859.

[8] ASKCOS, https://askcos.mit.edu/ (Accessed 10March 2024)

100a Framework and Results for Roadmapping to a Global Net-Zero Chemical Economy in an Uncertain Climate Future

Amrita Sen, Vyom Thakker, George Stephanopoulos, Bhavik Bakshi

The global transition to a net-zero carbon economy necessitates the transition of the chemicals and materials industry (CMI) as well. Products of the CMI, while hard to decarbonize, embody 70% of global greenhouse gas emissions.[1] Naturally, companies, policymakers and institutions worldwide have pledged and committed to enabling this solution.[2] The realization of such pledges, however, remains rife with uncertainty, unless a roadmap to carbon neutral chemicals is uncovered via systematic research.

Many low carbon (C) and net-negative C technologies are promising candidates to supplement or replace current methods of chemical production.[3] Recent studies show a combination of recycling, bio-based manufacturing, and carbon capture and utilization as the optimal means to a GHG emission free CMI.[4,5] However, in many cases, these technologies are in their nascency, and the associated research, development, commissioning, and operational costs are prohibitively high. Our preliminary results indicate possible "win-win" solutions that improve both value and abatement. Net-zero, however, demands higher operational costs under current policies. Therefore, strategic investments into and development of these technologies, in keeping with regional climate policy stipulations and government incentives, is critical.

The existing literature in the area designs net-zero value chains for the plastics industry and the required capital investments.[4,6] However, technologies with lower readiness levels are not considered in these analyses and the uncertainty associated with their development is not accounted for. Moreover, these studies do not connect with cross-sectoral decarbonization goals such as those imposed by sectors likely to consume chemical energy carriers, the effect of a changing climate, or the uncertainty in costs.

To bridge this gap, we devise a formulation to design a robust cost-optimal roadmap towards a net-zero chemical industry, accounting for the evolution of technologies as well as connections with integrated assessment models (IAMs). The problem is framed as a stochastic multi-period optimization that incurs the minimum costs and is constrained to reach net-zero by 2050 while satisfying the growing need for chemicals.

We utilize our open-access model of the chemicals and materials industry (CMI) as the background dataset.[7] This model provides an inventory of process data for conventional technologies that make up the CMI. Value-chain alternatives are added to this matrix to form a superstructure network of technologies. These alternatives include renewable energy driven process electrification, bio-based chemicals and plastics, point source carbon capture and utilization to value added chemicals, on-demand transformation technologies, mechanical and chemical recycling of polymers, pyrolysis and incineration of plastic wastes to produce primary chemicals and energy. The models used to accommodate circular flows and variable emission factors are adopted from earlier work.[8, 9, 10, 11] The costs for these technologies are estimated from techno-economic analyses of the same in literature and the distribution of costs is estimated from roadmapping literature in other sectors. The network incurs operational costs as well as those associated with research and development of emerging technologies. The formulation only allows fully evolved technologies to be deployed and models this evolution as sigmoidal curves. These sigmoidal models parametrically estimate the adoption of a technology and can thus incorporate qualitative expert knowledge about the adoption. The availabilities of biomass, renewable energy, and carbon capture are modelled according to scenarios from IAMs. The solution yields the investment decisions along the time horizon and the evolving value chain of processes and associated resource use and emissions.

Preliminary results indicate the need to invest in CCU technologies early on. This fulfills the downstream demand for olefins and aromatics via on-demand conversion reactions. Mechanical and chemical recycling scale up as soon as they reach the point of adoption. Around the milestone years, bio-based production and pyrolysis scale up to incur net negative emissions via a combination of carbon sequestration and product displacement. The remaining demands are met by electrification technologies powered by renewable energy. We find that the year of adoption of electrification is determined by the IAM scenario under consideration. Sensitivity analyses reveal the dependence of CCU technologies to steep adoption parameters. When these parameters are below threshold values, they are bypassed in lieu of direct electrification.

Our results help us identify the technologies which are robust to the climate policy, evolution parameters, climate evolution and cost parameters. We find the material circularity options are particularly robust and form a part of the solution in most cases. The use of a network model of the industry also inherently allows for the adoption of only those technologies which reduce emissions across the portfolio of products. This study is the first to combine realistic, comprehensive technology models, adoption models and the insight from IAMs, to roadmap to a sustainable and circular future for chemicals in a global context. The general framework and can be used for any subset of the global chemical product system such as a company's product portfolio.

References

1. Circle Economy (2020) The circularity gap report
2. Van Coppenolle, H., Blondeel, M., & Van de Graaf, T. (2023). Reframing the climate debate: The origins and diffusion of net zero pledges. Global Policy, 14(1), 48-60.
3. Gabrielli, P., Gazzani, M., & Mazzotti, M. (2020). The role of carbon capture and utilization, carbon capture and storage, and biomass to enable a net-zero-CO2 emissions chemical industry. Industrial & Engineering Chemistry Research, 59(15), 7033-7045.
4. Meys, R., Kätelhön, A., Bachmann, M., Winter, B., Zibunas, C., Suh, S., & Bardow, A. (2021). Achieving net-zero greenhouse gas emission plastics by a circular carbon economy. Science, 374(6563), 71-76.
5. Sen, A., Thakker, V., Stephanopoulos, G., & Bakshi, B. R. (2023). Designing roadmaps for transitioning to value chains with net-zero emissions: Case of the chemical industry. In Computer Aided Chemical Engineering (Vol. 52, pp. 2483-2488). Elsevier.
6. Zibunas, C., Meys, R., Kätelhön, A., & Bardow, A. (2022). Cost-optimal pathways towards net-zero chemicals and plastics based on a circular carbon economy. Computers & Chemical Engineering, 162, 107798.
7. Sen, A., Stephanopoulos, G., & Bakshi, B. R. (2022). Mapping Anthropogenic Carbon Mobilization Through Chemical Process and Manufacturing Industries. In Computer Aided Chemical Engineering (Vol. 49, pp. 553-558). Elsevier.
8. Thakker, V., & Bakshi, B. R. (2021). Toward sustainable circular economies: A computational framework for assessment and design. Journal of Cleaner Production, 295, 126353.
9. Thakker, V., & Bakshi, B. R. (2021). Designing Value Chains of Plastic and Paper Carrier Bags for a Sustainable and Circular Economy. ACS Sustainable Chemistry & Engineering, 9(49), 16687-16698.
10. Thakker, V., & Bakshi, B. R. (2023). Ranking Eco-Innovations to Enable a Sustainable Circular Economy with Net-Zero Emissions. ACS Sustainable Chemistry & Engineering.

11. Thakker, V., & Bakshi, B. R. (2023). Mapping the path to a net-zero chemicals industry by long-term planning with changes in technologies and climate. AIChE Journal, e18381.

100b Nylon 6 Depolymerization: Reactor Design, Catalysts, Kinetics, Process Modeling, and Data Analytics

Adam McNeeley, Niket Sharma, Y. Liu

Nylon 6 is a commonly produced engineering plastic that finds applications in textiles, electronics, and the automotive industry. Nylon 6 waste is attractive to chemically recycle, because of the high value of its monomer, caprolactam, and the ease of converting nylon 6 to caprolactam. Recovery of post-industrial scrap nylon 6 has been done for at least the last 50 years, and nearly every major nylon 6 manufacturer offers a nylon 6 product derived from recycled material. Recent industrial recycling efforts have focused on post-consumer wastes, such as carpeting or fiber reinforced automotive parts, to expand the scope of recoverable material and offer a larger volume of recycled products. This work presents an overview of the different nylon 6 depolymerization reaction configurations and catalysts, covers kinetics and modeling strategies and proposes new process innovations to reduce process energy demand.

The depolymerization of nylon 6 to caprolactam can be divided into three different reaction types; (1) liquid-phase hydrolysis where nylon 6 melt is mixed with water at a high enough pressure that the reaction is completely liquid phase to produce a mixture of caprolactam and oligomers. (2) steam stripping where nylon 6 melt is contacted with superheated water vapor and formed caprolactam is stripped into the vapor phase. (3) solvent-free depolymerization where nylon 6 melt is held under vacuum and formed caprolactam is removed in the vapor phase. Each reaction type has fundamental differences and offers certain advantages and disadvantages.

Nylon 6 depolymerization processes can be uncatalyzed but can also use various catalysts, such as phosphoric acid, sodium hydroxide, or can use new catalysts such as early transition metal metallocenes. We discuss the advantages and limitations of each catalyst. We describe common depolymerization byproducts. The primary depolymerization byproduct is the cyclic dimer of caprolactam, which is undesirable in the caprolactam product, but can be recovered and converted to caprolactam. Several other minor degradation byproducts are also known to form and must be separated in downstream purification processes.

We present a fundamental analysis of the depolymerization kinetics. The reaction kinetics of nylon 6 are unique because of the parallel reactions that occur during depolymerization. Two primary reversible reactions are a condensation reaction, which impacts the chain length of the melt and the average molecular weight, and the ring-opening reaction, which impacts the rate of caprolactam evolved directly from the melt. The reverse-ring opening reaction rate which forms caprolactam is proportional to the concentration of amine end groups in the melt, which is a function of the average molecular weight or chain length of the melt. Thus, the condensation reaction and ring-opening reaction rates must be well defined to properly model nylon 6 depolymerization.

We use the kinetic models built for nylon 6 polymerization in Aspen Plus as the basis for our depolymerization models, because the primary kinetics responsible for polymerization are the same for depolymerization. We describe some of the challenges and modifications required for the depolymerization kinetic model and demonstrate the ability of the model to reflect experimental data for all three reaction types and for different catalysts. We also indicate where uncertainties exist in the modeling assumptions and how future studies can be developed to resolve these uncertainties.

We employed advanced data analytics and machine learning techniques to preprocess experimental data, effectively identifying and excluding process outliers. This enhanced the data's integrity, which is crucial for accurate kinetic estimation. Our approach incorporated hybrid modeling to produce pertinent experimental data, culminating in the development of a data-driven hybrid model for the reactor, tailored to the available limited experimental dataset.

Our review of nylon 6 depolymerization patents and development of a robust kinetic models for uncatalyzed and catalyzed reactions allows us to develop comprehensive process models for the nylon 6 depolymerization process with different reaction types and reactor configurations. We demonstrate different process intensifications and reactor configurations that can reduce overall reactor size and process energy demand. Furthermore, we leveraged data-based reduced order models to refine the energy optimization of the complex depolymerization process, achieving a more precise and computationally efficient optimization outcome.

Consequently, our research synergistically fused elements of reactor design, catalysis, kinetics, process modeling, and data analytics to finely tune the nylon depolymerization process. This holistic integration paves the way for a more sustainable and efficient recycling of nylon polymers for different reactor configurations.

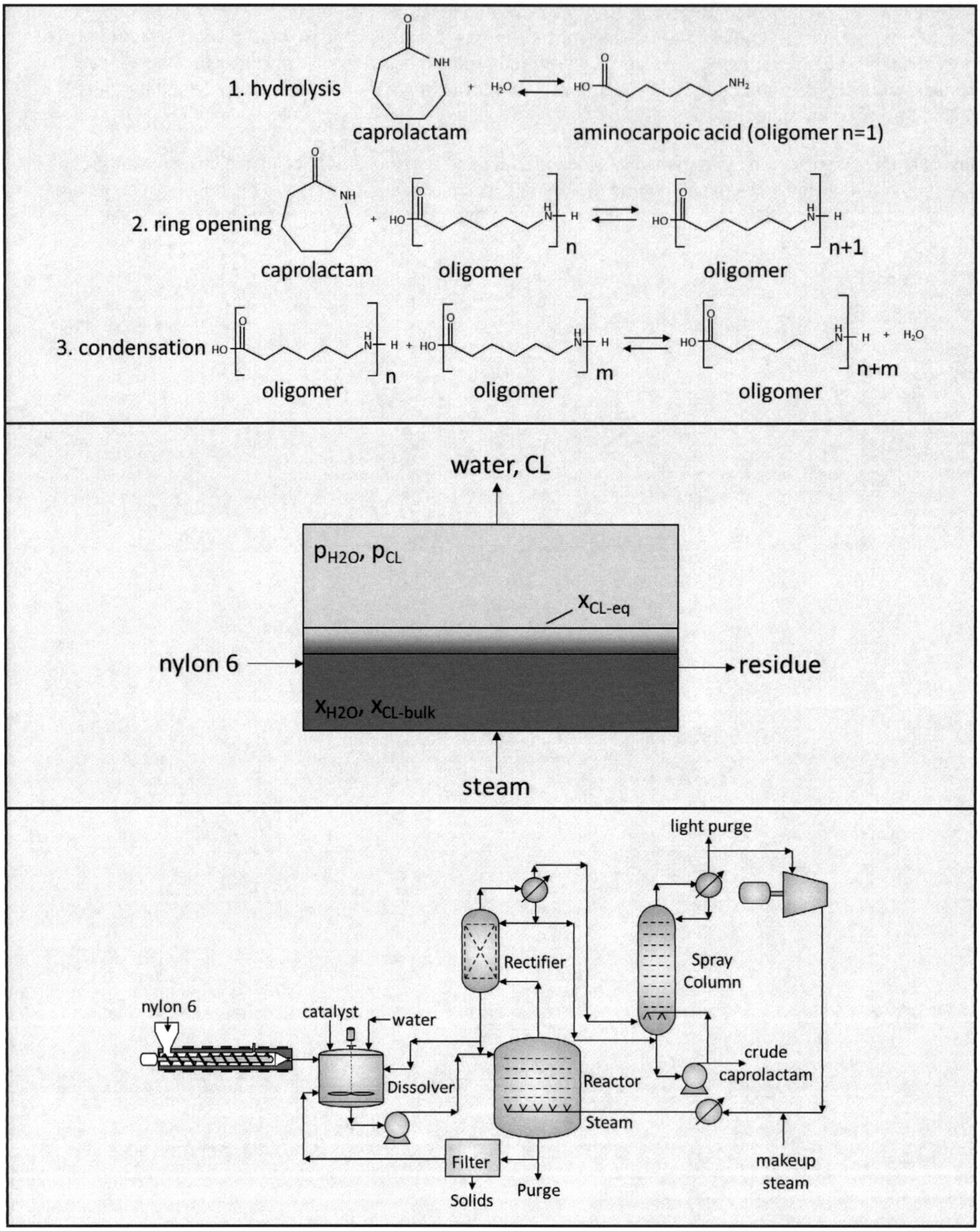
1. hydrolysis
caprolactam
aminocarpoic acid (oligomer n=1)
2. ring opening
caprolactam
oligomer
oligomer
n
n+1
3. condensation
oligomer
oligomer
oligomer
n
m
n+m
water, CL
p_{H2O}, p_{CL}
x_{CL-eq}
nylon 6
residue
x_{H2O}, $x_{CL-bulk}$
steam
light purge
Rectifier
Spray
Column
nylon 6
catalyst
water
Dissolver
Reactor
crude
caprolactam
Steam
Filter
Solids
Purge
makeup
steam

100c Comparative Study of Eco-Friendly Bioleaching Solvents for Sustainable Recovery of Critical Materials and Minerals from Spent Batteries

Haoran Wu, Meltem Urgun-Demirtas, Stina Daniel

Critical materials and minerals (CMMs) underpin both the electric vehicle (EV) revolution and the fight against climate change. However, surging demand from booming EV and clean energy sectors threatens CMMs supply in U.S., raising concerns about their shortages and price volatility. Black mass from spent battery represents a significant resource rich in CMMs like cobalt, nickel, and lithium. Valorization of spent batteries can offer environmental and economic benefits through reduced reliance on virgin materials and improved battery circularity.

Conventional CMMs extraction from black mass relies on energy-intensive pyrometallurgy or chemical-heavy hydrometallurgy. Bioleaching, utilizing biolixiviants (e.g., short-chain fatty acids or gluconic acid) derived from renewable resources, presents a promising alternative. Studies demonstrate remarkable potential with bioleaching, achieving over 90% recovery of critical metals like cobalt and lithium at small scales[1]. Further research is crucial to explore diverse biolixiviant options and optimize leaching conditions, ultimately reducing the operational and capital costs[2].

This study aimed to compare the effectiveness of various environmental friendly and cost-effective bioleaching solvents under different leaching conditions. The efficiency of biolixiviants production under different operation conditions (e.g., inoculum, temperature, pH) using renewable resources was explored. Subsequently, the black mass obtained from spent batteries served as the primary substrate for bioleaching experiments. Various leaching conditions (e.g., biolixiviants composition, pH, pulp density)were examined to establish a comparative dataset on CMMs recovery efficiency from spent batteries. The outcomes of this study helped the development of eco-friendly bioleaching processes and contributed to the creation of a sustainable end-of-life management strategy for the surge of battery waste.

Reference:

(1) Alipanah, M.; Reed, D.; Thompson, V.; Fujita, Y.; Jin, H. Sustainable Bioleaching of Lithium-Ion Batteries for Critical Materials Recovery. *Journal of Cleaner Production* **2023**, *382*, 135274. https://doi.org/10.1016/j.jclepro.2022.135274.

(2) Biswal, B. K.; Balasubramanian, R. Recovery of Valuable Metals from Spent Lithium-Ion Batteries Using Microbial Agents for Bioleaching: A Review. *Frontiers in Microbiology* **2023**, *14*.

100d Facile Remanufacturing and Re-Lithiation for Spent Lithium-Ion Battery Electrodes

Kasim Adewuyi

As the demand for Li-Ion batteries has increased, so have concerns about supply and environmental impact. From 2019 to 2020, global lithium demand increased by 18% to 56,000 tons. At that rate, a shortfall of 50 million tons would occur by 2050 based on proven global lithium reserves [1]. Additionally, the primary disposal method of spent batteries are landfills which could contaminate the environment with toxic heavy metals and organic electrolytes. To that end, we present a facile approach to remanufacturing applying Direct Recovery principles. Our approach maintains the underlying structure of old NMC-622 cathodes and combines it with fresh cathode slurry to fix the damage. A plasma treatment and washing process is used to remove degraded material before a fresh coat of slurry is applied. This allows us to restore performance to heavily degraded cells. To this end, we investigated different plasma treatments, treatment step order, slurry types and pressing to optimize our process. We determined that normal NMC-622 slurry with plasma treatment and pressing produced the best results in terms of Cell Viability (>80%) and Capacity Performance restoration on highly degraded cells. Furthermore, we also applied this approach to treating anode material where similar performance restoration of about 50% of pristine cells was all achieved.

The direct recovery remanufacturing approach has evolved and undergone optimization for NMC-622 cathodes and MCMB anodes, but basics were maintained. The general process is initial cell characterization where a baseline capacity is determined; disassembly where the constituent components (casing, electrodes, separator, spacer and spring) are separated; plasma treatment and washing where degraded material and electrolyte is removed; slurry treatment where fresh slurry fills in cracks and holes; a pressing process to make sure the new and old material has a strong connection before reassembly into Li half cells; Re-lithiation which reinserts Li ions into the treated crystalline structure of NMC622; and finally for cycle testing to determine performance.

The characterization process is critical because the old cells used for this process have been lying around the lab for indeterminate amounts of time. Determining the capacity performance of these cells allows for the establishment of two benchmarks, one based on the previous performance measured capacity, and one based on the practical capacity performance. The practical capacity for the NMC cells is 160 mAh/g and the initial mass of the NMC cathodes is assumed to be 10 mg (active material 8.55 mg). The previous performance benchmark is set by cycling the cells between their voltage limits of 2.8 V to 4.2 V for NMC for 5 charge and discharge cycles at a given C-rate, typically 0.1C. This characterization allows us to categorize cells broadly as Good (above 70% practical capacity), Poor (anything below 70% but above 0), and Zero (cells are flatlined or trapped because they never reach 4.2V).

The separation process is a delicate situation where the cell is opened using a crimping device. This allows for access to the electrode where the difficult process of removing the electrode occurs which can be difficult due to age and degradation. For the LMO half cells, the separator is stuck to the LMO and must be carefully cut and peeled away from each other without damaging the cathode. This process creates another level of classification based on how much damage the cathode underwent during the separation process. Currently referred to as Minimal, Moderate and Significant though this relies on the eye test instead of strict classification.

Plasma treatment should allow for the filling in of cracks to the electrode surface with fresh material.

Initially work was attempted with PlasmaTech Plasma Wand where treatment occurred in the air. However, the power and plasma application could not be effectively controlled.

Therefore, a switch occurred to fully self-contained PlasmaTech PE200 Plasma Processing Machine at 100 W for 1 min in an Oxygen environment. First, the switch occurred because the plasma wand must be raster, so the plasma application is highly variable. Second, while the wand has a maximum output of 30 W that power could not be guaranteed over the course of the entire process or from cathode to cathode as each cell must be treated individually. Third, unwanted side reactions could occur due to air exposure instead of pure Oxygen. This plasma treatment was shown to have no impact on pristine cathodes under Oxygen or Argon environments. Thus, it can be assumed that Oxygen catalyzes reactions that remove only degraded materials from the surface.

After plasma treatment, the electrodes are washed with DMC before the drying process occurs at 80oC in a Vacuum Oven, overnight. This process should remove the LiPF6 electrolyte, Li salt and side reaction material from the surface creating a clean surface to apply fresh slurry. DMC washing of pristine cathodes showed no loss of mass after even five minutes of washing. Drying for this was done on a hot plate to speed up the process.

Surface repair is the application of slurry to fill in any cracks or holes on the surface of the cathode after plasma treatment. The slurry types considered are normal slurry (NMC622, Carbon Black, PVDF, NMP) and salted slurry (NMC622, Carbon Black, PVDF, NMP and LiTFSi) where the salt amount is 10% wt of the NMC6222. The Anode slurry consists of MCMB, Carbon Black, PVDF and NMP. The slurry is carefully applied with a thin tipped paint brush. Cathodes are placed into the vacuum oven and heated at 120oC for 16 hours.

The pressing stage applies 1 MPa of pressure to the electrodes for 30 mins at room temperature. This pressure can be maintained for at least one hour without damaging the surface of the cathodes for pristine cells. However, raising the pressure to 2 MPa results in damage to pristine cells. Additionally, diminishing returns mean that 30 mins is optimal for compressing the fresh slurry material onto the old material. Finally, cycle testing can occur for these cells.

100f Chemical Recycling of Polycaprolactone Via Reactive Melt Processing

Anubhav Sarmah, Aristotle Zervoudakis, Michaela Pfau-Cloud, Marc A. Hillmyer, Christopher Ellison

An astounding ~300 million tons of plastic waste are dumped into our environment or landfills every year, making it one of the biggest crises in history. Among a broad range of potential solutions, chemical recycling is an exciting option which involves the depolymerization of end-of-life plastics into monomers or intermediates, which can then be purified by conventional means and used to synthesize virgin polymers. In principle, this mode of recycling can be used for multiple cycles without any deterioration of properties in the resynthesized polymer and can also be used for streams containing multiple plastic waste types. However, chemical recycling comes with a unique set of challenges: first, many chemical recycling processes involve the use of harsh solvents, thus introducing a new environmental hazard on the path to solving another. Secondly, and most importantly, chemical recycling processes are technologically and economically challenging to scale up, thereby slowing implementation in industry. In this work, we designed a modified twin-screw compounder which demonstrated successful chemical recycling of polyesters to their respective monomers. Inspired by industrial use of devolatilization extrusion set-ups, the modified compounder used herein simultaneously heated and mixed the polymer with a catalyst, allowing evolved monomer vapors to be removed by a vacuum system and collected in a cold trap. To explore the performance of this setup, we used commercially available polycaprolactone (PCL) for our experiments. Reactive distillation experiments were initially conducted in a heated round bottom flask with a distillation head to determine the optimum temperature and catalyst loading necessary for successful ring-closing depolymerization (RCD) . Recovered monomers were collected and assessed for yield and purity by ^{1}H nuclear magnetic resonance spectroscopy (NMR) and gas chromatography-mass spectroscopy (GC-MS). Compared to analogous distillation experiments, the compounder depolymerization experiments resulted in similar values of purity and yield. The results of this study establish the modified compounder as an easily scalable approach for polyester depolymerization.

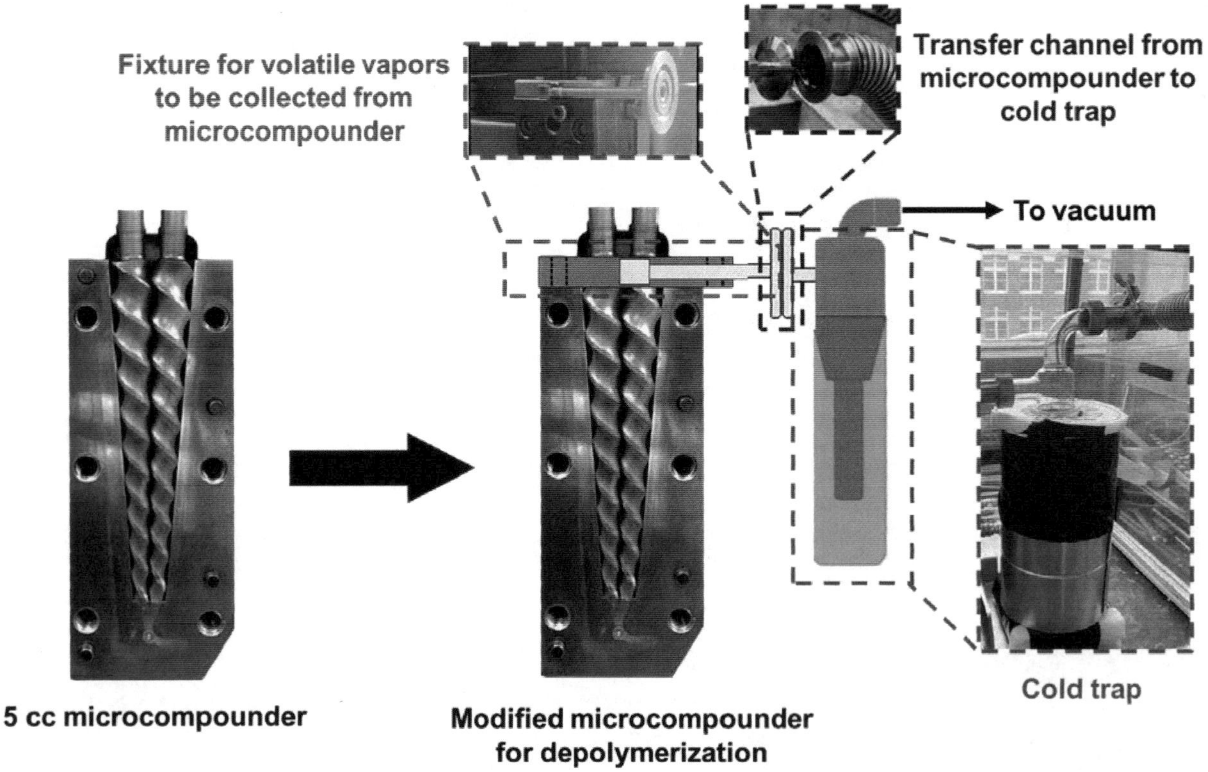

100g D o E - Assisted Development of Sustainable Mesoporous Silica Synthesis for Tailored Drug Delivery Applications.

Benjamin Roe, Carlos Brambila, Solomon F. Brown, Robert Milton, Siddharth Patwardhan

Porous silica has been widely researched as a material for use in drug delivery. The advantages of porous silica as a drug delivery system are pores of consistent size, uniform distribution and high volume which would make it optimal for drug loading. Unfortunately, existing silica materials have limited applications to drug delivery systems due to harsh conditions needed during synthesis, poor sustainability and an inability to effectively scale up the manufacturing process. Using a bio-inspired synthesis, we have been able to achieve a room temperature, rapid process to generate mesoporous materials suitable for drug delivery.

In the present work, we explore the parametric space of bio-inspired silica using a latin hypercube sampling methodology for establishing space filling variables and understand the relation between parameter, property and performance. The properties identified as critical quality attributes for drug delivery systems were the surface area, the surface charge and the yield of reaction. We successfully achieved precipitation of 300 materials with tailored properties by varying the synthetic parameters, including synthesis pH, precursor concentration, stoichiometry and purification conditions. We characterised all the materials for their porosity using nitrogen adsorption, zeta potential using light scattering and SI speciation (yield of reaction) using silicomolybdate assay. The results suggest a variety of outcomes can be achieved by varying our input parameters with surface areas ranging from 30-500m^2/g, surface charge ranging from -40 to +10 mV. Following the design and development of drug testing protocols. We have loaded our bio-inspired silica with the drug proxy calcein using wet impregnation of the material and measured the release profile and release kinetics. The presentation will show our results on developing a relationship between synthesis- property - performance, which will enable tailored drug delivery systems using a sustainable process designs.

160a Towards Resilient Pharmaceutical Supply Chains: Integration of Manufacturing Uncertainty

Miriam Sarkis, Nilay Shah, Maria M Papathanasiou

The pharmaceutical sector brings value to society through the discovery and deployment of life-saving medicines. In recent years, the market boom of next-generation therapies against life-threatening diseases and the urgent demands for vaccines during the pandemic have pressured the industry in rapidly scaling up manufacturing capacity to meet societal needs. Nevertheless, manufacturers catering for these markets reported shortages and delays due to unforeseen demand trends and uncertainty in manufacturing capabilities of platforms still under development [1, 2]. In this space, there is an increasing need for tools that help assess the impact of underlying manufacturing uncertainties on investment planning and mitigate against them.

We present a decision-support tool for the quantification of supply chain resilience for networks that require rapid scale-up for commercialization and operate under supply-demand uncertainty (Figure 1). Firstly, process uncertainty is quantified via uncertainty analysis and global sensitivity analysis conducted on the developed techno-economic models for the studied product platforms [3]. Given the quantified probability distributions of process performance, a sampling-based approach is proposed to generate process performance scenarios. Secondly, a stochastic mixed-integer linear problem is formulated for the optimization of investment planning under process and demand uncertainty. Specifically, the scenario-based supply-demand constraint is formulated as a chance constraint. The probability of meeting the constraint effectively approximates a resilience metric for the overall supply chain. The accuracy of this approximation is tested via Monte Carlo-based simulations of manufacturing-related outcomes. Candidate investment plans for single- and multi-product systems are assessed, and cost-resilience Pareto frontiers are constructed for a range of chance constraint probability levels.

Firstly, results of the uncertainty and sensitivity analyses suggest a volumetric propagation of uncertainty with respect batch size and scale, which is primarily due to uncertainty in upstream process productivity. Critical sensitivities for all key performance indicators with respect to input parameters are also identified and output uncertainty distributions are recorded. Secondly, the optimization results highlight that for each target demand mitigating against all process uncertainties outcomes requires larger manufacturing capacity and cost. As system reliability is decreased, cost decreases and smaller scales are selected, as they meet demands with lower probabilities and at lower cost. This analysis also allows the quantification of maximum reliability of the system under limited capacity. Finally, we showcase how the presented framework can be augmented to incorporate sustainability metrics, to support the quantification of trade-offs amongst cost, environmental footprint, therapy availability and financial risk.

[1] Lambrix E, Kent-Egan H (2022) Viral vectors: shortages and innovation. In: Biotech Review of the Year

[2] Capra E, Gennari A, Loche A, Temps C (2022) Viral-vector therapies at scale: today's challenges and future opportunities

[3] Sarkis M, Shah N, Papathanasiou MM (2023) Characterization of key manufacturing uncertainties in next generation therapeutics and vaccines across scales. J Adv Manuf Process. https://doi.org/10.1002/amp2.10158

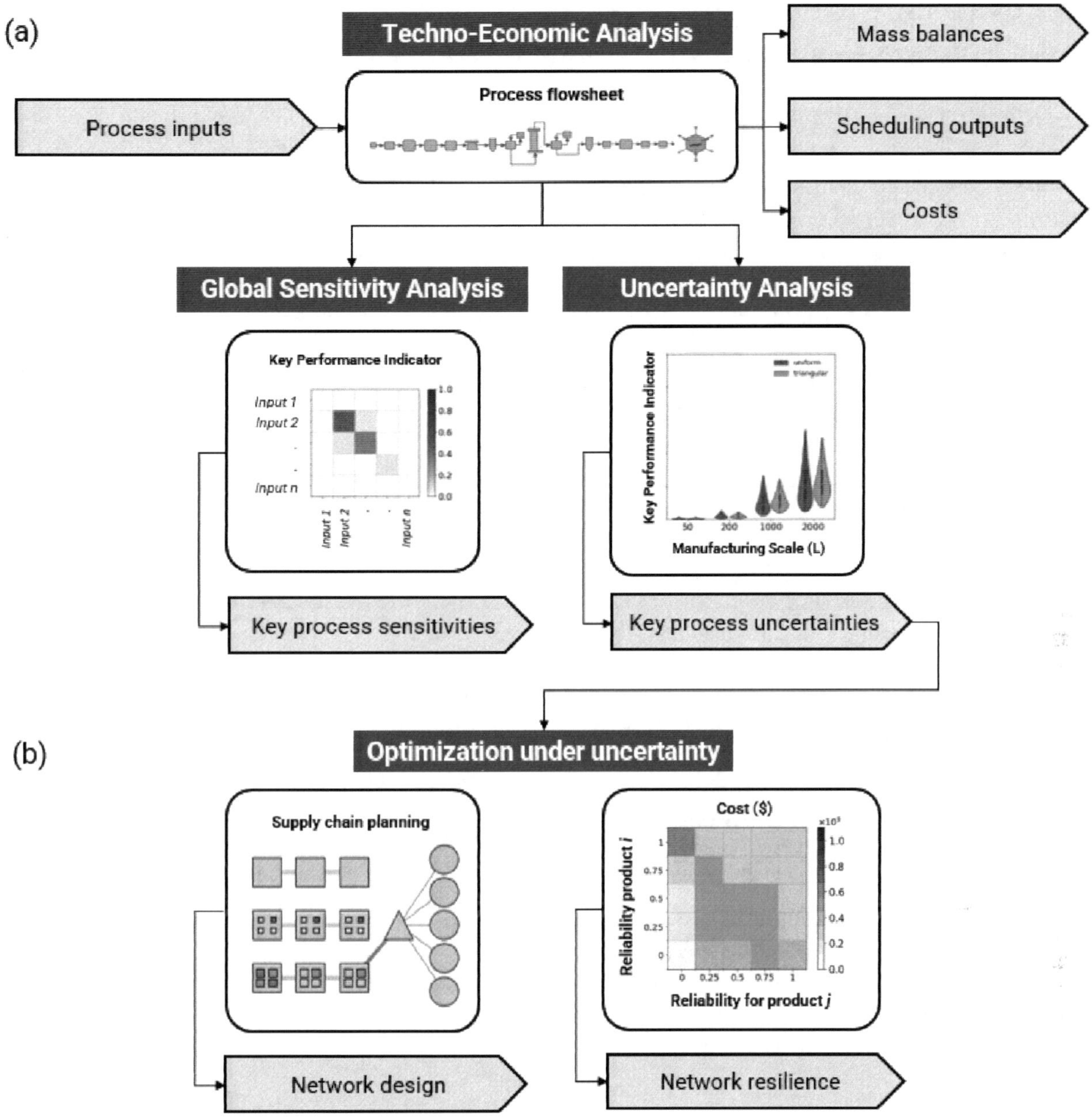

Figure 1: Framework for (a) uncertainty quantification and (b) supply chain optimization under uncertainty

160b Enhancing Resilience of the Energy-Water Nexus to Climate Change: A Holistic Framework and Dynamic Modeling Approach

Hesan Elfaki, Dhabia Al-Mohannadi

The impact of climate change on the interrelated energy and water systems is substantial, highlighting the immediate requirement for comprehensive initiatives to strengthen resilience within the energy-water (EW) nexus. This study aims to tackle the current vulnerabilities and complexities of the EW nexus by developing a dynamic modeling framework that considers the consequences of several climatic changes, such as rising temperatures, heatwaves, and sea-level rise.

The study examines the water and energy subsystems in the nexus, which include desalination plants, wastewater treatment facilities, and power generation plants. These subsystems are connected through the flow of water, power, and heat. A comprehensive resilience methodology is suggested to tackle the dynamic interactions among nexus components and uncertainties in climate change scenarios. The framework incorporates structured optimization model to guide decision-making processes by integrating resources within the nexus.

The dynamic properties of the models consider fluctuations in electricity and water demand and supply, as well as changes in climate variables over time. Specific case study on the state of Qatar is used to incorporate spatial factors. Representative indicators, such as economic and environmental metrics, are used to demonstrate the resilience of a system over various time periods. These indicators help guide in planning for long-term resilience and responding to short-term climate change stressors.

The study uses appropriate resilience indices to provide a quantitative measurement of the losses in system performance and the time it takes to recover, making it easier to evaluate the resilience of the system within the EW nexus.

This research highlights the crucial need to create strong resilience strategy integrating dynamic modeling approaches to protect the interconnected energy-water nexus from the complex challenges brought by climate change. This will enable the sustainable and adaptive management of essential resources in an increasingly uncertain future.

160c Cost-Effective Strategies Towards Net-Zero Plastic Packaging Based on a Resilient and Sustainable Circular Economy

Farshid Nazemi, Amrita Sen, Fabio Sporchia, George Stephanopoulos, Federico Pulselli, Bhavik Bakshi

Plastics constitute 80% of all carbonaceous chemical products and have the highest growth rate of all bulk materials, with global production reaching 400.3 MMT in 2022. Major sectors driving plastic demand include packaging, transportation, construction, electronics, consumer goods, machinery, and textiles. The production of plastics contributes significantly to greenhouse gas (GHG) emissions, estimated at 1.34 gigatonnes per year, primarily due to fossil resource consumption and inadequate recycling practices. Action on packaging should be prioritized as it comprises 37% of total plastic production, and since their short lifespan leads to substantial waste generation annually. Thus, addressing the fossil resource consumption and inadequate recycling practices associated with plastic packaging, represents a crucial opportunity to enable the net-zero transition of the carbon economy.

The technologies currently employed for manufacture of plastic packaging are dominantly fossil-based and the business models are not focused on waste recovery at the end-of-life of the products. Thus, emerging low-emission value-chain alternatives can significantly reduce the carbon footprint of the products. Their deployment however, is hindered by substantial capital investments and potentially higher operational costs. Consequently, meticulous planning to steer the plastic packaging economy towards achieving net-zero carbon emissions becomes imperative to guide decision-makers in making cost-effective investment decisions.

Prior research has focused on pathways to net-zero chemicals and plastics [1-3]. However, these studies have utilized aggregated models for plastics production and recycling, which might not be applicable for a multitude of end-use sectors. For example, the feasibility of mechanical recycling depends on chemical composition and quality of the plastic waste. Multilayer plastic films are one such packaging product which require alternative techniques for resin recycling beyond mechanical means such as the Solvent Targeted Recovery and Precipitation (STRAP) process, due to their heterogeneous chemical structure. In our investigation, we aim to address this gap by offering a more granular examination of technologies specific to the production and recovery of different types of plastic packaging.

Furthermore, prior research has predominantly focused on commercially available alternative technologies, neglecting the potential of options with lower technology readiness level (TRL) [1-3]. We address this gap by incorporating the evolution of technological innovations over time, with the goal of reaching an adoptable point characterized by the maximum TRL. In essence, our approach also accounts for the Research and Development (R&D) expenses associated with the scaleup of low TRL technologies, in addition to capital and operational costs. Thus, we project a realistic and holistic roadmap encompassing the possible bottlenecks towards a net-zero packaging industry.

Additionally, the current literature in the area uses commercial life-cycle inventories and costing datasets for their analyses, thereby hindering further dissemination behind steep paywalls. In contrast, our approach involves leveraging a publicly accessible database of the chemical industry developed by us earlier [5]. Our objective is to enhance transparency and accessibility by making all models and associated data openly available to the public since this is needed to jump start research to enable the urgent transition toward net-zero.

Our analysis indicates the infeasibility of current technologies to form a net-zero value chain for packaging. We therefore design cost-effective roadmaps for some future scenarios. The minimum emission value-chain for the BAU case serves as the baseline for our model, representing the continuation of current practices without significant technological interventions. The scenario for "Current Opportunity" makes commercially viable low-emission technologies available in the superstructure network. Investment decisions dictate the adoption of these technologies as needed. The scenario for "Future Opportunity" allows the scale-up of technologies at all value chain stages and TRLs, subject to the R&D costs and investment decisions.

This re-design of the packaging economy must ensure resilience and overall sustainability of the value chains. Potentially perverse solutions which are unendurable in the longer term, such as endless CCS and land-filling may be chosen as the primary value chain decisions, unless otherwise constrained. Moreover, cost-effective low emission technologies often shift environmental impact beyond metrics of immediate interest. We constrain the allowable biodiversity loss, water consumption, and land use to avoid such results.

In addition to prioritizing environmental sustainability, ensuring the resilience of packaging value chains is imperative. A resilient economy is characterized by its ability to withstand and adapt to disruptions, thereby preventing system collapse. For example, incorporating multiple technological options for waste treatment ensures that the system remains capable of waste recovery in the face of disruption to certain pathways. To assess the resilience of the economic network, we employ ecological resilience principles derived from Ecological Network Analysis (ENA).

We apply a multi-period optimization problem, developed for roadmapping to a net-zero chemical industry, to the packaging industry. We introduce additional constraints to ensure durable, resilient, and sustainable solutions. Our proposed methodology involves the generation of the following three roadmaps, each tailored to address the unique challenges and opportunities within the plastic packaging industry:

(1) Cost-effective transition solutions to achieve net-zero by 2050:

This roadmap prioritizes minimizing costs to achieve net-zero greenhouse gas (GHG) emissions by 2050. It does not consider constraints related to landfilling, Carbon Capture and Storage (CCS), circularity, sustainability, or resilience. The primary focus is solely on achieving a net-zero economy by 2050 without necessarily ensuring durable, sustainable, and resilient solutions for the plastic packaging supply chain.

(2) Cost-effective long-term solutions to achieve net-zero by 2050:

This roadmap aims to identify investment decisions necessary for achieving long-term net-zero solutions for plastic packaging. By imposing constraints on landfilling and addressing limitations associated with CCS, the goal is to develop solutions that are durable and capable of mitigating carbon emissions by and beyond 2050. Circularity constraints are also enforced, recognizing the finite nature of fossil resources and the importance of promoting circularity in the long term.

(3) Cost-effective long-term solutions to achieve a sustainable and resilient net-zero economy by 2050:

This roadmap outlines a path towards achieving a net-zero, sustainable, and resilient circular economy for plastic packaging. In addition to imposing constraints related to durable solutions, such as those discussed in the previous roadmap solution, this approach also incorporates constraints regarding environmental sustainability and supply chain resilience. For environmental sustainability, constraints are placed on other environmental impacts to ensure a net-zero impact across all environmental aspects, not just CO_2 emissions. For resilience, a constraint is applied based on the value of ecological resilience calculated for the network, aligning with the concept of the 'window of vitality' associated with resilient ecosystems [6].

Our initial findings show that a net-zero solution using the portfolio of current technologies is not feasible. Therefore, investing in innovation is necessary to achieve this goal. We see potential in both the current opportunity and future opportunity scenarios to create roadmapping solutions for both short and long-term planning. These scenarios reveal the large potential of innovative approaches for end-of-life treatment of materials, including mechanical, solvent-based, and chemical recycling to accelerate towards the goal. Additionally, the contribution of low-TRL solutions such as electrification of the industry to achieve cost-effective long-term net-zero solutions was captured. Bio-based technologies are shown to play a central role in achieving long-term sustainability. However, careful considerations are necessary to prevent environmental burden shifts, particularly regarding the high energy and water consumption of bio-based technologies and the potential conflicts deriving from the land required for biomass cultivation.

Reference:

[1] Stegmann, P., Daioglou, V., Londo, M., van Vuuren, D. P., & Junginger, M. (2022). Plastic futures and their CO2 emissions. *Nature*, *612*(7939), 272-276.

[2] Stegmann, P., Daioglou, V., Londo, M., & Junginger, M. (2022). The plastics integrated assessment model (PLAIA): Assessing emission mitigation pathways and circular economy strategies for the plastics sector. *MethodsX*, *9*, 101666.

[3] Zibunas, C., Meys, R., Kätelhön, A., & Bardow, A. (2022). Cost-optimal pathways towards net-zero chemicals and plastics based on a circular carbon economy. *Computers & Chemical Engineering*, *162*, 107798.

[4] Meys, R., Kätelhön, A., Bachmann, M., Winter, B., Zibunas, C., Suh, S., & Bardow, A. (2021). Achieving net-zero greenhouse gas emission plastics by a circular carbon economy. *Science*, *374*(6563), 71-76.

[5] Sen, A., Bakshi, B., Stephanopoulos, G. (Under Preparation). An open access network model of the global chemicals and materials industry towards systems analysis and design

[6] Chatterjee, A., & Layton, A. (2020). Mimicking nature for resilient resource and infrastructure network design. *Reliability Engineering & System Safety*, *204*, 107142.

160d Sustainable Supply Chain Design Under Uncertainty within the "Safe and Just Space": Modeling Framework and a Case Study of Li-Ion Battery

Ying Xue, Bhavik Bakshi

The global impact of human activities on the Earth's biophysical processes and resources is a matter of significant concern. Within the domain of supply chain (SC) management and enterprise-wide optimization, it is increasingly evident that stakeholders must prioritize the creation of sustainable value to maintain competitiveness [1]. This requires simultaneous optimization of environmental and social objectives. The 'Safe and Just' (SJS) framework delineates the ecological ceiling and social foundation for humanity, with the space in between referred to as the 'safe and just operating space'. This framework has garnered significant attention from policymakers and holds the potential to unite stakeholders around a comprehensive vision of sustainable development [2,3]. Uncertainties are inherent in supply chain design and quantitative measurements of environmental impacts, presenting a pervasive challenge [4]. Developing a supply chain design framework that operates within the SJS space while effectively addressing uncertainty is imperative. This work aims to bridge this research gap and contribute to the advancement of sustainable supply chain practices.

The planetary boundary framework has been used for ecological ceiling quantification however at product level it relies on subjective downscaling which has low geographical resolution and could lead to misleading results with high variability. The techno-ecological synergy framework which relies on scientific data and biophysical models has been used in defining the ecological ceiling for each process inside the supply chain. Social foundations have also been quantified through this framework [5]. This model has been applied at a life cycle scale for the supply chain meaning processes from raw materials extraction to the generation of final products are captured. In order to consider the effects of uncertainties, a two-stage mixed integer linear programming stochastic model is proposed to capture uncertainties in both raw material supply and final market demand. Stochastic optimization is a scenario-based approach for optimization under uncertainty and in this approach, uncertainties are captured by a number of discrete realizations [4]. In the first stage, where to produce which product and how much to produce is decided while the second stage determines the quantity of products to be shipped. The major constraints are due to capacity requirements and mass balance (supply chain and LCA network). While constraining the social foundations for each process, we looked into different ways of defining the objective functions and constraints to respect ecological ceiling: 1) minimizing the overall net emission of the whole supply chain (emission subtracts ecological threshold); 2) minimizing the percentage of overshooting (emission larger than ecological threshold) processes in different regions. The L-shaped method is applied for solving the design problem.

This framework is implemented and applied to a case study on the design of the Li-ion battery supply chain. Electric vehicles (EVs) have emerged as a promising solution to address sustainability challenges, particularly global warming [6]. The Li-ion battery, a crucial component of EVs, significantly contributes to greenhouse gas (GHG) emissions. Mineral production, a key input for Li-ion batteries, is susceptible to water supply and governmental policies. Additionally, market demands for Li-ion batteries are influenced by factors such as technological advancements and market competition, modeled here as stochastic variables. Scenarios are generated using Monte Carlo simulation based on historical mineral supply and battery demand data from the past decade. In the first stage, the model minimizes net production emissions, while in the second stage, it focuses on minimizing net transportation emissions. Social foundation constraints are applied in both stages, with penalty costs assigned to unsatisfied demands. To address risk aversion, we also formulated this it as a minimax problem. The results of this model are compared with the current Li-ion battery supply chain of Tesla, providing valuable insights into the effectiveness of the proposed framework.

The novelty of this work includes the development of a supply chain design model operating within the safe and just space while accounting for uncertainty alongside the algorithm to solve it. Notably, optimization occurs at the life cycle scale. By applying this framework to the Li-ion battery supply chain, it underscores the imperative of integrating the ecological ceiling and social foundation in facilitating the transition toward a more sustainable future.

Reference

1. Wheeler, Jonathan, et al. 'Designing biomass supply chains within planetary boundaries.' AIChE Journal 67.4 (2021): e17131.
2. Turner, Rachel A., and Jane Wills. 'Downscaling doughnut economics for sustainability governance.' Current Opinion in Environmental Sustainability 56 (2022): 101180.
3. Fanning, Andrew L., et al. 'The social shortfall and ecological overshoot of nations.' Nature sustainability 5.1 (2022): 26-36.
4. Gao, Jiyao, and Fengqi You. 'Modeling framework and computational algorithm for hedging against uncertainty in sustainable supply chain design using functional-unit-based life cycle optimization.' Computers & Chemical Engineering 107 (2017): 221-236.
5. Aleissa, Yazeed M., and Bhavik R. Bakshi. 'Possible but rare: Safe and just satisfaction of national human needs in terms of ecosystem services.' One Earth 6.4 (2023): 409-418.
6. Parmesan, Camille, Mike D. Morecroft, and Yongyut Trisurat. Climate change 2022: Impacts, adaptation and vulnerability. Diss. GIEC, 2022.

160e Towards Circular Food Supply Chains: Integrated Design of Food Packaging and Waste Management Infrastructures

Paola Alejandra Munoz Briones, Aurora del Carmen Munguía-López, Victor M. Zavala, Styliani Avraamidou

Plastic production, consumption, and waste have globally grown exponentially over the past decades.[1] Around 35-40% of the plastic produced worldwide is destined for packaging, a percentage that is projected to rise in the coming years. [2] Packaging plays a crucial role in food supply chains guaranteeing quality and food safety, extending the product's shelf-life while avoiding food waste, and facilitating food access. Nevertheless, most of the food packaging becomes waste after a single use contributing to multiple environmental impacts like water, land, and air pollution that threaten ecosystems, wildlife, and humans. Beyond traditional waste management technologies like mechanical recycling, incineration, and landfilling, new promising technologies are being explored to tackle complex plastic waste. Among the different packaging types, multilayer plastic films are commonly used, as they offer improved properties for preserving products using less material than the monolayer alternatives. However, multilayer plastic films cannot be recycled using conventional technologies due to their complexity, therefore, they require new technologies that are more energy and/or resource-intensive (e.g., dissolution-based processes like STRAP™, pyrolysis),[3] or emitting more greenhouse gases (e.g., incineration). The tradeoffs present, from the production to the waste management of packaging, lead to the need for the consideration of the entire food and packaging supply chain for the optimal design of food packaging and waste management infrastructure.

Food supply chain studies generally exclude packaging from their system of interest and within a few specific studies, there is a lack of a broader scope that analyzes different types of packaging along with different recycling technologies.[4] These studies analyze food packaging or recycling technologies focusing either on the packaging alternatives design or the comparison of different recycling technologies.[5] Some of them use a techno-economic analysis (TEA) to evaluate the feasibility of their alternatives and/or use a life-cycle assessment (LCA) as the methodology to determine the most environmentally friendly alternative. However, in our previous work, we have found that a further assessment is required to address a bigger array of environmental impacts beyond greenhouse gas emissions.[6]

To identify the optimal waste management infrastructure for food packaging, a complete analysis that includes existing waste management technologies and the potential implementation of new facilities and technologies is required. In this work, we aim to identify the best combination of packaging, waste management technologies, and the required infrastructure. Therefore, we propose a systems engineering framework based on multi-objective optimization that will help to analyze the different packaging alternatives and the potential waste management technologies for packaging and the food waste associated. We will illustrate the proposed framework on a case study focusing on the packaging of roasted coffee, involving a variety of packaging option types including monolayer and multilayer plastic films, rigid containers, glass, reusable glass, and metal cans. The integration of techno-economic analysis (TEA), life cycle assessment (LCA), and a circularity assessment enable a comprehensive analysis of packaging in the food supply chain from their production to end-of-life disposal.[7,8] This work aims to build the first steps in packaging and waste management infrastructure design considering economic and environmental aspects.

References

1. Wang, C., Liu, Y., Chen, W. Q., Zhu, B., Qu, S., & Xu, M. (2021). Critical review of global plastics stock and flow data. Journal of Industrial Ecology, 25(5), 1300-1317.
2. Stegmann, P., Daioglou, V., Londo, M., van Vuuren, D. P., & Junginger, M. (2022). Plastic futures and their CO2 emissions. Nature, 612(7939), 272-276.
3. Walker, T.W., Frelka, N., Shen, Z., Chew, A.K., Banick, J., Grey, S., Kim, M.S., Dumesic, J.A., Van Lehn, R.C. and Huber, G.W. (2020). Recycling of multilayer plastic packaging materials by solvent-targeted recovery and precipitation. Science advances, 6(47)
4. Baratsas, S. G., Pistikopoulos, E. N., & Avraamidou, S. (2021). A systems engineering framework for the optimization of food supply chains under circular economy considerations. Science of the Total Environment, 794, 148726.
5. Jeswani, H., Krüger, C., Russ, M., Horlacher, M., Antony, F., Hann, S., & Azapagic, A. (2021). Life cycle environmental impacts of chemical recycling via pyrolysis of mixed plastic waste in comparison with mechanical recycling and energy recovery. Science of the Total Environment, 769, 144483.
6. Munoz-Briones, P.A., Munguia-López, A. D. C, Sánchez-Riveraa, K.L., Zavalaa, V.M. Huber, G.W. & Avraamidou,S. (2024). Optimal Design of Food Packaging Considering Waste Management Technologies to Achieve Circular Economy. Submitted to Foundations of computer aided process operations/chemical process control.
7. Baratsas, S. G., Pistikopoulos, E. N., & Avraamidou, S. (2022). A quantitative and holistic circular economy assessment framework at the micro level. Computers & Chemical Engineering, 160, 107697.
8. Chialdikas, E., Munguia-López, A. D. C., Aguirre-Villegas, H., & Avraamidou, S. (2023). A framework for the evaluation of the circularity of plastic waste management systems: a case study on mechanical recycling of HDPE. Foundations of computer aided process operations/chemical process control: In-Press.

160f Assessing the Implications of Recycling Targets and Scale-up of Post-Consumer Plastics Recycling Infrastructure: A Reverse Supply Chain Optimization Approach

Tapajyoti Ghosh

As part of a broader initiative aimed at fostering sustainable waste management practices and mitigating greenhouse gas emissions associated with waste disposal, the Environmental Protection Agency (EPA) has recently established a national recycling objective: achieving a 50% recycling rate for municipal solid waste (MSW), including plastics, by 2050. This ambitious target is bolstered by increasing demand for commitments from brands to use recycled materials, creating both pressure and opportunities for advancing plastics recycling efforts. However, reaching these goals necessitates the development of a robust post-consumer plastics reverse supply chain infrastructure, surpassing current capacities. Achieving such a feat requires meticulous projections and analyses of the energy, economic, and environmental ramifications associated with transitioning towards a nearly closed-loop plastics economy. The main aim of this study is to explore the expansion required in recycling capacity, the pace at which this capacity needs to increase, optimal locations for additional capacity, and which materials would render recycling economically viable and environmentally advantageous, considering impact metrics that extend beyond carbon emissions and fossil energy use. Addressing these inquiries necessitates the implementation of a comprehensive regional or national system-wide reverse supply chain modeling, design, and optimization approach. The study will utilize the 4P (Plastic Parallel Pathways Platform) framework, developed by National Renewable Energy Laboratory (NREL), to formulate scenarios aimed at increasing recycling rates, leveraging the EPA PET target and commitments to recycled content as influential factors. Through 4P, valuable insights will be garnered regarding post-consumer plastics quality, distribution, available recycling technologies, associated costs, and material balance specifications. Although initially focused on PET, the 4P framework can be expanded to encompass other plastics and recycling streams. These insights will be integrated into the reverse logistics modeling and optimization tool (RELOG) developed by Argonne National Laboratory. This integration will facilitate decisions on the optimal location, timing, estimated capacities, and technology types for constructing plastics recycling plants, as well as determining the routes for post-consumer plastics. The analysis will encompass an assessment of costs, energy consumption, carbon emissions, and other environmental impacts across the life cycle of scaling up the recycling infrastructure.

160g Trojan Horse Strategy for Sustainable PET Plastics: A Comparative Evaluation of Techno-Economic and Lifecycle Impacts

Vedanti Sharma, Dhananjay Dileep, Mark Mba Wright, Eric Cochran

The recycling of Polyethylene Terephthalate (PET) is essential for achieving a circular economy. PET, derived from petroleum, necessitates recycling to promote sustainability and resource efficiency. Various recycling technologies, including mechanical and chemical processes, are crucial for establishing a circular economy for PET products, leading to reduced greenhouse gas emissions compared to linear production systems. The concept of chemically recyclable polymers underscores the importance of developing materials that can be efficiently recycled, contributing to supporting circular economy initiatives.

To overcome these challenges, a novel solution employing Trojan Horse units in PET is proposed. These units, integrated into the PET chain, facilitate targeted recycling processes, thereby conserving energy and preserving the physical and chemical properties of recycled PET akin to virgin counterparts. This research focuses on optimizing the implementation of Trojan Horse units and explores sustainable methods for producing Bio-PET. This paper underscores the significance of producing bio-based Polyethylene Terephthalate (PET) as a sustainable alternative to traditional petrochemical-derived PET. By utilizing biomass sources, bio-based PET reduces reliance on fossil fuels. It offers potential for a circular economy approach, especially with the development of bio-based PET products from recycled PET degradation products.

Assessing the scaling effects of circular economy strategies, particularly in the PET market, provides insights into the potential benefits of closed-loop recycling systems. The mature mechanisms of PET decomposition and recycling make it a suitable example for understanding the principles of recycling and depolymerization in the context of a circular economy. Recycling trojan horse incorporated PET and other plastic wastes on an industrial scale offers environmental benefits and aligns with the goals of a circular plastic economy.

Using the Techno-Economic Analysis (TEA) framework, this study compares the Trojan Horse approach with enzyme-based PET depolymerization technology. By employing a chemical process model based on the National Renewable Energy Laboratory (NREL) biorefinery design, the analysis evaluates capital and operational expenditures, energy consumption, potential product yields, and market viability. Initial findings indicate promising results for the Trojan Horse technology, sensitivity analysis is conducted to determine optimal Trojan Horse ratios and PET compositions for cost-effectiveness. The minimum selling price (MSP) of recycled PET is about $1.98/kg, whereas the price of the Trojan Horse incorporated PET is about $6.75/kg. This PET polymer consists of 5% Trojan horse, 25% bio-based TPA, and 70% recycled fossil TPA, thus naming it Bio-rPETTH. The TEA assessed the capital expenditure (CapEx) of $128 million, operational expenditure (OpEx) of $6,450 for Bio-rPETTH, and $2,180 for rPET.

Furthermore, a comprehensive Life Cycle Assessment (LCA) will be conducted to assess the environmental and socioeconomic impacts of both recycling methods. Considering the entire lifecycle of PET plastic using cradle to cradle approach. The LCA aims to inform decision-making towards the most sustainable PET waste management approach.

EFFECT OF AMBIENT CONDITION AND TERRAIN STRUCTURE ON THE DISPERSION OF ACCIDENTALLY RELEASED CHEMICALS

Sk Moumita Manjur & Sidney Lin
Lamar University, Beaumont, TX, USA.

Abstract

Communities in the regions having sophisticated big scale chemical industrial clusters have substantial risk to be exposed to constant air pollution by toxic or criteria air pollutant chemicals. They are also susceptible to severe catastrophic exposure to toxic or hazardous chemicals when an unplanned/accidental chemical release from a process unit of the facility. Besides, people working in the facility have high risks of exposure to the released toxic chemicals. Exposure to high concentration of toxic chemicals causes serious health hazard and even death. In the situation of an accidental release, emergency response and evacuation plan should be efficiently made and implied to minimize the dire consequence. Furthermore, chronic exposure to a hazardous chemical of a concentration which is low but higher than the permissible limit causes slow development of serious health problems such as cancer, lung disease etc. This kind of unpermitted release should also be addressed and minimized to ensure a safe and sustainable environment. A mathematical simulation can provide comprehensive understanding about the dispersion trend of a chemical depending on its properties, ambient condition, and terrain structure. A proper planning for reduction of pollution along with efficient emergency response and rescue action plan can be developed based on the understanding from mathematical simulation.

A finite element analysis model of released airborne toxic chemical from an industrial process unit such as chemical storage tank into surrounding ambience has been developed in this work using COMSOL Multiphysics. This is a high-resolution model using very small element size compared to the whole geometric domain. The diffusion and convection of the released chemical through ambient air is the prevailing driving mechanism of transport. The dispersion trend has been estimated by using Fick's 2nd law of diffusion integrated with convection by velocity field of air. Velocity field for different wind speeds have been calculated using compressible form of the Navier-Stokes and continuity equations. Transport and thermodynamic properties of air and chemical species are estimated at different temperatures and pressures. Dispersion trend of different pollutant species was calculated in different scenarios including i) ambient of different wind speeds, ii) different wind directions, and iii) different obstructed terrains. The calculated results are analyzed to study impacts of different wind speed, ambient temperature and terrain type on the propagation of chemical species over time.

Introduction

In October 2024, a serious chemical release incident happened in PEMEX oil refinery in Houston, TX. Two people were killed and 35 critically injured in this hydrogen sulfide leakage incident[1]. This is an obvious concern for the working people as well as the residents near the chemical facility. Hazardous or toxic chemical release can cause tremendous extent of loss to lives and assets. This life- threatening risk is accelerating as the economy growth nowadays. Therefore, a robust emergency rescue plan needs to be developed to prevent and minimize the loss during an accidental chemical release incident. Besides, proper planning should be deployed to combat the environmental pollution due to unplanned/unregulated chemical release from the chemical facilities. To address this problem, the dispersion trend of toxic/hazardous chemicals needs to be predicted properly. Based on the predicted concentration profile and history of released chemicals in a certain scenario, proper emergency response action or environmental protection plan can be undertaken. Dispersion trend of chemicals depends on several factors. For example, physical and transport properties of the chemicals determine the dispersion behavior; ambient conditions such as temperature, wind speed and direction are major factors to be considered. Wind speed and direction have strong dominance over the dispersion as the released chemicals undergo the convection driven by the wind. Terrain structure also plays a vital role on the dispersion trend. Chemicals propagation varies significantly in the complex terrain having building-like structures than in simple/plain terrain.

The study region of this research, Southeast Texas region is densely concentrated with complex cluster chemical industries. Most of them are petrochemical industries manufacturing plastics, rubber, basic chemicals, polymers and paints. 1,3- Butadiene is one of the common feedstocks used in these industries and is a toxic carcinogenic chemical by inhalation classified by US EPA[2]. The use of 1,3- Butadiene is increasing every year in this area[3]. There is a risk of accidental toxication by this chemical. Therefore, this chemical has been chosen for this study.

There are various approaches to predict the toxic gas dispersion trend. The dispersion trend and consequence scenario can be predicted within short period of time by DNV PHAST[4], ALOHA[5] and HAZDIG[6]. These software's are based on Gaussian dispersion and, Plume or Puff experience model. But these experience models are best for an open terrain without any complex structure. The CFD (Computational Fluid Dynamics) is another popular approach nowadays[7]. It can predict the dispersion trend and consequence scenario for complex exposure terrain where buildings and/or process structure are around the chemical release. The FEA (Finite Element Analysis) is a method by which precise high-resolution results can be attained. In this method, original domain is discretized into small elements. Equations are solved for each individual element and finally simulated results of all elements are combined back to the original domain for final result[8]. This study is carried out on the principle of FEA and CFD to understand the effect of exposure terrain structure on the dispersion trend in a precise manner.

Methods

Toxic gas release scenarios were considered inside a 50m x 50m x 50m domain containing a 2m x 2m x 3m storage tank of 1,3-Butadiene at its west end as shown in Figure 1. This work was conducted on both plain terrain (without obstacles) and obstructed terrain when two wind directions were used as shown in Table 1.

Table 1. Characteristics of Different Studies

Study	Terrain Type	Wind Speed & Direction
S-1	Plain	5 m/s. W to E
S-2	2 Obstacles	5 m/s. W to E
S-3	Plain	5 m/s. W to 30° NE
S-4	2-Obstacles	5 m/s W to 30° NE

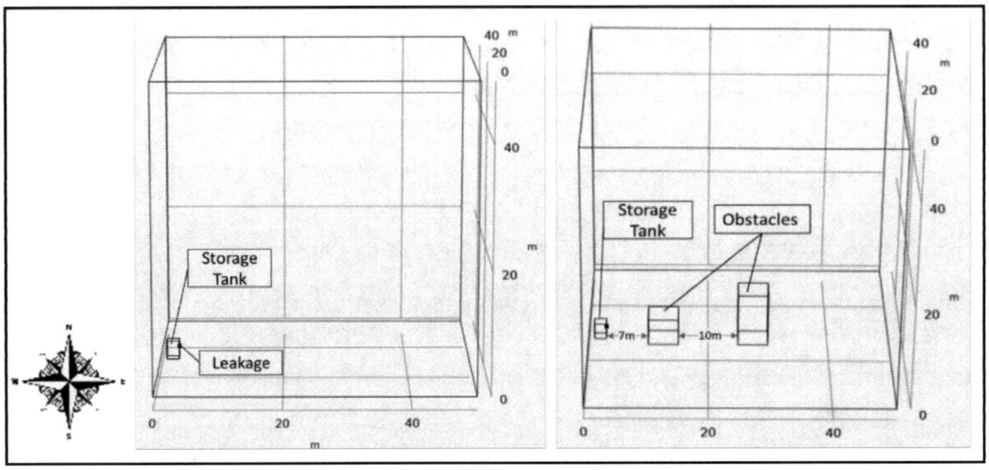

Figure 1: Geometry of the Model: Plain Terrain (Top) and Obstructed Terrain (Bottom)

COMSOL Multiphysics® 6.2 was used to simulate the gas release model. Based on CFD (Computational Fluid Dynamics), the wind velocity field was first calculated, and concentration profile and history of the released gas were calculated using the established wind velocity field afterwards.

Wind velocity field was estimated on the principle of Navier-Stokes and continuity equations as a compressible flow. Variation of wind along the height was taken into account in this work. The nominal height of boundary layer is set to be 274m[9] and the equation used for varying wind speed is:

$$V_z = V_g \left(\frac{Z}{Z_g}\right)^{\frac{1}{\alpha}}$$

where,

V_Z = mean wind speed at height Z aboveground

V_g = wind speed above the boundary layer

Z = height

Z_g = nominal height of boundary layer

α = power law coefficient

The concentration of the dispersed gas was calculated using Fick's 2nd law of diffusion. Physical properties of the dispersed gas obtained from literatures were used[10]. The Fluid Flow physics (for calculation of wind velocity field) and Transport of Diluted Species (for calculation of dispersed gas concentration) were fully coupled.

Size of mesh element ranging from 0.5 - 2.65m was used in the model. Maximum element growth rate is 1.13. Free tetrahedral elements were used with necessary corner refinement and boundary layer formation as shown in Figure 2.

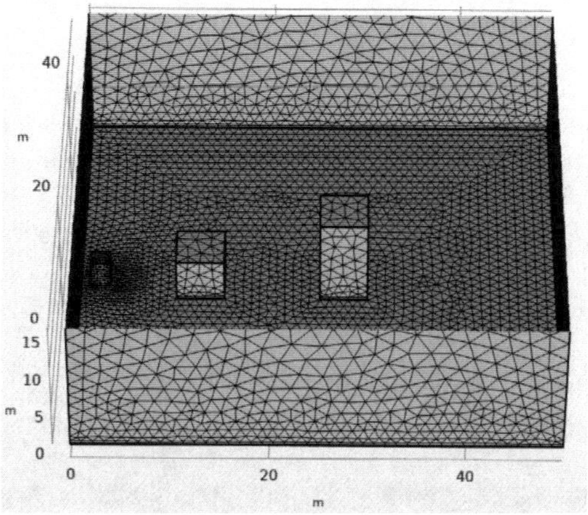

Figure 2: Mesh Used in the Model

The ambient temperature is set to be 25°C. The storage tank is assumed to be filled with pure 1,3- Butadiene gas and leaks from a pinhole at a rate of 0.5 kg/s.

Result & Discussion

The concentration of 1,3- Butadiene for all four studies listed in Table 1 have been calculated and analyzed. It was observed that the concentration profile of the dispersed gas varies notably in different exposure terrain and wind condition.

The contours for 1,3- Butadiene concentration at an elevation of 1.5m show the differences among all studies. The contours have been classified into four ranges i) 1 ppm; ii) 200 ppm; iii) 1000 ppm & iv) >2000 ppm. based upon the toxicity level of 1,3- Butadiene shown in Table 2.

Table 2. Toxicity Level of 1,3- Butadiene[11]

Reference Concentration Level	0.001 ppm	Chronic Exposure Limit
OSHA PEL	1 ppm	Exposure up to 8 hr. a workday
AIHA ERPG-2 Level	200 ppm	Exposure up to 1 hr. in any incident
NIOSH IDLH	2000 ppm	Immediately dangerous to life

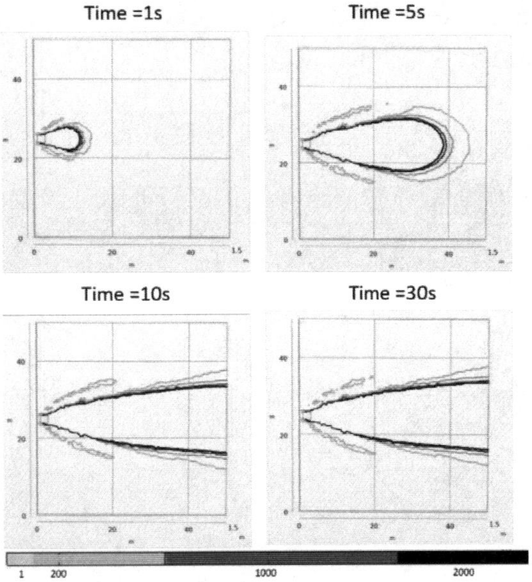

Figure 3: Contours of 1,3- Butadiene Concentration for Study S-1

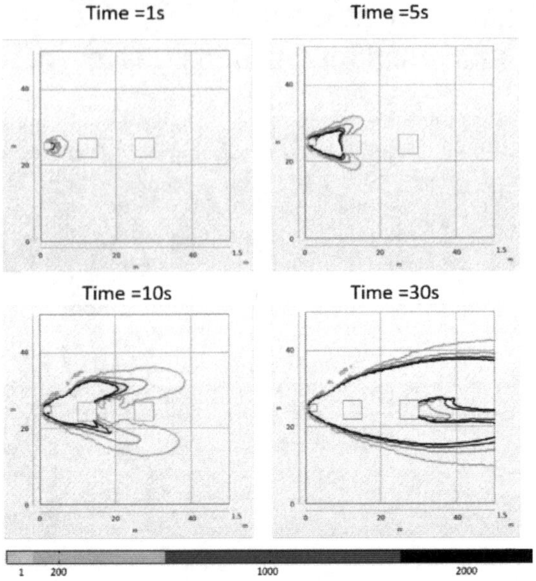

Figure 4: Contours of 1,3- Butadiene Concentration for Study S-2

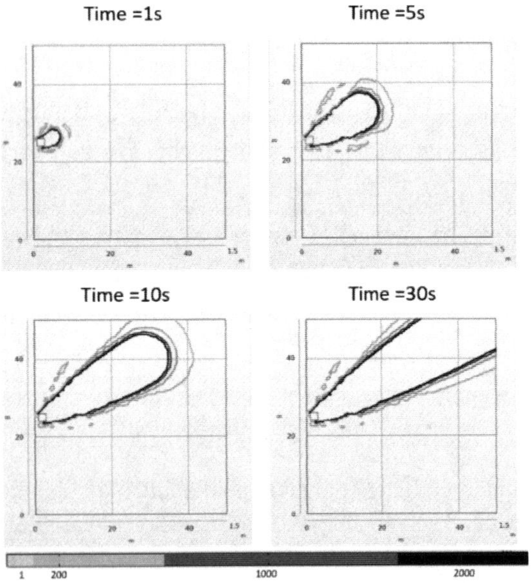

Figure 5: Contours of 1,3- Butadiene Concentration for Study S-3

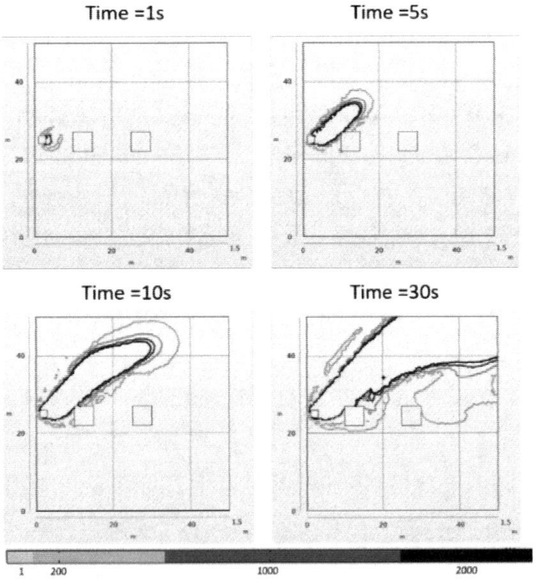

Figure 6: Contours of 1,3- Butadiene Concentration for Study S-4

The contours depicted in Figure 3 through 6 clearly show the difference in dispersion trend among the studies. The leaked gas covers wider space in obstructed terrain than in plain terrain as in Figures 3 and 4. Rather, the gas traverse straightly towards the wind direction in case of plain terrain. Difference in dispersion trend is also visible for the S-3 and S-4 shown in Figure 5 and 6.

To understand the impact of terrain type and wind direction on the concentration of dispersed gas, concentrations at two selected points shown in Figure 7 are calculated for all four studies.

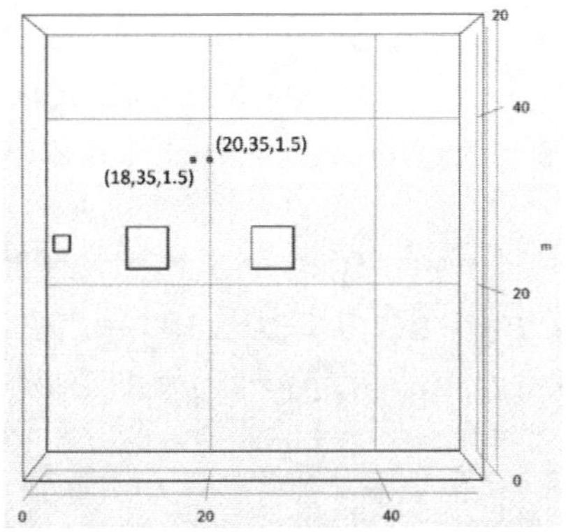

Figure 7: Points Selected for Observing Concentration Trend

The concentration histories of 1,3- Butadiene at these two points for four studies are presented in Figure 8 through 11.

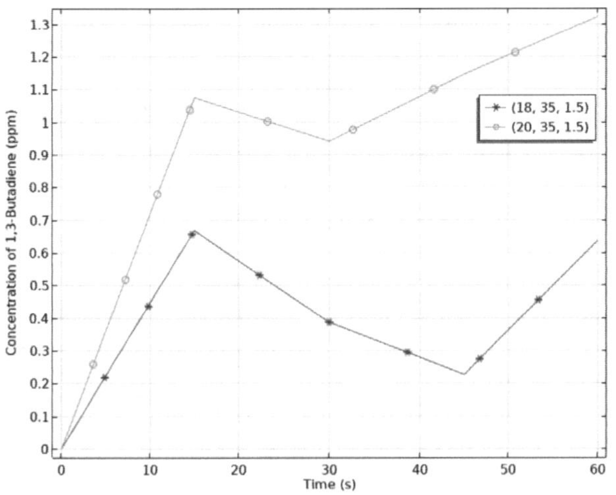

Figure 8: Concentration Profile of 1,3- Butadiene for Study S-1

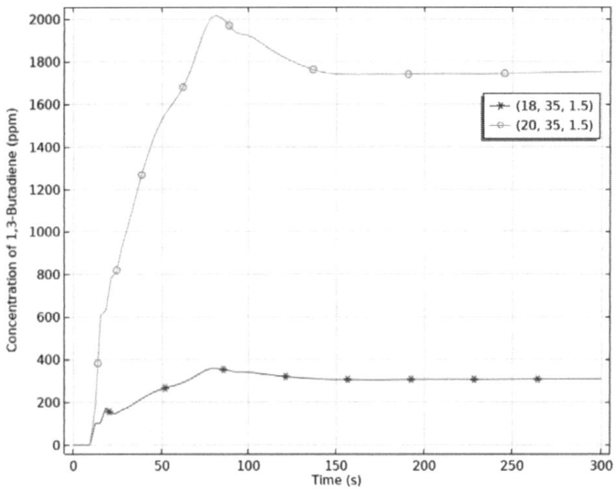

Figure 9: Concentration Profile of 1,3- Butadiene for Study S-2

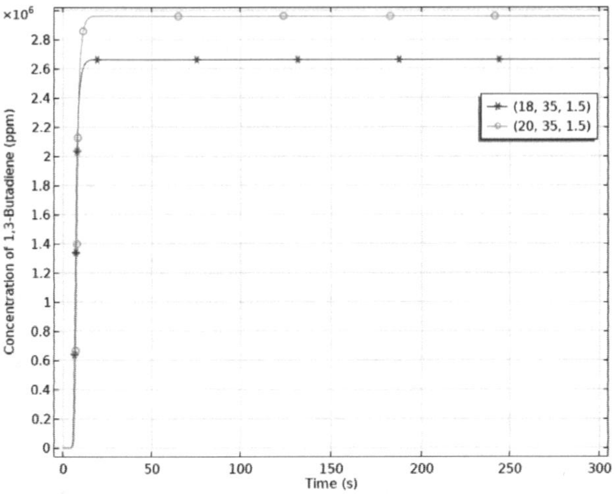

Figure 10: Concentration Profile of 1,3- Butadiene for Study S-3

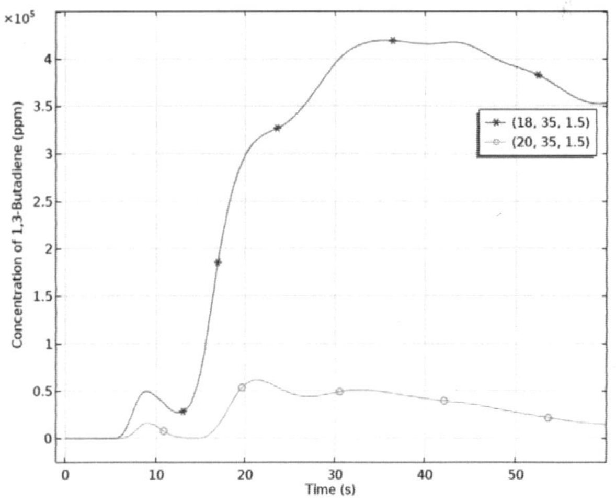

Figure 11: Concentration Profile of 1,3- Butadiene for Study S-4

It was observed that, for a same wind condition (speed & direction) concentration at a certain point varies with the terrain type. The maximum concentration can be almost 1500 times higher in study S-2 than that in study S-1. Gas concentration is tremendously high in study S-3 as the points fall within the main flow path of the gas following the wind to the northeast direction. Concentration is 10 times lower, and the trend is different in study S-4 than in S-3 due to the obstructions in the terrain.

To understand the extent of concentration buildup before and behind the obstacles concentrations along the selected lines before and behind each obstacle at 1.5m elevation as shown in Figure 12 are calculated and shown in Figures 13 and 14.

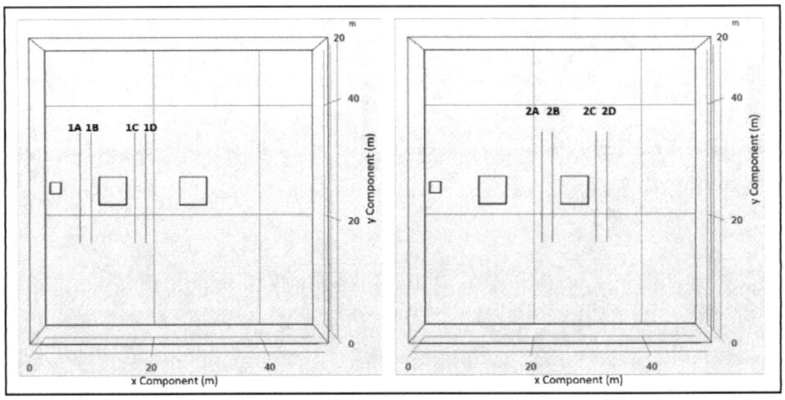

Figure 12: Selected Lines Around the i) 1st Obstacle (Top) & ii) 2nd Obstacle (Bottom)

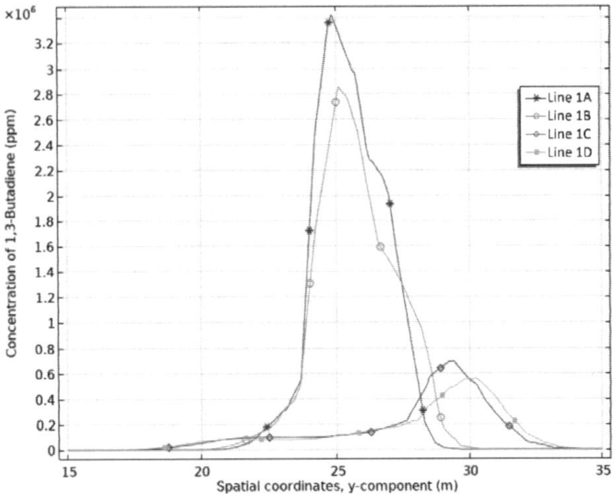

Figure 13: Concentration Profile along Selected Lines around the 1st Obstacle after 1 min.

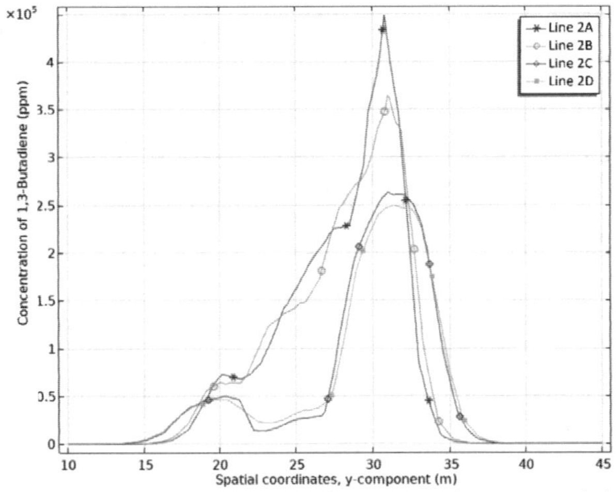

Figure 14: Concentration Profile along Selected Lines around the 2nd Obstacle after 1 min.

From Figure 13 and 14 it was observed that, the concentrations buildup before the obstacles are much higher than those behind the obstacles. Additionally, dispersed gas covers wider region after getting hindered by the obstacles when it moves forward.

Acknowledgement

This material is based upon work supported by the U.S. Department of Energy, Office of Science, Biological and Environmental Research Program under Award Number DE-SC0023216.

Conclusion

A mathematical model based on CFD and FEA were developed in this work. The effects of wind speed and direction along with terrain structure on the dispersion trend of the accidentally released gas were simulated and analyzed. Terrain type has notable influence over the dispersion trend. Gas travels along the straight path towards the wind direction and reaches on the other side of domain quicker in a plain terrain than in an obstructed terrain. The obstacles make hindrance to the gas pathway and cause very high concentration buildup around them. In an obstructed terrain, the gas disperses to a wider region than in plain terrain.

References

1. U.S. Chemical Safety and Hazard Investigation Board, PEMEX, Inc., PEMEX Deer Park Chemical Release, CSB, Washington DC, 2024.

2. Agency for Toxic Substances and Disease Registry (ATSDR) (1992). Toxicological Profile for 1,3-Butadiene. Public Health Service, U.S. Department of Health and Human Services, Atlanta, GA.

3. United States Environmental Protection Agency. (2023). *TRI Explorer* (2022 Dataset (released October 2023)) [Internet database]. Retrieved from https://enviro.epa.gov/triexplorer/, (November 08, 2024).

4. DNV, PHAST Software Introduction [Online]. https://www.dnv.com/software/services/plant/consequence-analysis/ (Accessed November 08, 2024)

5. EPA, ALOHA Software [Online]. https://www.epa.gov/cameo/aloha-software (Accessed November 08, 2024).

6. F.I. Khan and S.A. Abbasi (1999), "HAZDIG: A new software package for assessing the risks of accidental release of toxic chemicals", J Loss Prev Process Ind 12, March, pp.167–81.

7. Zhang, B., Liu, Y., & Qiao, S. (2018). "A Quantitative Individual Risk Assessment Method in Process Facilities with Toxic Gas Release Hazards: A Combined Scenario Set and CFD Approach," Process Safety Progress, 38(1), October, pp. 52–60.

8. Dr, J. N. R. (2005). "An Introduction to the Finite Element Method". McGraw-Hill Education.
9. Chen, Wai-Fah (1997). "Handbook of Structural Engineering". Boca Raton: CRC Press. pp. 12–50. ISBN 978-0-8493-2674-5.

10. Yaws, Carl L. (2003). "**Yaws' Handbook of Thermodynamic and Physical Properties of Chemical Compounds**". Knovel, February.

11. Occupational Safety and Health Administration (OSHA) (1998). Occupational Safety and Health Standards, Toxic and Hazardous Substances. Code of Federal Regulations. 29 CFR 1910.1000.

231b Enhanced Removal of Pharmaceutical Compounds from Aqueous Solution Using Deep Eutectic Solvents -Impregnated Graphene Oxide

Anjali Goyal, Tarek Lemaoui, Enas Nashef, Shadi Hasan, Hassan Arafat

The World Health Organization has designated antimicrobial resistance (AMR) as a 'silent pandemic,' with an annual mortality rate that could soar to 10 million by 2050. A prominent factor contributing to AMR is the persistence of antibiotics in aquatic environments, threatening marine ecosystems and human health.

Traditional water treatment methodologies actually not effectively mitigating these pollutants. This study innovates with deep eutectic solvents (DESs) as impregnating agent on graphene oxide (GO) to significantly enhance the removal of antibiotics from water bodies.

Graphene oxide, promising candidate among carbonaceous adsorbents, provides an extensive surface area and superior adsorption characteristics, serves an ideal framework for adsorption process. Nevertheless, the effectiveness of GO as a standalone adsorbent in eliminating organic pollutants from water is limited. This is primarily due to its pronounced hydrophilicity, stemming from the oxygen-containing groups on its basal plane and edges. It can be overcome by incorporating nanomaterials on the GO surface. However, the reported formulations include the consumption of highly volatile and toxic conventional solvents, presenting severe risks to human health and the environment. In response to this challenge, modifying GO with green solvents plays a pivotal role in environmental conservation. Ionic liquids (ILs) and deep eutectic solvents (DESs) are particularly noteworthy in this category as compared to organic solvents. ILs lose a liquid state when immobilized on the surface of GO by maintaining polarity and low volatility. Therefore, DESs, based on the wise selection of constituents, flourish with advantages, such as simple preparation, cost-effectiveness, low toxicity, and outstanding biocompatibility and biodegradability.

Most of the research has focused on applying DESs as extraction solvents. However, due to their liquid nature, DESs cannot be used as adsorbents independently due to instability in the water phase. This limitation can be overcome by immobilizing DESs onto the surface of GO materials. This immobilization or grafting not only enhances mass transfer but also improves adsorption interactions and the overall adsorption capacity of the materials. It is attributed by DESs tunable properties and environmental sustainability. This synergistic combination is relatively unexplored avenue in removal targeted pollutants in pharmaceuticals field.

A straightforward impregnation method used to synthesize DES-impregnated GO materials and subjected to employ various characterization techniques such as Fourier-Transform Infrared Spectroscopy (FTIR), UV-Vis Spectroscopy, Total Organic Content (TOC), Scanning Electron Microscopy (SEM), and X-ray Diffraction (XRD). A batch adsorption study was conducted to optimize contact time, pH, and adsorbent dosage, aimed to maximize removal efficiency.

The research findings are significant, demonstrating that both pristine and DES-impregnated GO materials exhibit remarkable adsorption efficiency, with removal rates of 86.75 mg/g and 98.40 mg/g, respectively, for a targeted antibiotic. The adsorption behavior closely follows a pseudo-first-order kinetic model and the Langmuir isotherm, indicative of monolayer adsorption on a uniformly distributed surface. Moreover, the thermodynamic parameters demonstrated that the adsorption process for selected pollutant is endothermic and spontaneous in nature. Chemical regeneration of adsorbent GO@ThyLevA using a ethanol provided significant adsorption capacities after 4 successive adsorption-desorption cycles for the antibiotics removal. These results underscore the potential of this novel approach in water treatment, contributing significantly to addressing the AMR challenge.

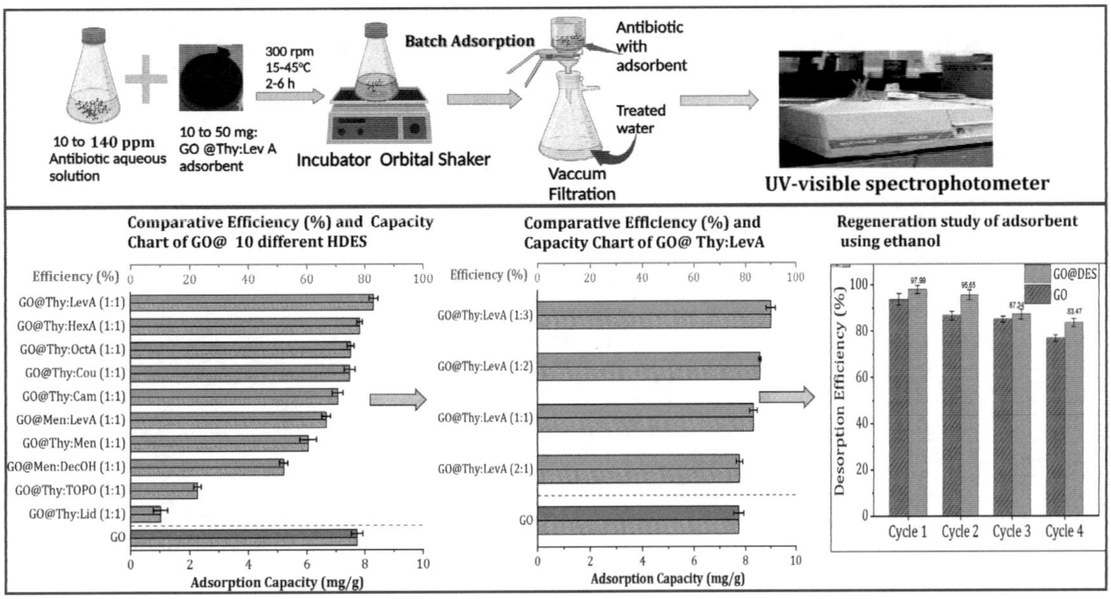

10 to 140 ppm
Antibiotic aqueous solution
10 to 50 mg:
GO @Thy:Lev A adsorbent
300 rpm
15-45°C
2-6 h
Batch Adsorption
Incubator Orbital Shaker
Antibiotic with adsorbent
Treated water
Vaccum Filtration
UV-visible spectrophotometer
Comparative Efficiency (%) and Capacity Chart of GO@ 10 different HDES
Efficiency (%)
GO@Thy:LevA (1:1)
GO@Thy:HexA (1:1)
GO@Thy:OctA (1:1)
GO@Thy:Cou (1:1)
GO@Thy:Cam (1:1)
GO@Men:LevA (1:1)
GO@Thy:Men (1:1)
GO@Men:DecOH (1:1)
GO@Thy:TOPO (1:1)
GO@Thy:Lid (1:1)
GO
Adsorption Capacity (mg/g)
Comparative Efficiency (%) and Capacity Chart of GO@ Thy:LevA
GO@Thy:LevA (1:3)
GO@Thy:LevA (1:2)
GO@Thy:LevA (1:1)
GO@Thy:LevA (2:1)
Regeneration study of adsorbent using ethanol
Desorption Efficiency (%)
GO@DES
GO
Cycle 1
Cycle 2
Cycle 3
Cycle 4

Exploring Hydrothermal Carbonization of Exotic Fruit Hydrochars for Acesulfame-K and Carbamazepine Removal from Water

*María Fernanda Lecaro[1], Yuliana Pullas[1], Andrés S. Lagos[1], Darío Niebieskikwiat[2], Luis Miguel Prócel[3], Marco León[4], Andrea C. Landázuri[1,5,6,7]**

[1] *Universidad San Francisco de Quito USFQ, Colegio de Ciencias e Ingenierías, Applied Circular Engineering & Simulation Group (GICAS), Chemical Engineering Department, Diego de Robles y Vía Interoceánica, P.O. Box 17-0901, Quito, Ecuador*

[2] *Universidad San Francisco de Quito USFQ, Colegio de Ciencias e Ingenierías, Departamento de Física, Diego de Robles y Vía Interoceánica, P.O. Box 17-0901, Quito, Ecuador*

[3] *Universidad San Francisco de Quito USFQ, Instituto de Micro y Nanoelectrónica, Diego de Robles y Vía Interoceánica, P.O. Box 17-0901, Quito, Ecuador*

[4] *Universidad San Francisco de Quito USFQ, Institute for Energy and Materials, Diego de Robles y Vía Interoceánica, P.O. Box 17-0901, Quito, Ecuador*

[5] *Universidad San Francisco de Quito USFQ, Instituto IBioMed, Diego de Robles y Vía Interoceánica, P.O. Box 17-0901, Quito, Ecuador*

[6] *Universidad San Francisco de Quito USFQ, Instituto de Investigaciones Biológicas y Ambientales (Biósfera), Diego de Robles y Vía Interoceánica, P.O. Box 17-0901, Quito, Ecuador*

*Corresponding author: alandazuri@usfq.edu.ec

Abstract

The sustainable management of waste presents a global challenge, requiring solutions to reduce the impact of organic waste that generates pollutants when ending up in landfills. In this regard, hydrothermal carbonization (HTC) stands out as a promising option to address this issue. This study explores the potential of hydrochars derived from exotic fruits like mango, mangosteen, African oil palm, and moringa to eliminate emerging contaminants such as Acesulfame-K and Carbamazepine from water through an adsorption process. The investigation prioritizes detailed characterization of these materials and the evaluation of their adsorptive capacities, considering variables such as dosage and particle size, alongside proposing potential alternative applications. Findings suggest that dosage does not significantly affect the process, yet both biomass type and particle size are relevant factors, with oil palm and mangosteen emerging as superior adsorbent materials. Physicochemical and electrical characterizations confirm the presence of cellulose in the treated material, indicating its utility for pollutant removal, while highlighting the potential of hydrochars as dielectric and insulating materials.

Introduction

Proper waste management presents a global environmental challenge. Specifically in Ecuador, organic waste often ends up in landfills, leading to the production of contaminating liquids and gases that pose risks to groundwater and soil [1], reports indicate non-compliance with waste management obligations, emphasizing the need for innovative approaches to address these issues. Hydrothermal carbonization emerges as a potentially effective solution. It involves a thermal and chemical process in which organic materials decompose in water at high temperatures and pressures; this process yields hydrochar (a solid fraction) and a liquid phase [2]. The applications of hydrochar span from biofertilizers [3] to construction materials [4], fuel production [5], dielectric materials [6], and water contaminant adsorption [7]. Studies demonstrate that hydrochars derived from agricultural residues can effectively remove contaminants such as diclofenac and triclosan [7]. As demand for efficient adsorbents grows, hydrochar offers a sustainable alternative to activated carbon in water treatment.

Water pollution, caused by human activity and exacerbated by industrialization and medication use, is a pressing environmental issue [8]. Current wastewater treatment methods struggle to efficiently and cost-effectively remove emerging pollutants [9], such as Carbamazepine and Acesulfame-K. Carbamazepine, used for treating epilepsy and mental conditions [10], has been found in high concentrations in Ecuadorian rivers [11], posing ecological and health risks, while Acesulfame-K, a calorie-free artificial sweetener, remains unchanged in water treatment plants, and its presence can lead to the formation of harmful byproducts during ozone treatment [12]. Evaluating hydrochar as a sustainable alternative to activated carbon for water treatment is proposed to address industrial solid waste challenges.

This study seeks to promote circular economy practices by utilizing the byproducts of introduced exotic plants and fruits like mango, mangosteen, African palm, and moringa. Mango byproducts, obtained through second-generation biorefinery, extract valuable phytochemicals and bioactive compounds for cosmetic, nutraceutical, and food applications [13]. Mangosteen, despite generating waste during its processing, offers antioxidant properties and potential for value-added product development [14]. African palm byproducts, rich in lignin, hemicellulose, and cellulose, find use in energy generation, paper production, and fertilizers [15]. Moringa, known as the "miracle tree," provides healing properties and diverse applications in pharmaceuticals, cosmetics, and food industries [16], with its leaves and seeds serving as raw materials for biofuel and water purification processes [17], showcasing the potential of fruit and plant byproducts in various sectors, aiming to reduce waste and promote sustainability in Ecuador's fruit industry.

Methodology

Experiments on the removal of emerging contaminants (Carbamazepine and Acesulfame-K) were conducted using UV-Vis spectroscopy for exploratory analyses and High-Performance Liquid Chromatography (HPLC) for compound separation. For both methods, doses of 0.5 and 1.5 grams of hydrochar per liter of solution were utilized, with variations in the particle size of hydrochars. For the physicochemical characterization of materials, Fourier Transform Infrared (FTIR) spectroscopy, X-ray Diffraction (XRD), and Scanning Electron Microscopy (SEM) were performed to determine the composition, crystallographic structure, and morphology of all native materials and hydrochars produced at 175°C for 1 hour. Additionally, after the HPLC experiments, the most effective materials were used to assess their specific surface area through Brunauer-Emmet-Teller (BET) analysis. Finally, pellets for electrical characterization underwent frequency sweeps using a Semiconductor Device Parameter Analyzer. This last characterization was used to determine the relative permittivity, conductivity, and loss tangent of the samples.

Results and Discussion

The assessment of contaminant removal was conducted in experiments using artificially contaminated water with Carbamazepine and Acesulfame-K respectively, utilizing exploratory methods with UV-Vis spectrometry. Subsequently, HPLC coupled with UV-Vis was also utilized as a method to verify that the detected absorbances correspond solely to the presence of the pharmaceuticals and not to other extractives originating from the adsorbent materials. In the figure below, examples of chromatograms from five experiments are presented.

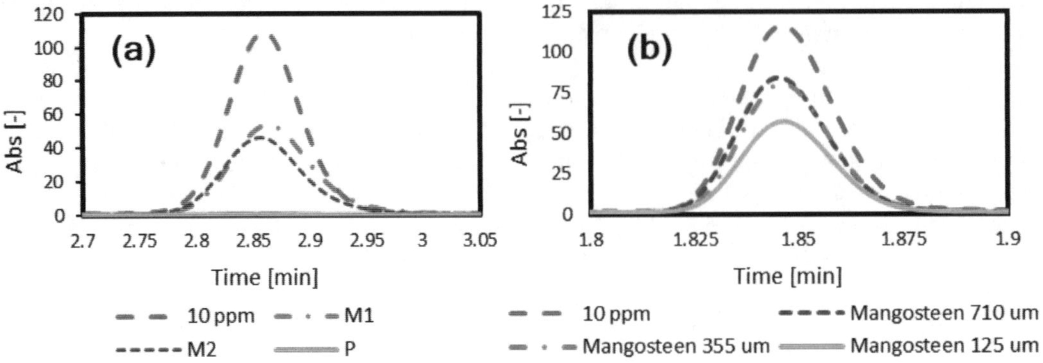

Figure 1. Chromatograms for (a) Carbamazepine removal with Mango and Palm hydrochar, and (b) Acesulfame-K removal with Mangosteen hydrochar.

For Figure 1 (a), M1 represents Mango hydrochar with particle size ranging from 125 μm to 250 μm and a dosage of 1.5 g/L. M2 refers to the same native material. P corresponds to Palm hydrochar with particle size ranging from 125 μm to 250 μm and a dosage of 0.5 g/L. In contrast, to understand the impact of particle size on the treatment of artificially contaminated water, the removal of Acesulfame-K is observed in Figure 1 (b); through the four chromatograms, which include absorbance data from three removal experiments using Mangosteen hydrochar, it becomes evident that a smaller particle size leads to a proportional reduction in contaminant levels as indicated by the absorbance. All aliquots are taken at 5-minute intervals.

In Figure 1 (a), it can be observed that Palm hydrochar successfully reduced the concentration of Carbamazepine below the equipment's detection limit, which would be less than 0.1 ppm. Consequently, this material presents the possibility of serving as an adsorbent. It is worth noting that the studied materials have not undergone an acid or base activation process, which has been shown to enhance the removal capacity of various organic contaminants in wastewater [18].

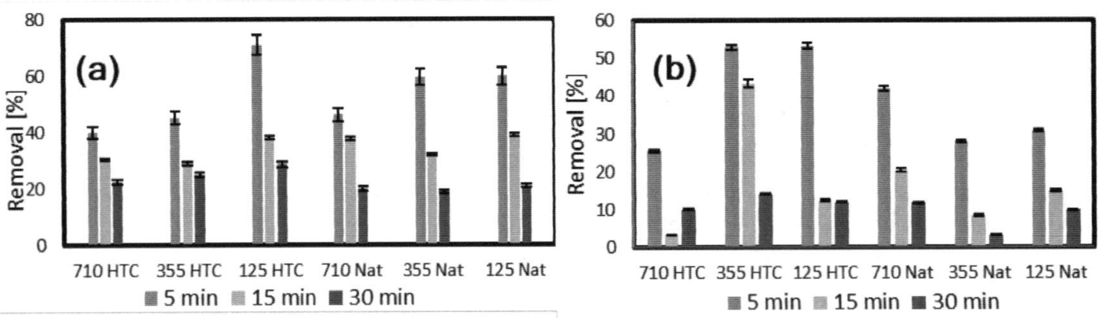

Figure 2. Removal rate for (a) Carbamazepine, and (b) Acesulfame-K with palm hydrochar.

To understand the impact of particle size, as well as the treatment and adsorption behavior, Figure 2 illustrates the removal of (a) Carbamazepine and (b) Acesulfame-K in six experiments. Each experiment involves aliquots taken at 5, 15, and 30 minutes. For Carbamazepine, a smaller particle size leads to an increased percentage of contaminant removal, while over time, the contaminant dissociates from the material surface, resulting in an elevated concentration in the water. Conversely, regarding Acesulfame-K, extending the experiment duration does not enhance the removal percentage; instead, it reduces the overall removal. Consequently, the optimal duration for improved removal appears to be short, specifically 5 minutes.

A substantial amount of data was collected, and therefore, for the purpose of conducting a more thorough analysis, it was processed using Design Expert. As a result, statistical significance of factors was obtained, and the summary is presented in Table 1.

Table 1. Results of the ANOVA statistical analysis.

Factors	Treatment (T), biomass (B), contaminant (C), dose (D), and size (S).
Responses	Removal at 5, 15, and 30 minutes.

Analysis	**Time [min]**		
	5	15	30
R^2	0,640	0,616	0,533
R^2 adj.	0,573	0,549	0,424
R^2 pre.	0,482	0,460	0,270
Ad. Pre. > 4	12,9	13,8	10,6
Relevant factors	T, B, and C	T and C	T and C
Relevant interactions	TS, TB, BC y STC	Ct, BC y TBDS	CS, BC, and CTS

Notably, the hydrocarbon exhibited higher contaminant absorbance compared to the native mango material, suggesting carbonization might not be necessary, although further statistical analysis is required. The multicategorical ANOVA analysis in table 1 indicates the treatment significantly impacts removal, with smaller particle sizes favoring hydrocarbon use within specific ranges. While initial sampling showed rapid removal within 5 minutes followed by contaminant concentration increase, indicating desorption, the lack of extended data points hinders fitting to an adsorption isotherm. Figure 2 (a) supports the greater removal efficiency of smaller particle sizes, with ANOVA highlighting the interaction of particle size with other factors. On the other hand, dosage did not influence contaminant removal significantly, considering both contaminants and biomasses.

Various characterizations were completed to complement the removal analyses, starting from FTIR spectra shown in figure 3 was used to understand the composition of hydrochar and native materials.

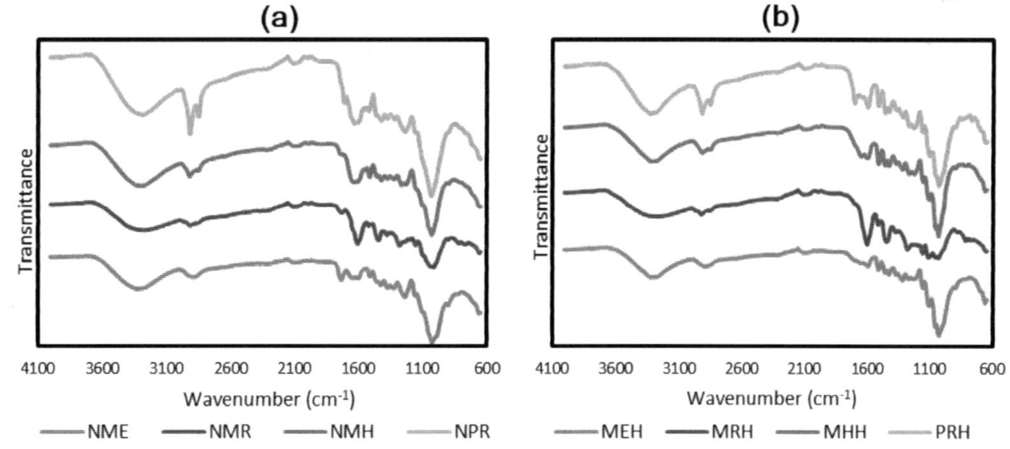

Figure 3. FTIR spectra of (a) native materials, and (b) hydrochars.

Figure 3 presents the FTIR spectra in the range of 1580-1600 cm^{-1}, the folding of NH vibrations is observed. Finally, in the range of 1100 to 1300 cm^{-1}, the folding of (CH) in cellulose, hemicellulose, and lignin is visualized [19]. The peak located at 1730 cm^{-1} corresponds to the vibrations of hemicellulose with the stretching of CO. Similarly, the peak at 1520 cm^{-1} exhibits lignin vibrations, along with the peak around the wavelength of 1200 cm^{-1} showing the asymmetric stretching of cellulose, and finally, the peak near 1107 cm^{-1} belongs to the asymmetric stretching of COC [20]. However, it can be observed that there are slight variations around 1600 cm^{-1}, resulting in a more pronounced peak than in the native sample, which can be explained by the presence of nitrogen compounds.

Furthermore, to comprehend the morphology of native materials, SEM images are presented in Figure 4 below. From these images, it can be observed that the native materials of mangosteen, palm, and moringa exhibit porosities on their surfaces, whereas mango appears in the form of fibers.

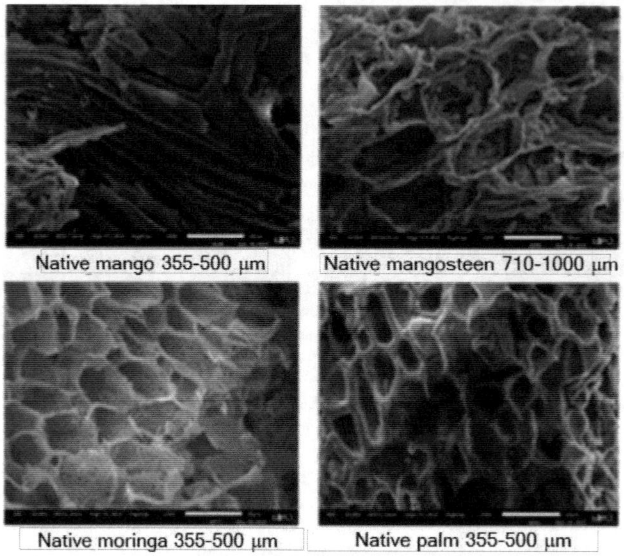

Figure 4. SEM images of (a) Mango endocarp, (b) Mangosteen rinds, (c) Moringa husk, and (d) Palm residues native materials

Then, X-Ray Diffraction Analysis was performed to provide information about the crystallographic structure and physical properties of the different mango samples.

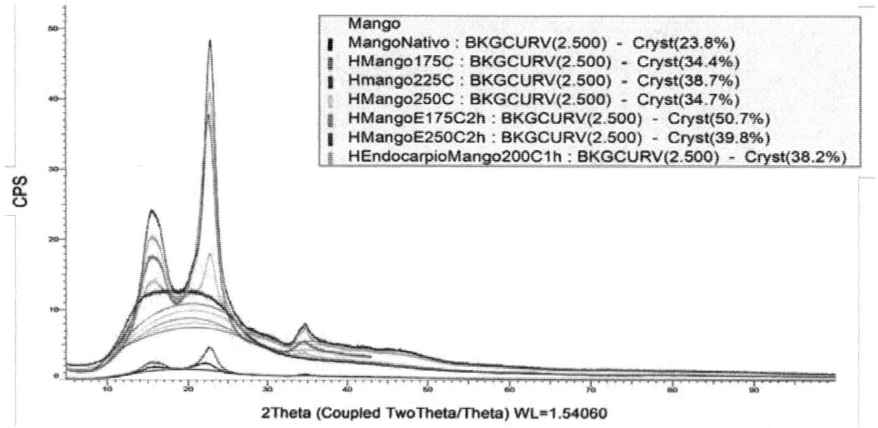

Figure 5. XRD spectra of native mango endocarp and hydrochars.

These diffractograms demonstrate the presence of cellulose. The native mango contains a significant percentage of crystalline cellulose (with a crystallinity index, CI, of 23%). The CI remains constant at around 35% up to 225°C, with no difference between 1 hour and 2 hours of hydrothermal treatment. At

250°C, the CI decreases to approximately 15% after 1 hour. After 2 hours at the same temperature, the crystalline cellulose is completely decomposed.

The most promising samples from further study are the ones treated for 1 hour at 175°C, with mango endocarp hydrochars containing significant amounts of cellulose. X-ray diffraction analysis shows that samples treated at temperatures up to 225°C maintain their relative crystallinity index, while samples treated at 250°C do not contain crystalline cellulose, indicating significant degradation kinetics. The presence of cellulose in water treatment materials is important due to its adsorption and contaminant removal capabilities, with lignocellulose-based materials like bioadsorbents and cellulose-based materials like cellulose acetate membranes used for removing emerging water contaminants like pharmaceuticals and chemicals, and for heavy metal removal, dye adsorption, and pharmaceutical removal. Given that cellulose is a renewable and abundant resource, it is an attractive option for water treatment applications, with its adsorption properties and capacity to eliminate contaminants making it a promising choice for water treatment applications.

The FTIR spectra analysis reveals that hydrochars maintain a consistent chemical structure with subtle changes, contrasting with more significant structural alterations reported at temperatures above 210°C and longer HTC treatment durations compared to the native material. Notably, altered hydrochars exhibit reduced peak intensities in the 1100 to 2000 cm^{-1} range, indicating structural degradation affecting lignin, cellulose, and hemicellulose components. Increasing the surface area of adsorbent materials is known to enhance contaminant removal through adsorption processes, yet this study found that such removal processes also facilitate mass transfer phenomena, leading to the diffusion of additional compounds like furfurals, sugars, phenolic compounds, and volatile fatty acids into the liquor. Understanding the impact of conductivity on contaminant removal is crucial, emphasizing the need for further research in this area to elucidate these complex interactions.

The relative permittivity and conductivity of native mango endocarp and hydrochars has been measured to provide information on the samples' properties at different temperatures.

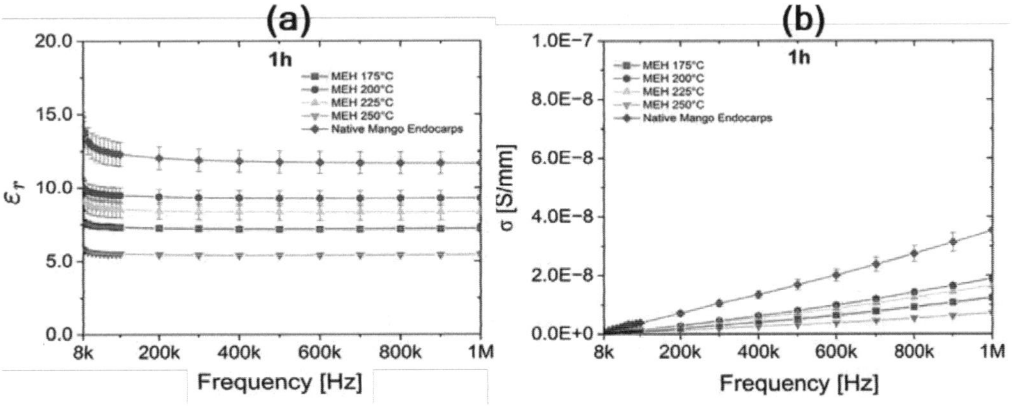

Figure 6. (a) Relative permittivity, and (b) conductivity of native mango endocarp and hydrochars.

These materials demonstrate themselves to be insulating and dielectric, meaning they can store energy in the form of an electromagnetic field. At higher temperatures and longer treatment times, conductivity decreases while dielectric capacity increases. In general, all the studied materials exhibit dielectric properties, insulation, and low conductivity, behaving similarly to previous studies [6].

Studies have investigated the preparation and characteristics of adsorbents derived from lignocellulosic biomass and their activated carbons, but limited information is available on the correlation between the removal of emerging organic contaminants from water using lignocellulosic hydrocarbons and their dielectric and conductive properties. Although relative permittivity and conductivity are independent, a material's conductivity and capacity to remove organic contaminants from water can exhibit an inverse relationship. Specifically, as the concentration of organic carbon in a sample increase, conductivity tends to decrease, with practical applications for water quality monitoring and assessing organic carbon removal methods [21]. The electrical conductivity of water is also crucial in wastewater treatment processes, such as electrocoagulation, which is influenced by water's electrical conductivity [22]. Therefore, the conductivity of a material is essential in removing organic contaminants from water, with significant implications for enhancing water quality in various applications and industries [21]. In the present study, the electrical properties of adsorbent materials were determined. It would be interesting to analyze the conductivity of artificial water during the batch process of contaminant removal and understand the influence of adding these materials during the adsorption process in the liquid phase, along with potential changes in the conductivity of the treated water.

Conclusions

The removal of emerging contaminants from wastewater poses a potential challenge, and it is essential to explore options for enhancing current water treatment processes. Through statistical analysis, it was determined that the hydrothermal carbonization process improves the removal of Carbamazepine and Acesulfame K. The highest contaminant removal occurred within 5 minutes, leading to a focus on this time frame for further analysis. Interestingly, the dosage was not statistically significant within the studied range of 0.5 g/L to 1.5 g/L. However, both biomass and particle size were relevant factors. Decreasing particle size significantly enhanced Acesulfame removal and impacted hydrocarbon treatment. Notably, Palm and Mangosteen were identified as effective adsorbent materials. Additionally, physicochemical characterizations using infrared spectroscopy and X-ray diffraction confirmed the presence of cellulose in the native material, with minimal degradation observed in the material treated at 175°C for 1 hour. This suggests that cellulose could be a useful factor for contaminant removal. Finally, hydrochars can function as dielectric and insulating materials, and future research can reveal more information about the relation between conductivity and water contaminants removal.

References

[1] J. Ulloa, "LOS RELLENOS SANITARIOS," *Revista de Ciencias de la Vida*, vol. 4, pp. 2–17, 2006.

[2] A. L. Pauline and K. Joseph, "Hydrothermal carbonization of organic wastes to carbonaceous solid fuel – A review of mechanisms and process parameters," *Fuel*, vol. 279, no. April, p. 118472, 2020, doi: 10.1016/j.fuel.2020.118472.

[3] L. Wang, Y. Chi, K. Du, Z. Zhou, F. Wang, and Q. Huang, "Hydrothermal treatment of food waste for bio-fertilizer production: Formation and regulation of humus substances in hydrochar," *Science of the Total Environment*, vol. 838, no. 1, 2022.

[4] M. M. Santos *et al.*, "Reducing cement consumption in mortars by waste-derived hydrochars," *Journal of Building Engineering*, vol. 75, no. February, p. 106987, 2023, doi: 10.1016/j.jobe.2023.106987.

[5] E. Song, S. Park, S. Han, E. Lee, and H. Kim, "Characteristics of Hydrothermal Carbonization Hydrochar Derived from Cattle Manure," *Energies (Basel)*, vol. 15, no. 23, 2022, doi: 10.3390/en15239195.

[6] A. C. Landázuri *et al.*, "Valorization of ripe banana peels and cocoa pod husk hydrochars as green sustainable 'low loss' dielectric materials," *J Clean Prod*, vol. 426, no. March, 2023, doi: 10.1016/j.jclepro.2023.139044.

[7] L. Delgado-Moreno, S. Bazhari, G. Gasco, A. Méndez, M. El Azzouzi, and E. Romero, "New insights into the efficient removal of emerging contaminants by biochars and hydrochars derived from olive oil wastes," *Science of the Total Environment*, vol. 752, no. December 2019, p. 141838, 2021, doi: 10.1016/j.scitotenv.2020.141838.

[8] M. Janet Gil, A. María Soto, J. Iván Usma, and O. Darío Gutiérrez, "Emerging contaminants in waters: effects and possible treatments Contaminantes emergentes em águas, efeitos e possíveis tratamentos," vol. 7, no. 2, pp. 52–73, 2012, [Online]. Available: http://www.scielo.org.co/pdf/pml/v7n2/v7n2a05.pdf

[9] N. Martínez Garrido, "Estudio de la problemática de los contaminantes emergentes en depuradoras de aguas residuales: tecnologías de tratamiento," pp. 1–43, 2021, [Online]. Available: https://repositorio.upct.es/bitstream/handle/10317/9923/tfm-mar-est.pdf?sequence=1&isAllowed=y

[10] J. S. Maan, T. vi H. Duong, and A. Saadabadi, *Carbamazepine*. 2023.

[11] A. Voloshenko-Rossin *et al.*, "Emerging pollutants in the Esmeraldas watershed in Ecuador: Discharge and attenuation of emerging organic pollutants along the San Pedro-Guayllabamba-Esmeraldas rivers," *Environmental Sciences: Processes and Impacts*, vol. 17, no. 1, pp. 41–53, 2015, doi: 10.1039/c4em00394b.

[12] F. T. Lange, M. Scheurer, and H. J. Brauch, "Artificial sweeteners-A recently recognized class of emerging environmental contaminants: A review," *Anal Bioanal Chem*, vol. 403, no. 9, pp. 2503–2518, 2012, doi: 10.1007/s00216-012-5892-z.

[13] E. M. Yahia, J. de J. Ornelas-Paz, J. K. Brecht, P. García-Solís, and M. E. Maldonado Celis, "The contribution of mango fruit (Mangifera indica L.) to human nutrition and health," *Arabian Journal of Chemistry*, vol. 16, no. 7, 2023, doi: 10.1016/j.arabjc.2023.104860.

[14] N. A. Hamid and J. J. You, "Mangosteen peel-derived hydrochar prepared via hydrothermal carbonization for methylene blue removal," *IOP Conf Ser Earth Environ Sci*, vol. 765, no. 1, 2021, doi: 10.1088/1755-1315/765/1/012114.

[15] T. da S. Agostini-Costa, "Bioactive compounds and health benefits of some palm species traditionally used in Africa and the Americas – A review," *J Ethnopharmacol*, vol. 224, no. May, pp. 202–229, 2018, doi: 10.1016/j.jep.2018.05.035.

[16] A. C. Landázuri-Rojas, J. S. Villarreal, E. R. Núñez, M. M. Pico, A. S. Lagos, and M. Caviedes, "Experimental evaluation of crushed Moringa oleifera Lam. seeds and powder waste during coagulation-flocculation processes," *J Environ Chem Eng*, vol. 6, no. 4, pp. 5443–5451, 2018, doi: 10.1016/j.jece.2018.08.021.

[17] G. D. Asensi, A. M. D. Villadiego, and G. R. Berruezo, "Moringa oleifera: Revisión sobre aplicaciones y usos en alimentos," *Arch Latinoam Nutr*, vol. 67, no. 2, pp. 86–97, 2017.

[18] E. Danso-Boateng, E. Nyktari, A. D. Wheatley, R. G. Holdich, and A. S. Mohammed, "Removal of Organic Pollutants from Effluent of Anaerobic Digester Using Hydrochars Produced from Faecal Simulant and Sewage Sludge," *Water Air Soil Pollut*, vol. 231, no. 5, 2020, doi: 10.1007/s11270-020-04557-3.

[19] M. Dahiru, Z. Uba Zango, and M. A. Haruna, "Cationic dyes removal using low-cost banana peel biosorbent," *Am. J. Mater. Sci.*, vol. 8, no. 2, pp. 32–38, 2018, [Online]. Available: https://doi.org/10.5923/j. materials.20180802.02.%0A

[20] L. Leng *et al.*, "A review on nitrogen transformation in hydrochar during hydrothermal carbonization of biomass containing nitrogen," *Science of the Total Environment*, vol. 756, p. 143679, 2021, doi: 10.1016/j.scitotenv.2020.143679.

[21] J. Gáfaro and G. Navarro, "Demostrar la Influencia de la Aplicación de Carbón Activado en Polvo (CAP) Sobre la Calidad del Agua Potable para la Remoción De GSM Y MIB en dos Condiciones de Operación (Con y Sin Ayudante de Floculación) de la Planta de Tratamiento de Agua Potable de B," 2019.

[22] L. Oviedo, V. Medina, and M. Beltan, "TRATAMIENTO DE AGUAS RESIDUALES EN LA INDUSTRIA DE ALIMENTOS EN COLOMBIA MEDIANTE LA ELECTROCOAGULACIÓN," 2023. [Online]. Available: https://www.ncbi.nlm.nih.gov/books/NBK558907/

231d Managing PFAS-Contaminated Adsorbents with a Life-Cycle Perspective

Margaret O'Connell, Petros Stamoulis, William Dichtel, Rosa Gwinn, Jennifer Dunn

Background: Adsorbents can remove PFAS from contaminated wastewater and drinking water. These PFAS-contaminated adsorbents must be dealt with in an environmentally responsible manner. They can be disposed of in landfills, but this poses leakage risks and requires manufacturing of increasing amounts of adsorbents. Incineration can release toxic compounds and is disallowed in many regions. Options that destroy PFAS and allow for adsorbent recycling include ball milling and low-temperature incineration. In our analysis, we evaluate different PFAS destruction and adsorbent recovery options in the context of full fuel cycle energy consumption, greenhouse gas emissions, and solid waste generation to identify the options with promising environmental performance.

Methods: Using the properties and preparation methods for biomass-derived adsorbents [Ching et al., 2020], we consider three routes for PFAS disposal or destruction and, when applicable, adsorbent recovery. These include landfilling, low-temperature incineration, and ball-milling. We generate material and energy balances for these three options and develop full fuel cycle energy consumption, emissions, and waste generation using background data and calculations in the Greenhouse gases, Regulated Emissions, and Energy use in Technologies (GREET) model [Argonne National Laboratory, 2023].

Results: Per m^3 of water treated, energy consumption and life-cycle greenhouse gases are nearly identical across all treatment options. This result arises because the share of energy or greenhouse gas emissions of the treatment method is less than 5% of overall impacts. For example, regardless of whether PFAS is landfilled (with adsorbents undergoing single use), or destroyed with ball milling or low-temperature incineration, life-cycle greenhouse gas emissions are approximately 0.2 kg CO_2e. Ongoing analysis is evaluating additional destruction techniques such as supercritical water oxidation, exploring different adsorbent performance, and investigating the dependence of type of treated water on life cycle assessment results.

Implications: Preliminary results indicate that treating PFAS rather than landfilling contaminated adsorbents saves large volumes of solid waste without an increase in greenhouse gas emissions. Every effort should be made to use adsorbents that can be regenerated while PFAS is destroyed to limit solid waste impacts of PFAS management.

Citations:

Argonne National Laboratory. Greenhouse gases, Regulated Emissions, and Energy use in Technologies (GREET) model, 2023 release. https://greet.anl.gov/

Ching, C., Klemes, M. J., Trang, B., Dichtel, W. R., Helbling, D. E. β-Cyclodextrin Polymers with Different Cross-Linkers and Ion-Exchange Resins Exhibit Variable Adsorption of Anionic, Zwitterionic, and Nonionic PFASs. *Environ. Sci. Technol.* 2020. 54: 12693-12702.

231e Breakthrough Extraction of Aqueous Perfluorooctanoic Acid Using Natural Deep Eutectic Solvents for Sustainable Water Remediation

Sana Eid, Tarek Lemaoui, Ahmad Darwish, Maguy Abi Jaoude, Fawzi Banat, Shadi Hasan, Enas Nashef

Addressing the urgent need for environmentally sustainable solutions in water remediation, particularly in tackling persistent emerging contaminants such as perfluorooctanoic acid (PFOA), this work reports for the first-time deep eutectic solvents (DESs) for the extraction of PFOA from aqueous medium. The design framework of the DESs was based on multiple metrics, including extraction performance, density, viscosity, environmental impact, and hydrophobicity. Through the multi-criteria evaluation of ten DESs, trioctylphosphine oxide and lauric acid (TOPO:LauA 1:1) emerged as the superior choice with an exceptional 99.7% extraction efficiency in a single stage only –far outperforming the toluene benchmark's 82.3%– and showing favorable characteristics across all other assessed criteria. Remarkably, this exceptional efficiency of TOPO:LauA was sustained >98% across diverse operational conditions, including a broad pH and temperature spectrum (3–9 pH; 15–100°C), solvent-to-feed ratios down to 1:7, concentrations as low as 0.1 ppm, rapid equilibration within 1 minute, and reusability throughout seven cycles without any observable degradation, as verified by spectroscopic analysis. TOPO:LauA also demonstrated exceptional selectivity for PFOA even when mixed with other per- and polyfluoroalkyl substances (PFAS) and under a complex synthetic wastewater environment. Moreover, computational quantum chemistry modeling was applied, providing insights into the extraction's molecular mechanism, and showing a strong agreement with the experimental findings. The development of this new DES for the extraction of PFAS signifies a critical breakthrough in water treatment, presenting an alternative that surpasses traditional solvents and conventional methods of PFAS removal, such as adsorption and membranes, in terms of efficacy, sustainability, and scalability.

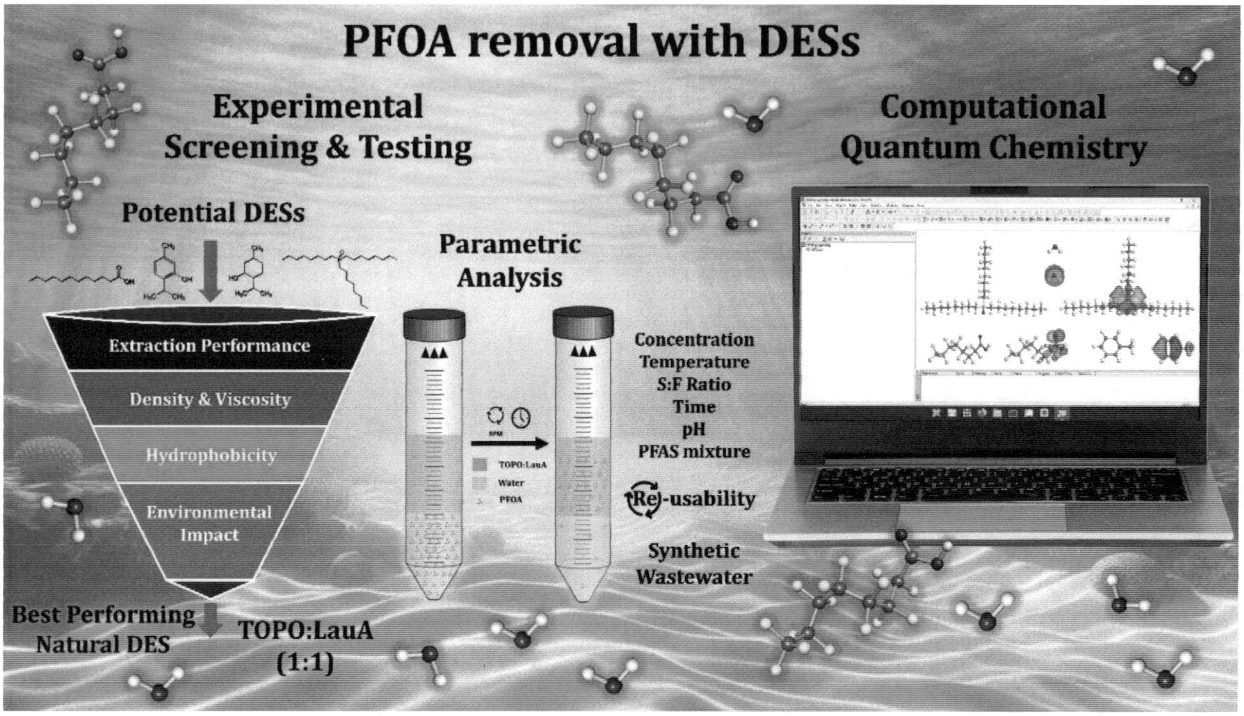

231f Understanding the Fate and Transport of Perfluorooctanesulfonic Acid (PFOS) and Microplastics in Aquatic Environment

Jasneet Pala, Gabriel Barbosa, Christoffer Turner, Milad Esfahani

Microplastics have emerged as a growing environmental concern, attracting considerable attention due to the complexity of their removal and their capacity to transport accompanying pollutants. Owing to their mobility, small size, and propensity to accumulate, microplastics pose toxicity risks to both terrestrial and aquatic organisms, as well as to humans, triggering inflammation and oxidative stress. Additionally, the presence of per- and polyfluoroalkyl substances (PFAS) in water can cause health effects on humans and animals over time. In this work, we delved into the intriguing interplay between these two emerging water contaminants at different conditions (temperature, contact time and pH). Our primary focus centered on unraveling the interaction dynamics between perfluorooctanesulphonic acid (PFOS) and prevalent microplastics, predominantly comprising polyethylene (PE) (20 μm), polypropylene (PP) (17 μm), and polyvinyl chloride (PVC) particles found in water. Our findings revealed PFOS adsorption onto PE, PP, and PVC, reaching peaks of approximately 9%, 12%, and 6% after 24 hours of contact and escalating to 14%, 20%, and 20% within 8 days, respectively. Following this, a noticeable trend of desorption was observed over the subsequent month for each type of microplastic, prompting further investigation through the characterization of PFOS-adsorbed microplastics. Moreover, as temperatures increased from 25^0 C to 45^0 C, a notable increase in PFOS adsorption occurred (approximately 41%, 52%, and 48% within 8 days in PE, PP, and PVC, respectively), indicative of an endothermic reaction during the adsorption process. Concurrently, molecular dynamics simulations were employed to model the adsorption mechanism of three different perfluorooctyl PFAS including 30 molecules on PE, PP, and PVC microplastic particles. The molecular simulation results suggest a pronounced dispersive interaction between the polymer and PFAS molecules for PE and PP; while the dispersive interaction dominates the PVC systems, as compared to PE and PP, a much stronger coulombic PFAS-PVC interaction is found.

231g Bioengineering Under Siege: The Unseen Threat of Co-Existing PFAS and Microplastics on Ecosystem Health

Mileisha Velazquez, Katherine Deliz-Quiñones

Aquatic ecosystem health, and consequently, human well-being, is compromised by contamination associated with anthropogenic activities. These pollutants are transferred to aquatic systems where they become bioavailable for organism uptake. Exposure to harmful chemicals such as the emergent contaminants per- and polyfluoroalkyl substances (PFAS) and microplastics (MPs) can negatively impact ecological functions of aquatic systems and pose harmful ecotoxicological effects to aquatic organisms. Both types of pollutants are introduced to the environment through runoff, product breakdown or wastewater discharge. These are more likely to co-exist during product manufacturing, thus polluting and interacting within aquatic systems. Combined effects of interacting pollutants have the potential to be additive and cause further harm to those exposed, which is pertinent for further research. Persistent and widespread, PFAS have high chemical and heat resistance and are hydrophobic with strong carbon-fluorine bonds, charged with hydrophilic functional groups. Similarly, MPs have a high molecular weight, strong carbon-carbon bonds, hydrophobic surface, and large surface area. While these properties are useful for product longevity and resistance, they promote bioaccumulation and physiological damage to exposed organisms, thus affecting their health and ecological roles. Examining the ecotoxicological effects of PFAS and MPs combined rather than separately provides deeper insights, revealing their combined impact on aquatic life and their co-occurrence in natural environments.

The co-existence of these pollutants and their combined effects on biofilms, an ecological engineering community, have not yet been studied. The ubiquitous biofilms are composed of diverse microbial communities, which contribute to nutrient and biogeochemical cycling, substrate stabilization, and are used for bioremediation and restoration effort strategies. Microbial community composition may be altered by the co-existence of these pollutants, and/or the mechanisms by which they provide ecological functions as ecosystem engineers. Further understanding of these complex interactions between PFAS, microplastics, and biofilms is essential to address co-contamination research gaps such as changes in ecological processes and human health risks and acquire novel knowledge regarding fate, transport, and behavior of associated contaminants, thus addressing co-contamination research gaps.

663a Sustainability Assessment in Additive Manufacturing End-of-Life Material Management Practices

John D. Chea, Gerardo Ruiz-Mercado, Raymond Smith, David E. Meyer, Michael Gonzalez, William Barrett

Additive manufacturing (AM) offers a variety of novel material manufacturing techniques for a wide range of applications across many industries, allowing for on-demand production of complex, customized materials. Most system engineering evaluation and improvement efforts for AM are centered at the manufacturing process stage. However, AM create unintentional environmental releases in end-of-life (EoL) pathways, affecting overall life cycle sustainability. Traditional sustainability assessment tools often fail to enable a comprehensive, risk-based approach to chemical safety decision-making. This research finds potential material allocation and occupational exposure in EoL management, then analyzed the overall sustainability of AM EoL material management practices using resource-efficiency, environmental, energy, and economic indicators from the GREENSCOPE (Gauging Reaction Effectiveness for the Environmental Sustainability of Chemistries with a multi-Objective Process Evaluator) tool. The trade-off analysis finds potential process modifications for material/energy source reduction, pollution prevention, risk mitigation, and overall sustainability improvements. The AM EoL-specific sustainability analysis serves as a resource to offer insights and empower policymakers and stakeholders to enhance pollution prevention strategies and optimize the AM EoL supply chain.

663b Using Laser Irradiation to Separate Polymer Adhesives from Silicon in End-of-Life PV Panels for Recycling Purpose

Touhid Bin Anwar, Christopher J. Bardeen

The active semiconductor component (Silicon) of a solar photovoltaic (PV) panel is encapsulated by an ethylvinylacetate (EVA) adhesive and protected with a top glass layer. The detachment of this glass-EVA layer from the silicon (Si) is a significant challenge in recycling end-of-life PV panels. To tackle this issue, a novel impulsive light-debonding technique was devised and tested on both model and commercial poly-crystalline PV panels. Nanosecond laser pulses can effectively induce transient melting at the EVA-Si interface, severing the adhesive bond. The debonding efficiency, influenced by laser pulse fluence (energy per area), wavelength, applied pressure, and scanning speed, was systematically evaluated. For model PV panels, the necessary single-pulse laser fluences for gravity-driven separation were found to be 0.2, 0.3, and 0.7 J/cm^2 at wavelengths of 355 nm, 532 nm, and 1064 nm, respectively. Shorter wavelengths concentrate heat near the silicon surface, minimizing the fluence required for debonding, whereas higher fluences offset increased scanning speeds. Post-irradiation analysis of the silicon surfaces, using optical and electron microscopy, revealed that the textured antireflection coating was eliminated, yet the silver metal grid remained unaffected. Initial tests with 532 nm laser pulses successfully detached the glass-EVA layer from sections of decommissioned commercial PV panels, despite extensive cracking in the top glass layer. The nanosecond laser debonding method provides a new method for disassembly of end-of-life solar panels that avoids the use of chemical solvents or heating, or mechanical grinding.

663c Electrochemical Recovery of Rare Earth Elements from Neodymium-Iron-Boron Based Magnets Coupled with Iron Recovery

Alamgir M Haque, Christian E Alvarez-Pugliese, Gerardine Botte

Rare earth elements (REE) have widespread applications in many commodities including wind turbines, electric vehicles, batteries, cell phone displays, etc. Hence, the market demand for REEs is expected to increase in the future. However, the shortage of natural deposits and environmental risks associated with mining create concerns for the sustainable supply of REEs. A prospective secondary source of these elements could be recycling REE-containing products once these reach their end of life (EOL). Neodymium-iron-boron (NdFeB) permanent magnets could be a prominent example of such a product. These magnets have 27-32% rare earth elements (REEs) (Nd, Pr, Dy), 67-73% iron, 1% boron, and other minor metals. These magnets have high energy density and are widely used in many applications, especially in wind turbines. Due to the increasing global wind power capacity, a large stockpile of EOL magnets is predicted to be available in the near future.

Researchers have developed several processes including the hydrometallurgical and pyrometallurgical route for recycling NdFeB magnets. However, recovering iron, the major component of these magnets, poses a significant challenge. The conventional hydrometallurgical and pyrometallurgical recovery processes extract REEs leaving behind a solid waste normally as Fe(III) hydroxide/ Fe(III) oxide hydroxides, or as jarosite.

The Institute for Sustainability & Circular Economy at Texas Tech University has developed an electrochemical recycling route for NdFeB magnets. A novel method for REE recovery via acid leaching coupled with the electrowinning of iron will be presented. The following aspects will be included: the REE extraction process, the effect of iron speciation on REE recovery, and iron electrowinning as a means of metallic iron recovery from acid waste.

663d Polymer-Degrading Potential of a Diverse Set of Environmentally Ubiquitous Microorganisms Using High-Throughput Biodegradation Testing

Alex Zappi, Katharina Fransen, Gabrielle Godbille-Cardona, Natalie Mamrol, Kristala Prather, Bradley Olsen

With the growing accumulation of plastic waste, great focus has been directed towards the development of polymers with short lifespans in the natural environment. One solution is the use of biodegradable polymers, functionally-useful structures with the ability to be enzymatically-broken down by microorganisms. To determine the effectiveness of these new polymers, a biodegradation test must be performed; however, these tests often take many months and are often unstandardized. Recently, a high-throughput method to determine biodegradability was developed to address this need using clear-zone assays, but this method uses a single species, *P. lemoignei*. To better reflect natural conditions and encompass the environmental enzymatic scope, this work expanded the method to new species. Initially, ununiform colony morphology made this expansion difficult, but it was found that colony shape can be controlled using media composition and richness. An exploration of culturing conditions and media compatibility was required, leading to the detailed screening of a set of minimal media on each species. In some cases, limited carbon availability encouraged polymeric breakdown, emphasizing the effect of culturing conditions on depolymerase expression. Using these new conditions, this method was used to determine the unexplored degrading capability of 20 polymer-degrading species on 13 commercial polymers. These results emphasize the difference in degrading-capability of key degraders while emphasizing the importance of microbial diversity in polymer biodegradation. Trends between polymer structure and specific microbial species are clearly evident with certain species favoring specific functional groups over others, and particular polymeric structures and classes are shown to be generally favored for degradation.

663e Carbon Sequestration Potential of Stabilized Hydrochar from Cattle Manure

Cordel Bever, Chuck Coronella

Hydrothermal carbonization (HTC) is a thermochemical treatment process with applications to upgrade or convert diverse biomass feedstocks to a carbonaceous solid product. Recent discoveries in soil science and climate mitigation suggest that processes such as HTC may also provide a means to produce carbonized materials for long-term carbon sequestration in soil. Carbonized organic matter produced by fires has been found to be extremely resilient to natural decomposition and can help improve soil health and fertility under some conditions. These materials, often referred to as pyrogenic carbon, can act as a means for long-term carbon sequestration by first allowing plants to pull carbon from the air via photosynthesis, then stabilizing the plant carbon as char and storing it in soil. Thus engineers are seeking scalable methods to produce similarly stable soil amendments from organic matter for carbon sequestration. Pyrolysis, a traditional process used to make charcoal, is one such method. Pyrolysis has been demonstrated to create pyrogenic carbon, termed biochar, for use in soil. Biochar has been produced by pyrolysis of diverse feedstocks, including wood, corn stover, food waste, municipal solid waste, biosolids, and manure. In pyrolysis, the biomass is slowly heated in an inert gaseous environment to temperatures between approximately 400 and 600 °C. When applied to biomass with high moisture content, a significant amount of energy is consumed by first evaporating water to dry the biomass before the temperature reaches reaction temperature. Thus, pyrolysis is inefficient when applied to wet biomass feedstocks. On the other hand, HTC is performed in an environment of liquid water, obviating the requirement for drying.

HTC excels at converting wet feedstocks into hydrochar, a carbonaceous material with some chemical and physical similarity to biochar but is somewhat less stable in soil. When placed in soils, hydrochar has been found to release much of its carbon to the atmosphere. Thus, we seek to produce a pyrogenic carbon by way of HTC but with stability of the biochar produced by pyrolysis. The research presented here aims to address the concerns surrounding hydrochar's limited soil stability and proposes the use of secondary processing to further stabilize hydrochar and wet agricultural waste for sequestering carbon.

We use HTC as the primary process step for thermochemical conversion of cow manure into hydrochar, and evaluate three secondary treatments for the potential to produce a stable char. Two oxidation processes were evaluated, including treatment in weak and strong chemical oxidizers (hydrogen peroxide and potassium dichromate, respectively.) The goal of these treatments is to oxidize the relatively labile carbonaceous materials, leaving behind only relatively inert material which might resist rapid decomposition in soils. Treatment with hydrogen peroxide was found to decrease the hydrochar's inherent soil stability and treatment with potassium dichromate provided a minimal increase to its carbon sequestration potential.

We also evaluated pyrolysis for the potential to produce a stable carbon. The hydrochar produced by HTC is substantially hydrophobic and does not suffer the inefficiency of moist biomass when subjected to pyrolysis. The biochar produced from pyrolysis of hydrochar exhibited substantial improvements in markers for increased soil stability, including reduced oxygen and hydrogen content. In fact, the biochar produced from pyrolysis of hydrochar exhibits a higher carbon content as well as a higher carbon yield relative to the biochar produced from direct pyrolysis of manure. soil sequestration properties over biochar derived from pyrolysis of cattle manure. We offer several suggestions of how this two-step process can be utilized at scale for a significant impact of carbon sequestration.

663g End-of-Life Options for Materials in Wind Turbine Blades

Caroline Cameron

The wind industry provided 17.8% of US electricity in 2022 and installations are expected to double in coming years, which increases both material demand and the subsequent waste stream as wind turbines reach their end-of-life (EoL). The glass fiber-reinforced polymer composites (FRPC) that comprise approximately 70% of a wind turbine blade's mass are difficult to separate into their component parts for recycling. Yet, these materials could be recovered and recycled to keep them in use and out of the landfill, thus increasing their value and longevity. In this work, we analyze the increased circularity in the material economy when glass recovery is implemented as compared to current practices of landfill and cement production. Reprocessing waste wind turbine blades into a secondary raw material comes at a cost of increased energy utilization, global warming impact, and fine particulate matter formation compared to landfill; however, utilization of the waste materials also creates benefits of reduced human and eco toxicities, land use, and use of mineral and fossil resources. This work represents an integrated framework for assessment and trade-off analysis of EoL pathways as research and infrastructure efforts continue to expand the recycling industry.

430a Sustainable Pathways to Acrylic Acid - the Impact of Biomass Source on the Catalytic Dehydration of Lactic Acid to Acrylic Acid

Christopher Nicholas

Acrylic acid and the four primary acrylate derivatives are the cornerstone of a $10B market with broad application across superabsorbent polymers, paints, coatings, and adhesives industries. Producing these chemicals necessary for modern life from sustainable pathways is of key interest. We focus on the dehydration of feeds derived from lactic acid, a product sustainably fermented from a range of bio-derived sugar sources.

Utilizing an inert and site-selective chemical titrant to form an inorganic/ organic composite can provide desirable selectivity tuning. Here, a selectivity enhancement from 60 to upwards of 96 C% was demonstrated for methyl lactate dehydration to methyl acrylate and acrylic acid over a composite Na-FAU based zeolite catalyst using amines as titrants to suppress side reactions on in situ-generated Bronsted acid sites.

How does the selectivity and catalytic process affect the sustainability of bio-based chemicals production? For transition of petrochemicals to biochemicals, two necessary hurdles must be met: 1) cash cost of production by the bio-based method should be close to that of the petrochemical and 2) sustainability, on a life-cycle basis, should be significantly better than that of the petrochemical. Here we present results on life cycle analysis of the lactic acid to acrylic acid dehydration and show how the biomass sugar source impacts the life cycle assessment.

Our results indicate, on a cradle to gate of the factory basis, that acrylic acid production from sugarcane is a typically net carbon sequestration and that beets and corn, two widely used sugar sources, offer significant improvement versus today's oil derived process for acrylic acid production.

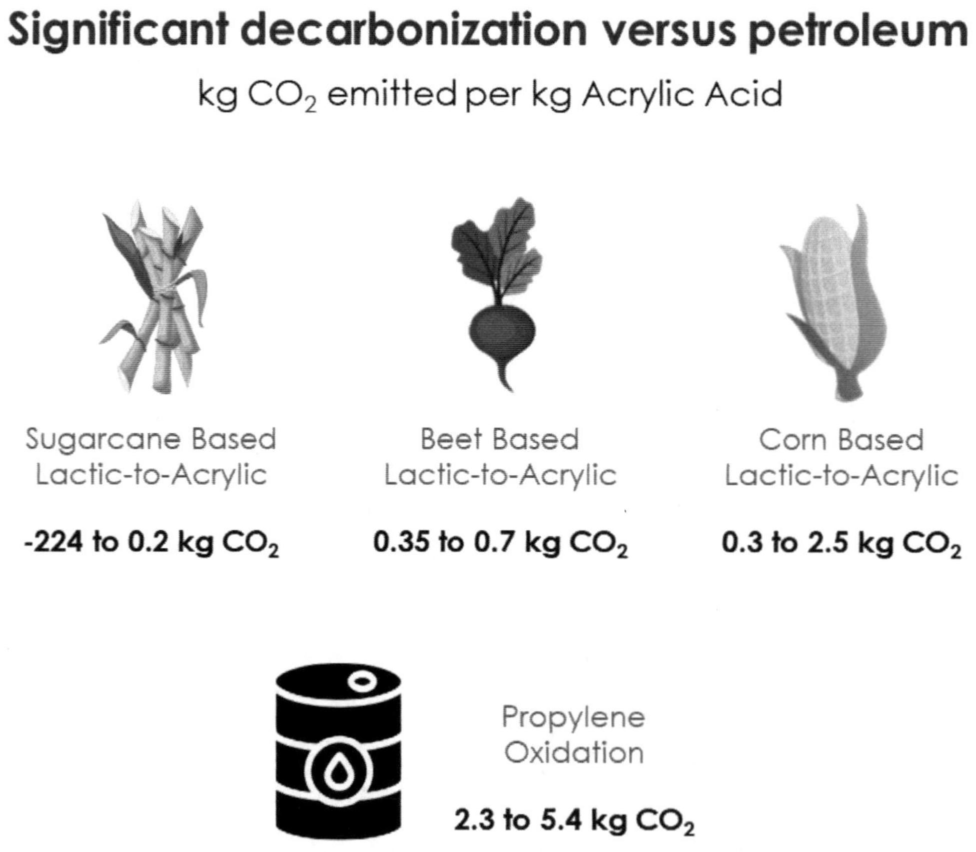

430b Directed Mineralization of Waste Carbon to Develop Negative Emission Cements

David Kenney, Andrew Charlebois, Nima Rahbar, Shuai Wang, Michael Timko, Andrew R Teixeira

Direct air capture (DAC) and point source capture (PSC) are current state-of-the-art efforts to close the industrial carbon balance. These processes fight against thermodynamics, as they capture dilute atmospheric carbon dioxide (400 ppm) for it to be concentrated and upcycled to consumer chemicals.

Alternatively, the United States produces 292 million tons of municipal solid waste (MSW) with nearly 80% organic biomass. Traditional waste management produces nearly 8% of the national anthropogenic emissions. This study aims to leverage the organic fraction of MSW as a pre-concentrated source of carbon and optimize reaction thermodynamics and kinetics to develop a competitive net-negative alternative cement. This net energy-positive, carbon-negative process relies on hydrothermal mineralization (HTM), a patent-pending process that converts wet organic waste into thermally stable calcite (CaCO3). First, the use of supercritical water oxidation (SCWO) is a high-temperature, high-pressure process that allows for the combustion of organic carbon under supercritical conditions (374°C, 22.4 MPa). Waste has a significant energy content (24-40 MJ/kg) and can operate as a source of bioenergy. In some cases, energy can be exported from the process. The second oxidative step drives the high-pressure CO_2 produced from the SCWO and mineralizes calcium hydroxide ($Ca(OH)_2$) to the most stable calcite ($CaCO_3$) form.

Ordinary Portland Cement (OPC) is the current industrial standard for cementitious materials. It relies on the mining and calcination of CaCO3 to form calcium oxide (CaO), which is a key component of the primary strength source. This work then looks at the capability to integrate HTM with trailblazing work to develop a competitive alternative cementitious material (10 MPa) that acts as a calcite sink and relies on polymer binding for strength, instead of traditional cement that uses $CaCO_3$ as a feed. This work was modeled in ASPEN/HYSYS and GREET, a lifecycle analysis software, and found that in offsetting traditional cement production, this pathway offers a substantial net-negative alternative to two of the largest global polluters.

430c CO_2 Capturing Technology through Lignin-Based Polymers

Hoyong Chung

Climate change, driven by excessive CO2 emissions, poses a pressing threat to global ecosystems and human well-being. CO2 capture technology has emerged as a promising approach to mitigate this challenge, enabling the capturing of CO2 from ambient air for storage and conversion into valuable products. In this presentation, we report an innovative and cost-effective polymer for CO2 capture and utilization for the first time, using abundant and non-food-based biomass lignin. The modified ionic polymer synthesized through the reaction of ion containing reagent with lignin under alkaline conditions. Subsequently, the modified lignin-based polymer is employed for CO2 capture from both direct air and concentrated CO2 sources at room temperature and atmospheric pressure. Structural characterization of the polymers is accomplished through 1H, 13C, and 2D-HSQC NMR, and FT-IR spectroscopy. The CO2 capture process is established through the formation of specific functional ions alongside the presence of CO2. The captured CO2 is precisely quantified by using inverse-gated proton decoupled 13C NMR with an internal standard. Remarkably, the amount of captured-CO2 of the ionic lignin polymer per one gram are 1.06 mmol (47 mg) from concentrated CO2 source and 0.60 mmol (26 mg) of CO2 from ambient air exposure. Significantly, the captured CO2 in the new polymer is utilized to synthesize cyclic carbonate as an example. The cyclization reaction in a unique reaction setup, where the CO2-captured lignin-based polymer as an active CO2 source. The reaction was successfully conducted by supplied CO2 from the new lignin-based polymer. Moreover, the fully controlled recycling of ionic lignin polymer achieved via repeated CO2 release ↔ CO2 capture. Hence, this novel methodology offers a promising pathway towards sustainable, recyclable CO2 capture and utilization and represents a significant step forward in the quest for effective carbon capture technologies.

430d Large-Scale Decarbonization through Synergetic Utilization of Forest Biomass Resources

Yuan Yao

Forest biomass is one of the most abundant biomass resources used by modern industries. Forests also offer many nature-based carbon dioxide removal (CDR) solutions, such as afforestation/reforestation and converting woody biomass to biochar or other long-lived wood products. Previous studies have used life cycle assessment (LCA) to quantify the process-level decarbonization potential of individual forest biomass utilization pathways. However, there is increasing interest in understanding the large-scale, synergetic effects of those nature-based solutions related to forest systems. Furthermore, LCA often does not account for the variability in environmental conditions and forest management approaches over time and space, affecting the design and operability of engineering systems utilizing biomass as feedstock. A comprehensive understanding of the carbon dynamics associated with the interlinked forest-industrial systems across land and product life cycles at various scales is essential for system-level optimization to combat climate change.

This study aims to address the methodological and knowledge gaps by developing a multi-scale life cycle modeling framework that merges machine-learning-assisted, process-based LCA with dynamic forest ecosystem modeling. This approach facilitates the analysis of carbon fluxes in forests and associated wood product lifecycles with spatial and temporal dynamics. This framework is demonstrated by a case study in the southeastern United States. Our analysis explores how diverse engineering wood product selections, forest management practices, temporal considerations, and geographical variability influence the overall carbon balances of the system. Our results suggest that the effectiveness of combined CDR strategies is significantly influenced by time and location. This presentation will further delve into the details of the conditions where integrating various forest-based CDR methods can effectively mitigate greenhouse gas emissions while advancing a resilient bioeconomy. Finally, based on the results, the presentation will discuss policy implications and regional CDR opportunities.

430e Multiscale Analysis for the Substitution of Natural Gas Boilers Using Biomass

Roberto Cifuentes García, Guillermo Galán Iglesias, Mariano Martin

The progress of humankind has been tightly bound to the intense use of fossil fuels, resulting in large greenhouse gas emissions mainly composed of CO_2 whose levels are upwards of 36.8 Gt CO_2/year (IEA, 2022). Over the last decades, natural gas was positioned as the main fossil fuel to obtain energy in the combined cycle, cogeneration plants as well as in domestic heating systems. Nonetheless, the interruption of Europe's natural gas supply from Russia via the Nord Stream pipeline due to the Ukrainian War has resulted in substantial fluctuations in global energy prices, which leads to the need to develop alternative strategies towards energy security using local resources to obtain energy and chemicals. This fact, along with the transition to renewable and clean energy sources places biomass as one of the most promising alternatives for the future to guarantee energy security. It could be used through anaerobic digestion to obtain biomethane yet this process to reduce energy dependence is not enough to satisfy the demand (Martín-Hernández et al., 2020; Taifouris and Martín, 2018). In addition, the European Union agreed to phase-out boilers powered by fossil fuels in a gradual manner, whose installation will not be allowed as of January 1st 2025 (European Commission, 2023). Concerning this scenario, the exploitation of lignocellulosic material can be carried out from X-to-power via generating electricity burning the biomass waste to supply heat pumps or power-to-X-to-power through pellet production.

The problem is approached as a two-level analysis bound by a multiscale techno-economic study. First, the process level includes the optimization of the performance of pellet factories, previously evaluated in the reference work (Cifuentes et al., 2024), and TPS, which includes a Rankine cycle and a cooling tower that is modelled are based on previous work (Guerras and Martín, 2020). Input-output models are developed to predict their yield and economics (CAPEX and OPEX). At the second level, the strategy decision-making one, an extended facility location problem (FLP) is formulated with the purpose of determining the optimal combination of the pair of technologies, satisfy the demand, as well as the spots to install the facilities to process the biomass through a supply chain, performed having regard to economic, social, and environmental factors. The structure of the model consists of five different layers, corresponding to biomass availability, biomass transformation factories to produce pellets, thermal power stations (TPS), electric knots, and customers. Before being supplied to the customers, it is first distributed to the electric knots located in transformer substations, and from there, electricity arrives at population centers. The pellets manufactured, as the objective is energetic, have to be transformed into energy with their experimental heating value outlined (Yancey et al., 2013). In order to validate the methodology, Spanish peninsular soil is selected which is divided into spatial units between the province and the municipality called agricultural shires (MAPAMA, 2024).

To meet the demand for heating within Spain, the optimal outcome is the installation of 18 TPS and 3 pellet factories based on corn stover and eucalyptus wood, with the consumption of 12.2% of the total lignocellulosic waste. The sum to be invested is up to 2,200 M€, where up to 13% corresponds to pelletization and the remaining 87% to electricity generation. Another 2 case studies were performed as the budget overpasses the Spanish Governmental fund for energy, 1,150 M€ (Gobierno de España, 2022), as the aggregate of the 15 regional ones (From 40 to 450 M€, being 1,600 M€ the total). Using the State budget, 2 TPS and 6 pellet plants are the results, however, by limiting the investment per region the obtained facilities are 10 TPS added to 7 pellet plants.

Acknowledgments

The authors would like to acknowledge Salamanca Research for the optimization licenses, Europa Union's Horizon research and innovation program under the Maria Skłodowska-Curie grant agreement no. 778168 (IProPBio), the funding received from the Junta de Castilla y León and the European Commission under the program Contratación de Personal Técnico de Apoyo a la investigación por las Universidades Públicas de Castilla y León and the FPU PhD fellowship from the Spanish Ministry of Universities.

References

Cifuentes-García, R., Galán, G., Martín, M., 2024. Multiscale analysis for the valorization of biomass via pellets production towards energy security. J. Clean. Prod., in press.

European Commission, 2023. New rules to boost energy performance of buildings [WWW Document]. URL https://ec.europa.eu/commission/presscorner/detail/en/IP_23_6423 (Accessed 3.21.24).

Gobierno de España, 2022. BOE-A-2022-22128 Ley 31/2022, de 23 de diciembre, de Presupuestos Generales del Estado para el año 2023. [WWW Document]. URL https://www.boe.es/eli/es/l/2022/12/23/31 (Accessed 3.21.24).

Guerras, L.S., Martín, M., 2020. On the water footprint in power production: Sustainable design of wet cooling towers. Appl. Energy 263, 114620. https://doi.org/10.1016/j.apenergy.2020.114620

IEA, 2022. CO2 Emissions in 2022 [WWW Document]. URL https://www.iea.org/reports/co2-emissions-in-2022 (Accessed 3.21.24).

MAPAMA, 2024. Comarcas Agrarias [WWW Document]. URL https://www.mapa.gob.es/es/cartografia-y-sig/ide/descargas/agricultura/ (Accessed 3.21.24).

Martín-Hernández, E., Guerras, L.S., Martín, M., 2020. Optimal technology selection for the biogas upgrading to biomethane. J. Clean. Prod. 267, 122032. https://doi.org/10.1016/J.JCLEPRO.2020.122032

Taifouris, M.R., Martín, M., 2018. Multiscale scheme for the optimal use of residues for the production of biogas across Castile and Leon. J. Clean. Prod. 185, 239–251. https://doi.org/https://doi.org/10.1016/j.jclepro.2018.03.018

Yancey, N., Tumuluru, J.S., Wright, C., 2013. Drying, Grinding and Pelletization Studies on Raw and Formulated Biomass Feedstock's for Bioenergy Applications. J. Biobased Mater. Bioenergy 7, 549–558. https://doi.org/10.1166/jbmb.2013.1390

430f Effect of Zeolite on Hydrothermal Carbonization of Loblolly Pine and Its Impact on Carbon Capture

Swarna Saha, Toufiq Reza

In recent times, the release of CO_2 has become a contentious topic due to its link to global warming and other natural calamities. The idea of producing an environmentally friendly adsorbent for enhanced carbon capture by introducing high porosity and functionalities through incorporating catalytic impact of zeolite from loblolly pine (LP) was explored in this study. The research work aims to perform a systematic hydrothermal carbonization (HTC) of LP at different operational temperatures varying from 200°C to 260°C using 0 to 10 wt.% of zeolites. The produced hydrochars after HTC were further superactivated using KOH in a ratio of 2:1 at 700°C to 900°C to synthesize ultraporous superactivated zeolite composites (SAZs). The SAZs were evaluated for their elemental composition, surface functionalities, thermogravimetric behavior, BET surface area, and scanning electron microscopy morphology to understand the catalytic effect of zeolites during HTC. The catalytic effect of zeolite has improved the carbonization and resulted in thermally stable SAZs. Compared to control runs without any zeolite, the SAZs showed less aromatic and thermally stable carbon but higher surface area and pore volume. The ultraporous superactivated zeolite composites. The enhanced CO_2 capture efficiency was a result of the influence of increased surface porosity, which improved the physical adsorption mechanisms.

Biobased Films Using Ecuadorian Agro-industrial Residues - In the Framework of a Circular Bioeconomy, Circular Engineering, and Sustainability

Briggitte León Intriago[1], Lourdes M. Orejuela-Escobar[1,2,3,4,7*], Ana Gabriela Andrade Sosa[1], Inés Hernández Guerra[1,5], Arleth Gualle Brito[1], Andrea C. Landázuri [1,2,3,4], Alfredo Valarezo[4], Edgar Javier Guerrón-Orejuela[6,7]

[1] *Applied Circular Engineering Group, Department of Chemical Engineering, College of Science and Engineering, Universidad San Francisco de Quito USFQ. Quito, Ecuador*

[2] *Institute of Biomedicine Research, USFQ, Ecuador*

[3] *Institute of Biological and Environmental Research BIÓSFERA, USFQ, Ecuador*

[4] *Institute for Energy and Materials, Mechanical Engineering Department, Colegio de Ciencias e Ingenierías, Universidad San Francisco de Quito, (USFQ), Quito, Ecuador*

[5] *Universidad de Carabobo, UC. Valencia, Venezuela*

[6] *University of South Florida, Tampa, FL.USA*

[7] *Geocircular Consulting Group, Tampa, FL. USA*

*Corresponding author email: lorejuela@usfq.edu.ec

Abstract

The valorization of forestry, agricultural, agro-industrial, and municipal bioresidues and by-products aligns with the United Nations Sustainable Development Goals, driving the emergence of Green Chemistry, Circular Bioeconomy (CBE), and Circular Engineering (CEng) approaches. These novel solutions favor the application of CBE in Chemical Engineering to replace non-renewable materials and fossil-based products with sustainable use of renewable bioresources. Tons of bioresidues and by-products from forest, agriculture and food industries are generated annually worldwide, and their improper disposal negatively affects the environment. Efforts have made to add value to this residual biomass by reincorporated them into production chains as important feedstocks of bioproducts and biopolymers. Proper agro-industrial residues management plays a significant role in achieving several SDGs. In Ecuador, agro-industrial waste exceeds 2 million tons per year, providing a considerable raw material source for generating new bio-based products and materials through green technologies. In this study, we produced four biofilm prototypes using representative Ecuadorian agro-industrial residues such as avocado seeds (*Persea americana*, Fuerte variety), green plantain peels (*Musa paradisiaca*), mango cotyledons (*Mangifera indica* L.), cocoa pod husks (*Theobroma cacao* L.), and vegetable ivory (*Phytelephas aequatorialis*) meal residues; and characterized them through TGA, XRD, tensile strength and biodegradation techniques. The results showed that these bioresidues could be used to generate biobased films. From the resulting prototypes, the avocado seed film turned out to be the most resistant and with the best physical properties, representing a starting point for future studies and applications. Challenges need to be overcome to improve the mechanical properties of these materials, to be suitable for industrial applications.

Bioresidues as a sustainable source of high value-added bio-based products and bio-based materials to achieve Sustainable Development Goals

In the need for more sustainable production systems in accordance with the Sustainable Development Goals of United Nations, Green Chemistry, Circular Bioeconomy (CBE) and Circular Engineering (CEng), which favors the application of CBE in Chemical Engineering, come into the scene with novel and viable solutions [1], [2]. In this context, industrial processing seeks the replacement of non-renewable materials and fossil-based products with a sustainable use of renewable bioresources [3], [4] thus reducing the emission of greenhouse gases and waste. In fact, millions of tons of forestry and bioresidues from food and agriculture industries are generated every year worldwide [5]. After being separated from the value chain and not properly disposed, these wastes affect negatively the environment (increased temperatures, greenhouse gases, pollutants, etc.) [6]. In recent years, efforts have been strengthened to improve the management of agro-industrial waste, adding value to them; with the possibility of reincorporating them into the production chain since many of them are an important source of bioproducts and biopolymers of industrial and commercial interest [4]. Additionally, agro-industrial waste management plays a crucial role in achieving some United Nations Sustainable Development Goals (SDGs)[7]. Goal 11 (Sustainable cities and Communities) aims improving municipal waste management to reduce the adverse environmental impacts of cities. Goal 12 (Responsible Consumption and Production) targets to substantially reduce waste generation through prevention, reduction, recycling, and reuse by 2030, as well as achieve environmentally sound management of chemicals and wastes throughout their life cycle. Furthermore, proper waste management contributes to Goal 3 (Good Health and Well/being); Goal 6 (Clean Water and Sanitation); Goal 7 (Affordable and Clean Energy); Goal 8 (Decent Work and Economic Growth); Goal 13 (Climate Action); Goal 14 (Life below Water); and, Goal 15 (Life on Land). As these SDGs pursue pollution reduction, recycling increase, renewable energy generation, biofuels and biomaterials production, jobs creation, and climate change impacts mitigation.

In Ecuador, agro-industrial waste can exceed 2 million tons per year, many of which are a source of starch and lignocellulose [8], [9]. This indicates that there is a considerable amount of raw material in this country that applying green technologies can generate new bio-based products and bio-based materials. On the other hand, the high consumption of petroleum-based plastics has resulted in the massification of its production[10]. Many of these plastics are single-use, and their inadequate disposal constitutes a threat to existing ecosystems as they disintegrate very slowly[11] [12], [13]. Therefore, it has been imperative to find ecological alternatives that meet the current demand for single-use plastic. The use of polymers from agro-industrial residues and by-products is an alternative to the use of polymers derived from petroleum, which has allowed the production of biofilms intended for conventional consumption [14] [15].

Biofilms from representative Ecuadorian agro-bioresidues

The objective of this study was to explore the potential of the direct use of residual biomass, from Ecuadorian agroindustries, for the production of biobased films. Bioresidues including avocado (*Persea americana*, Fuerte variety) seeds, green plantain (*Musa paradisiaca*) peels, mango (*Mangifera indica* L.) cotyledons, cocoa (*Theobroma cacao* L.) pod husks, and vegetable ivory (*Phytelephas aequatorialis*) meal residues. Given the chemical composition of the bioresidues; avocado seed, mango cotyledon and green plantain peel were considered as sources of starch. Alternatively, cocoa pod husk was considered a source of cellulose polysaccharide and vegetable ivory meal residue as a source of mannan polysaccharide (hemicellulose). Before being used, the bioresidues samples were subjected to a Soxhlet extraction with 70% ethanol to obtaining extractive-free residual biomass using a modified NREL standard procedure[16], subsequently dried and ground until achieving particles of approximately size 63 μM.

Four prototypes of biobased films were formulated and produced, and then were analyzed for their thermal stability, crystallinity, tensile strength, and biodegradation rate. Each biobased film consisted of 90% starch biomass, 5% cellulose biomass and 5% mannan biomass. For the mixed prototype, the starch sources were divided into three equal parts (30% avocado seed; 30% mango cotyledon; 30% green plantain peel). The thermal stability was determined through a Thermal Gravimetric Analysis (TGA); and the crystallinity index by X-ray Diffraction (XRD) following the procedures of Orejuela et al. 2023 [17]; the tensile strength was preformed through a universal testing machine; biodegradation was determined by the difference in mass of the films in a controlled environment with soil [18].

Films of 0.36-0.41mm thickness, flexible in texture, and opaque were produced (Figure 1A). From the TGA results, a very similar behavior was observed among all the generated curves, with 100-200°C being the range in which these prototypes have adequate thermal stability to be used (Figure 1B). Regarding crystallinity, an absorption peak between 2θ = 15-28° was observed for all prototypes (Figure 1C), which indicates the presence of starch, of which an amorphous and semicrystalline structure has been reported [19]. The avocado seed film presented greater crystallinity in relation to the others. This difference could be related to the amylose content of the avocado seed [20]. Regarding tensile strength, variable measurements are observed between the prototypes (Figure 1D), with the avocado film being the one with the greatest resistance (Figure 1C). Finally, from the biodegradation test it was determined that after 28 days, the mass loss was above 50% for all cases (avocado seed prototype=54.71%; mango cotyledon prototype=77.98%; green plantain peel prototype=52.3%; mixed prototype=59.16%), being the mango cotyledon film the one with the highest rate (Figure 1E). Both the physical and the chemical composition of biopolymer-based materials affect their biodegradability, as well as factors such as humidity, pH, oxygen content, nutrient availability, and others [21].

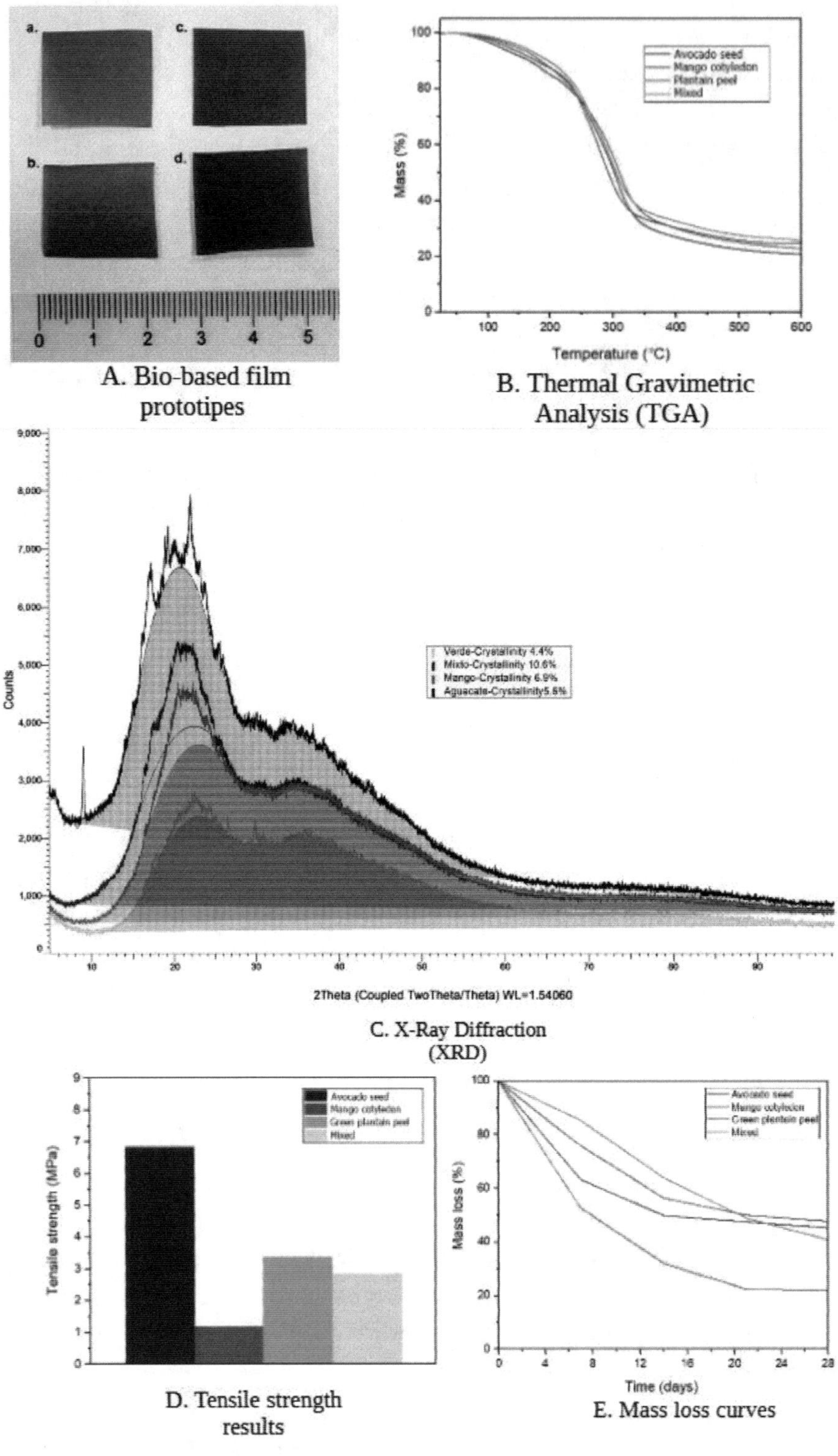

Figure 1. Bioplastics prototypes from agro.residues: A. Bio-based film prototypes; B. TGA curves of bio-based prototypes; C. XRD diffractograms of bio-based prototypes; D. Tensile strength results of bio-based prototypes; E. Biodegradation results of bio-based prototypes.

Conclusions

Through this study, it is determined that it is possible to use residual biomass to produce biobased films without prior isolation of its polymeric components. However, challenges arise to improve the mechanical properties of these materials, to be suitable for their intended application. From the rcsulting prototypes, the avocado seed film turned out to be the most resistant and with the best physical properties, representing a starting point for future studies and applications. Hence, there is much more to explore and take advantage of agro-industrial waste in Ecuador, which opens the route to new opportunities to create products for everyday consumption, in a more sustainable way.

The utilization of agro-industrial wastes for bio-based film manufacture aligns with the principles of CBE and CEng. It promotes the valorization of waste streams, reduces the dependence on fossil-based resources, and contributes to waste reduction and environmental sustainability. Moreover, these bio-based films can be designed to be biodegradable and/or compostable, further supporting the CE and reducing plastic pollution. In summary, the development of bio-based films using Ecuadorian agro-industrial residues and by-products represents a promising approach within the framework of CBE, CEng, and sustainability. It enables the valorization of waste resources, promotes the transition towards renewable and biodegradable materials, and contributes to environmental protection and sustainable development.

References

[1] A. Ncube, S. Mtetwa, M. Bukhari, G. Fiorentino, and R. Passaro, "Circular Economy and Green Chemistry: The Need for Radical Innovative Approaches in the Design for New Products," *Energies (Basel)*, vol. 16, no. 4, Feb. 2023, doi: 10.3390/en16041752.

[2] F. Ramonet, "Circular bioeconomy applications in chemical engineering," Technische Universität Wien, Austria , 2023.

[3] M. Voss, C. Valle, E. Calcio Gaudino, S. Tabasso, C. Forte, and G. Cravotto, "Unlocking the Potential of Agrifood Waste for Sustainable Innovation in Agriculture," *Recycling*, vol. 9, no. 2, p. 25, Mar. 2024, doi: 10.3390/recycling9020025.

[4] L. M. Orejuela-Escobar, A. C. Landázuri, and B. Goodell, "Second generation biorefining in Ecuador: Circular bioeconomy, zero waste technology, environment and sustainable development: The nexus," *Journal of Bioresources and Bioproducts*, vol. 6, no. 2, pp. 83–107, May 2021, doi: 10.1016/j.jobab.2021.01.004.

[5] F. C. Lopes and R. Ligabue-Braun, "Agro-Industrial Residues: Eco-Friendly and Inexpensive Substrates for Microbial Pigments Production," *Front Sustain Food Syst*, vol. 5, Mar. 2021, doi: 10.3389/fsufs.2021.589414.

[6] P. R. Yaashikaa, P. Senthil Kumar, and S. Varjani, "Valorization of agro-industrial wastes for biorefinery process and circular bioeconomy: A critical review," *Bioresour Technol*, vol. 343, p. 126126, Jan. 2022, doi: 10.1016/j.biortech.2021.126126.

[7] Department of Economic and Social Affairs, "U. Nations, "The 17 Goals. The 2030 Agenda for Sustainable Development."

[8] F. Torres, F. Guerrero, C. Varas, and T. Solis, "View of Circular economy vision for local development focused on waste derived from two agricultural crops in Ecuador," *Journal of Business and Entrepreneurial Studies*, vol. 6, 2022.

[9] S. Aguiar, M. Enríquez Estrella, and H. Uvidia Cabadiana, "Residuos agroindustriales: su impacto, manejo y aprovechamiento," *AXIOMA*, vol. 1, no. 27, pp. 5–11, Dec. 2022, doi: 10.26621/ra.v1i27.803.

[10] "Plastic Consumption: Understanding Its Global Impact and Challenges," Better Future Factory .

[11] S. Dey, G. T. N. Veerendra, P. S. S. A. Babu, A. V. P. Manoj, and K. Nagarjuna, "Degradation of Plastics Waste and Its Effects on Biological Ecosystems: A Scientific Analysis and Comprehensive Review," *Biomedical Materials & Devices*, May 2023, doi: 10.1007/s44174-023-00085-w.

[12] L. O. Palma Gallardo and J. Alba Flores, "Do all Plastics Damage the Environment?," *Front Young Minds*, vol. 10, Sep. 2022, doi: 10.3389/frym.2022.722532.

[13] P. Li, X. Wang, M. Su, X. Zou, L. Duan, and H. Zhang, "Characteristics of Plastic Pollution in the Environment: A Review," *Bull Environ Contam Toxicol*, vol. 107, no. 4, pp. 577–584, Oct. 2021, doi: 10.1007/s00128-020-02820-1.

[14] M. E. Orqueda *et al.*, "Feasibility of active biobased films produced using red chilto wastes to improve the protection of fresh salmon fillets via a circular economy approach," *Food Hydrocoll*, vol. 133, p. 107888, Dec. 2022, doi: 10.1016/j.foodhyd.2022.107888.

[15] J. X. Chan, J. F. Wong, A. Hassan, and Z. Zakaria, "Bioplastics from agricultural waste," in *Biopolymers and Biocomposites from Agro-Waste for Packaging Applications*, Elsevier, 2021, pp. 141–169. doi: 10.1016/B978-0-12-819953-4.00005-7.

[16] A. Sluiter *et al.*, "Determination of structural carbohydrates and lignin in biomass, in: Laboratory Analytical Procedure (LAP)," *National Renewable Energy Laboratory*, Jun. 2008.

[17] L. F. Zambrano-Mite *et al.*, "Optimization of microfibrillated cellulose isolation from cocoa pod husk via mild oxalic acid hydrolysis: A response surface methodology approach," *Heliyon*, vol. 9, no. 6, p. e17258, Jun. 2023, doi: 10.1016/j.heliyon.2023.e17258.

[18] Md. R. Amin, M. A. Chowdhury, and Md. A. Kowser, "Characterization and performance analysis of composite bioplastics synthesized using titanium dioxide nanoparticles with corn starch," *Heliyon*, vol. 5, no. 8, p. e02009, Aug. 2019, doi: 10.1016/j.heliyon.2019.e02009.

[19] O. S. Jangong, P. L. Gareso, I. Mutmainna, and D. Tahir, "Fabrication and characterization starch/chitosan reinforced polypropylene as biodegradable," *J Phys Conf Ser*, vol. 1341, no. 8, p. 082022, Oct. 2019, doi: 10.1088/1742-6596/1341/8/082022.

[20] E. Indarti, Nurlaila, M. Muzaifa, S. Noviasari, Z. F. Rozali, and E. M. Yusup, "Characteristics of avocado (Persea americana) and kluwih (Artocarpus camansi) seeds starch with different extraction methods," *IOP Conf Ser Earth Environ Sci*, vol. 951, no. 1, p. 012095, Jan. 2022, doi: 10.1088/1755-1315/951/1/012095.

[21] A. Folino, A. Karageorgiou, P. S. Calabrò, and D. Komilis, "Biodegradation of wasted bioplastics in natural and industrial environments: A review," *Sustainability (Switzerland)*, vol. 12, no. 15. MDPI, Aug. 01, 2020. doi: 10.3390/su12156030.

97a Techno-Economic Analysis of Liquid Absorbent-Based Carbon Capture for Achieving Net-Zero Natural Gas Power Generation

Nicholas Cross, Mengyao Yuan, Wenqin Li

To effectively meet climate goals for decarbonizing the electric grid, urgent action is required to mitigate CO2 emissions from fossil fuel power plants while transitioning to a sustainable electricity future. Natural gas currently serves as a major power source for the US electric grid, with numerous additional natural gas power plants slated for construction in the near future. Integrating carbon capture and storage (CCS) processes into existing natural gas-fired power plants is recognized as a pivotal strategy for decarbonizing these facilities, therefore aligning with climate objectives while ensuring reliable electricity supply.
Historically, CCS technology has aimed to capture 90% of the CO2 from power plant flue gases (i.e., 90% capture efficiency). Recent advancements indicate that liquid absorbent-based technologies can achieve capture rates as high as 95% at reasonable costs (e.g., <5% increase relative to 90% capture).[1] However, to meet the ambitious US target of carbon-free electricity by 2035, further exploration into the technical and economic feasibility of >95% CCS is imperative.
A new generation of carbon capture liquid absorbents are under development to target capture efficiencies exceeding 95% while minimizing energy consumption and cost increases. This study presents a technoeconomic analysis of a >95% carbon capture process utilizing validated experimental data obtained from a pilot-scale carbon capture system employing an innovative liquid absorbent. Pilot experimental work has demonstrated that this liquid absorbent can achieve over 99% capture efficiency with minimal reboiler duty increase. The analysis aims to predict capture performance and economics at commercial-scale natural gas power plants. Comparative assessments between the novel and traditional liquid absorbents, such as monoethanolamine, targeting a similar range of capture efficiencies are conducted to evaluate performance and costs.
Additionally, recognizing the upstream emissions associated with the leakage of natural gas during its extraction, transmission, and distribution, this study explores the feasibility and costs of achieving net-zero electricity production from natural gas power plants. To inform the comprehensive assessment of emissions reduction strategies, state-of-the-art measurements of methane emissions from US oil and gas systems are incorporated, providing crucial insights into the total emissions reduction required for achieving net zero emission electricity from natural gas power generation.

(1) Du, Y.; Gao, T. Y.; Rochelle, G. T.; Bhown, A. S. Zero- and negative-emissions fossil-fired power plants using CO2 capture by conventional aqueous amines. Int J Greenh Gas Control 2021, 111. DOI: 10.1016/j.ijggc.2021.103473. Schmitt, T.; Leptinsky, S.; Turner, M.; Zoelle, A.; Woods, M.; Shultz, T.; James, R. Fossil Energy Baseline Revision 4a; National Energy Technology Laboratory, 2022.

This work was performed under the auspices of the U.S. Department of Energy by Lawrence Livermore National Laboratory under Contract DE-AC52-07NA27344. LLNL Release number LLNL-ABS-862506.

97b Comparative and Predictive Analyses of Different Adsorption Processes for Carbon Capture from Cement Plants Flue Gas Streams

Abdullah Al Moinee, Fateme Rezaei

Cement and steel industries are among those sectors that are hard to decarbonize in the near future. As such, more efforts should be dedicated to mitigating CO_2 emissions from these industries. Specifically considering cement power plant, their flue gas streams are usually at elevated temperature (>100 ºC), which makes the capture process challenging and complex. This study explores modeling and optimization of different adsorption processes such as pressure swing adsorption (PSA), temperature swing adsorption (TSA), and electric swing adsorption (ESA) tailored to the unique requirements of CO_2 capture and utilization offering precise insights into key performance indicators (KPIs) crucial for cement production and decarbonization. The process systems engineering (PSE) approach involved optimizing KPIs such as purity, recovery, productivity, and energy consumption with a keen focus on maximizing the production of hybrid supplementary cementitious materials (SCM), while minimizing environmental footprints. The effects of operational key variables on the KPIs and overall capture performance were systematically analyzed for each adsorption scenario. Through rigorous conceptual design, simulation, and empirical validation, we conducted sensitivity analyses and optimized KPIs, with the Kiln's cooler end emerging as a promising CO_2 capture source. Assuming a precooling step, the room temperature PSA process acheived purity levels exceeding 90%, surpassing ESA and TSA (85-90%), with a targeted recovery rate potentially exceeding 95% (>90%), providing higher productivity rates of around 80-85%. Targeting energy consumption of 0.8-1.2 kWh/kg_{CO2} captured, the actual capture rate varied around 1.5 kWh/kg_{CO2}. By bridging performance projections with process engineering solutions, this research contributed to the advancement of decarbonization initiatives in the cement industry, with iterative testing, predictive modeling, and a robust framework, driving towards a sustainable future.

97c Deploying an *in Situ* Raman Methodology for Online, Live Solvent Composition Analysis for Process Monitoring of CO_2 Capture

Dushyant Barpaga, Abhishek Kumar, Po Ki Tse, Amanda Lines, Samuel Bryan, David Heldebrant

Solvent-based carbon capture continues to represent a key pathway to meet climate change mitigation goals. For both, the continued adoption of these capture processes in industrial process and for development work by researchers, the optimization and testing of this process would be significantly aided by integrating *in situ* solvent monitoring capability.[1] Real-time analysis for feedback on varying water, carbon dioxide, and dissolved gas contaminants offers the operator the potential to respond more quickly in maintaining a balance of steady operation. It also allows insight into solvent integrity and life cycle analysis. Using unique data acquisition techniques targeted for this purpose, Raman spectra of a single-component water-lean solvent, N-(2-ethoxyethyl)-3-morpholinopropan-1-amine (EEMPA) were collected at known and controlled CO_2, H_2O, SO_2, and NO concentrations via a custom-built vapor-liquid equilibrium cell coupled with an *in situ* Raman cell. The quantification of CO_2, H_2O, SO_2, and NO loadings in EEMPA was performed by principal component regression and partial least squares methods with analysis of uncertainties. An algorithm was developed using this training set data that calculates and outputs the 5-component composition of any unknown EEMPA-based solvent sample from the capture process. This model has been deployed and demonstrated for use in an industrial scale capture process at the National Carbon Capture Center in Alabama, US. We describe and showcase the development of this capability and the results from its deployment in realistic conditions.

[1] Lines et al. Anal. Chem. 2023, 95, 15566-15576. DOI: 10.1021/acs.analchem.3c02281

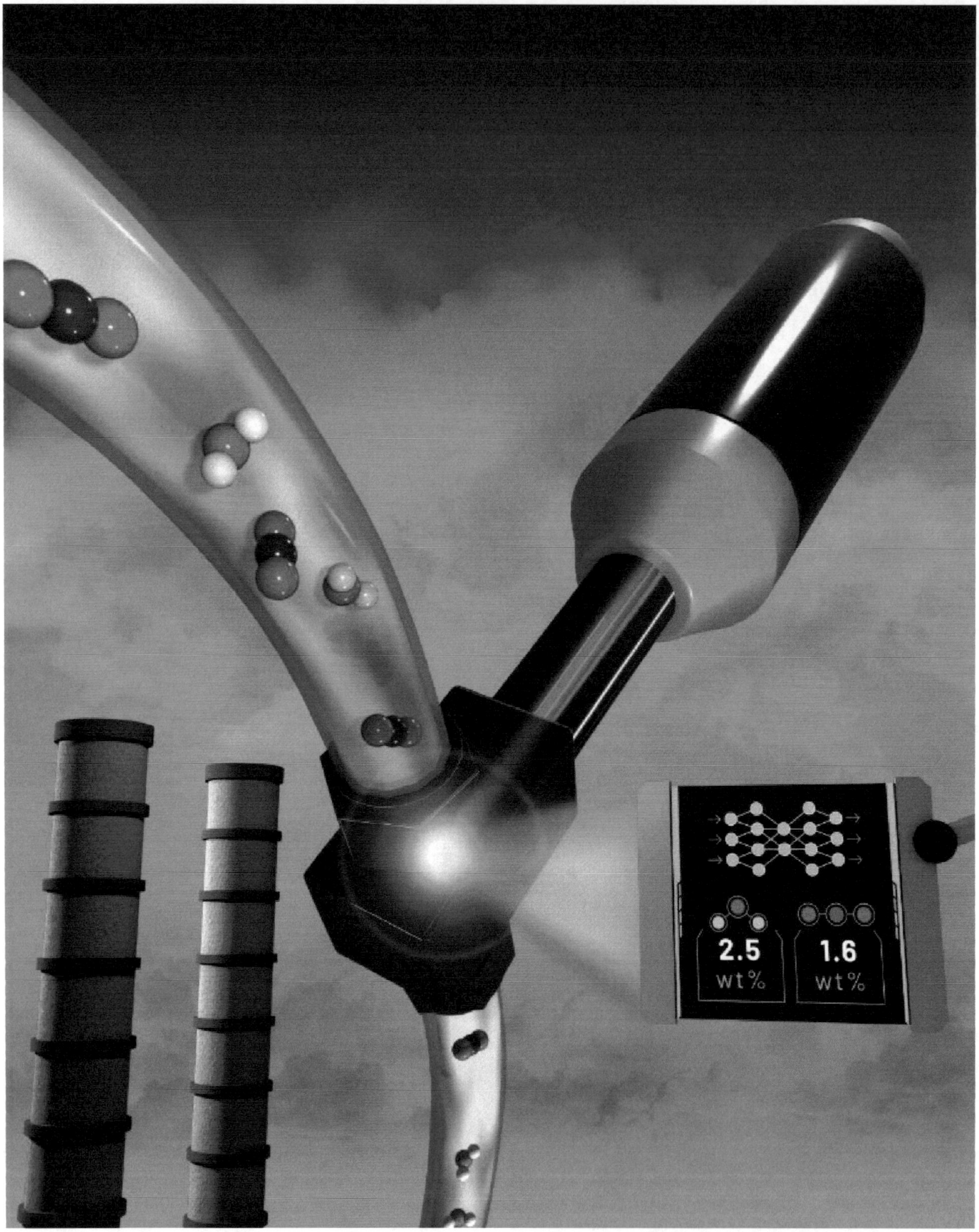
2.5
wt%
1.6
wt%

97d Pilot Testing of Frostcc™, a Novel Cryogenic Carbon-Capture Technology, at the National Carbon Capture Center

Jonathan Stickel, Jeffrey Young, Widad Yossri, Michael McNelis, Miles Abarr

FrostCC™ is a patented and transformational carbon-capture technology that separates CO_2 from industrial or power-plant flue gas using a cryogenic process. It consists of a highly recuperated Brayton refrigeration cycle, employing the flue gas as an auto-refrigerant to generate all required cooling to separate the CO_2 from the flue gas. This process entails a series of steps, including compression, cooling, and expansion of flue gas, leading to the desublimation of CO_2 on heat-exchanger (HX) surfaces. Periodic cycling of these HXs induces the transformation of the collected solid CO_2 to a liquid state, allowing for underground storage of the captured CO_2 or conversion into valuable products. FrostCC can capture up to 99% of the CO_2 from a wide range of industrial and power sources, offering a cost-effective approach to significantly curb greenhouse gas emissions and combat climate change.

Carbon America has built and operated a pilot plant with FrostCC technology at the National Carbon Capture Center (NCCC), which is the topic of this presentation. Turbomachinery components, specifically a scroll compressor and radial expander, along with several shell-and-tube heat exchangers, were used to provide compression, cold recuperation, and expansion to provide sufficient cooling to frost and capture >95% of the CO_2 from the flue gas. Integrated operation of all the components of FrostCC were demonstrated for the first time. The FrostCC pilot system captured CO_2 from flue gas generated by a natural-gas boiler. Additional accomplishments were developing robust control techniques, exploring the capture efficiency of FrostCC for a range of operating conditions, and validating physics-based computational models. Technical achievements and key lessons learned from pilot-scale operations at NCCC will be presented along with next steps for FrostCC development and commercial deployment.

97e Thin-Film Composite Membranes Based on Poly(1,3 dioxolane) and Uio-66-NH_2 for CO_2/N_2 Separation

Gengyi Zhang, Narjes Esmaeili, Fathy Mohammed Eldesoky Attia, Haiqing Lin

Membrane technology has been pursued in gas separations, such as post-combustion carbon capture, and a key challenge is to fabricate thin film composite (TFC) membranes containing a defect-free selective layer (< 500 nm) with state-of-the-art manufacturing processes. For example, poly(1,3 dioxolane) (PDXLA) has been synthesized via photo-polymerization, which presents the CO_2 permeability of 200 Barrer and CO_2/N_2 selectivity of 60. The highly crosslinked structure makes it hard to be prepared as TFC membranes. Herein, we successfully synthesized high molecular weight PDXLA via reversible addition–fragmentation chain-transfer (RAFT) polymerization. The water-soluble PDXLA remains amorphous with a macular weight of 200 kDa. We successfully coated PDXLA selective layer (as thin as 140 nm) on the gutter layer of polydopamine (PDA) modified polydimethylsiloxane (*d*PDMS). The TFC membranes exhibited CO_2 permeance of 700 GPU and CO_2/N_2 selective of 35 at room temperature. Given the mixed matrix materials as a promising avenue for achieving the dual objective of high selective and permeance, the metal-organic frameworks (UiO-66-NH_2) were incorporated into the PDXLA selective layer. The smooth mixed matrix selective layer benefited from hydrogen bonding between amine in UiO-66-NH_2 and PDXLA. The effect of UiO-66-NH_2 size (25 to 160 nm) on the separation performance of PDXLA membranes was also investigated. For example, adding 5 mass% of UiO-66-NH_2 (25 nm) increases CO_2 permeance from 700 to 880 GPU and CO_2/N_2 selectivity from 35 to 70, in contrast, adding 5 mass% of UiO-66-NH_2 (160 nm) increases CO_2 permeance from 700 to 890 GPU and CO_2/N_2 selectivity from 35 to 52. This showcases the smaller particles with strong interaction potential benefits of fabricating high-performance TFC membranes.

97f Development of All-Polymer Sorbents for High Capacity CO_2 Capture

Ankana Roy, Hannah Holmes, Lisa S. Baugh, David C. Calabro, Johannes Leisen, Yi Ren, Simon C. Weston, Ryan Lively, M.G. Finn

Natural gas combined cycle (NGCC) power plants are more energy efficient than their coal counterparts, and NGCC can be combined with CO_2 capture to further reduce emissions. CO_2 capture materials tailored for NGCC flue gas will be necessary for the advancement and widespread utilization of the technology. Adsorbents are often used in structured contactors to increase productivity and reduce pressure drop compared to traditional packed beds. However, structured contactors typically contain inactive binders or polymeric materials in addition to the adsorbent, which can limit the amount of active material in the contactor. The development of polymers with high CO_2 affinities eliminates the need for a binder by enabling contactors composed entirely of the active material.

One promising class of materials for this purpose is polymers of intrinsic microporosity (PIMs) due to their high surface area, permanent interconnected microporosity and ability to fine tune through chemical and morphological modifications. However, PIM-1 can only weakly bind CO_2 through physisorption in the micropores, leading to low uptake in dilute CO_2 feeds. To resolve this, we chemically modified the PIM-1 backbone with guanidine, a Brønsted base, to create a novel polymeric CO_2 adsorbent.

Here, we present the synthesis and characterization of PIM-guanidine. The adsorbent was evaluated extensively for CO_2 capture from dry and humid streams, including isobar, breakthrough, and cycling experiments. Solid state (13CP MAS) NMR spectroscopy after $^{13}CO_2$ adsorption was used to further probe the sorption mechanism and distinguish between physisorbed and chemisorbed CO_2. The results reveal guanidine's role in the CO_2 adsorption mechanism, the advantages of functionalizing PIM-1, and the effect of humidity on PIM-guanidine's performance. Finally, the powder is transformed into polymeric contactors composed completely of the active material. Overall, these results advance our understanding of amine-CO_2 chemistry and demonstrate PIM-guanidine's potential for scale-up and industrial application in NGCC processes.

97g CFD Modeling of High-Flux Plate-and-Frame Membrane Modules for Carbon Capture at Industrial and Power Generation Plants.

Cheick Dosso, Hector Pedrozo, Thien Tren, Lingxiang Zhu, Victor A. Kusuma, David Hopkinson, Lorenz Biegler, Grigorios Panagakos

Industrial emissions of carbon dioxide (CO_2) constitute a significant portion of global emissions. To address the urgent need to reduce atmospheric CO_2 levels and achieve net-zero emissions by 2050, carbon capture technologies (CCT) have emerged as promising solutions. Among these, membrane processes have gained traction as energy-efficient methods for CO_2 capture over the past two decades.

This study investigates the application of CO_2-selective flat sheet membranes for capturing CO_2 emissions from point sources. Utilizing Computational Fluid Dynamics (CFD) models, we design high-flux plate-and-frame membrane modules to ensure uniform flow distribution among membrane elements, minimize dead-end zones, and alleviate concentration polarization issues commonly encountered in gas separation membranes. Our objective is to drive advancements in membrane technology by optimizing module designs based on specific membrane properties and operating conditions.

Our modeling approach encompasses Multiphysics processes, incorporating fluid flow and transport of concentrated species within the membrane module. Previously, we showed that our numerical results demonstrate excellent agreement between experimental findings and model predictions regarding CO_2 recovery, CO_2 mole fraction in the retentate and permeate streams, and stage-cut.

Recently, we successfully scaled up our design to a larger, stacked version featuring additional membranes and a greater surface area, capable of handling higher flow rates. While our earlier results were based on a single 24 cm^2 membrane operating at a maximum flow rate of 600 sccm, our current models accommodate up to 5 membranes for a total surface area of 120 cm^2 with flow rates reaching 3200 sccm. These results were validated for the new scaled up design with experimental data establishing further the reliability of our models.

Moreover, whereas previous work assumed the gas feed stream was comprised solely of CO_2 and N_2 for simplicity, our latest models, currently undergoing validation, incorporate CO_2, N_2, and O_2. These models can accommodate up to 10 membranes, totaling a surface area of 960 cm^2. The integration of computational fluid dynamics (CFD) with experimental data enhances our understanding of the process, facilitating the development of efficient CO_2 separation through membrane systems both from the view point of optimizing the module design and the overall process.

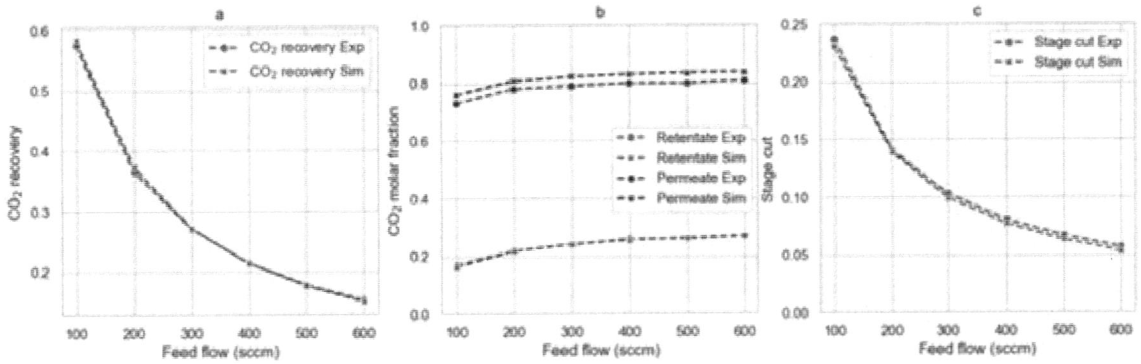

CO_2 concentration (mol/m³) profiles

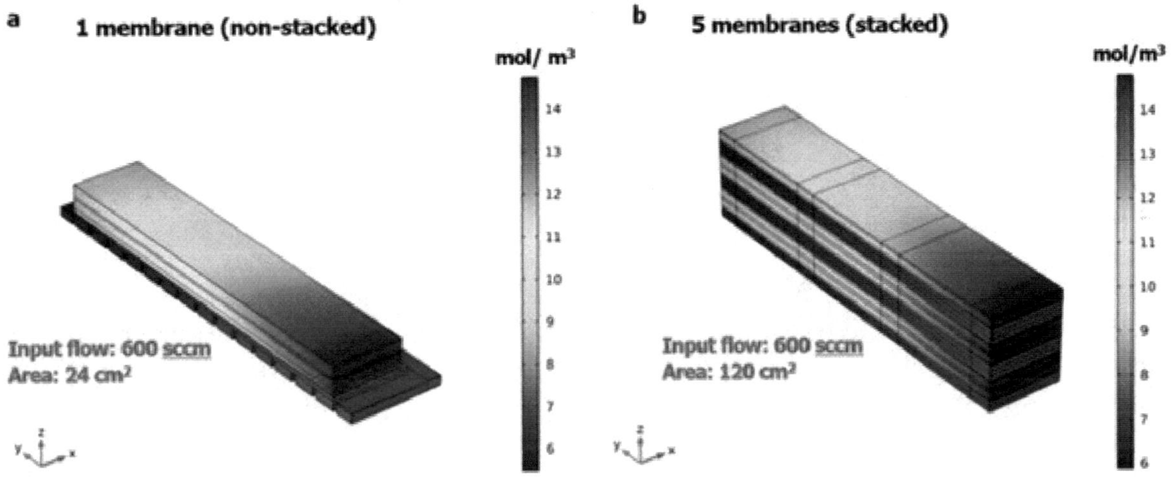

97h Corrosion Analysis of Non-Aqueous Solvent for Carbon Capture

Katy Jessop, Paul Mobley, Jak Tanthana, Vijay Gupta, Marty Lail

RTI International is developing a novel non-aqueous solvent (NAS) based process for post-combustion CO_2 capture. The process has the potential of substantially reducing the thermal energy required for solvent regeneration compared to the NETL baseline cases.[1] RTI has already shown the reduction of thermal energy using NAS in a Bench-scale Gas Absorption System (BsGAS) at RTI with simulated flue gas, a pilot-scale unit at SINTEF's Tiller facility in Trondheim, Norway with coal-derived flue gas, and testing at the National Carbon Capture Center (NCCC) in Wilsonville, Alabama with coal-derived flue gas. Further NAS testing was completed at Technology Centre Mongstad's flexible amine-based capture plant (TCM DA) in Mongstad, Norway during March-September 2022 with 2752 hours of time on stream. This plant can treat up to 60,000 Sm^3/hr flue gas supplied from a combined heat and power (CHP) plant and residual fluidized catalytic cracker (RFCC).

The capital cost of CO_2 capture plants is one of the largest drivers in the cost of CO_2 captured. As a result, efforts to reduce capital costs are one of the most effective ways to minimize the cost of CO_2 capture technologies. Most capture systems are currently built with stainless steel due to the corrosive nature of MEA and other aqueous solvents, but cheaper alternatives including carbon steel or plastic liners may be suitable for solvents like NAS. In addition to testing the performance of NAS under various operating conditions at TCM DA, corrosion due to the use of NAS was also monitored to identify suitable materials of construction (MOC) for future capture plants and to generate an accurate techno-economic analysis (TEA). Corrosion coupons made of a variety of materials (carbon steel, stainless steel, elastomer, resin) were inserted in the piping downstream of the absorber and reboiler and in the regenerator overhead condenser vapor stream. At the end of testing, the metal coupons were analyzed for corrosion rates, pitting, and stress cracking; elastomer and resin coupons were analyzed for changes in material hardness and mass. Solvent samples were also collected to determine dissolved metal, heat stable salt (HSS), water content, and conductance.

NAS (0.01 μm/yr) exhibited virtually no corrosion for CS 1018 under absorber conditions compared to MEA (4170 μm/yr)[2] due to the lower water content of the solvent. Reported literature values for aqueous solvent corrosion rates and NAS corrosion results for previous testing campaigns (NCCC and SINTEF) showed corrosion rate can be well predicted by a power law correlation with the conductivity of the solvent. The corrosion rate for stainless steel was about one order of magnitude lower than for the carbon steel. However, this correlation greatly overpredicted the NAS corrosion rates observed at TCM.

Solvent was routinely sampled at various points in the system during steady state operation to track system operation and detect any changes in solvent composition. Samples were analyzed for amine, water, and CO_2 content from both the absorber and regenerator side of the system. In addition, the concentrations of metals in the samples were analyzed using inductively coupled plasma mass spectrometry (ICP-MS) for metal content. Chromium (Cr), iron (Fe), and nickel (Ni) are present in the TCM DA MOC and are not present in the flue gas in significant quantities, therefore the presence of these metal ions indicates corrosion in the process equipment. In a comparable TCM DA test campaign with MEA, concentrations of Fe, Cr, and Ni ions were an order of magnitude higher than concentrations measured in NAS.[3] For instance, a maximum of ~17 ppm Fe was seen in the MEA campaign, whereas ~1.1 ppm Fe was observed in the NAS campaign.

Fischer et al. (2017) suggested that a positive linear correlation exists between the rate of corrosion and the rate of formate ion production.[4] Formate concentration was monitored throughout the TCM DA campaign but remained below detectable levels. Comparatively, studies of lean and rich 7 m MDEA (45.5 wt%) showed formate production of >400 ppm and >150 ppm respectively after 14 days.[5] NAS' ability to inhibit production of formate and other HSS may be a key contributor to reduced corrosivity.

The low corrosion rate of NAS indicates that CO_2 capture plants could use low-cost MOC, lowering capital costs significantly and reducing the cost of CO_2 capture. A TEA will also be included to show the reduction in cost of capture from the use of cheaper materials.

1. Schmitt, T.; Leptinsky, S.; Turner, M.; Zoelle, A.; White, C. W.; Hughes, S.; Homsy, S.; Woods, M.; Hoffman, H.; Shultz, T.; James III, R. E. *Cost and Performance Baseline for Fossil Energy Plants Volume 1: Bituminous Coal and Natural Gas to Electricity*; National Energy Technology Laboratory (NETL): United States, 2022.

2. Gunasekaran, P.; Veawab, A.; Aroonwilas, A., Corrosivity of Single and Blended Amines in CO2 Capture Process. *Energy Procedia* **2013,** *37*, 2094-2099.

3. Hjelmaas, S.; Storheim, E.; Flø, N. E.; Thorjussen, E. S.; Morken, A. K.; Faramarzi, L.; de Cazenove, T.; Hamborg, E. S., Results from MEA Amine Plant Corrosion Processes at the CO2 Technology Centre Mongstad. *Energy Procedia* **2017,** *114*, 1166-1178.

4. Fischer, K. B.; Daga, A.; Hatchell, D.; Rochelle, G. T., MEA and Piperazine Corrosion of Carbon Steel and Stainless Steel. *Energy Procedia* **2017,** *114*, 1751-1764.

5. Closmann, F.; Nguyen, T.; Rochelle, G. T., MDEA/Piperazine as a solvent for CO2 capture. *Energy Procedia* **2009,** *1* (1), 1351-1357.

499a Particles As Heat Transfer Medium to Drive Endothermal Chemical Reactions By Solar Energy

Martin Roeb

Particles as Heat Transfer Medium to drive endothermal chemical reactions by solar energy

Keywords: particles, receiver, solar thermal, redox reaction, catalyst

Martin Roeb, Juan Pablo Rincon, Clarisse Lorreyte, Dimitrios Dimitrakis, Christos Agrafiotis, Dennis Thomey, Christian Sattler

High-temperature solar receivers with particles as a heat transfer medium have the potential to increase the operating temperature far above 900 °C and improve efficiency in comparison to technologies operating with molten salt at significantly lower temperature [1]. In addition to using the high-temperature heat for electricity generation, the heat can also be used directly for several industrial applications. One area of those applications is solar-chemical processes. In the particle receiver technology solid particles made from e.g. sintered bauxite, calcined flint clay, or quartz sand are heated directly by concentrated sunlight and act as a heat transfer medium. Various designs of particle receivers, e.g. free-falling particles, with braked particle flow, centrifugal receivers and fluidized-bed particle receivers are currently under development.

In order to further pursue the advantages of particle technology in concentrating solar technology, fundamental questions on material resistance, heat transfer processes, possible receiver alternatives and the use of chemically active particles have been scientifically investigated. In addition to the bauxite particles used to date as an inert material and as pure heat transfer medium, chemically active particles can also be used in solar receivers. Concentrating solar technology can be used not only in electricity generation, but also via generation of fuels and commodities in the industrial and transportation sectors. Chemically active means, for example, that they can be used as a catalyst material in gas-solid reactions or they can be reduced by means of a reduction-oxidation cycle at higher temperatures using solar radiation and oxidized at lower temperatures. As the reduction step is endothermic and the oxidation step is exothermic, they can be used for thermochemical storage. Other important fields of application are air separation, use as an oxygen pump, water splitting to produce hydrogen and CO2 splitting to produce CO.

Particle temperatures of up to 1500 °C are sometimes required for the chemical reactions described. This is well above the previous operating temperature range of the CentRec receiver [2]. Due to the operating temperature and the chemical reactions, different receiver-reactor concepts or a combination of several components such as receiver, heat exchanger and chemical reactor are used in these processes. Particles can be used both in receivers, where they directly absorb solar radiation, and as storage materials. In some cases, even the particles themselves represent the material to be treated and converted to the desired product like in the case of solar lime or cement production [3, 4].

A number of receivers and reactor concepts for thermochemical applications are described and analyzed with respect to their operation performance and scaling potential. A distinction is made depending on the application and its operating conditions. Different receiver concepts are compared. The advantages and disadvantages of those concepts are compared using an evaluation matrix. In addition to the technical aspects, the criteria of cost, scalability and sustainability are also considered.

In a first example the investigation of a novel thermochemical cycle in which the solid sulfur cycle is combined with DLR´s next generation particle receiver. The particles are heated in the centrifugal particle receiver CentRec with the help of solar irradiation and the hot particles are fed into the sulfuric acid splitting reactor. The heat energy of the particles is used to evaporate sulfuric acid to SO3 and finally decomposes into SO2 with the help of a catalyst (e.g. Fe2O3) at a temperature above 800 °C [5]. A lab scale particle heated reactor test setup for sulfuric acid splitting was built and tested as a proof of concept. The lab scale reactor is a new concept of counter-current flow shell-and-tube heat exchanger with particles and sulfuric acid mass flow rate of 10 and 1.8 kg/h, respectively. The particles which are electrically heated up to 900 °C acts as heat transfer medium and move downward on the shell side while the sulfuric acid flows upward inside acid resistant tubes. A one-dimensional heat transfer model was developed based on the correlations of flow boiling heat transfer coefficients, particle bed heat transfer coefficient and sizing of the shell-and-tube heat exchanger.

Another application example deals with an innovative approach of biomass conversion. The thermal energy derived either from particle heat carriers heated up in a solar receiver [6] or through electrical energy from renewable energy sources. The target products of the thermochemical process are bio-oil, biochar and syngas. The thermal energy required for the pyrolysis process is provided by the particles, which are heated using concentrated solar energy, so less biomass is wasted for the provision of energy. During periods of low or no solar irradiation, excess renewable energy from the grid can be converted into thermal energy for the pyrolysis process. To achieve the required temperature of approximately 800°C for the particles, a solar receiver is developed based on a rotary kiln concept.

And a third example is a particle based thermochemical process for the generation of another commodity: Nitrogen is required as a feedstock for the production of ammonia and nitrate fertilizers. Oxygen is removed from air by a thermochemical reaction to such an extent that the resulting nitrogen purity is sufficient for the generation of ammonia using the Haber-Bosch process. Key element of such process is the thermochemical reactor, where the air separation takes place and which contains a redox material that can reversibly incorporate and release oxygen without major changes of its crystal structure. When it is heated oxygen is released. This oxygen is flushed out of the reactor with a carrier gas. When then air is fed into the reactor, the oxygen oxides the oxygen-lean redox material and pure nitrogen remains as the gas phase. Lastly the material is heated again, the oxygen is released and the process starts all over again [7].

References:

[1] C.K. Ho, A new generation of solid particle and other high-performance receiver designs for concentrating solar thermal (CST) central tower systems, in: Manuel J. Blanco, Lourdes Ramirez Santigosa (Eds.), in Woodhead Publishing Series in Energy, Advances in Concentrating Solar Thermal Research and Technology, Woodhead Publishing (2017), Pages 107-128, ISBN 9780081005163.

[2] M. Ebert, L. Amsbeck, R. Buck, J. Rheinländer, B. Schlögl, S. Schmitz, M. Sibum, R. Uhlig. (2019) Operational Experience of a Centrifugal Particle Receiver Prototype. In: 24th SolarPACES International Conference on Concentrating Solar Power and Chemical Energy Systems, SolarPACES 2018, 2126 (030018), 030018-1. AIP Publishing. SolarPACES 2018, 2.-5. Oct. 2018, Casablanca, Morocco.

[3] G. Moumin, S. Tescari, P. Sundarraj, L. de Oliveira, M. Roeb, C. Sattler. (2019) Solar treatment of cohesive particles in a directly irradiated rotary kiln, Solar Energy 182, 480-490.

[4] J. P. Rincon Duarte, D. Kriechbaumer, B. Lachmann, S. Tescari, T. Fend, M. Roeb, C. Sattler.

(2022) Solar calcium looping cycle for CO2 capturing in a cement plant. Definition of process parameters and reactors selection, Solar Energy, vol. 238, 189–202.

[5] V.K. Thanda, D. Thomey, L. Mevißen, H. Noguchi, C. Agrafiotis, M. Roeb, C. Sattler. (2022) Solar thermochemical energy storage in elemental sulphur: design, development and construction of a lab-scale sulphuric acid splitting reactor powered by hot ceramic particles. AIP Conference Proceedings (2445). American Institute of Physics (AIP) 2445, 130008.

[6] S. Rodat, S. Abanades, G.Flamant. (2010) Experimental Evaluation of Indirect Heating Tubular Reactors for Solar Methane Pyrolysis, International Journal of Chemical Reactor Engineering, 8 (1), Article A25.

[7] L. Klaas, M. Pein, P. Mechnich, A. Francke, D. Giasafaki, D. Kriechbaumer, C. Agrafiotis, M. Roeb, C. Sattler. (2022). Controlling thermal expansion and phase transitions in Ca1-xSrxMnO3-δ by Sr-content, Phys. Chem. Chem. Phys. 24, 27976-27988.

499b Experimental Evaluation of a Continuous Oxidation Reactor for High-Temperature Discharge of Thermochemical Energy Storage Particles

Juvenal Ortiz-Ulloa, Owen Ramsey, Lucas Freiberg, Fuqiong Lei, Michael Hayes, Philipp Schimmels, Like Li, Andre Benard, Joerg Petrasch, Kelvin Randhir, Nesrin Ozalp, James F. Klausner, Nick AuYeung

Thermochemical Energy Storage (TCES) using metal oxide particles has emerged as a promising technology to facilitate dispatchability of renewable energy. TCES stores thermal energy in the form of chemical bonds in metal oxide particles for short or long-term periods, which can subsequently be released as heat when needed. Apart from providing energy storage, TCES has other advantages, including high energy density, decoupled charging and discharging steps for long-term storage, and the ability to deliver high-temperature heat (~1000 °C). On the one hand, the "charging" step involves exposing the particles to high temperature heat and an inert sweep gas where they undergo a reduction step releasing oxygen. Numerous studies recommend that the input energy should come from sustainable and low-cost sources such as concentrated solar or renewable excess electricity. On the other hand, the "discharging" or oxidation step requires that the particles are maintained at certain operating conditions of temperature and atmosphere that ensure that the exothermic oxidation step proceeds at a high reaction rate, and effective heat transfer to a heat transfer fluid that will be used downstream for applications such as power generation or industrial heat. Such operating conditions are often constrained by the need of maintained a coupled reduction and oxidation steps with intermediate hot storage and low efficiencies due to the intrinsic heat and mass transfer limitations of gas-particle contacting patterns. A proposed reactor concept has been put forward to facilitate the effective extraction of high-temperature heat from TCES particles while maintaining high oxidation reaction rates. The design involves a particle-gas counter-flow reactor with heat exchanger located between two moving beds. The moving beds allow for efficient transfer of sensible heat between the gas and solids, thereby enabling complete preheating and cooling of the gas and particles with minimal sensible heat losses. The construction of this reactor involved the use of high-temperature resistant materials such as superalloy Hastelloy X for the reactor and heat exchangers, top and bottom moving beds. Experiments using N_2 as a heat transfer fluid demonstrate that the heat exchangers can achieve sustained HTF outlet temperatures of ~ 900 °C with particle flowrates of 2.2 g/s and a reactor temperature of ~ 1000 °C. The main performance metrics are thermal efficiency and conversion. In this study, the thermal efficiency is defined as the useful energy extracted by the HTF divided by all the energy inputs at steady state such as heat of reaction, auxiliary external heating, and pumping work. Our experiments so far have obtained thermal efficiencies ranging from 30 to 50% with oxidation conversions of approximately 66%. Future work will involve testing key operating parameters such as HTF flowrate to maximize efficiency and conversion while maintaining adequate thermodynamic conditions for the oxidation reaction.

The presented reactor development for TCES using metal oxide particles represents a potential advancement in the field of energy storage. The implementation of the system in various thermal applications is enhanced by the decoupled charging and discharging steps, integration of different HTFs, and the ability to feed and remove solids and gas at ambient temperature. With further development and refinement, this reactor concept has the potential to advance the way we store and utilize energy from renewable sources.

499c Experimental Evaluation of an Indirectly-Irradiated Packed-Bed Solar Thermochemical Reactor for Calcination–Carbonation Chemical Looping

Lifeng Li, Alireza Rahbari, Mahdiar Taheri, Roelof Pottas, Bo Wang, Morteza Hangi, Leanne Matthews, Lindsey Yue, José Zapata, Peter Kreider, Alicia Bayon, Chi-Hwa Wang, Terrence W. Simon, Joe Coventry, Wojciech Lipiński

The two-step calcium oxide based calcination–carbonation cycle is studied for carbon dioxide capture and solar thermochemical energy storage applications. An indirectly-irradiated packed-bed solar thermochemical reactor is experimentally evaluated using simulated high-flux solar irradiation provided by a multi-source solar simulator. Experimental runs include a single calcination reaction step as well as single and multiple (up to four) consecutive calcination–carbonation cycles. The samples are characterised using scanning electron microscopy (SEM) and X-ray diffraction (XRD). The reactor temperature peaked at 1,035°C. The average solar-to-chemical conversion efficiency, defined as the ratio of heat consumed by the reaction to radiant heat supplied to the reactor, was found to be between approximately 1.3% and 8.6% for the five performed experimental runs. The necessary advancements to the presented reactor design identified during the experimental campaign include improvements in thermomechanical characteristics of ceramic and metallic parts of the reactor to prevent fast mechanical and chemical degradation, application of more robust high-temperature reaction chamber seals, and optimisation of reactor geometry and gas flow patterns towards spatially more uniform thermal conditions and chemical reaction rates.

499d Counter-Current Regenerative Redox Chemical Looping for Syngas Production: Packed Bed Reactor Design, Evaluation, and Feasibility Study

Zachary Hart, Alon Lidor

Solar radiation provides a semi-infinite amount of energy to Earth, enough to meet annual global demands in just 1 hour utilizing minimal surface area [1]. Even with the limited efficiency of solar conversion technologies, solar energy may still be one of the most promising sources of clean energy, necessary to decarbonize the energy sector. The diurnal and proximal variation of renewable resources in general, however, make it necessary that cost-effective, long-term storage solutions be investigated. Current efforts in solar energy storage systems span thermal, mechanical, chemical, electrochemical, and hybrid systems [2].

Of those, chemical storage as fuels is attractive due to its easy transportation and compatibility with current energy infrastructure. Recent solar-to-fuel research has focused largely on thermochemical conversion of carbon dioxide and steam to syngas, a mixture of carbon monoxide and hydrogen, which can be further upgraded to liquified hydrocarbon fuels via downstream chemical processes, such as Fischer-Tropsch synthesis. One feasible mechanism for this transformation is regenerative redox chemical looping utilizing nonstoichiometric metal oxides as an oxygen exchanger. In this process, metal oxide is first exposed to a reducing environment deficient in oxygen at high temperatures, allowing oxygen to be released thus creating oxygen vacancies in the lattice. Following reduction, the metal oxide is exposed to an oxidizing environment containing carbon dioxide and/or steam, allowing the metal oxide to return to its initial state while producing carbon monoxide or hydrogen [3]. There are ongoing efforts on improving the reactor efficiency which to date has been limited to 5%, with various proposed designs aimed at recovering the sensible heat losses between reduction and oxidation and minimizing losses. A significant part of these efforts includes the development of new redox materials that can operate at lower reduction temperatures. One aspect that is not widely pursued is the feedstock conversion, which is inherently low in these systems, ranging from less than 0.1% up to only 20% for some materials/reactors. Low conversion of CO_2 and/or H_2O results in large and energy/capital intensive separation processes, jeopardizing the commercial viability of the process.

Recently, the use of nonstoichiometric oxides in a redox reactor with alternating flow directions during reduction and oxidation has been proposed, achieving similar performance to counter-current membrane systems due to the gradient of chemical potential stored in the oxide. This has been demonstrated for the water-gas shift (WGS) [4] and reverse water-gas shift (RWGS) [5] reactions using a packed bed reactor. Such a system overcomes the design, fabrication, and operating limitations of high-temperature membrane reactors, with demonstrated conversion of over 90% for these thermodynamically-limited reactions. It also negates the need to design complex reactors with moving oxide materials at high temperatures. These findings call for further investigation of the counter-current configuration, optimizing reactor sizing and operating conditions, along with an updated system-level technoeconomic analysis.

Therefore, this study aims to evaluate the economic feasibility of the newly proposed counter-current thermochemical regenerative redox chemical looping reactor utilizing nonstoichiometric metal oxides, evaluate reactor productivity via performance indicators, and conduct a system-level technoeconomic analysis on the process.

Physics-based models have been developed using the commercially available finite element software

COMSOL to simulate the fluid flow through porous media, mass transfer, and kinetics of the system. 1D models have been developed for both co-current and counter-current configurations for comparison and validation against available data, allowing for a quick parametric study to establish the dependencies between the main parameters and reactor performance. A 2D axisymmetric model will consequently be developed for the counter-current configuration for a more detailed analysis. Key model assumptions include: (1) the oxidizer is in equilibrium with its corresponding thermolysis reaction at the entrance of the reactor, (2) fluid properties are evaluated for the bulk inlet species, and (3) reaction kinetics are limited by the metal oxide and rapidly approach equilibrium.

Reactor performance is evaluated over a wide range of reactor dimensions and operating conditions via performance indicators used in the field of solar thermochemistry [6]. Because both the oxidation and thermolysis reactions produce the same products, selectivity and yield are not needed to characterize the reactor performance. Instead, only cycle conversion and solar-to-fuel efficiency are required.

Cycle conversion is defined as the ratio of total unreacted reactant to total available reactant in the feedstock. Solar-to-fuel efficiency $\eta_{solar\text{-}to\text{-}fuel}$ represents the fraction of input energy that can be recovered by product combustion over the different energy input terms: the sensible heat required to increase the reactor temperature from the oxidation to reduction temperature, the endothermic heat required to drive the reduction reaction, the work required to pump sweep gas during reduction and oxidizer during oxidation, and the work required to separate released oxygen contaminant from inert gas. In practice, $\eta_{solar\text{-}to\text{-}fuel}$ is the reactor efficiency, neglecting losses due to solar collection and downstream processes such as reactant/product separation and gas-to-liquid synthesis.

Simultaneously optimizing cycle conversion and efficiency is a challenge, often trading off between each other. Achieving high conversion requires long cycle times and diminished kinetic rates near complete conversion, thus decreasing solar-to-fuel efficiency. Contrarily, infinitesimally small cycle times maximizes fuel production rates, but inefficiently utilizes reactants thus decreasing conversion [7]. However, when considering the effects of reducing large downstream separation units, a different optimal operation regime is expected compared to previous works to date.

The developed models will demonstrate the improved performance of a counter-current regenerative redox reactor, maximizing the difference between oxidizer concentration and metal oxide vacancies across the reactor length, resulting in increased conversion with minor losses in efficiency due to increased reduction energy demands and downstream separation work. Preliminary results from the 1D model confirmed this, showing achievable conversions up to 90% using the counter-current configuration. Utilizing shorter time cycles will increase the reactor efficiency while obtaining much higher conversion than demonstrated values to date. It is not expected that reactor diameter will have a dramatic effect on reactor performance but increasing it may allow for higher oxidizer flowrates while maintaining high conversion and decreasing the required pumping work.

By performing rigorous numerical investigation, the conversion-efficiency tradeoff will be evaluated in detail, informing further efforts with guidance on reactor sizing, optimal operating conditions, and expected results for both experimental testing and computational validation. The subsequent system-level technoeconomic analysis will determine the system feasibility and identify technological gaps and recommended directions for future research efforts to advance the field of solar thermochemical hydrogen and fuel production.

References:

[1] Steinfeld A, Meier A. Solar fuels and materials. in: Cleveland C (ed) Encyclopedia of energy, vol

5. Elsevier, Amsterdam, (2004) pp 623–637. https://doi.org/10.1016/B0-12-176480-X/00322-3.

[2] J. Mitali, S. Dhinakaran, A.A. Mohamad, Energy storage systems: a review, in: Energy Storage and Saving, Volume 1, Issue 3 (2022) pp 166-216. https://doi.org/10.1016/j.enss.2022.07.002.

[3] A. Meier, A. Steinfeld, Solar Energy in Thermochemical Processing, in: C. Richter, D. Lincot, C.A. Gueymard (Eds.), Sol. Energy, Springer New York, New York, NY (2013) pp. 521–552. https://doi.org/10.1007/978-1-4614-5806-7_689.

[4] Metcalfe, I.S., Ray, B., Dejoie, C. et al. Overcoming chemical equilibrium limitations using a thermodynamically reversible chemical reactor. Nat. Chem. vol 11, (2019) pp 638–643. https://doi.org/10.1038/s41557-019-0273-2.

[5] B. Bulfin, M. Zuber, O. Gräub, A. Steinfeld, Intensification of the reverse water–gas shift process using a countercurrent chemical looping regenerative reactor, Chem. Eng. J. 461 (2023). https://doi.org/10.1016/j.cej.2023.141896.

[6] B. Bulfin, M. Miranda, A. Steinfeld, Performance Indicators for Benchmarking Solar Thermochemical Fuel Processes and Reactors, Frontiers in Energy Research (2021). https://doi.org/10.3389/fenrg.2021.677980.

[7] Timothy C. Davenport, Chih-Kai Yang, Christopher J. Kucharczyk, Michael J. Ignatowich, Sossina M. Haile, Maximizing fuel production rates in isothermal solar thermochemical fuel production, Applied Energy (2016). https://doi.org/10.1016/j.apenergy.2016.09.012.

499e Numerical Heat Transfer Simulation and Optimization of Indirectly Heated Porous Foams in Thermochemical Hydrogen Generation

Durgesh Ranjan, Elizabeth Gager, Caroline Hill, Juan C. Nino, Jonathan Scheffe

Solar-driven high temperature thermochemical fuel production has been seen as a potential solution towards the decarbonization of various energy intensive industries such as automotive iron and steel, and aviation industries. One of the major approaches towards sustainable fuel production is utilizing the solar heat flux to drive a high temperature redox reaction of metal oxides for hydrogen production. In recent times, reticulated porous foams with varying pore size and geometry have been investigated for the effective heat transfer and improved solar to fuel efficiency. Heat transfer is one factor that governs how much of the sample undergoes redox cycling, and therefore has a significant impact on the overall performance of hydrogen-producing reactors. Porosity in foam samples improves radiative absorption within the volume, resulting in an increased heated zone and uniform temperature gradient. However, parameters such as the size of pores, overall porosity and topology of such samples need to be optimized to maximize the benefits offered by foam structure. In this study, we are carrying out a parametric analysis on the heat transfer characteristics of porous foam samples of varying geometry and boundary conditions using Discrete Ordinate (DO) radiation model.

The model considers all three types of heat transfer, as well as necessary reactor components like the foam sample, work tube, and insulator. The performance of the modeling is evaluated based on the minimization of the temperature gradient (both axial and radial) in the foams as it can severely affect reduction rates because of the inherent thermal resistance. DO model uses the radiative transfer equation (RTE) considering absorbance, emittance, and scattering for a user defined finite number of discrete polar and azimuthal angles. Unlike the typically used Monte Carlo ray tracing method, this method does not use ray tracing and is computationally inexpensive and fast, especially if multiple geometric models are being evaluated. The model offers flexibility in the choice of 3D geometric models i.e., whether a 3D computed tomography scan of a fabricated foam or a CAD model. It is therefore possible to investigate how foam designs affect its thermal performance, which could guide future advancements in structural design. Additionally, the heating source used in the modelling is indirect type i.e., using heat transfer fluid (HTF) such as steam from a solar receiver to drive the reaction rather than concentrated solar flux as is commonly observed in ray tracing models. The thermal effects of endothermic reduction and exothermic oxidation reactions are represented as heat sinks and heat sources, respectively based on the oxygen non-stoichiometry and partial molar enthalpy. Other parameters that influence temperature distribution in foam have been investigated, including the temperature and velocity of process gas and HTF, as well as physical characteristics of outer insulator. The simulation data is verified with experimental results in literature and a description of the solution dependence on angular discretization, as well as on boundary conditions have also been presented.

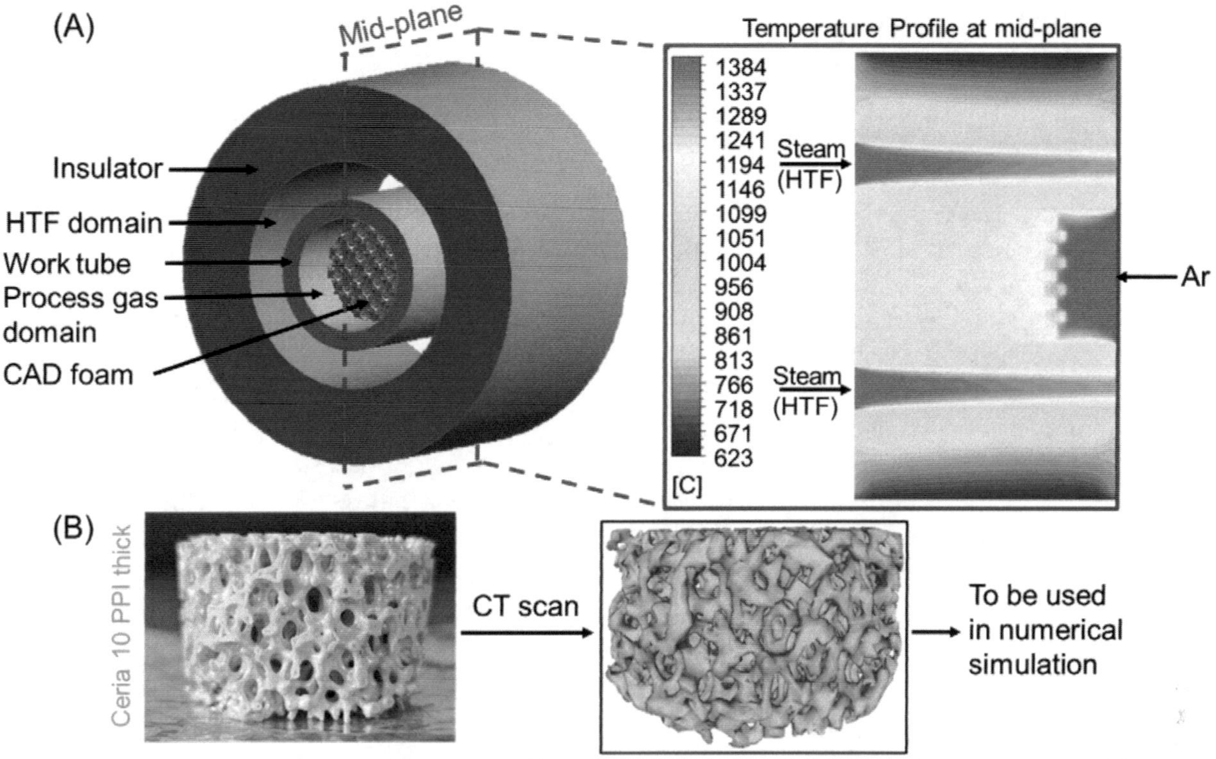

Figure 1. Discrete Ordinates radiation model (A) Model geometry showing the outer insulator, heat transfer fluid (HTF) flow domain, work tube, process gas flow domain and the CAD foam model. The temperature profile of all the domains is shown for a particular case of steam inlet at 1400 °C (4.5 kg hr^{-1}), Ar at 1400 °C (2100 sccm) and foam sample porosity = 0.82. (B) Ceria foam replica and its computed tomography scan that is used in place of CAD foam model to analyze its heat transfer performance.

499f A Photothermal Reactor with a Selective Solar Absorber to Produce Liquid Fuels Via Ethylene Oligomerization

Aisulu Aitbekova, Magel Su, Matthew Salazar, Harry Atwater

Utilizing solar thermal energy for thermochemical processes has attracted substantial recent interest due to its capacity to provide a sustainable alternative for the synthesis of fuels, chemicals, and materials compared to traditional industrial methods. Furthermore, thermocatalytic processes allow operation at reduced temperatures and pressures achievable with direct solar-to-thermal energy conversion alone. Here, we present a scalable photothermal reactor to exploit a selective solar absorber that converts sunlight into thermal energy for sustainable fuel synthesis. We report a selective solar absorber consisting of an anti-reflective coating and an optically transparent metal layer, achieving a calculated maximum temperature of 249 °C under one sun illumination and 125 °C under ambient operating conditions (25 °C, 1 atm). The photothermal reactor was employed in batch and flow operation using homogenous and heterogenous Ni-catalyzed ethylene oligomerization reactions. In batch operation, homogenous oligomerization yielded $C_6 - C_{24}$ hydrocarbon products, and in flow operation, heterogenous oligomerization yielded 25 % ethylene conversion to butene and hexene products. Finally, we present simulated results for a large-scale photothermal reactor, which predict spatially uniform maximum potential temperatures up to 120 °C and 210 °C under one and three sun illumination, respectively, demonstrating the potential of this system for generating fuels from sunlight using scalable devices.

499g Decarbonizing Ethylene Production Using Concentrating Solar Power

John Fawcett, Abhaya K. Datye, Andrew T. DeLaRiva, Christopher Riley, H. Evan Bush, Erik Spoerke, Dresean Abeyta, Benjamin Harvey

Ethylene, the world's highest volume petrochemical, is produced using steam cracking of alkanes. This reaction is carried out at high temperatures (>900°C) by the burning of natural gas which produces around a metric ton of CO_2 per ton of ethylene produced and requires energy inputs of up to 16 GJ per ton of produced ethylene [1-2]. In industry this heat and energy transfer occurs by the combustion of natural gas or other fossil fuels. The purpose of this research is to develop an alternative process that uses solar heating of a catalytic reactor to drive the dehydrogenation of ethane to ethylene. As has been previously demonstrated, ethane dehydrogenation can be performed at temperatures reachable by concentrated solar technologies (CST) [2]. The challenge then becomes what to do about the inherent intermittence of solar power. Thermal storage media already used in CST applications provide an intriguing solution. Quartz tubes were filled with solid thermal storage media, then heated to temperatures required for the dehydrogenation reaction while an ethane-carrier gas mixture was flowed through the system. Quartz tubes were chosen as they have been previously demonstrated to achieve ethylene yields of up to 68% using the open tube setup [2-4] This experimental setup demonstrated that thermal storage media can transfer sufficient energy to drive the dehydrogenation of ethane. This study also demonstrates that the packed tube is more efficient than an empty quartz tube in transferring heat to the reactant gases.

- "Net-zero carbon ethylene production via recovery of CO2 from cracking furnace flue gas." n.d. S&P Global. https://www.spglobal.com/commodityinsights/en/ci/products/net-zero-carbon-ethylene-production.html.
- R. Riley, A. De, I. L. Ibarra, A. K. Datye, and S. S. Chou, "Achieving high ethylene yield in non-oxidative ethane dehydrogenation," Applied Catalysis A: General, vol. 624, pp. 118309–118309, Aug. 2021, doi: https://doi.org/10.1016/j.apcata.2021.118309.
- Scott A. Steinmetz, Andrew T. DeLaRiva, Christopher Riley, Paul Schrader, Abhaya Datye, Erik D. Spoerke, and Christopher J. Kliewer *The Journal of Physical Chemistry C* **2022** *126* (6), 3054-3059, DOI: 10.1021/acs.jpcc.1c09955
- Efficient Low-temperature, Catalyst free Dehydrogenation Of Alkanes, by C. R. Riley. (2022, Jul. 14). *US20230037379A1*. [Online]. Available: https://ppubs.uspto.gov/pubwebapp/

98a An Integrative Framework for Optimal Decarbonization Strategies in the Chemicals Sector

Sheikh Mahdi Ahmed, Joan Cordiner

In response to climate change, industries are pivoting towards low-carbon energy systems, navigating through a maze of economic, technical, and environmental considerations to mitigate carbon emissions effectively. Our approach introduces a dynamic decision-making framework employing Mixed Integer Linear Programming (MILP) to evaluate and optimize decarbonization technologies. This tool not only assesses the viability of such technologies but also explores diversified bidding strategies and participation in ancillary markets to enhance investment value and mitigate financial risks. Central to this framework is its ability to adapt to the fluctuating energy markets, evolving global emissions trading schemes and carbon border mechanisms, ensuring industries can make informed, long-term investment decisions. By applying this model to an industrial case study, we demonstrate how it can predict market dynamics' influence on investment outcomes, thereby guiding industries to minimize risks and maximize returns. This ensures competitiveness and profitability in a transitioning low-carbon economy, providing a comprehensive roadmap for industrial decarbonization.

98b Design of a Supply Chain for the Production of Biodegradable Plastics Based on the Revalorization of Sewage Sludge

Jose Enrique Roldán-San Antonio, Mariano Martin

The growth in the production of synthetic plastics since the 20th century has promoted the development of the industry. The wide range of polymers and their properties make these materials promising alternatives for the manufacture of products used in the automotive, electronics and construction sectors, as well as for the production of containers and packaging used in the food industry [1]. Among all plastic applications, packaging represents the main market, accounting for 44% of global plastic demand [2]. The most commonly used polymers for packaging are polypropylene (PP), polystyrene (PS), polyvinyl chloride (PVC), polyethylene terephthalate (PET) and polyethylene (PE) [3], the later representing 63% of all packaging waste [4]. A possible alternative to mitigate the environmental impact of this fossil – based polymers is the development of polymeric materials with renewable origin, biodegradables and with mechanical properties similar to polyethylene, as can be found in starch – based plastics [5,6].

To evaluate the polyethylene substitution potential through the production of starch – based biodegradable blends, a multiscale analysis is performed including process scale up and supply chain conceptual designed. The supply chain is composed by the wastewater treatment plants (WWTP), CO_2 and sawdust sources and polymers productive centres. These facilities are integrated processes where the biopolymer is produced from renewable and waste sources such as sawdust, sludge, and CO_2, obtaining all the intermediates required (e.g., methanol, glycerol, starch) [5]. The supply of sludge to the production centres has been considered through pipelines directly connected to the WWTP to facilitate the transport. In this way, the installation of these centres will be limited to the locations of the WWTP. For the case of sawdust, the supply to the production centres has been considered by land transport in trucks from locations with availability of shredded lignocellulosic material after forestry maintenance of pine populated regions. Finally, CO_2 is obtained from DAC and/or MEA – based CO_2 capture methods, being supplied to the production centre by pipeline. The integrated facility was optimized [5] and scaled up. Next, the supply chain model was formulated including linear algebraic constraints based on the availability of sawdust and CO_2 at the raw material locations together with capacity limits set by the availability of sludge at each location of the production centres. Additionally, economic aspects such as investment and operational costs of the production network and environmental impact have been included. In this way a mixed – integer linear model (MILP) is formulated to determine the optimal network which maximizes the fraction of demand for polyethylene replaced by starch-based biopolymer. The methodology presented has been applied to the case study of peninsular Spain to show the approach. Agricultural county has been considered as discretization level of the problem, accounting for 343 regions, solving a total of 1M equations, 1.4M continuous variables and 1400 binary variables solved in GAMS® using CPLEX.

Considering a total budget of 2200M$, corresponding to the tax imposed by the European Union [7] over the use of non-renewable plastic manufacturing in Spain between 2024-2027, the optimization problem was solved focusing on the maximization of PE substitution by starch – based polymers. The results showed the ability to satisfy the 40% of the polyethylene demand, capturing a total of 1.5 Mt of CO_2 per year and promoting a decarbonization of 1% with respect to the year 2023. In addition, the total biodiesel production of the proposed network is 0.47 Mt/year, corresponding to a 1.47% of the diesel consumption in Spain.

Acknowledgments

This work was supported by funding to José Enrique Roldán San Antonio under the call for predoctoral contracts USAL 2021, co-funded by Banco Santander. This work was supported by projects PID2019-105434RB-C31 and TED2021-129201B-I00 of the Spanish Government

References

(1) Andrady, A. L.; Neal, M. A. Applications and Societal Benefits of Plastics. *Philos. Trans. R. Soc. B Biol. Sci.* **2009**, *364* (1526), 1977–1984. https://doi.org/10.1098/rstb.2008.0304.

(2) Plastics Europe. *Plastics - the Facts 2022*; 2022.

(3) Beswick, R.; Dunn, D. J. *Plastics in Packaging: Western Europe and North America*; iSmithers Rapra Publishing, 2002.

(4) Kunwar, B.; Cheng, H. N.; Chandrashekaran, S. R.; Sharma, B. K. Plastics to Fuel: A Review. *Renew. Sustain. Energy Rev.* **2016**, *54*, 421–428. https://doi.org/10.1016/j.rser.2015.10.015.

(5) Roldán-San Antonio, J. E.; Martín, M. Optimal Integrated Plant for Biodegradable Polymer Production. *ACS Sustain. Chem. Eng.* **2023**, *11* (6), 2172–2185. https://doi.org/10.1021/acssuschemeng.2c05356.

(6) Vanderreydt, I.; Rommens, T.; Tenhunen, A.; Mortensen, L. F. Greenhouse Gas emissionsand Natural Capital implicationsof Plastics (Including Biobased Plastics). *Eur. Top. Cent. Waste Mater. a Green Econ.* **2021**, No. May, 1–68.

(7) WTS Global. *Plastic Taxation in Europe: Update 2023*; 2023.

98c Transition Toward a Decarbonized Steel Industry: A Supply Chain Analysis for the United States

Yeonseo Jeong, Qi Zhang

The iron and steel industry, traditionally known for its substantial energy use, is now navigating a pivotal transition towards net-zero production. The combined use of direct reduction (DR) and electric arc furnaces (EAF) with green hydrogen—namely the green H-DRI-EAF system—emerges as a key strategy for decarbonizing the steel industry. It has been demonstrated that emissions from green H-DRI-EAF could be as low as only 2.8% of those from the energy-intensive blast furnace-basic oxygen furnaces (BF-BOF) route [1].

The shift to green H-DRI-EAF requires a holistic supply chain analysis to effectively incorporate novel technologies and capitalize on renewable energy and regional assets, such as abundant wind energy and iron ore reserves. However, there is a notable gap in green steel supply chain analysis for the United States, the world's fourth-largest steel producer [2]. Given its iron and steel industry's reliance on mini-mills—small-scale EAFs primarily using scrap steel—for nearly two-thirds of production [3] and the need to consider unique iron ore qualities [4], a customized approach to integrating the green H-DRI-EAF system is essential.

In this work, we developed a supply chain planning model to provide a roadmap for integrating the green H-DRI-EAF system and phasing out traditional, carbon-intensive processes. The model spans multiple time periods to reflect changes in parameters such as steel demand and emission reduction goals over time. It incorporates critical parameters including capital, transportation, and purchasing costs, resource availability, and current production capacities. Results from solving the proposed supply chain optimization model for a planning horizon of up to the year 2050 provide insights into the strategic co-location of processes, such as the placement of iron production in regions with high renewable energy capacity, as well as the trade-off between production and transportation costs. The study reveals trends in the phase-out of existing BF-BOF plants and the expansion of EAF operations, making use of existing mini-mills. Furthermore, the sensitivity of the results on key parameters, including emission reduction targets and carbon tax rates, has been investigated to assess their effects on the overall long-term strategy for the transition towards green steel production in the United States.

References

[1] V. Vogl, M. Åhman, and L. J. Nilsson, "Assessment of hydrogen direct reduction for fossil-free steelmaking," *J. Clean. Prod.*, vol. 203, pp. 736–745, Dec. 2018, doi: 10.1016/j.jclepro.2018.08.279.

[2] World Steel Organization, "2022 World Steel in Figures."

[3] Congressional Research Service, "Domestic Steel Manufacturing: Overview and Prospects," R47107, May 2022.

[4] L. Bloomquist, "Iron Range endowed with more than enough ore," *Mesabi Tribune*, Oct. 30, 2019. Accessed: Aug. 28, 2023. [Online]. Available: https://www.mesabitribune.com/mine/iron-range-endowed-with-more-than-enough-ore/article_34104296-f9df-11e9-81ef-7be29e3f69cb.html

98d Supply Chain Model for Bioenergy with Carbon Capture and Storage (BECCS) in Iowa.

Mrinal Manoj Redij, Mark Mba Wright, Lisa Schulte

Bioenergy with Carbon Capture and Storage (BECCS) is a crucial technology for achieving net-zero emissions in the global environment, as underscored by the International Energy Agency's climate change models, which suggest that at least 2 Gt CO_2 per year removal by BECCS should be implemented by 2050 to keep the global temperature rise below 2°C (IEA, 2009; National Academies of Sciences, Engineering, and Medicine, 2019). This study proposes a supply chain model for BECCS in Iowa, utilizing the state's unique agricultural and geological resources to address the need for sustainable and scalable carbon-negative energy solutions. Our approach utilizes advanced feedstock optimization, and innovative carbon capture innovations, based on Iowa's potential for bioenergy production and CO2 sequestration. Iowa's rich biomass resources and suitable geological formations provide an exceptional opportunity for BECCS deployment. Researchers have identified Iowa as having significant potential for underground carbon dioxide storage, indicating that the state's geological formations could safely sequester substantial amounts of CO_2. However, optimizing the supply chain to balance economic viability and environmental sustainability is a significant challenge. Our model builds upon the Roads to Removal biorefinery logistics model to evaluate the economic viability of BECCS. Distinctively, our model is developed using the Julia programming language, renowned for its high-performance capabilities in technical computing, and utilizes comprehensive data from the US Billion Ton reports, ensuring an accurate and robust foundation for analysis. Key to our model is the advanced analysis of biomass feedstocks, utilizing optimization algorithms to determine the most efficient and sustainable combinations for bioenergy production.

The research develops a nuanced model that accurately outlines biorefinery operations, their feedstock preferences, and the bioenergy products they produce. This model utilizes data from the US Billion Ton Reports to ensure optimal feedstock utilization without compromising sustainable limits. It is designed to achieve a minimum of 60% of Iowa's potential for carbon sequestration, utilizing biorefinery configurations that are economically viable and environmentally sustainable. Through strategic resource use and severe constraints, our approach aligns economic viability with environmental sustainability, balancing bioenergy production with the goal of maximizing carbon sequestration efficiency. Our research underscores the critical role of innovative supply chain models in achieving sustainable bioenergy production and effective carbon management, highlighting Iowa's potential as a leader in BECCS deployment and that motivates the further research.

98e The Energy and Materials Transition Nexus– Multiscale Modeling and Optimization

Betsie Montano Flores, Marco De Sousa, Rahul Kakodkar, Shayan Niknezhad, Efstratios Pistikopoulos

The energy transition is propelled by the motivation to decarbonize processes and obtain viable costs-effective low-carbon technologies. This has led to a gradual shift from conventional fossil fuel sources to renewable energy. However, the realization of the energy transition necessitates a robust infrastructure, which relies heavily on the availability of resources and materials. Materials play a critical role in empowering the energy transition, serving for energy storage, electric vehicle production, grid infrastructure, and construction of wind and solar power systems [1] amongst others. As the production and procurement of these materials generate emissions, introduce additional supply chain costs and constraints, the consideration of the energy and materials transition nexus becomes essential to fully address sustainable energy challenges.

In this work, we propose a multiscale modeling and optimization decision-making framework capable of coordinating the energy and materials nexus. Based on a mixed integer optimization formulation, the framework involves various key aspects, including (i) an integrated resource-task-material-network design and scheduling model to simultaneously account for material and energy requirements for the material and energy transition supply chains, and (ii) tracking of disparate emissions originated from both materials and energy systems components. A case study focusing on a mobility transition scenario in Texas towards electric vehicles (EVs) is used to illustrate the key features and capabilities of the proposed framework, including: (i) carbon value vector utilization to produce polymeric materials, (ii) sustainable production of hydrogen [2], (iii) power generation through intermittent renewables with storage considerations, and (iv) establishing grid infrastructure for charging the EVs. The study, implemented in energiapy [3], reveals intricate interconnections between energy supply, materials production, and vehicle production, emphasizing the importance of an integrated energy-materials-mobility nexus approach.

References

[1] T. Watari, B. C. McLellan, D. Giurco, E. Dominish, E. Yamasue, K. Nansai, Total material requirement for the
global energy transition to 2050: A focus on transport and electricity, Resources, Conservation and Recycling 148
(2019) 91–103.

[2] C. R. Allen, S. G. Baratsas, R. kakodkar, S. Avraamidou, C. D. Demirhan, C. F. Heuberger-Austin, M. Klokken-
berg, E. N. Pistikopoulos, A multi-period integrated planning and scheduling approach for developing energy
systems, Optimal Control Applications and Methods 44 (2023) 355–372. doi:https://doi.org/10.1002/oca.2866.

[3] R. Kakodkar, E. Pistikopoulos, Energiapy-an open source python package for multiscale modeling & optimization
of energy systems, in: 2023 AIChE Annual Meeting, AIChE, 2023.

98f Strategic Supply Chain Optimization for Bio-Jet Fuel Production in the Brazilian Sugarcane Industry

Madelynn Watson, Aline Veronese da Silva, Pedro Gerber Machado Sr., Celma O. Ribeiro, Claudio Augusto Oller do Nascimento, Alexander Dowling

Brazil is a global leader in renewable energy and sustainability efforts. For example, in 2019, 44.1% of energy in Brazil came from renewable sources, compared to only 10.4% in the US [1]. Additionally, Brazil was one of the first countries to use biofuels for transportation due to multiple incentive programs the Brazilian government has supported since the 1970s [2]. They are currently the second-largest producer of the world's bioethanol and the first-largest producer of bioethanol from sugarcane [3-4].

Biofuels, especially those produced at large scales, have gained increasing interest, specifically from the aviation industry. The aviation industry currently accounts for ~3% of global CO_2 emissions. Moreover, aviation cannot rely on common decarbonization solutions (e.g., electrification or hydrogen) in the short- to medium-term (2030-2050) as they require two or three decades of development and capital-intensive investments [5]. Bio-jet fuels appeal to aviation companies to reduce CO_2 emissions due to their economic feasibility and drop-in compatibility with existing aircraft and fuel systems [6]; however, several challenges currently hinder commercial-scale bio-jet fuel development. One notable challenge is underdeveloped biomass-to-bioenergy supply chains, which result in high biomass and bio-jet fuel costs [6-7].

Biofuel supply chains face many unique challenges, for example, time-sensitive storage and transportation from biomass degradation, supply uncertainties from biomass seasonality, and infrastructure incompatibilities [7-8]. Transportation limitations and infrastructure incompatibilities force biofuel supply chains (i.e., bioethanol) to rely on trucks, barges, or trains, which can be expensive and environmentally intensive at high volumes and long distances [8]. Furthermore, high transportation costs may offset the expected benefits of "economy of scale" [8]; thus, it is crucial to strategically design biofuel supply chains that balance transportation and capacity investment costs.

Bioethanol from sugarcane can be upgraded to bio-jet fuel via the ASTM-certified pathway alcohol-to-jet (ATJ) [5-8]. Brazil's experience in renewable energy development, historical government support of biofuels, and existing biomass infrastructure create a unique opportunity for bio-jet fuel development.

In this work, we strategically design a supply chain model to optimize bio-jet fuel capacity distribution in Brazil to meet 10% of the country's jet fuel demand with bio-jet fuel from sugarcane. Specifically, we proposed a mixed-integer linear programming (MILP) superstructure model to optimize the location of sugarcane mills that invest in bio-jet fuel capacity, assignment of airports to integrated biorefineries, and transportation links that connect sites across the supply chain network. The optimal supply chain network minimizes the cost of bio-jet fuel from sugarcane based on existing biomass infrastructure (sugarcane mill and airport locations), bio-jet fuel policy incentives, and market prices of sugar, ethanol, and jet fuel. Finally, we identify opportunities to quantify multi-objective trade-offs of costs and environmental impact with sensitivity studies to guide biofuel infrastructure and policy development.

References

[1] IEA (2022), Energy Statistics Data Browser, IEA, Paris https://www.iea.org/data-and-statistics/data-tools/energy-statistics-data-browser

[2] Castro REN, Alves RMB, Nascimento CAO, Giudici R. Assessment of sugarcane based ethanol production. In: Basso TP, Basso LC, editors. Fuel ethanol production from sugarcane Rijeka: IntechOpen; 2019. https://doi.org/10.5772/intechopen. 78301. Ch. 1

[3] Mutran, V. M., Ribeiro, C. O., Nascimento, C. A. O., & Chachuat, B. (2020). Risk-conscious optimization model to support bioenergy investments in the Brazilian sugarcane industry. *Applied Energy*, 258. https://doi.org/10.1016/j.apenergy.2019.113978

[4] López-Ortega, M. G., Guadalajara, Y., Junqueira, T. L., Sampaio, I. L., Bonomi, A., & Sánchez, A. (2021). Sustainability analysis of bioethanol production in Mexico by a retrofitted sugarcane industry based on the Brazilian expertise. *Energy*, 232, 121056.

[5] Dahal, K., Brynolf, S., Xisto, C., Hansson, J., Grahn, M., Grönstedt, T., & Lehtveer, M. (2021). Techno-economic review of alternative fuels and propulsion systems for the aviation sector. *Renewable and Sustainable Energy Reviews*, 151. https://doi.org/10.1016/j.rser.2021.111564

[6] Watson, M. J., Machado, P. G., da Silva, A. v., Saltar, Y., Ribeiro, C. O., Nascimento, C. A. O., & Dowling, A. W. (2024). Sustainable aviation fuel technologies, costs, emissions, policies, and markets: A critical review. *Journal of Cleaner Production*, *449*. https://doi.org/10.1016/j.jclepro.2024.141472

[7] Doliente, S. S., Narayan, A., Tapia, J. F. D., Samsatli, N. J., Zhao, Y., & Samsatli, S. (2020). Bio-aviation Fuel: A Comprehensive Review and Analysis of the Supply Chain Components. *Frontiers in Energy Research*, *8*. https://doi.org/10.3389/fenrg.2020.00110

[8] Yue, D., You, F., & Snyder, S. W. (2014). Biomass-to-bioenergy and biofuel supply chain optimization: Overview, key issues and challenges. *Computers and Chemical Engineering*, *66*, 36–56. https://doi.org/10.1016/j.compchemeng.2013.11.016

98g State-Task Network Representation of Early-Stage Process for Identification of Best Processing Route – in Application of Reactive Carbon Capture

Wonsuk Chung, Ung Lee

Superstructure has been in limelight for conceptualization method of a large-scale chemical system due to its excellence of capturing characteristics from multiple processes. A proper representation of superstructure allows identification of all the mathematically feasible pathways by mixed integer programming (MIP). Combination State-task network (STN) representation with superstructure optimization can find the best pathways.[1] Here, state is information of a stream (compound flowrate, temperature, and pressure) and task is a process which relates the inlet streams and the outlet streams, obeying mass and energy balance. Hence, a superstructure as STN representation commonly represents tasks as nodes and states as edges. There is one branch of STN representation, a generic model, where four tasks (mixing, reaction, waste separation, and product separation) are incorporated into one node allows is proposed.[2] This generic model is adopted to develop a framework to construct, analyze, and optimize different superstructures.[3,4]

Herein, we introduce our previous work, where a versatile superstructure framework which can represent various types of chemical processes is proposed.[5] STN representation in this framework does not limit any number of states, tasks, inlets, and outlets of a process. Logic-based outer approximation (LOA)[6] which reformulates a strategy-finding problem from a complicated system into MIP is adopted, allowing quick optimization of a given superstructure. Any nonlinear equations can be incorporated into each task, thus flexibility to represent a chemical process in a superstructure is guaranteed. Compared with conventional superstructure representation as mixed-integer nonlinear programming (MINLP), this framework outperforms the conventional one in terms of both optimization time and solvability.

Application of this framework can stretch identification of promising pathways from a system in its technological infancy. An illustrating example is reactive carbon capture (RCC), a strategy that captures CO_2 from flue gas or air by amine scrubbing and utilizes it into chemicals (e. g. methanol, CO, and formic acid) without CO_2 release (Figure 1). As energy to produce pure CO_2 can be saved, RCC can be a true solution for carbon neutrality;[7] however, all the RCC processes are in early stage. This study construct RCC superstructure using the framework, by representing the process as STN representations, keeping the first-principles of the processes. As it is known that the price of renewable energy varies by climate, this study identifies which route is best among various RCC technologies for each country. Compared to other carbon capture and utilization (CCU) pathways, the competitiveness of RCC routes in terms of both profitability and carbon reduction will be measured.

Figure 1. Reactive carbon capture (RCC). a) RCC by using amine scrubbing and electrochemical CO_2 conversion to syngas[7] and b) superstructure representation of CCU system with RCC.

References

1. Yeomans and Grossmann, 1999, A systematic modeling framework of superstructure optimization in process synthesis, *Chem. Eng.* 23, 709-731.
2. Quaglia et al., 2012, Integrated business and engineering framework for synthesis and design of enterprise-wide processing networks, *Chem. Eng.* 38, 213-223.
3. Bertran et al., 2017, A generic methodology for processing route synthesis and design based on superstructure optimization, *Chem. Eng.* 106, 892-910.
4. Chung et al., 2022, Computer-aided identification and evaluation of technologies for sustainable carbon capture and utilization using a superstructure approach, *CO_2 Util.* 61, 102032.
5. Chung et al., 2023, Identification of sustainable carbon capture and utilization (CCU) pathways using state-task network representation, *Chem. Eng.* 178, 108408.

6. Raman and Grossmann, 1994, Modelling and computational techniques for logic based integer programming, *Chem. Eng.* 18, 563-578.
7. Langie et al., 2022, Toward economical application of carbon capture and utilization technology with near-zero carbon emission, *Commun.* 13:e7482.

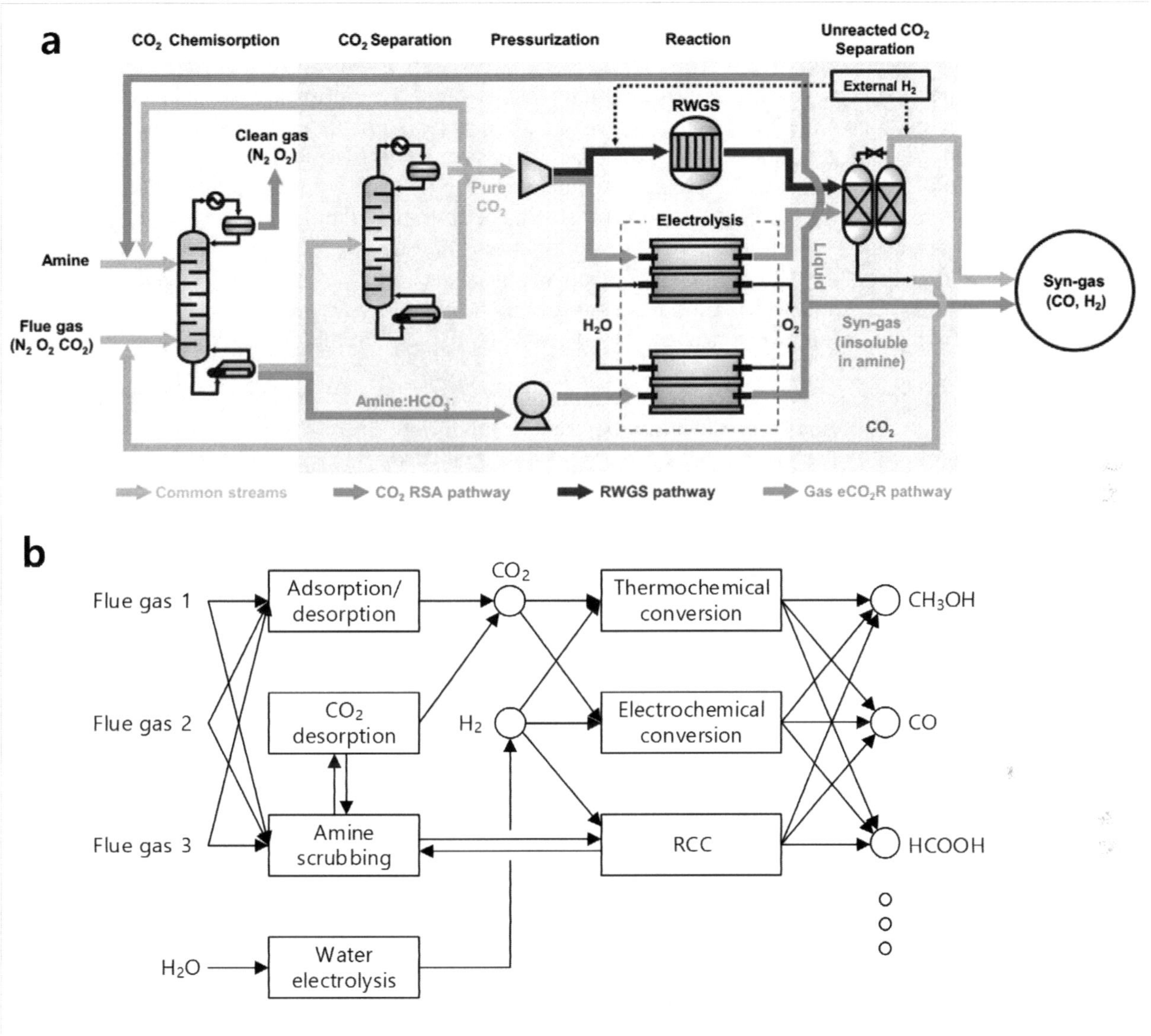

98h Integration of Temporal Factors into Spatially Explicit Optimization of Second Generation Biofuels Supply Chains

Phuc Tran, Eric O'Neill, Christos Maravelias

Motivated by the need to alleviate the heavy dependence on fossil fuels and address the climate change crisis while maintaining energy security, biomass-derived liquid transportation fuels and energy products have received attention as an alternative. Particularly, there has been great interest for the development of biofuels produced from non-food sources, commonly known as second-generation biofuels. These products require ligno-cellulosic feedstock materials which are less land and water intensive, and do not contribute to the food-versus-fuels debate. Ligno-cellulosic crops are great feedstocks for biofuels production due to their low costs, net carbon emission, and ability to be grown on marginal or non-arable lands without displacing food crops [1][2][3]. With the goal of better understanding the potential of this sector, a mixed-integer linear programming (MILP) optimization framework is developed for the design and operation of the supply chain (SC) of ligno-cellulosic biofuels over a given time horizon. This work aims to integrate temporal factors into traditional SC models in order to identify key decision variables regarding the SC network, obtaining not only the lowest cost, but also environmental impact.

A typical biofuels supply chain network design (SCND) problem consists of an underlying network of harvesting sites, plants, distribution centers, and transportation modes and infrastructure linking those entities [4][5]. While the cultivation of biomass closely relates to landscaping decisions and the agricultural sector, the production of fuels depends on the technology and operation of a conversion plant. The distribution network serves to establish connections between facilities and harvesting sites and delivering materials from one place to another. Mathematical optimization models can identify the strategic and operational planning of the entire SC network. Appropriate optimization models for the design and operation of biofuel SC are critical to increasing the economic viability and decreasing the environmental impact of the system [6].

While many optimization approaches have been developed and applied to the design of second-generation biofuels SCs, there are no approaches integrating the spatial and temporal dimensions. Most spatially explicit studies look into the system for one representative year and do not account for changes in, for example, energy demands. Thus, the predicted designs may not be suitable in the future [7][8]. On the other hand, models that account for temporal aspects mainly focus on aggregate capacity and production planning, overlooking the spatial dimension [9].

To address this limitation, we develop an integrated optimization model that considers simultaneously time-varying and spatially explicit aspects, thereby ensuring that the proposed networks are robust to the ever-changing energy market. Facilities can be established at any time point throughout the horizon and once established, they can be expanded to accommodate more production. In addition to the system configuration, time-specific operational decisions, such as harvesting and production planning, are obtained.

The regional area of study includes the eight states in the US Midwest: Ohio, Michigan, Indiana, Illinois, Wisconsin, Missouri, Iowa, and Minnesota. This region is the topic of interest since switchgrass, a cellulosic feedstock, and other native grasses grown on marginal lands in the region can facilitate large amounts of carbon-negative biofuels. Marginal lands are lands less suitable for food production. They also have varying crop yields and sequestration potentials of soil-organic carbon (SOC). The availability and distribution of marginal lands depend significantly on the geographical location and the definition of marginal lands used. In this study, we specifically chose "historically abandoned" lands, which are defined as lands that have been taken out of food production for at least 5 years, and not converted to urban or water use. Crop simulation models are used to attain realistic, field-specific, high-resolution data on biomass yield and soil carbon potential, using land quality, local weather, and management options [10] [11].

In terms of energy demand evolution, the Energy Information Administration (EIA) has come up with 8 different scenarios for the biofuels market [12]. We use the "High Oil Price" projection case. The projection assumes higher global crude oil prices, making biofuels more price competitive with petroleum-based fuels. As a result, a sharp rise in biofuels domestic demands and consumption is predicted. Furthermore, we consider the

greenhouse gasses (GHG) balance of the entire system. We adopt the concept of "social cost of carbon" (SCC) so that both environmental and economic metrics can be considered. The term is defined as the economic cost by an additional ton of carbon dioxide emission according to the Environmental Protection Agency (EPA) and climate scientists [13]. By utilizing SCC.

Our results show that the inclusion of temporal aspects into spatially explicit optimization models is feasible. We show that the optimal solution varies significantly with the weight placed to environmental outcomes. The developed model helps shed light into how the integrated land-crop-logistics-biorefinery systems should be developed over time.

References:

[1] G. P. Robertson, S. K. Hamilton, B. L. Barham, B. E. Dale, R. C. Izaurralde, R. D. Jackson, D. A. Landis, S. M. Swinton, K. D. Thelen, and J. M. Tiedje, "Cellulosic biofuel contributions to a sustainable energy future: Choices and outcomes," Science, vol. 356, 6 2017.

[2] S. N. Naik, V. V. Goud, P. K. Rout, and A. K. Dalai, "Production of first and second generation biofuels: A comprehensive review," Renewable and Sustainable Energy Reviews, vol. 14, pp. 578–597, 2 2010.

[3] R. E. Sims, W. Mabee, J. N. Saddler, and M. Taylor, "An overview of second generation biofuel technologies," Bioresource Technology, vol. 101, pp. 1570–1580, 3 2010.

[4] M. Eskandarpour, P. Dejax, J. Miemczyk, and O. P´eton, "Sustainable supply chain network design: An optimization-oriented review," Omega, vol. 54, pp. 11–32, 7 2015.

[5] H. Yildiz, J. Yoon, S. Talluri, and W. Ho, "Reliable Supply Chain Network Design," Decision Sciences, vol. 47, pp. 661–698, 8 2016.

[6] C. Papapostolou, E. Kondili, and J. K. Kaldellis, "Development and implementation of an optimisation model for biofuels supply chain," Energy, vol. 36, pp. 6019–6026, 10 2011.

[7] R. T. Ng, D. Kurniawan, H.Wang, B. Mariska, W.Wu, and C. T. Maravelias, "Integrated framework for designing spatially explicit biofuel supply chains," Applied Energy, vol. 216, pp. 116–131, 4 2018.

[8] L. He-Lambert, B. C. English, D. M. Lambert, O. Shylo, J. A. Larson, T. E. Yu, and B. Wilson, "Determining a geographic high resolution supply chain network for a large scale biofuel industry," Applied Energy, vol. 218, pp. 266–281, 5 2018.

[9] A. Azadeh, H. Vafa Arani, and H. Dashti, "A stochastic programming approach towards optimization of biofuel supply chain," Energy, vol. 76, pp. 513–525, 11 2014.

[10] B. Maestrini and B. Basso, "Subfield crop yields and temporal stability in thousands of US Midwest fields," Precision Agriculture, vol. 22, pp. 1749–1767, 2021.

[11] E. G. O'Neill, R. A. Martinez-Feria, B. Basso, and C. T. Maravelias, "Integrated spatially explicit landscape and cellulosic biofuel supply chain optimization under biomass yield uncertainty," Computers & Chemical Engineering, vol. 160, p. 107724, 4 2022.

[12] EIA, "Annual Energy Outlook 2023."

[13] EPA, "The Social Cost of Carbon: Estimating the Benefits of Reducing Greenhouse Gas Emissions," tech. rep., EPA, 2017.

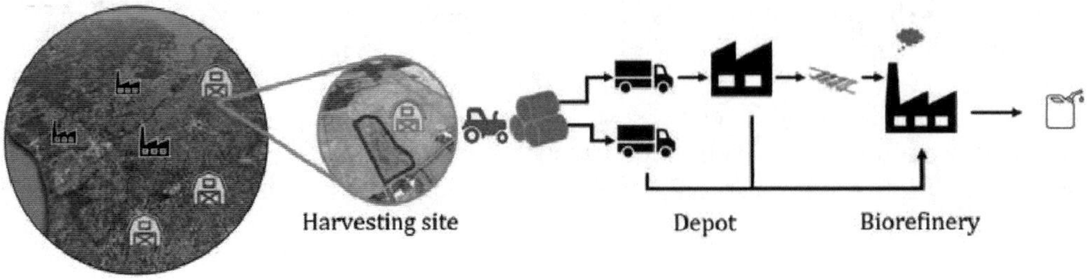

Figure 1: Structure of a Biofuels Supply Chain

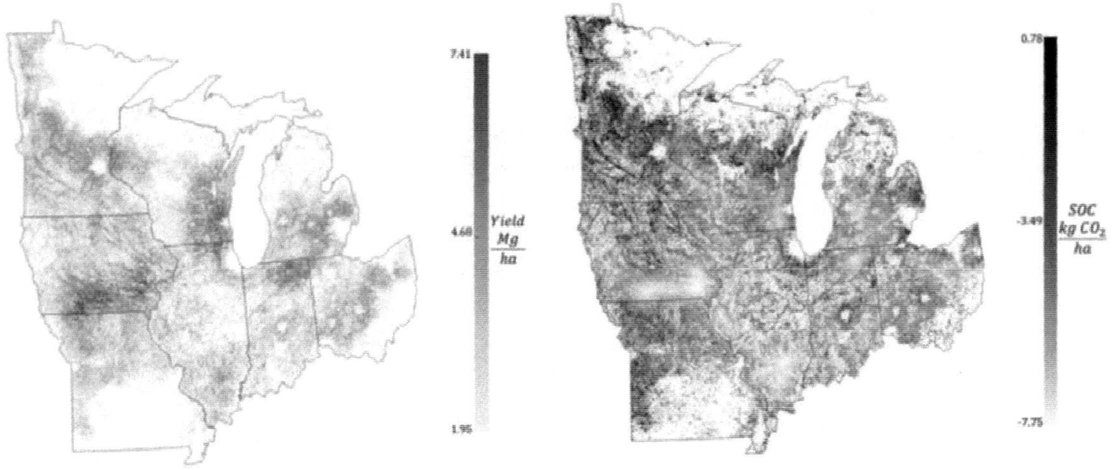

a) b)

Figure 2: The distribution of potential biomass fields weighted by: a) the potential available biomass (Mg/ha), b) the potential SOC (kgCO$_2$/ha)

662a Techno-Environmental Analysis of Olefins Production from Syngas Via Sorption-Enhanced Synthesis

Aurora Cavaliere, Massimiliano Materazzi, Paola Lettieri, Andrea Paulillo, Nilay Shah, Semra Bakkaloglu

Nowadays, the most widely used method for producing olefins is from fossil fuels through the cracking process. A staggering 400 Mt of CO_2 are emitted by light olefin plants, accounting for 27% of all CO_2 emissions in the industry. In recent years, the transformation of methanol into olefins has been gaining popularity, due to the possibility of generating methanol from renewable feedstock. However, the conventional process still presents several challenges, including high energy consumption and significant direct CO_2 emissions. Moreover, the current dependence on fossil raw materials, such as coal and natural gas, transformed into syngas to produce methanol raises further environmental concerns. A promising solution to address these challenges is the re-optimization of the entire production chain, implementing a circular model, where waste and biomass are used as sustainable feedstock for a new process. Indeed, advanced thermochemical technologies (ATTs), like pyrolysis and gasification, have an important role in converting waste into valuable chemical feedstock, hence promoting the green energy transition.

This project focuses on a novel olefin synthesis process specifically tailored for operating with CO_2-rich feedstock, as in the case of syngas obtained from the gasification of waste and biomass. In particular, the sorption-enhanced olefin synthesis (SEOS) technology helps us to overcome thermodynamic limitations associated with methanol formation and subsequent dehydration, thus improving yields and selectivity towards light olefins, while reducing energy consumption and CO_2 by-products.

A preliminary experimental campaign has identified the best catalyst and water sorbent combination for performing SEOS from a simulated biomass syngas stream. This has demonstrated the feasibility of combining two different processes (i.e., methanol synthesis and methanol-to-olefins) into a single reacting system, thus significantly simplifying the process layout and integration potential.

A sophisticated set of models has been developed using Aspen Plus to simulate the performance of the fully integrated biomass-to-olefins system, as well as to produce the mass and energy balance of the full plant. Experimental data have been directly used to inform the product distribution of direct syngas-to-olefins in the presence or absence of water removal steps and validate the model results. An energetic optimization and a sensitivity analysis of different parameters were done to probe and increase the efficiency of the plant. Furthermore, a techno-economic analysis, of the proposed plant layout was conducted to understand the operating cost of this technology and the levelised cost of olefin products.

Finally, the optimised data, mass, and energy balances will be used to perform a comprehensive Life Cycle Assessment (LCA) analysis to unveil the sustainability performance but also the bottlenecks of the process, comparing with other olefin synthesis processes, and to analyze different impact categories, such as climate change, resource use, energy carriers, freshwater eutrophication, etc., using a cradle-to-gate approach.

662c Automatic High Throughput Reaction Pathway Identification for More Sustainable Chemicals Production

Dion Jakobs, Lucas Francisco dos Santos, Gonzalo Guillen-Gosalbez

The chemical industry is a known hard-to-abate industry due to its high dependence on fossil resources as sources of carbon-rich raw materials and energy for high-temperature industrial heating [1]. Unlike other sectors, such as the power sector where renewables can decouple electricity generation from carbon, the production of chemicals is intrinsically bound to carbon. This is due to the wide range of carbon-based chemical products (*i.e.*, vastly different number of carbon atoms in the molecules) [1], and the dependence on high-temperature thermal energy [2]. To achieve a more sustainable chemicals production, we could transition towards a net zero emission (NZE) chemical industry by closing the so-called "carbon loop". NZE pathways that present promising options for achieving a closed carbon loop include those that couple fossil feedstock technologies with carbon capture and storage (CCS), and others based on renewable resources, e.g., carbon capture and utilization (CCU), carbon recycling, and biomass [1].

Amongst these potential alternatives, CCU stands out as it utilizes captured CO_2, which is potentially emitted at the chemical's production process as a direct emission and at the end-of-life phase via incineration or degradation. This CO_2 could replace the traditional carbon-rich fossil feedstocks, enabling an intrinsically circular production. CCU shows some advantages over other NZE pathways for the chemical sector, such as not relying on CO_2 underground storages and the avoidance of leakage risks, in contrast to CCS [3], and the lack of competition for the carbon feedstock [4] unlike biomass. According to many projections, it is likely that most, if not all, forms of NZE technologies will be implemented in some optimized combination in chemicals production, taking advantage of regional aspects, resource availability, and socio-political acceptance [5]. Despite the recognized potential of CCU as a necessary part of achieving a sustainable chemical industry [5, 6], this pathway still faces the challenges of having a high energy and electricity demand [7] and, critically, many of the CO_2-based production processes are still at low technology readiness level (TRL), only being proven in the lab scale [8]. Although many researchers have made great progress in the study of reaction feasibility, reaction mechanisms, catalysts, etc., there are still large gaps in the understanding of what potential these new technologies have for profitably reducing the reliance on fossil resources and CO_2 emissions [5, 8].

At present, experimental groups develop new potentially appealing routes, whose sustainability performance is seldom investigated in-depth in the original publications. Based on such articles, modeling groups develop techno-economic and environmental studies, often focusing on the most promising or mature technologies. This decoupled approach implies that the discovery of promising new sustainable processes relies often on an iterative, and manual method involving chemical intuition, simulations, and experimentation – often costly and time-consuming. Alternatively, we could employ chemical reaction databases such as Reaxys, CAS SciFinder, and USPTO to identify new promising chemical reaction pathways from sustainable feedstocks , which would then be analyzed more in-depth using process simulation and LCA.

A number of works have already utilized the said chemistry datasets for different purposes, including creating and analyzing the network of organic chemistry [10] (NOC), which could enable the rapid search for innovative reaction pathways or key chemicals necessary to the NZE transition in the chemical industry. The vast application of the NOC includes statistical network analyses [11], identification of strategic molecules for future circular supply chains [12], and investigations on the cyclic nature of organic chemistry [13]. Additionally, the NOC allows the identification of sustainable multistep chemical reaction pathways for fine chemicals within simplified small sub-networks [14]. Based on these datasets, large-scale screening of new reaction pathways can be completed either manually combined with human chemical intuition [8] or programmatically via high throughput screening to drive the direction of more extensive research [9]. The main limitation of these approaches is the availability of data. Manual exploration approaches are typically bounded by the intuition and knowledge of the human users. Alternatively, programmatic exploration is severely hindered by the prevalence of data gaps and unprocessed data.

To contribute towards the automatic exploration and identification of promising sustainable chemical routes for bulk and fine chemicals production, we introduce a high throughput reaction pathway screening tool based on the Reaxys chemistry database. This work innovates on similar works by introducing methods to automatically fill in critical data gaps, such as stoichiometry, and a greedy search algorithm to search the NOC for promising sustainable reaction pathways. In this exemplary case applied to CCU, the high throughput tool generates a localized bipartite graph network representation of the NOC by querying the Reaxys API. The gathered data is processed and any critical data gaps are filled using optimization or other property estimation techniques. Incomplete stoichiometric reactions are completed using optimization models that propose chemically valid completed stoichiometric equations. Green chemistry-inspired key performance indicators (KPIs), such as atom economy, hydrogen efficiency, etc. are calculated for each reaction. To traverse the dense graph from CO_2 to a desired target bulk or fine chemical, a greedy, depth-first, graph path search algorithm is developed. The algorithm explores along the best-performing reaction pathway until it reaches the target chemical and restarts, always searching along the next best partially explored or unexplored reaction pathway until the desired number of reaction pathways have been collected.

Based on the KPI selected, the high throughput screening tool delivers a ranked list of the most promising full reaction pathways, i.e. pathways connecting CO_2 and other selected cutoff raw materials to the target chemical. The ranked list can be further characterized by the type of reactions occurring within the reaction pathways, e.g. electrochemical or thermocatalytic reactions, that can be specifically analyzed or independently compared. With this work, we enable the broader audience of researchers investigating sustainable chemistry and in particular, CCU, to have access to relevant and curated data so that they might identify and focus efforts on, by design, promisingly sustainable reaction pathways for our transition to a future NZE chemical industry. Our results show that the method proposed can identify pathways performing well in a range of sustainability metrics that would be very hard to find using simple iterative methods based on heuristics.

References

[1] P. Gabrielli *et al.*, “Net-zero emissions chemical industry in a world of limited resources,” *One Earth*, vol. 6, no. 6, pp. 682–704, Jun. 2023, doi: 10.1016/j.oneear.2023.05.006.

[2] C. Chung, J. Kim, B. K. Sovacool, S. Griffiths, M. Bazilian, and M. Yang, “Decarbonizing the chemical industry: A systematic review of sociotechnical systems, technological innovations, and policy options,” *Energy Research & Social Science*, vol. 96, p. 102955, Feb. 2023, doi: 10.1016/j.erss.2023.102955.

[3] E. A. Quadrelli, G. Centi, J. Duplan, and S. Perathoner, “Carbon Dioxide Recycling: Emerging Large-Scale Technologies with Industrial Potential,” *ChemSusChem*, vol. 4, no. 9, pp. 1194–1215,

Sep. 2011, doi: 10.1002/cssc.201100473.

[4] A. Muscat, E. M. De Olde, I. J. M. De Boer, and R. Ripoll-Bosch, “The battle for biomass: A systematic review of food-feed-fuel competition,” *Global Food Security*, vol. 25, p. 100330, Jun. 2020, doi: 10.1016/j.gfs.2019.100330.

[5] P. Markewitz *et al.*, “Worldwide innovations in the development of carbon capture technologies and the utilization of CO2,” *Energy Environ. Sci.*, vol. 5, no. 6, p. 7281, 2012, doi: 10.1039/c2ee03403d.

[6] Á. Galán-Martín, V. Tulus, I. Díaz, C. Pozo, J. Pérez-Ramírez, and G. Guillén-Gosálbez, “Sustainability footprints of a renewable carbon transition for the petrochemical sector within planetary boundaries,” *One Earth*, vol. 4, no. 4, pp. 565–583, Apr. 2021, doi: 10.1016/j.oneear.2021.04.001.

[7] I. Ioannou, Á. Galán-Martín, J. Pérez-Ramírez, and G. Guillén-Gosálbez, “Trade-offs between Sustainable Development Goals in carbon capture and utilisation,” *Energy & Environmental Science*, vol. 16, no. 1, pp. 113–124, 2023, doi: 10.1039/D2EE01153K.

[8] A. Otto, T. Grube, S. Schiebahn, and D. Stolten, “Closing the loop: captured CO 2 as a feedstock in the chemical industry,” *Energy & Environmental Science*, vol. 8, no. 11, pp. 3283–3297, 2015, doi: 10.1039/C5EE02591E.

[9] J. M. Weber, Z. Guo, C. Zhang, A. M. Schweidtmann, and A. A. Lapkin, “Chemical data intelligence for sustainable chemistry,” *Chem. Soc. Rev.*, vol. 50, no. 21, pp. 12013–12036, 2021, doi: 10.1039/D1CS00477H.

[10] C. M. Gothard *et al.*, “Rewiring Chemistry: Algorithmic Discovery and Experimental Validation of One-Pot Reactions in the Network of Organic Chemistry,” *Angewandte Chemie*, vol. 124, no. 32, pp. 8046–8051, Aug. 2012, doi: 10.1002/ange.201202155.

[11] P.-M. Jacob and A. Lapkin, “Statistics of the network of organic chemistry,” *React. Chem. Eng.*, vol. 3, no. 1, pp. 102–118, 2018, doi: 10.1039/C7RE00129K.

[12] J. M. Weber, P. Lió, and A. A. Lapkin, “Identification of strategic molecules for future circular supply chains using large reaction networks,” *React. Chem. Eng.*, vol. 4, no. 11, pp. 1969–1981, 2019, doi: 10.1039/C9RE00213H.

[13] M. D. Bajczyk, P. Dittwald, A. Wołos, S. Szymkuć, and B. A. Grzybowski, “Discovery and Enumeration of Organic-Chemical and Biomimetic Reaction Cycles within the Network of Chemistry,” *Angew Chem Int Ed*, vol. 57, no. 9, pp. 2367–2371, Feb. 2018, doi: 10.1002/anie.201712052.

[14] A. A. Lapkin *et al.*, “Automation of route identification and optimisation based on data-mining and chemical intuition,” *Faraday Discuss.*, vol. 202, pp. 483–496, 2017, doi: 10.1039/C7FD00073A.

662d A Computational Toolbox Supporting the Development of Safe and Sustainable By Design Chemicals and Materials

Dimosthenis Sarigiannis, Fotini Nikiforou, Achilleas Karakoltzidis, Anna Agalliadou, Tomas Rydberg, Maja Halling, Chiara Laura Battistelli, Emilio Benfenati, Cecilia Bossa, Evert Bouman, Émilien Bourgé, Milena Brouwer-Milovanovic, Annabel Hill, Eleni Iacovidou, Ivo Iavicoli, Tomi Kanerva, Therese Kärnman, Veruscka Leso, Jenny Lindén, Magnus Lofstedt, Bernd Nowack, Araceli Sánchez Jiménez, Susanne Resch, Gianluca Selvestrel, Kirsi Siivola, Anezka Sharma, Vrishali Subramanian, Rosella Telaretti Leggieri, Martijn van Bodegraven, Joanke Van Dijk, Jaco Westra, Ziye Zheng, Antonis Gypakis, Spyros Karakitsios

The European Union (EU), through the Chemicals Strategy for Sustainability (CSS) and the Zero Pollution Action Plan, has emphasized the imperative for an environment and economy that are both climate-neutral and toxic-free. This emphasis underscores a transition towards a Safe and Sustainable by Design (SSbD) approach, by integrating safety and sustainability considerations into the early design stages of chemicals and materials, focusing on the whole life cycle. SSbD aims to foster the chemical innovation process by developing and assessing novel and existing chemicals. This is achieved by adhering to circular economy principles, all while mitigating the potential negative impacts on human health, the environment and society in the short and long term. The EC introduced a framework for SSbD, which is a comprehensive five-step approach for the safety and sustainability assessment of chemicals and materials. The first three steps involve safety assessment including hazard, occupational exposure, environmental exposure, and consumer exposure. The last two steps are related to the sustainability assessment, including environmental and socioeconomic aspects.

Within the EU partnership on chemical risk assessment (PARC), the PARC SSbD toolbox is being developed. The PARC toolbox aims at the operationalization of the EC framework, by providing an integrative and innovation toolbox for both innovators and regulators. It will include all the relevant data, methods, and tools for SSbD, along with automated pipelines that encompass every step of the framework across the stages of innovation. The tools will be structured and mapped in accordance with the stages of chemical or material development into a phased commercial product development (stage-gate model) and the five steps of the EC framework. The PARC toolbox is based upon a methodological framework (in development) that leverages advanced methods for hazard, exposure, human health risk assessment, and sustainability assessment. Initiating from user-provided data, including chemical structure parameters and a designated parameter list, a range of tools with varying levels of complexity will be employed to facilitate the SSbD assessment across the innovation process.

The development of the alpha version of the toolbox is currently underway. As part of this development process, a detailed case study was conducted to explore the toolbox's full potential and to discern any disparities in the findings between the two stages as well as between the different tools used. The case study included the assessment of Bisphenol-A (BPA) and two potential alternative substances: Bisphenol-AP and Isosorbide. The selected substances underwent evaluation in two distinct applications: as a BPA alternative for polycarbonate bottles and as a BPA replacement in epoxy resin paints. To determine the efficiency of the toolbox, the case study was conducted in both the early and late stages of innovation. In early innovation, only the structure and the potential application of the substance were considered as input information. In the context of late innovation, the assessment process was informed by data derived from pertinent databases (e.g. ECHA Database, EFSA, PubChem, CompTox Chemical Dashboard) or from relevant studies. In both scenarios, a range of tools were used to perform the SSbD assessment.

In early innovation, Quantitative structure-activity relationship (QSAR) models, such as VEGA, OECD QSAR toolbox, Janus, Oncologic, Mistra SafeChem *in silico* Toolbox and Danish (Q)SAR database, were applied in Step 1 to predict the hazard endpoints (e.g., carcinogenicity, mutagenicity, reproductive toxicity, endocrine disruption) requested in the EC JRC framework. In Steps 2 (manufacturing and processing phase) and 3 (use phase), a number of models, including ECETOC TRA, ProScale, ART, Stoffenmanager, INTEGRA, ConsExpo, SimpleBox, CEM and Vermeer FCM, were employed to forecast potential risks relating to occupational, environmental, and consumer exposure. These models were used in both early and late innovation, and it was found that there were differences in their outcomes. The observed difference in the results between the early and late stages of innovation can be attributed to the source of input data used. Specifically, the former relied on QSAR predictions, while the latter was based on experimental measurements. In Step 4, a preliminary Life Cycle Assessment (LCA) was conducted to evaluate the environmental sustainability in early innovation. This assessment was performed using the GaBi LCA software, which is a widely recognized tool for conducting LCAs. Lastly, the exploration of Step 5, the socioeconomic analysis step, was very limited.

Overall, QSAR models, played a vital role in predicting hazards and physicochemical properties during early innovation when data availability was limited. In addition, variations in SSbD scores for Steps 1-3 were observed, with the selection of QSARs and the available data level influencing the results. Therefore, it is of great importance to evaluate the reliability of predictions resulting from QSARs and to perform an uncertainty analysis of the results from different tools. As a result, integrating New Approach Methodologies (NAMs) with QSARs to assess hazard potency is highly significant. Moreover, a need for further development of models for endocrine disruption and immunotoxicity prediction is highlighted. Finally, it is crucial to develop, test, and incorporate reliable predictive models for toxicity and tools for prospective LCA and socioeconomic assessment. In particular for LCA in early innovation, the lack of data regarding up- and downstream life cycle stages contributes to results uncertainty, and better guidance on how to populate the model is required.

In conclusion, the concept of SSbD is characterized by a high level of complexity arising from the diverse types of assessments essential for its implementation. This study has established a basis for the further refinement of the SSbD toolbox, aiming to enhance its efficacy and applicability and eventually foster cohesion across diverse policies and strategies. The SSbD toolbox will be a guidance tool that will include all the relevant data, tools, and methods for the operationalization of SSbD while focusing on the integration of risk and sustainability assessment.

662e Optimal Solvent Selection Using Computer-Aided Molecular Design (CAMD) to Obtain Octacosanol from Sugarcane Rind

Nistala Venkata Subrahmanyam, Sharad Bhartiya, Vinay A Juvekar, Urmila Diwekar

Introduction

India stands as the world's second-largest sugar producer and a significant consumer. With over 700 sugar factories contributing to an annual production of approximately 340 lakh metric tonnes of sugar [1]. This sector's milling and clarification processes generate substantial solid wastes, including bagasse, rind, press mud, and bagasse fly ash. Nearly one-third of each ton of crushed sugarcane is collected as sugarcane waste and leaves, comprising esters, aldehydes, carboxylic acids, and alcohols [2]. Press mud, a byproduct, contains valuable nutraceuticals such as policosanols, which are aliphatic alcohols with carbon atoms ranging from C20 to C32. Nutraceuticals, a fusion of 'nutrition' and 'pharmaceuticals,' are natural health supplements. The United States accounted for the largest market share for Nutraceuticals in the North American region, owing to consumers' rising interest in preventive healthcare and self-medication and an increase in the aging population's demand for functional food and beverages to prevent or mitigate aging-related conditions. In addition, increasing healthcare spending and rising consumer disposable income, coupled with changing consumer perception towards nutraceuticals due to their proven health benefits, further propels the market's growth [3].

Policosanols, primarily composed of octacosanol (66% - 67%) [4], can be derived from sources like plant waxes (sugarcane wax, soybean wax, wheat extract), fruits, leaves, seeds, and animal wax like beeswax [5]. These policosanols exhibit various health benefits, including reducing platelet aggregation, preventing low-density lipoprotein (LDL) peroxidation, and addressing cardiovascular diseases [5], [6]. Additionally, Octacosanol (C28), a prominent component, has been noted for its potential anti-Parkinsonism effect [7]. Being hydrophobic, it is insoluble in water but partially soluble in edible oils [8]. Given its significance, there has been a growing interest recently in extracting octacosanol from sugarcane.

1-octacosanol is a long-chain alcohol available in plant waxes and is a member of the broader family of policosanols (20-36 carbon). It is used as a dietary supplement and is known to have the potential for reducing blood cholesterol, stress-related sleep disorders, and benefits for Parkinson's disease. It contains antioxidants. A detailed review of the health benefits of octacosanol is available [9]. Irmak et al. (2006) report the number of dietary supplements containing policosanols that are commercially available in the US market [5]. Most of these products are prepared from beeswax or sugar cane extracts. While octacosanol is found in many plant waxes, its separation from filter mud produced after sugarcane juice clarification using extraction with hot ethanol reflux has been reported [10]. A US patent by Chinen (2009) [11] also reports a method of extracting octacosanol from sugarcane skin and extracting it from an organic solvent such as benzene, isopropanol, and hexane. The extract is then eluted using benzene and further purified using thin-layer chromatography. Analysis of octacosanol has been reported using GC at 320°C and an HP-5 capillary column with helium as carrier gas. The extraction step needs to be followed by a purification step that will fractionate the desired compound from the vast number of phytochemicals available in the filter press mud.

The primary goal of current research is to develop an economically viable process to obtain octacosanol from sugarcane wax or filter press mud. This involves starting from optimizing solvents used in extraction and adsorbents used in adsorption steps to complete process design and operations. In this paper, we address the first step of optimizing solvents used for extraction of octacosanol.

GCM for Solvent Selection

In the first stage of the process, we propose a framework for computer-aided molecular design (CAMD) to generate optimal solvents and adsorbents for efficient and cost-effective purification of octacosanol from sugar cane filter mud and wax.

A group contribution method (GCM) is a technique to estimate and predict thermodynamic and other properties from molecular structures. CAMD is the reverse use of the group contribution method to generate molecules with desirable properties.

CAMD for solvent selection involves consideration of various properties like distribution coefficient, and solvent selectivity. These properties can be defined in terms of infinite dilute activity coefficients as given below.

Distribution coefficient (m) is a measure of solvent capacity that provides an equilibrium relation between extract and raffinate [12]. The relation between infinite dilute activity coefficient γ^{∞} and m is defined as shown in Eq. 1, [13]

$m \approx \gamma^{\infty}{}_{B,A}/\gamma^{\infty}{}_{B,S}$ (1)

where A, B, S represent raffinate, solute, and solvent, respectively.

Solvent selectivity (β) is defined as the measure of effectiveness of solvent to separate solute from the rest [12]. The relation between infinite dilute activity coefficient γ^{∞} and β is shown in Eq. 2, [13]

$\beta = m_B/m_A \approx \gamma^{\infty}{}_{A,S}/\gamma^{\infty}{}_{B,S}$ (2)

The group contribution method of UNIFAC can be used to obtain the infinite dilute activity coefficients.

CAMD for Solvent Selection

CAMD methods are successfully applied to separations such as distillation, extraction, and absorption. However, this is the first time such an approach has been taken to extract and purify a nutraceutical from biomass.

An optimization approach to CAMD involves solving the following problem, Eq.3, using a combinatorial optimization method, as the resulting problem involves the combinatorial explosion of alternatives.

Maximize $m = (\gamma^{\infty}{}_{B,A}/\gamma^{\infty}{}_{B,S})(MW_A/MW_S)$: Maxime distribution coefficient for solvent design (3)

Subject to

$\beta = (\gamma^{\infty}{}_{A,S}/\gamma^{\infty}{}_{B,S})(MW_B/MW_A) \geq S$: Solvent selectivity constraint for solvents (4)

$\sum\sum V_{ij} \leq \varepsilon$: Structural Constraint (5)

where MW is molecular weight

UNIFAC model for calculation of activity coefficients

This poses a challenge to optimization techniques. The possibility of obtaining local minima or maxima is very high in such problems. Some of the biomimetic optimization techniques, like the ant colony optimization, show a higher probability of obtaining global solutions. However, these techniques can be computationally intensive. In earlier work, we have increased the efficiency of ant colony optimization by using the Hammersley Sequence Sampling (HSS) to generate random numbers needed for the algorithm [14], [15]. We have used this Efficient Ant Colony Optimization for our work here.

Results and Discussions

The groups used for the generation of molecules are tabulated in Table 1.

Using the above resources, the solvents generated are tabulated in Table 2 along with distribution coefficient values and solvent selectivity values.

Theoretically, for such a system of compounds, solvents were generated. Also, using this method, we found m and β of some of the common solvents like n-hexane (2.336 & 0.488), toluene (5.601 & 0.915), and ethanol (0.060 & 1.479), respectively. It has been found experimentally that these three solvents do require little heating for wax to dissolve and form a clear solution. Hence, the model used here has a good validation for separating the desired component from the rest. It can be seen from Table 2 that the solvents generated are better solvents in terms of distribution coefficient and selectivity than the commonly used solvents.

In the next step, we will validate these solvents using experiments.

Conclusion

Octacosanol is one of the policosanols which has wide-range of health benefits. This can be obtained from sources like rice-bran, beeswax, sugarcane, seeds, leaves, etc. Achieving this may be done in two stages: extraction (or leaching) and adsorption. Our present focus would be on selecting the solvent for extraction. This has been achieved using a computational technique called CAMD incorporating the UNIFAC model. Various solvents were generated with respect to certain properties like distribution coefficient and solvent selectivity.

Table 1 The groups used for the solvent generation using UNIFAC model

Sno.	1	2	3	4	5	6	7	8	9	10
Sub-group	CH_3	CH_2	CH	C	CH_2=CH	CH=CH	CH=C	CH_2=C	OH	H_2O

Sno.	11	12	13	14	15	16	17	18	19	20
Sub-group	ACH_3	ACH_2	ACH	AC	ACCH	ACOH	CH_3CO	CH_2CO	CHO	COOH

Table 2 Distribution coefficient and solvent selectivity values of generated solvents

Name	m	β
Ethane	383.61	1.079
Ethanal	142.51	9.436
Propanal	65.689	7.275
Propane	37.521	0.671
2-methyl propanal	34.232	5.941
Butanal	34.055	5.942
2/3-methyl butanal	20.217	5.048
Pentanal	20.142	5.049
2,2-dimethyl propanal	18.305	5.082
2,3-dimethyl butanal	13.302	4.407
2/3/4-methyl pentanal	13.266	4.408

Butane	9.856	0.560
Hexanal	13.229	4.408
2,2/3,3-dimethyl butanal	12.337	4.437
2-methyl propane	9.932	0.560
2,3/4-dimethyl pentanal	9.447	3.924
2/3/4/5-methyl hexanal	9.427	3.924
Heptanal	9.407	3.925
2,2/3,3/4,4-dimethyl pentanal	8.927	3.949

References

[1] Ramesh Chand, "Report of the Task Force on Sugarcane and Sugar Industry," 2020. Accessed: Apr. 06, 2024. [Online]. Available: https://www.niti.gov.in/sites/default/files/2023-02/SugarReport.pdf

[2] R. Sharma, "Microbial-assisted remediation of food processing industry waste," in *Microbes and Microbial Biotechnology for Green Remediation*, Elsevier, 2022, pp. 391–403.

[3] Mordor Intelligence, "North America Nutraceutical Industry Size & Share Analysis - Growth Trends & Forecasts (2024 - 2029) Source: https://www.mordorintelligence.com/industry-reports/north-america-nutraceutical-market," North America.

[4] EFSA Panel on Dietetic Products Nutrition and Allergies (NDA), "Scientific Opinion on the substantiation of health claims related to policosanols from sugar cane wax and maintenance of normal blood LDL-cholesterol concentrations (ID 1747, 1748, 1864, 1951, 1954, 4693) and maintenance of normal blood HDL-cholesterol concentrations (ID 1747, 1748, 1864, 1951, 1954, 4693) pursuant to Article 13(1) of Regulation (EC) No 1924/2006," Wiley-Blackwell Publishing Ltd, Jun. 2011. doi: 10.2903/j.efsa.2011.2255.

[5] S. Irmak, N. T. Dunford, and J. Milligan, "Policosanol contents of beeswax, sugar cane and wheat extracts," *Food Chem*, vol. 95, no. 2, pp. 312–318, Mar. 2006, doi: 10.1016/j.foodchem.2005.01.009.

[6] A. Srisaipet, S. Phromchan, and T. Jaipaeng, "Policosanol extraction from beeswax and improvement of the purity," *Fluids and Chemical Engineering Conference (FluidsChE 2017)*, vol. 111, no. 02005, 2017.

[7] S. Oliaro-Bosso *et al.*, "Regulation of HMGCoA reductase activity by policosanol and octacosadienol, a new synthetic analogue of octacosanol," *Lipids*, vol. 44, pp. 907–916, 2009.

[8] S. Sen Gupta and M. Ghosh, "Octacosanol educes physico-chemical attributes, release and bioavailability as modified nanocrystals," *European Journal of Pharmaceutics and Biopharmaceutics*, vol. 119, pp. 201–214, Oct. 2017, doi: 10.1016/j.ejpb.2017.06.020.

[9] J. C. Taylor, L. Rapport, and G. B. Lockwood, "Octacosanol in Human Health." Nutrition, Vol. 19, No. 2, 2003, pp. 192-195. doi:10.1016/S0899-9007(02)00869-9

[10] S. Ou, J. Zhao, Y. Wang, Y. Tian, and J. Wang, "Preparation of octacosanol from filter mud produced after sugarcane juice clarification," *LWT*, vol. 45, no. 2, pp. 295–298, 2012, doi: 10.1016/j.lwt.2011.08.011.

[11] Isao Chinen, "Compositions and Foods and Drinks Containing Higher Fatty Acid Derivative," Dec. 09, 2004

[12] K.-J. Kim, U. M. Diwekar, and K. G. Joback, "Greener solvent selection under uncertainty," in *ACS Symposium Series*, Washington, DC; American Chemical Society; 1999, 2002, pp. 224–237.

[13] K. J. Kim and U. M. Diwekar, "Efficient combinatorial optimization under uncertainty. 2. Application to stochastic solvent selection," *Ind Eng Chem Res*, vol. 41, no. 5, pp. 1285–1296, Mar. 2002, doi: 10.1021/ie0101691.

[14] B. H. Gebreslassie and U. M. Diwekar, "Efficient ant colony optimization for computer aided molecular design: Case study solvent selection problem," *Comput Chem Eng*, vol. 78, pp. 1–9, Jul. 2015, doi: 10.1016/j.compchemeng.2015.04.004.

[15] B. H. Gebreslassie and U. M. Diwekar, "Homogenous multi-agent optimization for process systems engineering problems with a case study of computer aided molecular design," *Chem Eng Sci*, vol. 159, pp. 194–206, 2017, doi: 10.1016/j.ces.2016.05.026.

662g Navigating the Route of Polyester Textiles to Net-Zero Emissions

Aniket Mali, Bhavik Bakshi

The rise in emissions and consequent global warming pose significant challenges, urging nations worldwide to take decisive action, as demonstrated by the commitments made in the Paris Agreement [1]. Recognizing industries, particularly the textile sector, as major contributors to emissions underscores the imperative of addressing this issue [2]. Adding to the emissions, the impact of plastic pollution originating from the textile industry extends across vast distances. Specifically, discarded plastic clothing is estimated to have dispersed at least 1.4 quintillion (or one million trillion) plastic microfibers into the ocean [3]. Hence, textiles present a significant environmental threat. Amongst these textiles, polyester fiber stands out as the most utilized, significantly contributing to emissions. This is primarily due to its heavy reliance on fossil-based resources and energy-intensive production processes [4]. Many industries are aiming to achieve net-zero emissions owing to these factors [5]. This underscores the urgent need for mitigation strategies that simultaneously address environmental impact and economic feasibility.

In response to these pressing challenges, this research aims to chart a pathway toward achieving net zero emissions while considering economic factors. Initially, we conduct a comprehensive life cycle assessment (LCA) and techno-economic analysis of the polyester fiber value chain, delineating a cradle-to-gate system boundary. This
approach captures key manufacturing stages where significant improvements can be made. By employing comparative LCAs, we evaluate various alternative polyester molecules, such as Polyethylene terephthalate (PET), Polytrimethylene terephthalate (PTT), and Polylactic acid (PLA), alongside different fossil-based and biobased raw materials, thereby comparing their environmental footprints and economic efficiencies. Our findings indicate that PLA exhibits minimal emissions, while the conventional polyester manufacturing process from PET proves to be cost-efficient.

Utilizing multi-objective optimization and Pareto Analysis, we endeavor to identify optimal combinations of alternatives that strike a balance between emission reduction and cost efficiency. Our investigation reveals that a synthesis of PLA and PET production processes yields valuable trade-off insights for stakeholders, facilitating progress toward net-zero emissions. We also find that excluding PLA as an alternative in the optimization problem impedes the path to achieving net-zero emissions and increases the cost of the optimal combination. To address these issues, we identify cost and emission hotspots within the conventional route and propose solutions to mitigate them. Our analysis identifies electricity production from fossil routes and energy production from fossil resources as major emitters, contributing significantly to the life cycle cost. Therefore, we propose the incorporation of renewable energy alternatives and carbon capture and utilization processes as decision variables in the problem, offering promising avenues for achieving net-zero emissions in a financially sustainable manner. By examining trade-offs involved in incorporating these technologies, we find that utilizing wind energy, a combination of PLA and PET, carbon capture by direct air capture, and chemical looping, and converting the captured CO2 to ethylene are part of the trade-off solutions that lead to more cost-efficient and environmentally friendly outcomes.

In addition, we plan to extend the life cycle to encompass the use phase and recognizing consumers as significant stakeholders. We will explore diverse end-of-life scenarios to attain zero waste, emphasizing circularity. Furthermore, we aim to integrate emerging technologies and project future scenarios to develop a roadmap to achieve net-zero emissions [6]. Proposed multiperiod planning strategies provide a comprehensive roadmap toward achieving net-zero emissions, considering economic factors alongside environmental objectives. Ultimately, this research aims to provide comprehensive solutions to enhance the environmental sustainability of the polyester fiber value chain, contributing to the broader goal of fostering a greener future for the textile industry while ensuring economic viability.

References:

1. Savaresi, A. (2016). The Paris Agreement: A New Beginning? Journal of Energy & Natural Resources Law, [online] 34(1), pp.16–26. doi: https://doi.org/10.1080/02646811.2016.1133983.

2. Farhana, Kaniz, et al. "Energy Consumption, Environmental Impact, and Implementation of Renewable Energy Resources in Global Textile Industries: An Overview towards Circularity and Sustainability." Materials Circular Economy, vol. 4, no. 1, 29 Mar. 2022, https://doi.org/10.1007/s42824-022-00059-1.

3. UNEP (2019). Fashion's tiny hidden secret. [online] UN Environment. Available at: https://www.unep.org/news-and-stories/story/fashion's tiny hidden secret.

4. Palacios-Mateo, C., van der Meer, Y. and Seide, G. (2021). Analysis of the polyester clothing value chain to identify key intervention points for sustainability. Environmental Sciences Europe, [online] 33(1). Available at: https://enveurope.springeropen.com/articles/10.1186/s12302-020-00447-x.

5. www.formula4media.com. (n.d.). Achieving Net-Zero - Textile Insight. [online] Available at: https://www.formula4media.com/articles/achieving-net-zero.

6. Thakker, V. and Bakshi, B.R. (2024). Mapping the path to a net-zero chemicals industry by long-term planning with changes in technologies and climate. AIChE journal. doi:https://doi.org/10.1002/aic.18381.

289a Application of Modified Meshgraphnets for Subsurface Prediction during CO_2 Sequestration

Paul Holcomb, Alex Sun, Chung Shih, Guoxiang Liu, Hema Siriwardane

In the face of the increasingly dire consequences of anthropogenic climate change, capturing and storing carbon dioxide is paramount. However, several impediments exist to the safe and effective subsurface storage of CO2, such as cost of transport, identification of suitable sites for subsurface storage, and assessment of long-term risk from storage in subsurface aquifers. Accurate subsurface modeling is necessary to ensure that CO2 storage is both safe and effective. Still, such modeling has traditionally required either substantial time and computational power (numerical simulation) or a substantial amount of pre-existing data for training (machine learning models). Additionally, these models lack flexibility in dealing with both changes in discretization of the input data and generalizability beyond the data on which they are trained. In order to address these issues, this research applies graph neural networks (GNNs) to predict subsurface saturation and pressure during CO_2 injection in a model of the Illinois Basin-Decatur Project (IBDP).

GNNs provide a flexible, intuitive method for representing and manipulating complex unstructured data, which is often found in many practical domain problems such as fluid flow and subsurface characterization. These unstructured grids are easily represented in GNNs by representing spatially localized features such as permeability, porosity, saturation, and pressure as nodes in a graph and relationships between these properties as edges connecting these nodes. This research applies a specific GNN model called MeshGraphNets (MGN) to model the change in CO2 saturation and pressure over a 50-month time period (36 months of injection, 14 months post-injection). The MGN model leverages a message passing process that allows the network to learn both the spatial and temporal dynamics of this system simultaneously. Additionally, training on a limited dataset (64 realizations, 20 time points each) resulted in a high degree of accuracy in saturation prediction both within the same timeframe as the training (20 months, 0.039 average RMSE for saturation) and when projecting out to the end of injection (36 months, 0.053 average RMSE for saturation).

Temporal predictions such as those generated by MGNs and other similar models are prone to accumulated error over time; in order to address this, a multi-step rollout (MSR) training process was applied to calculate training loss. This method mimics the forward prediction during inference by "rolling out" multiple time points in a single training step using the previous prediction as input to the MGN model. By calculating the loss several time steps forward from the current prediction, the model is forced to find a more stable state over time. Application of MSR to the MGN model resulted in an average 15% reduction in inference error over time during forward prediction.

This study showcases the immense potential of GNNs as a game-changing methodology for predicting pressure and saturation evolution in CCS projects, ultimately paving the way for more sustainable and effective carbon storage solutions.

Figure 1: Expected (left), predicted (center), and difference between expected and predicted (right) CO_2 saturation at 20 months (top) and 36 months (bottom) for the MGN model.

IBDP Saturation

Time Point: 20 months

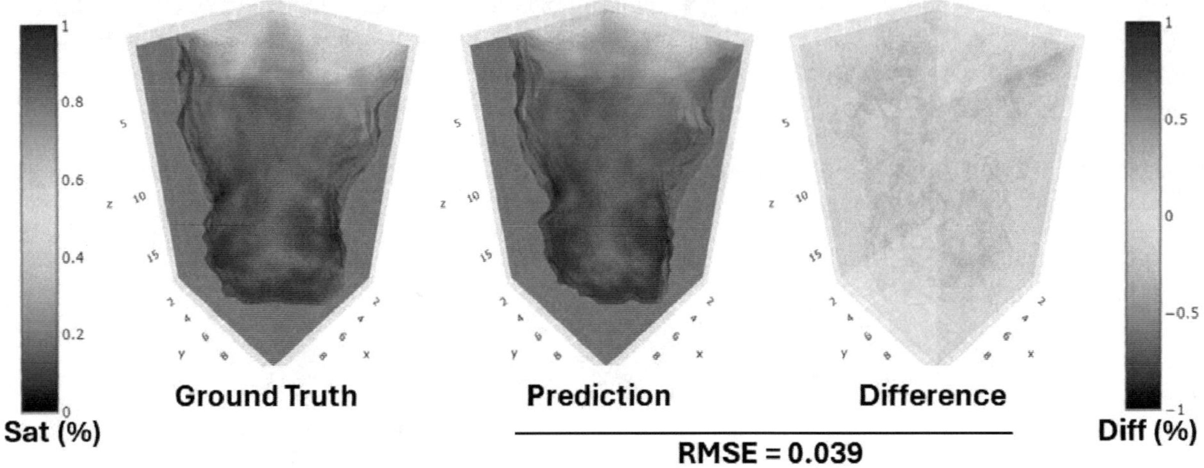

RMSE = 0.039

Time Point: 36 months

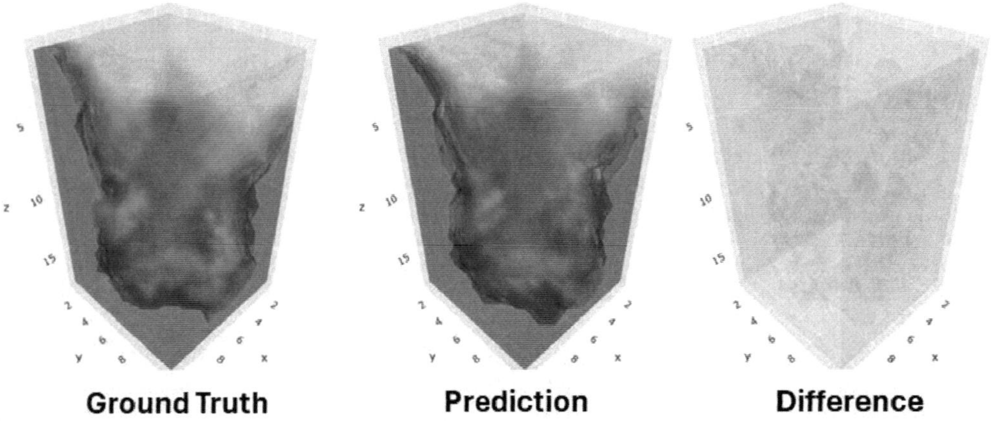

RMSE = 0.053

289b Monitoring Carbon Storage Sites Using Electromagnetics: Recovering a Baseline Conductivity Model at the Kemper Carbonsafe Site

Colton Kohnke, Yaoguo Li, Richard Hammack

When supercritical CO_2 is injected into the subsurface reservoir for storage, it needs to be monitored to ensure containment. Geophysics is one component of a successful monitoring strategy for carbon storage sites. Seismic surveys are a common element of these strategies, with repeat surveys occurring about every 5 years. However, seismic is expensive, the sensitivity to CO_2 saturation is limited, and it requires a large footprint on the surface. Electromagnetic (EM) surveying methods offer repeat surveys at a significantly lower cost, direct sensitivity to the CO_2 saturation in the storage formation, and the ability to utilize airborne measurements to increase spatial coverage. EM can potentially provide enough information to increase the time between seismic surveys, thereby lowering the overall cost to monitor the reservoir. EM methods are useful because injected CO_2 is resistive. When the CO_2 is injected into a conductive saline reservoir, it causes the conductivity of the area containing the CO_2 plume to drop. The drop in conductivity will alter the preferential path of current flow from an electric dipole transmitter, and then an image of the CO_2 plume can be recovered from measured magnetic field data. Specifically, we focus on semi-airborne EM, with a grounded dipole transmitter, and airborne magnetic field receivers.

To validate previous simulation results and determine the suitability of EM for monitoring efforts, we conducted a test EM baseline survey at the Kemper County CarbonSAFE site in Mississippi. A map of the survey area is shown in Figure 1. Supercritical CO_2 injection is planned at the marked 19-2 well location. Two transmitter dipoles were implemented, one oriented North-South along the highway, and the other forming a U-shape along county roads and forming an East-West dipole between the injection electrodes. The transmitters operate at a base frequency of 2.50 Hz. A high-temperature SQUID 3-component magnetometer receiver was flown along North-South flight lines spaced 75 m apart and 20 m altitude. The SQUID magnetometers were used due to their ability to detect magnetic fields lower than 0.1 pT with a low level of internal noise.

We performed a 3D inversion on the measured magnetics data to recover a baseline conductivity model. Nearby well logs were used to create a 1D reference model that is resistive in the top 900 m and conductive below. The geologic stratigraphic column of the region from previous studies suggests a sandstone aquifer at approximately 300 m depth (not present on the well logs due to logging starting depth) and 330 m of saline aquifer starting at 900 m. Figure 2 shows a cross-section depicting the recovered conductivity using the North-South transmitter. The inversion recovers a conductivity model with the conductive near surface reservoir despite not being present in the reference model. The saline aquifer is shown to start at about 1 km, be laterally continuous, and is approximately 300 m thick. Both results are consistent with known geology.

EM surveying methods successfully delineates the injection zone. Future surveys will use this recovered baseline model as a point of comparison so that the shape of the CO_2 plume and reservoir health can be monitoring as the plume moves in the subsurface. Now that a baseline conductivity model for the area has been recovered and validated against known geology, simulations of different injection scenarios can be performed and analyzed to determine the usefulness of EM surveying methods for monitoring predicted CO_2 plumes within the context of the full reservoir monitoring plan. While these results are promising for monitoring efforts at the Kemper CarbonSAFE site, the usefulness of EM elsewhere is dependent on local geology and the presence of anthropogenic sources of EM noise. For instance, EM will struggle with recovering the CO_2 plume in deep storage reservoirs (depths greater than 2 km), conductive geologic overburden, or sites near urban areas. Therefore, modeling studies and test surveys are important to determine the feasibility of the monitoring strategy.

Figure 1: Map of the study area with flight area in orange, North-South transmitter in yellow and the East-West transmitter in red.

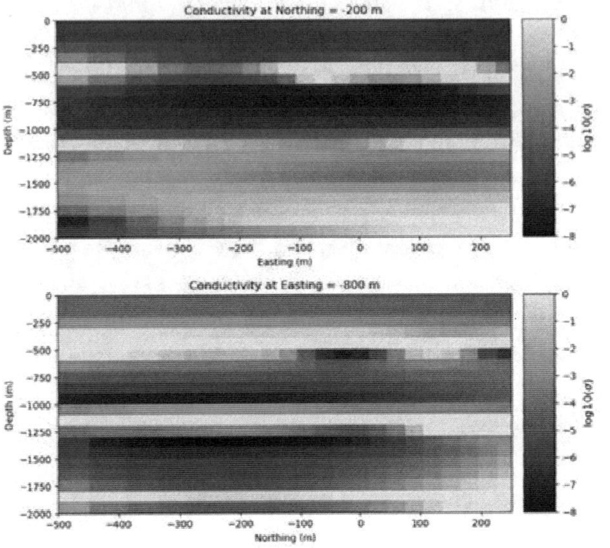

Figure 2: Inversion results using the North-South transmitter.

289c Permeability and Porosity of Tuscaloosa Mudstone at CO_2 Injection Site

Vyacheslav Romanov, Yee Soong, Igor Haljasmaa

Saline formations are promising targets for large-scale geologic carbon storage – likely representing billions of tons of potential resource in the conterminous United States, alone. One such candidate target is the Lower Tuscaloosa Formation into which the Southeast Regional Carbon Sequestration Partnership's (SECARB) Cranfield Project has injected more than 1 million metric tons of CO_2, into the saline part of the field comprised of primarily sandstones. The Tuscaloosa Formation is generally heterogeneous and composed of gravel, sand, and clay lithologies. In this work, composition, porosity, and permeability of the mudstone samples from the NETL-supported Cranfield Project were studied, before and after exposure to CO_2–brine environment. Specifically, a core sample was exposed to CO_2-saturated brine at the sequestration temperature and pressure for a period of six months. Powdered Tuscaloosa mudstone samples were also studied using optical microscopy and X-ray diffraction (XRD) to observe the mineralogical, textural, and geochemical changes resulting from CO_2-brine exposure.

Two pieces of core extracted from the depth of 3104-3105 m measured 10 cm in diameter and about 9 cm and 13 cm in length, respectively. Core sample densities were measured to be 2.39 g/cc and 2.51 g/cc for the first and second samples, respectively. Porosity and permeability of both samples were characterized prior to their exposure to CO_2-saturated brine; the second sample was subsequently used in the geochemical studies. Exposure of this sample to CO_2-saturated brine led to some chemical alterations and a significant increase in the permeability. Because of the relatively low initial permeability and relative thickness of the Tuscaloosa mudstone, the observed increase in permeability in response to CO_2-brine exposure is likely insufficient to substantially impact the interval's effectiveness as a natural seal.

289d Insight into the CO2 Pipeline Transportation Cost: New Cost Calculation Model with Regional Variation and CO2 Pipeline Network.

Kwang Hoon Baek, Sheik Tanveer, Pingping Sun, Amgad Elgowainy

Mapping of existing carbon dioxide (CO_2) point sources and potential storage sites suggests that a long distance exists between them. CO_2 delivery from the capture point to the storage point is an important process in the carbon capture and storage (CCS) system. While most studies focus on the CO_2 delivery cost by a pipeline distance, there are other aspects, which are normally overlooked, i.e., determining a realistic CO_2 pipeline design between existing sources and storage sites to calculate the optimum CO_2 delivery cost. In this study, we introduce a newly developed CO_2 pipeline transportation model and emphasize those key aspects with example cases based on existing industrial CO_2 sources and potential CO_2 storage locations.

We developed a python-based CO_2 pipeline transportation cost model by leveraging Argonne's 15-year modeling experience on pipeline networks and detailed cost data for natural gas pipelines in the United States. The model calculates a CO_2 pipeline delivery cost in two steps. First, it draws a pipeline route upon the existing road network, so that it captures a more realistic pipeline route and distance to calculate the pipeline length. Then, for a given CO_2 delivery conditions (CO_2 throughput, pressure, temperature, pipeline length, region, etc.), it calculates the CO_2 pipeline delivery cost by optimizing the pipeline diameter with the number of booster pumps. As a unique feature, this new model simulates a pipeline network to calculate the optimum CO_2 delivery cost by connecting multiple CO_2 sources to destination(s) or storage sites. This pipeline network suggests a more practical CO_2 delivery configuration because the CO_2 storage potential of a selected site is normally much greater than CO_2 amounts from individual CO_2 sources.

In this study, we will provide key insights into the cost of CO_2 pipeline transportation. It is known that the levelized CO_2 delivery cost (dollar per metric tons of CO_2, $/MT) will be reduced with a larger throughput of CO_2 transported and a shorter pipeline distance. In our calculations, we applied for up-to-date costs for material, labor, right of way, and other construction costs as well as the regional cost variations based on reported natural gas transmission pipeline costs. In Figure 1 and Figure 2, we compared CO_2 delivery costs in different regions. There are six groups with nine regions, and each group resulted in different CO_2 delivery costs. For example, when one million metric tons of CO_2 is delivered annually by a 200-mile pipeline, the levelized cost could be as low as $10/MT in Group 4, or as high as $35/MT in Group 1. This large difference (more than three times) indicates that we should consider the regional aspect of the CO_2 delivery cost.

Figure 3 (Case 1) and Figure 4 (Case 2) are the preliminary cases to demonstrate the capability of our model that captures realistic pipeline routes by a CO_2 pipeline network. In Case 1, we made a scenario connecting six CO_2 sources (ethanol plants) to a storage destination in Minnesota. The model calculates the optimized pipeline network connecting the CO_2 sources to the destination. The detailed pipeline information of each segment is provided in the table below the figure. The levelized cost to collect and deliver 1.17 million metric tons annually is $9/MT. In Case 2, there are four CO_2 sources (ethanol plants) connected to the destination in Ohio. The levelized cost to collect and deliver 0.59 million metric tons annually is $19/MT. Compared to Case 1, even though the total pipeline length is shorter, the CO_2 delivery cost is higher for two reasons: 1) the delivered CO_2 amount is smaller, and 2) the regional cost is higher.

We are currently developing CO_2 delivery scenarios from actual industrial CO_2 sources to potential storage locations (saline formations) using our model. The results will provide better understanding of CO_2 delivery cost tied with storage sites as a function of CO_2 throughput, distance, region, CO_2 sourcing (multiple small sources vs. single large source), and pipeline network. Our analysis will be an important link between studies on CO_2 capture and CO_2 storage.

(Fig. 1) 9 Regions and 6 Groups

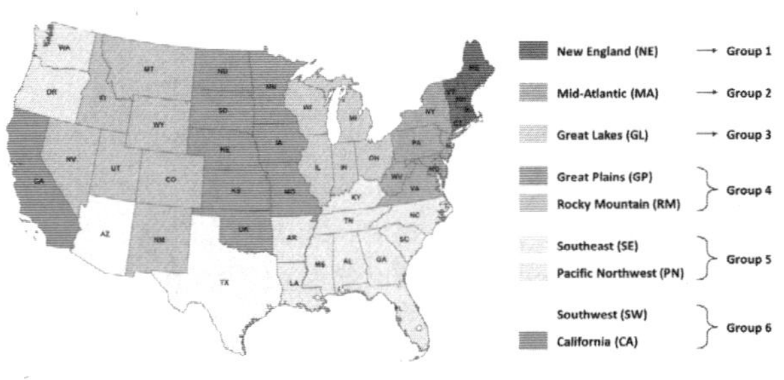

(Fig. 2) Regional Cost Variation

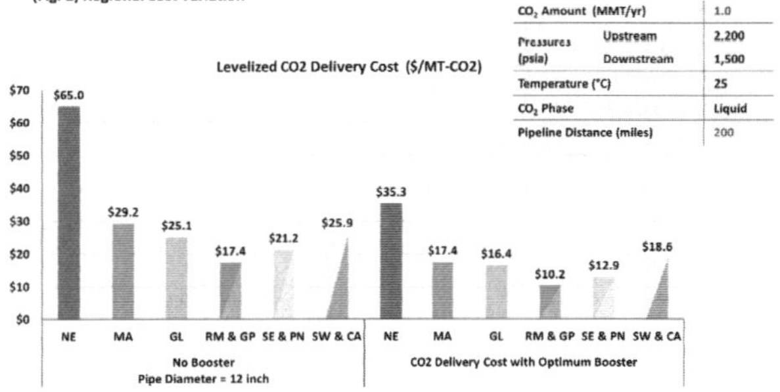

(Fig. 3) Pipeline Network – Case 1

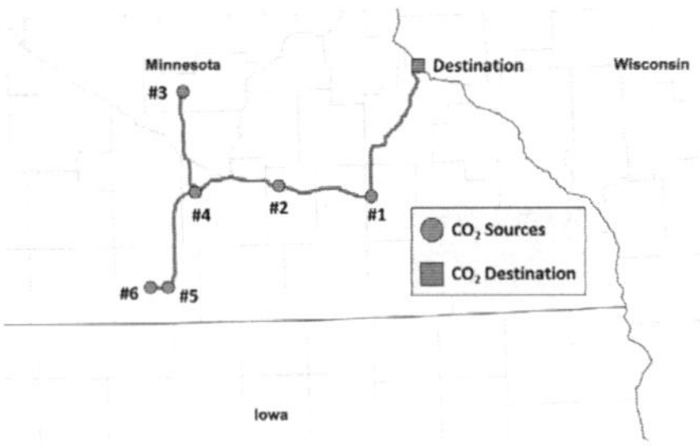

Pipeline Route	CO_2 Amount (metric tons/year)	Distance (mile)	Pipe Diameter (inch)	# Boosters
#6 → #5	152,420	5.3	4	1
#3 → #4	223,851	36.4	4	2
#5 → #4	355,765	37.3	4	3
#4 → #2	695,813	33.5	6	2
#2 → #1	952,130	34.9	8	1
#1 → Destination	1,174,272	55.1	8	1

Levelized CO_2 Delivery Cost for This Network: $9.1/Metric tons of CO_2

(Fig. 4) Pipeline Network – Case 2

Pipeline Route	CO_2 Amount (metric tons/year)	Distance (mile)	Pipe Diameter (inch)	# Boosters
#3 → #2	124,741	32.6	4	1
#4 → #1	263,153	43.1	4	2
#2 → #1	278,532	33.0	4	2
#1 → Destination	587,089	41.1	6	0

Levelized CO_2 Delivery Cost for This Network: $19.0/Metric tons of CO_2

289e Unveiling the CO_2 Storage Resources of the St. Peter Sandstone and Everton Formation in the Illinois Basin

Mansour Khosravi, Zohreh Askari, Roland Okwen

This study explores the Middle Ordovician St. Peter Sandstone and the Everton Formation within the Illinois Basin, highlighting their potential as significant geological storage reservoirs. In particular, the St. Peter Sandstone and Everton Formation emerge as promising targets for CO2 storage in southwestern Illinois, especially in areas where the Mt. Simon Sandstone is absent or has a low thickness. Through a detailed examination of their porosity, permeability, and spatial variation in thickness and modeling to support volumetric estimations, this research underscores the formations' resources for CO_2 storage and their role in climate change mitigation.

The St. Peter Sandstone Formation, characterized as a mature pure quartz arenite, spans the entirety of the Illinois Basin. Within the basin, the thickness of the St. Peter Sandstone varies significantly, ranging from approximately 33 to 656 feet (10 to 200 meters). This formation has undergone extensive diagenetic changes, including mechanical compaction and cementation. The Everton Formation, deposited in the early Middle Ordovician period, primarily consists of dolomite and sandstone intervals. The thickness of the Everton Formation varies, extending from less than 100 to about 125 feet (30 to 38 meters), and its presence is primarily confined to the southern regions of Illinois.

Porosity estimations and trends were derived from an analysis of approximately fifty wells in southern Illinois and over twenty wells in Indiana. To estimate permeability, a complex approach was adopted, involving the creation of a supervised machine learning algorithm. This algorithm was trained utilizing a dataset comprising core sample data alongside permeability estimations inferred from Nuclear Magnetic Resonance (NMR) log data and additional petrophysical logs, enabling the prediction of permeability across more than thirty wells. Subsequently, a rigorous geocellular model was developed and populated with porosity and permeability values that were geostatistically distributed. This modeling approach enabled the calculation of both bulk and pore volumes, facilitating a thorough evaluation of the potential storage resources within the St. Peter Sandstone and Everton Formation.

The porosity and permeability analyses conducted across these formations revealed a broad spectrum of values, with porosity figures ranging from less than 1% to upwards of 20%, and permeability rates spanning from less than 0.01 millidarcies to over 500 millidarcies. Significantly, this research uncovered a pronounced correlation between increasing depth and reduction in both porosity and permeability that can be attributed to compaction. Adapting methodologies from the DOE/NETL Carbon Storage Atlas (2015), the potential CO_2 storage resources within the St. Peter Sandstone and Everton Formation were calculated, factoring in geographical extent, stratigraphic thickness, and formation porosity, alongside adjustments for temperature and pressure variations. This analysis suggests a substantial sequestration potential, projecting prospective storage resources to range from 680 million tonnes to 7,300 million tonnes of CO_2 in southern Illinois for worst-case and best-case scenarios, respectively.

This research emphasizes the role of geological storage as a technology to lower atmospheric CO_2 concentrations, marking a critical step forward in combating climate change. Moreover, it highlights the St. Peter Sandstone and Everton Formation as indispensable components within a portfolio of carbon storage targets in the Illinois Basin. Significantly, the modeling indicates the southwest-central part of Illinois, particularly in Washington, Madison, Clinton, and Bond Counties, possesses reservoir properties within these formations that are suitable for CO_2 storage. This insight is critical for providing optionality in the development of CO_2 storage projects in the Illinois Basin that are not only geologically suitable, but also commercially attractive and publicly acceptable.

289f Offshore Stored CO_2 Residence Time: Accurate Assessment with Geochemical, Seismic, and Modelling Integration

Richard Coffin, Richard Coffin, Maria Vasilyeva, Hussain Abdulla

Coastal geologic carbon sequestration focuses on injection of supercritical CO2 into deep sediment coastal reservoirs. Seal integrity above these reservoirs is critical for safe long-term storage of CO2. It is generally assumed that if a reservoir was able to contain hydrocarbons for millions of years, it should be suitable for CO2 storage. However, many hydrocarbon reservoirs are leaking, probably best documented by the ubiquitous natural oil slicks observed e.g., in the Gulf of Mexico. Despite realization that many reservoirs will leak CO2, we propose, abiotic and biogeochemical reactions during CO2 migration from the injection point to the ocean floor surface provide diverse traps that result in efficient sediment CO2 sequestration. First, during the migration concentration in porewater becoming supersaturated can result in a carbonate formation. This carbonate formation may cause a trap and with the CO2 build up there can be pressurization that results in horizontal migration to a new vertical path or pressurization that results in fractures of carbonate or clay layers. This transport can occur at multiple intervals during migration to the surface sediment and this step-by-step trapping creates long term vertical migration. With final transport to the shallow sediment there is a complex series of biogeochemical cycles that will sequester deep sediment sourced CO2 that are controlled by the presence of electron acceptors. Moving down the biogeochemical eH gradient to the absence of sulfate, CO2 is a near final electron acceptor and can be cycled by bacteria into methane or organic material.

This potential effectiveness of the seafloor to act as a buffer depends strongly on localized flux rates of CO2 and organic carbon loading that controls microbial metabolism rates. Highly concentrated CO2 flux, at or approaching advection, would overwhelm the biogeochemical systems and result in a direct flow of CO2 to the overlying water column. In contrast, with lower vertical migration rates due to trapping and low permeability through clay seals would slow down any biogeochemical reactions as well as move these reaction fronts closer to the seafloor. These parameters need to be understood and constrained above individual reservoirs for an evaluation of the seafloor as potential secondary seal for any leakage of CO2 from deeper reservoirs. We predict that careful selection of deep sediment CO2 sequestration sites will provide long term trapping with a thorough seismic data review of fractures and vertical fluid and gas migration focus points integrated with assessment of shallow sediment biogeochemical trapping potential. This presentation will review characteristics of sediment that result in vertical migration of injected CO2 and an overview of trapping events during the vertical migration. With a thorough integration of seismic and geochemical data and fate-transport modeling there is capability to select efficient storage locations in the deep coastal sediment and provide accurate assessment of CO2 residence time.

289g Research and Development of Carbon Capture and Storage in the Illinois Basin

Carl Carman

The Illinois Basin has hosted numerous research and commercial carbon capture and storage (CCS) projects over the last two decades, many led by the Illinois State Geological Survey (ISGS). These projects include the world-renowned Illinois Basin – Decatur Project (IBDP), multiple U. S. DOE-funded Carbon Storage Assurance Facility Enterprise (CarbonSAFE) projects across Phases I, II, and III, and several commercial projects of varying scale. Current CCS research in the Illinois Basin is focused on further geospatial and petrophysical characterization of recognized storage reservoirs and sealing horizons, as well as identification of additional units with potential storage and sealing properties. While the geological and industrial setting appears favorable for commercial-scale CCS development, remaining challenges include underdeveloped transportation infrastructure, fledgling regulatory frameworks at the municipal, county, and state levels, and negative public perception.

The Illinois Basin is a structural and sedimentary basin underlying approximately 80,000 square miles of Illinois, southwestern Indiana, and western Kentucky. It is elliptical, with the long axis trending NW-SE and deepest in southeastern Illinois and southwestern Indiana. The Illinois Basin is situated favorably for CO_2 supply, being in the heart of ethanol production in the Midwest U. S. and is near three major metropolitan areas and their associated industries: Chicago, St. Louis, and Indianapolis. However, capture and storage studies are not limited to ethanol. For example, the UIC Class VI permits issued to Wabash Carbon Services (WCS) are for storage of CO_2 from the Wabash Valley Resources (WVR) ammonia gasification plant near Terre Haute, IN. The ISGS is also currently studying storage potential for sites at or near coal-fired power plants and cement production.

Geologically, the Illinois Basin comprises multiple storage complexes suitable for carbon storage, particularly in the Cambrian and Ordovician strata deposited atop the Precambrian basement rock. These storage complexes leverage several units with favorable reservoir characteristics, including the Mt. Simon Sandstone, the Potosi Dolomite, and the combined St. Peter Sandstone and Everton Dolomite. The Illinois Basin also has multiple sealing formations, including the Eau Claire Formation, Maquoketa Group, and New Albany Shale Group. Additional units are being characterized for storage or sealing properties, such as the New Richmond Sandstone as a potential storage reservoir.

To date, the U. S. EPA has granted four Underground Injection Control (UIC) Class VI permits to construct injection wells for the geologic storage of CO_2 in the entire U. S., all four of which are in the Illinois Basin and in EPA Region 5. Two permits were issued to Archer Daniels Midland Company (ADM) in Decatur, IL. The first was issued for CCS1, the CO_2 injection well for the IBDP, a demonstration project which successfully injected and stored one metric megaton of CO_2 in the Mt. Simon Sandstone from 2011 to 2014.

The IBDP site is now in its post-injection site care (PISC) period, continuing with a prescribed monitoring program detailed in the Class VI permit. The Illinois Industrial Carbon Capture and Storage (IL-ICCS) project built off the learning from the IBDP and was issued a Class VI permit to install an injection well (CCS2) roughly 0.6 miles from CCS1. The IL-ICCS project is active and to date has stored over two metric megatons of a permitted six metric megatons of CO_2 in the Mt. Simon Sandstone. The primary seal for both wells is the Eau Claire formation, and the Maquoketa Group, a regionally continuous and thick caprock unit which acts as a secondary seal.

Two Class VI permits were issued by the U. S. EPA in January 2024 to WCS to allow the construction of injection wells for the injection and permanent storage of up to twenty metric megatons of CO_2 in Vermillion and Vigo counties in Indiana. Now in its "pre-operation" phase, the WCS project intends to inject into the Cambrian Potosi Dolomite, which demonstrates significant vugular porosity and is a known potential lost circulation zone in the Illinois Basin. The primary seal is the Maquoketa Group. In addition to the permits granted to ADM and WCS, twenty-two additional permit applications are under review by the U. S. EPA, e. g., three Class VI permits applied for by One Earth Sequestration for CO_2 storage near their ethanol production facility in Gibson City, IL.

The geological and industrial advantages of the Illinois Basin present a uniquely promising environment to develop commercial-sale CCS to an extent that can help mitigate climate change. However, challenges to acceptance and implementation remain, particularly in the areas of regulation and public approval. Ongoing research by the ISGS is focused on identifying and characterizing new geologic formations suitable for the permanent storage and containment of CO_2, providing risk-based assessment of capture, transport, and storage potential, and increased stakeholder education and interaction.

289h Novel Carbon Storage Monitoring Methods

Trevor Richards, John Hunt, Cesar Barajas-Olalde, Justin Kovacevich, Kyle McBride, Bethany Kurz

We present a high-level overview of past and ongoing research efforts led by the Energy & Environmental Research Center (EERC) at the University of North Dakota to demonstrate novel and sustainable carbon storage-monitoring methods at multiple commercial-scale (i.e., >1 million tonnes of carbon dioxide sequestered annually) carbon capture and storage (CCS) sites. Research with our project partners, which has received support from the U.S. Department of Energy (DOE) and Plains CO_2 Reduction (PCOR) Partnership, is aligned with DOE's goal to advance monitoring technology for carbon dioxide (CO_2) plume tracking and leakage (assurance) monitoring as well as the mission of the PCOR Partnership, a government–industry consortium with over 200 members, to accelerate deployment of commercial-scale CCS technologies.

The EERC has conducted field testing associated with its scalable, automated, sparse seismic array (SASSA) 2.0 technology at a permitted CCS site supported by the PCOR Partnership. A sparse surface array of seismic nodes was deployed around the project site to collect seismic data from midpoints set at multiple azimuths and distances around the injection well for tracking the injected CO_2 plume within the target zone (i.e., storage reservoir) through time-lapse comparison of data acquired between the baseline (preinjection) and the first year and a half of operations. The seismic data were acquired with a set of static seismic sources that were operated remotely after installation. EERC field personnel retrieved and redeployed the seismic nodes about once every 6 weeks to obtain the raw field data for processing. SASSA 2.0 benefitted the local community by providing a low-impact, semiautonomous solution to conducting seismic survey work. Initial processing of the results indicated satisfactory signal-to-noise within the storage reservoir.

The EERC recently completed baseline research activities at a permitted CCS site with DOE support. The overall goal of the project is to demonstrate the commercial readiness of several novel carbon storage-monitoring technologies for the benefit of future CCS projects. Survey planning benefitted from drone surveillance work conducted prior to survey acquisition, which generated a digital elevation model that the planning team used in the geophysical survey designs.

Potential fields methods, such as electromagnetic (EM) surveys, are underutilized in CCS monitoring. A project objective is to use EM methods to quantify saturation changes through electrical resistivity measurements of the near-surface and at-reservoir depths. The near-surface measurements of electrical resistivity can be used to corroborate with nuclear magnetic resonance-logging data planned to be collected from a group of shallow groundwater wells within the area surveyed. This method will provide a novel approach to monitoring the near-surface environment for leakage (assurance monitoring) at a CCS site. To date, the baseline near-surface and at-reservoir EM surveys have been acquired, and initial results indicate satisfactory signal-to-noise.

The EERC also deployed three-component (3C) seismic sensors in strategic surface locations as part of the baseline activities associated with the research overlay supported by DOE. The purpose of the 3C sensors was to measure ambient noise at the site to generate seismic images of the subsurface from the passive (noise) seismic data collected. The method involves estimating the orientation of maximum horizontal compressive stress (SHmax); therefore, measurements of SHmax from core available within the project area were used, and a set of mechanical earth models were created as part of this research. This method will demonstrate a very low-impact method for CO_2 plume monitoring, as no intrusive seismic sourcing is required.

One of the objectives for improving the economics of future seismic surveys is to demonstrate the ability to image a CO_2 plume from repeat 2D seismic lines sourced with conventional seismic sources with varying degrees of decimation. Low-impact mobile seismic sources were also tested at high-density spacing along those 2D seismic lines for testing viability of imaging to the target storage depths. These methods will improve the sustainability of future carbon storage monitoring by reducing impact and cost.

In a similar effort at another site, the EERC is planning to collect a large (>200-square-mile) 3D seismic survey associated with a DOE-supported CCS project. This 3D survey design will implement compressive sensing concepts and serve as a baseline for future carbon storage monitoring. The objective of this task will be to convert the conventional survey design supplied by the selected seismic acquisition contractor into a compressive sensing design that produces superior data density for the same in-field effort and, therefore, reduces costs for future repeat monitor surveys.

As an alternative to acquiring large-scale 3D seismic surveys for CCS projects, the EERC has been involved in past underground injection control Class VI projects where the design of a grid of 2D seismic lines was selected to track the migration of the injected CO_2 plume. Design patterns have varied based on the site-specific needs of the CCS project. In cases where active mining operations are a concern for land use and access to roads, a sparse 2D grid along roads can be used to limit impact to site operations. Another approach involved acquiring 2D seismic lines in a radial pattern around CO_2 injection to create a pseudo-3D image of the injected CO_2 plume.

We presented a high-level overview of many novel and sustainable carbon storage-monitoring methods that are at various stages of planning or demonstration to accelerate the deployment of CCS technologies at future CCS sites across the United States. Our work impacts the general CCS industry by providing novel and more sustainable methods for tracking CO_2 plume migration and performing assurance monitoring. Our work primarily benefits three parties: 1) CCS community members through knowledge sharing of lessons learned; 2) CCS operators through commercialization of additional methods, including improvements on workflows and simplification of fieldwork; and 3) stakeholders of CCS projects through implementation of low-impact and more autonomous monitoring solutions.

701a A Multi-Objective Decision-Making Framework for Renewable Energy Transportation

Halil Iseri, Funda Iseri, Eleftherios Iakovou, Efstratios Pistikopoulos

Energy transition is gaining pace in the world with heightened awareness of the environmental effects and health impacts of fossil fuels, as well as concerns on security of energy supply along with decreasing cost of renewable energy deployment. In this regard, hydrogen economy, referring to the wide use of hydrogen and derived products as the main source of fuel, is regaining attention as hydrogen is regarded as a potential solution to address renewable intermittency, de-fossilization of hard-to-abate industries and long-term energy storage challenges [1,2]. However, the path towards de-fossilization presents significant additional challenges for many industrialized economies, as their domestic renewable energy generation is projected to be insufficient to meet their energy and de-fossilization demands in the foreseeable future [3]. On the other hand, the imbalance of renewable energy generation potential, levelized cost and demand in between different geographies demonstrates the potential of a future global green energy economy, through the trade of green fuels. This potential and need call for modeling frameworks that capture the specific needs of decision-makers to make informed decisions.

In this research, we present a framework for modeling and optimizing decisions for the selection of an optimal hydrogen-based energy carrier for transporting renewable energy considering different locations, transportation mediums and often conflicting objectives. The important features of the potential energy carriers of the future are determined and weighted employing a survey addressed to energy executives. The resulting decision-making problem is modelled as a multi-objective optimization problem through Mixed-Integer Linear Programming (MILP), while the weights of different objectives are determined with respect to Analytical Hierarchy Process (AHP) [4]. A case study of transporting renewable energy from different locations with low levelized cost of electricity to a highly industrialized region with insufficient resources is considered to determine the suitable energy carrier and transport mechanism with respect to conflicting objectives.

References

[1] Veziroglu, T N, and Barbir, F. Hydrogen: Its comparison with fossil fuels and its potential as a universal fuel. United States:1993.

[2] International Energy Agency, The Future of Hydrogen, 2019

[3] The National Hydrogen Strategy, BMWi, 2020

[4] Saaty, TL., The analytic hierarchy process (AHP), The Journal of the Operational Research Society 41(11)

701b Decarbonizing Natural Gas Sector: Natural Gas/Hydrogen Blends and Synthetic Gas Production Pathways

Kyuha Lee, Vincenzo Cappello, Pingping Sun, Amgad Elgowainy, Kwang Hoon Baek, Pallavi Bobba

Natural gas (NG) is widely used across all energy sectors. Combustion of fossil NG leads to significant greenhouse gas (GHG) emissions. It is estimated that NG was responsible for approximately 30% of the total U.S. GHG emissions in 2021. Replacing NG with cleaner alternatives, such as hydrogen (H_2) produced from renewable or low-carbon sources, could have substantial impacts on decarbonizing U.S. energy sectors. While H_2 can play an important role in such decarbonization, the infrastructure for H_2 transportation is costly in the early markets and is yet to be fully developed. One promising option to decarbonize the NG supply chain and mitigate the lack of hydrogen delivery infrastructure is to inject clean H_2 into the NG pipeline networks to deliver a blended NG/H_2 gas. In this work, we conducted a quantitative analysis to estimate life cycle GHG emissions associated with delivering blended gas in an existing NG transmission pipeline. While a significant reduction in life cycle GHG emissions can be achieved when delivering the same NG baseline volume throughput of blended gas but at a reduced energy flow, maintaining the energy throughput will likely limit the maximum life cycle emissions reduction to about 6% as the H_2 blending ratio will likely be constrained to approximately 30 mol% due to pipeline operating limitations. Further reduction in GHG emissions of gas blend while maintaining energy throughput would likely require modification of the NG pipeline infrastructure. Alternative to use of H_2 for blending with NG, H_2 may be used to produce synthetic natura gas (SNG) utilizing captured CO_2, thus bypassing the infrastructure hurdles associated with gas blending. In this study, we developed an engineering model of SNG production at an industrial scale to investigate its levelized cost and life cycle GHG emissions. Depending on the CO_2 supply chain, the levelized cost of SNG is estimated to be in the range of $45–76 per million British thermal units (mmBtu) on a higher heating value basis. SNG can also reduce life cycle GHG emissions by 52–88% compared with fossil NG.

701c A Technoeconomic Analysis and Life Cycle Assessment for Long Distance Hydrogen Transportation Via Hydrogen Carriers

Gasim Ibrahim, Bosong Lin, Haoxiang Lai, Guiyan Zang

NOTE: This abstract is tentative in terms of authors and content as it goes through a vetting process with industrial collaborators.

Hydrogen is often considered a promising source of clean energy. However, its low volumetric energy density makes its efficient transportation challenging, especially across long distances. Hydrogen carriers have been proposed as a means of overcoming some of the challenges associated with hydrogen transport. Several studies have been conducted on specific hydrogen carriers along specific supply chains; comparing costs, GHG emissions, mass efficiency, and energy efficiency. However, the hydrogen carrier supply chain has many different variables and design choices that influence the costs and emissions associated with the transport, making comparisons between the different carriers, and between different studies for the same carriers, challenging.

In this work, we conduct a technoeconomic analysis (TEA) and lifecycle analysis (LCA) of widely studied hydrogen carriers, including ammonia, synthetic natural gas (SNG), methanol, and methylcyclohexane (MCH), in addition to liquid hydrogen as a benchmark. The TEA and LCA cover the conversion of hydrogen into the carrier, export terminal storage, transportation, import terminal storage, and final reconversion of the carrier into hydrogen. Furthermore, the analyses include multiple supply chain variables such as: origin of hydrogen being transported, type of ship and fuel used in transportation, method used in land transportation, fuel used in reconversion step, and inclusion of carbon capture and storage in the reconversion stage. We leverage combined industrial and academic expertise to develop a tool for a comprehensive and accurate TEA and LCA of the supply chains, while incorporating common variations in assumptions and design choices. Through the tool we answer important questions related to long distance hydrogen transportation, such as: which hydrogen carries are lowest in carbon intensity and cost for a given supply chain, and under which conditions can a given carrier be economically feasible. Answering these questions will play an integral role in the current efforts to scaleup hydrogen use as a clean fuel, or a feedstock for the production of other low-carbon fuels.

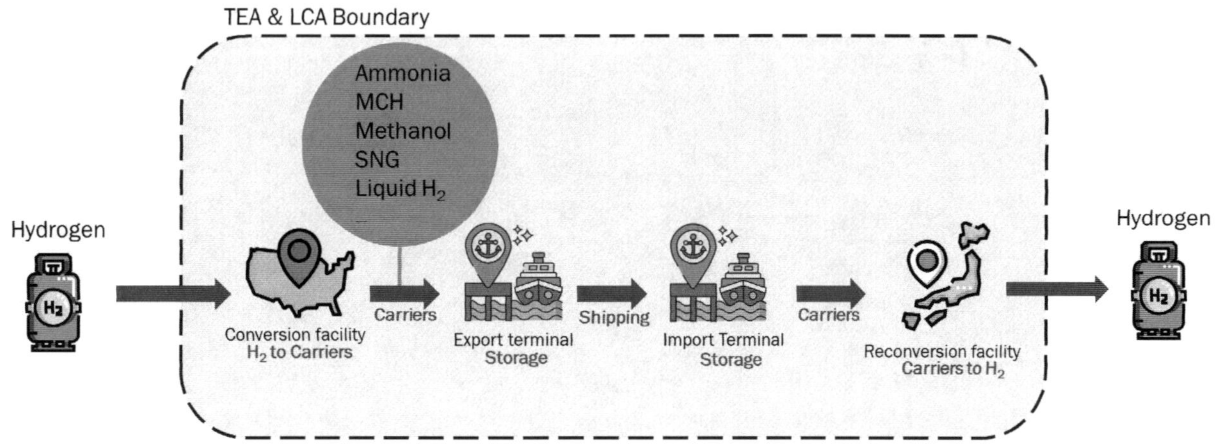

701d Modeling and Analysis of Electrofuel Production Using CO_2 and Nuclear Energy Via Methanol Intermediate

Hernan E. Delgado, Vincenzo Cappello, Pingping Sun, Clarence Ng, Kwang Hoon Baek, Amgad Elgowainy

Electrofuels (e-fuels), which are low carbon fuels produced with carbon dioxide (CO_2) and low carbon hydrogen (H_2), can be a potential way to reduce CO_2 emissions in hard-to-decarbonize sectors, such as aviation or heavy duty vehicles and heavy machinery. In this work, we modeled and analyzed an e-fuel production process from that converts CO_2 to electro fuels using methanol as an intermediate; furthermore, the process uses nuclear energy to satisfy plant consumption and to produce H_2 via high temperature electrolysis. Relative to the commonly studied e-fuels from Fischer Tropsch process that only has distillate yield of 60-70% with significant naphtha yield of 30%, the present work increases the distillate (jet and diesel cut) yield to 98% with only 2% naphtha. The modeled process has a carbon conversion of 96%, a process energy efficiency of 50% relative to total thermal and electricity inputs, and life cycle GHG emissions of 5-7 gCO_2e/MJ liquid fuel estimated with the Greenhouse gases, Regulated Emissions, and Energy use in Technologies (GREET) model. Additionally, a techno-economic analysis was performed for three scales corresponding to nuclear power of 100, 500, and 1000 MW_e, where minimum fuel selling prices (MFSP) of $6.9-8.6/gal were obtained with current H_2 costs of $3.8-4.0/kg assuming market electricity price of $70/MWh and SOEC stack cost of $155/kw. The e-fuels cost can be reduced to $4.2-6.4/gal with an H_2 cost of $2.0-2.5/kg, generated from low cost electricity of $25-34/MWh; in both cases, the MFSP does not consider tax credits. Finally, the impact of potential tax credits (e.g., 45Q, 40B, 45Z from the Inflation Reduction Act) is also evaluated.

701e Techno-Economic Analysis and Life Cycle Assessment of Production, Storage and Transportation of Hydrogen Carriers

Maanasa Bhat, Gasim Ibrahim, Sili Deng, Guiyan Zang

Hydrogen demand is expected to increase by six times in the next decade and the development of hydrogen supply chains all over the world is of utmost importance. Due to its inherent volatile nature, hydrogen poses considerable risks and expenses in storage and transportation. A strategy to overcome these challenges is utilizing hydrogen carriers, which are liquid or solid materials that can store hydrogen energy and can be safely and inexpensively transported. The carriers are subsequently used directly in applications such as power generation or dehydrogenated back to hydrogen. While individual studies on the costs and emissions of producing the carriers are available in literature, the variability in production methods, properties and end-use of the carriers creates challenges in analyzing and comparing them fairly.

In this paper, we discuss four hydrogen carriers in tandem - methane, methanol, ammonia, and liquid hydrogen. Techno-economic analysis and life cycle analysis are performed along a common framework to estimate the costs and emissions of production, storage and oceanic transportation of the carriers. Low carbon production pathways are considered with utilization of green hydrogen and captured carbon dioxide. In the production stage, ammonia is found to have the lowest levelized cost and second to lowest greenhouse gas (GHG) emissions, marginally higher than liquid hydrogen. For transportation costs, a fixed supply route from Australia to Japan is considered and several port-to-port routes are analyzed for economic and environmental feasibility. Transportation costs and GHG emissions are found to be significantly affected by port-to-port distance via ship charter costs and the type of fuel used in the ship.

We expect the results from the current paper to provide a guideline towards identifying the suitable hydrogen carrier for a given application and supply route and to contribute towards fastened development of robust hydrogen supply chains.

701f CO_2-to-Sustainable Synthetic Fuel Process: Scenario-Based Techno-Economic and Carbon Footprint Analysis

Jinsu Kim, Yanhui Yuan, Ben McCool, Matthew Realff

Carbon dioxide can be efficiently utilized through liquid hydrocarbons as an energy carrier, making it a viable solution for transitioning from a linear carbon flow to a circular one. Synthetic fuel is crucial due to challenges in decarbonizing the sector, such as aviation transport, and ensuring fuel production regardless of infrastructure availability.

This work discusses the sustainable synthetic fuel (SSF) production system, considering various aspects, including technical readiness, energy, economics, carbon intensity, and carbon footprint. A conceptual process flow was developed by integrating mature and promising chemical operational units. The Fischer-Tropsch-based synfuel-making route was studied by incorporating different types of electrolyzers. The first process was designed by the proton exchange membrane electrolysis cell (PEMCE) to produce hydrogen, which was technically viable. The second process was designed by the high-temperature solid oxide electrolysis cell (SOEC) to coproduce hydrogen and carbon monoxide, which can replace the water-gas shift reactor. After deriving the process flowsheet, multiple scenarios for SSF production were illustrated by comparing carbon footprint analysis results based on energy origin. The study parametrically investigated factors such as the cost of electricity and natural gas, scale of CO_2 utilization, origin of the captured CO_2, and internal recycling flow. Through the work, we revealed the bottlenecks of the SAF-making process to be economically viable and environmentally benign.

THE PRIMARY AQUEOUS ALUMINUM-COPPER BATTERY

Dr. Sirshendu Guha, Engineers India Limited, New Delhi, India

Abstract

A novel aqueous primary battery has been developed, comprising a cathode made of copper (Cu) or a copper alloy, and an anode made of aluminum (Al) or an aluminum alloy [Al/Al^{3+}, SO_4^{2-} // Cu^{2+}, SO_4^{2-}/ Cu (Guha S, inventor; 2022 November 06. AQUEOUS COPPER-ALUMINIUM (Cu-Al) PRIMARY BATTERY. Indian Patent Application No. 202231063340)].

The cathode compartment contains an aqueous solution of copper sulfate ($CuSO_4$) as the electrolyte, while the anode compartment contains an aqueous solution of aluminum sulfate ($Al_2(SO_4)_3$). These compartments are separated by a ceramic, wood, or inert separator that facilitates ion transport.

To prevent cupric ions (Cu^{2+}) from migrating to the aluminum (Al) or aluminum alloy anode compartment, the catholyte and anolyte are supplemented with aqueous potassium nitrate (KNO_3) as a supporting electrolyte. This innovation results in a primary battery with exceptional performance, boasting a specific discharge capacity exceeding 2500 mAh per gram of anode material, and demonstrating highly stable electrochemical behavior.

The battery exhibits an open circuit voltage (OCV) exceeding 0.6 volts per unit cell. It can be constructed in various sizes and excels in self-discharge properties. Notably, it does not generate hydrogen (H_2) gas at the aluminum (Al) or aluminum alloy anode, nor does it deplete the aqueous solvent during discharge.

Furthermore, it avoids parasitic corrosion of the aluminum (Al) or aluminum alloy anode.

With an energy density surpassing 600 Wh/kg of anode material and an anodic efficiency reaching 86% or higher, the battery also demonstrates the ability to deliver high electric current densities, exceeding 5 mA per gram of anode material.

Keywords: Galvanic Cell; Aqueous Primary Battery; Aluminum-Copper Battery; High Specific Discharge Capacity

Summary of the Invention

This invention relates to a novel aqueous primary battery particularly an aqueous primary battery possessing aluminum (Al) or an alloy of aluminum (Al) anode. There is a greater need today to introduce sources of clean energy that do not cause environmental degradation and global warming. The conventional Aluminum-air battery with aqueous electrolyte has severe limitations such as parasitic corrosion of aluminum anode, passive aluminum hydroxide layer formation on the anode, evolution of hydrogen gas, and severe self-discharge rate [1,2]. Aluminum (Al) is environmentally friendly and a very low-cost metal, any battery with Aluminum (Al) anode is considered a highly attractive candidate for the production of large-capacity batteries that can be used in electrical applications to reduce dependence on fossil fuels. To develop a successful aluminum anode or an alloy of aluminum (Al) anode-based battery, one needs to address the serious shortcomings associated with a conventional Aluminum-air battery.

The invented aqueous primary battery will not be associated with (a) parasitic corrosion of aluminum or an alloy of aluminum anode, (b) passive aluminum hydroxide layer formation on aluminum or an alloy of aluminum anode, (c) evolution of hydrogen gas from aluminum or an alloy of aluminum anode, (d) severe self-discharge rate and (e) consumption of water which is a solvent during discharge.

This invention has given birth to a primary battery which is composed of a positive electrode (cathode), made of pure copper (Cu) or an alloy of copper (Cu), and a negative electrode (anode), essentially made of pure aluminum (Al) or an alloy of aluminum along with a separator, primarily made of ceramic or wood or any other inert material which allows transport of ions through the separator. The copper or its alloy electrode compartment in the battery uses an aqueous copper sulfate ($CuSO_4$) solution possessing a concentration of Cu^{2+} ions of at least 1(M) (1.0 mole/lit) or more as catholyte. The aluminum or its alloy electrode compartment in the battery uses an aqueous aluminum sulfate ($Al_2(SO_4)_3$) solution possessing a concentration of Al^{3+} ions of at least 1(M) (1.0 mole/lit) or more as anolyte. The catholyte and anolyte should be supplemented with aqueous potassium nitrate (KNO_3) solution of at least 1(M) (1mole/lit) concentration or more as supporting electrolyte to prevent cupric ion (Cu^{2+}) crossover to the aluminum or its alloy electrode compartment.

The battery reaction can be given by the equation: $3\ Cu^{2+}(aq) + 2\ Al\ (s) = 3\ Cu\ (s) + 2\ Al^{3+}(aq)$

The Nernst Equation for the battery can also be depicted as,

$$E_{cell} = E^0_{cell} - (RT/6F)\ \ln\ (a^2Al3+ / a^3Cu2+)$$

E_{cell} is the electromotive force (EMF) of the battery; E^0_{cell} is the electromotive force (EMF) of the battery at standard conditions, R is universal gas constant, T is battery operating temperature, F is Faraday's constant and aAl3+, aCu2+ are activities of Al^{3+} and Cu^{2+} ions respectively.

$$E^0_{cell} = E^0_{(cathode)} - E^0_{(anode)} = [0.337 - (-1.662)]\ \text{Volt} = 1.999\ \text{volt} \sim 2.0\ \text{volt}$$

Examples

Experimental Results and Discussion for Battery-1

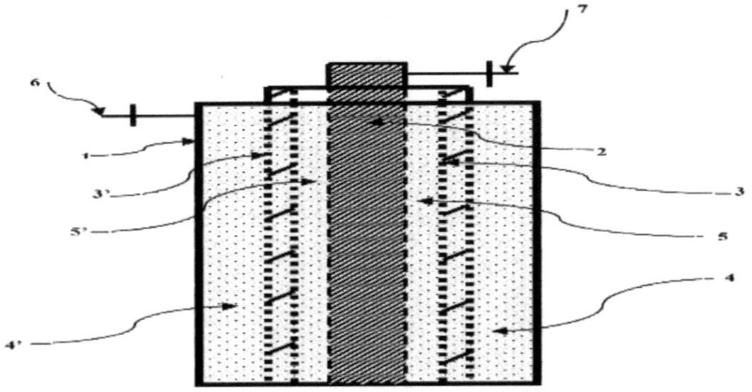

FIG.1: Battery-1 Front View [3]

LEGEND FOR FIG.1, FIG.2

1 – Cu or its alloy Case (Cathode)
2 – Al or its alloy Anode
3, 3' – Ceramic Separator
4, 4' – Catholyte (aq. $CuSO_4$)
5, 5' – Anolyte (aq. $Al_2(SO_4)_3$
6,7 – External Electrode Terminals

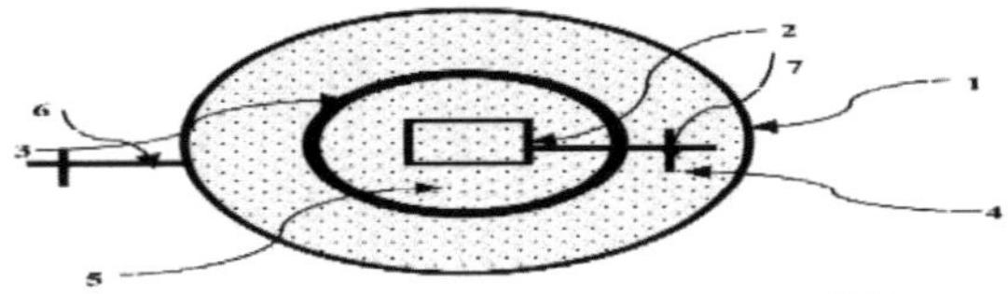

FIG.2: Battery-1 Top View [3]

In Fig.1, Fig.2, the Effective area of Al Anode is greater than or equal to 15872 mm^2, and the Effective area of the Cu Cathode is greater than or equal to 21566 mm^2 [3].

In Fig.1, Fig.2, the ceramic separator thickness is 4mm which can be reduced to lower the internal resistance of the battery. The catholyte and anolyte liquid levels were greater than or equal to 62 mm from the BTL of the cylindrical Cu or its alloy case [3].

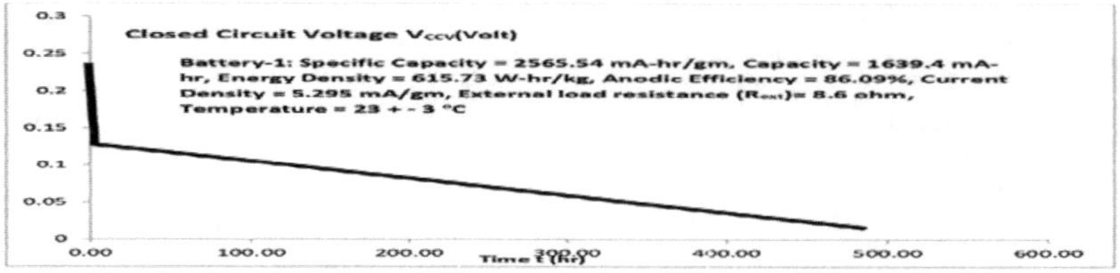

Fig.3: Discharge Curve for Battery-1 [3]

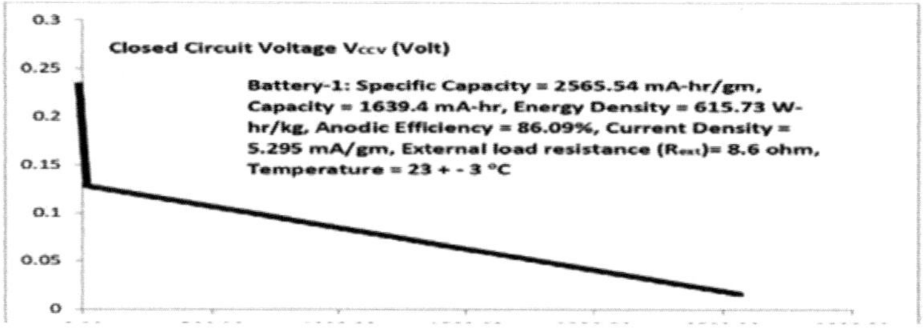

Fig.4: Voltage versus Specific Discharge Capacity Plot for Battery-1 [3]

Fig.3, and Fig.4 reveal that Al-Cu Battery (Battery-1) delivered a very high specific discharge capacity which can exceed 2500 mAh per gm of anode material [3].

These curves also reveal that the Al-Cu Battery (Battery-1) possesses reasonably high energy density which can exceed 600 W-hr/kg and has good anodic efficiency which can be as high as 86% or more concerning the theoretical specific discharge capacity of Al anode being 2980 mAh/gm [3].

Moreover, the Al-Cu Battery (Battery-1) has the capability of delivering large electric current density exceeding 5 mA/gm approximately [3].

Fig.5 is showing output voltage or open-circuit voltage V_{OCV} profile of the Al-Cu Battery (Battery-1) before the start of discharge through the external resistive load of 8.6 ohms [3].

Fig.5 indicates that the maximum output voltage or open-circuit voltage V_{OCV} of the Al-Cu Battery (Battery-1) is greater than 0.6 volts at an environment temperature of 23 + - 3 ^{o}C [3]. Fig.5 also reveals that the Al-Cu Battery (Battery-1) possesses highly stable electrochemical behavior having output voltage or open-circuit voltage V_{OCV} plateau at around 0.61 volts [3]. Fig.6 shows that a red light-emitting diode (LED) can be powered by the invented Al-Cu Battery (Battery-1). Fig.7 indicates that the measured open circuit voltage of the invented Al-Cu Battery (Battery-1) can exceed 0.6 volts per unit cell.

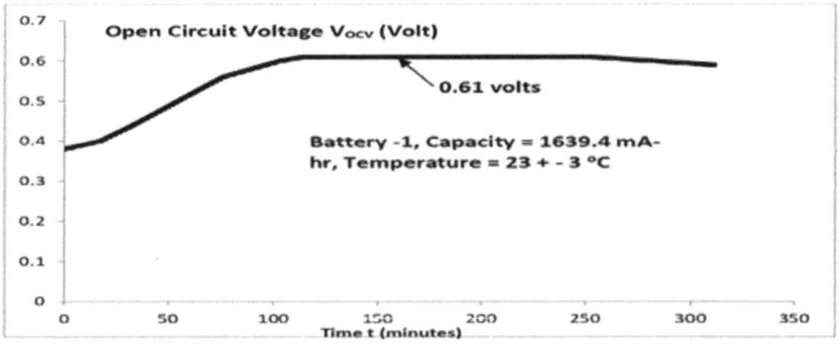

Fig.5: Voltage versus Time before Discharge Plot for Battery-1 [3]

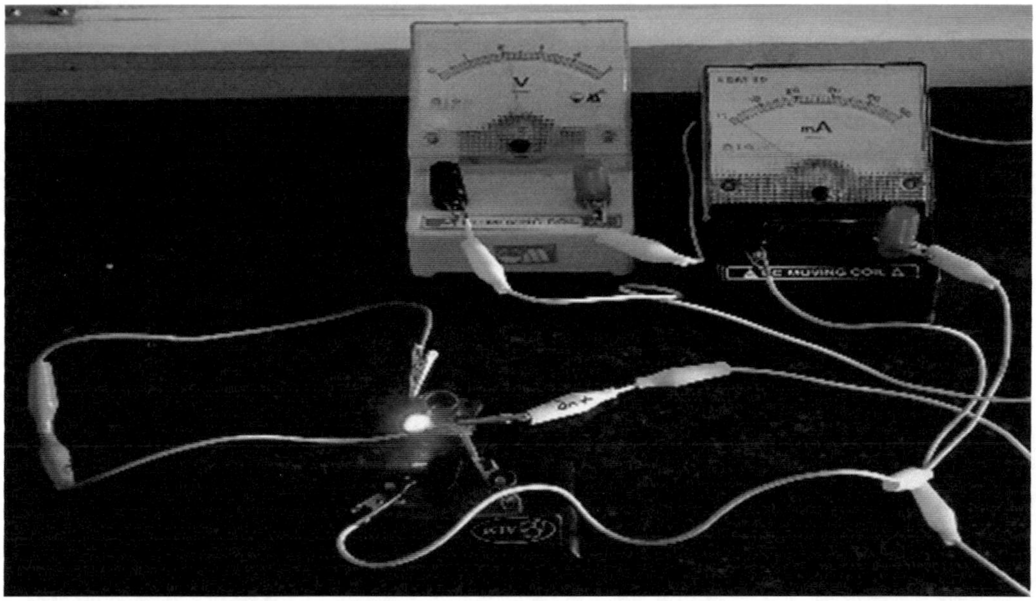

Fig.6: Red LED powered by Battery-1

Fig.7: Measured Open Circuit Voltage of Battery-1

Conclusions

A novel aqueous primary battery has been invented, comprising a cathode made of Cu or a Cu alloy, and an anode made of Al or an Al alloy. The cathode compartment contains an aqueous solution of $CuSO_4$ as the electrolyte, while the anode compartment contains an aqueous solution of $Al_2(SO_4)_3$. These compartments are separated by a ceramic, wood, or inert separator that facilitates ion transport. The catholyte and anolyte are supplemented with aqueous KNO_3 as a supporting electrolyte to prevent Cu^{2+} crossover to the anode compartment and consequent electroplating of Cu on the anode. The battery exhibits an open circuit voltage exceeding 0.6 volts per unit cell and possesses a specific discharge capacity exceeding 2500 mAh per gram of anode material, has energy density surpassing 600 Wh/kg of anode material, and has an anodic efficiency reaching 86% or higher. It does not generate H_2 gas at the anode, nor does it deplete the aqueous solvent during discharge. Furthermore, it avoids parasitic corrosion of the Al or Al alloy anode and can deliver high electric current densities, exceeding 5 mA per gram of anode material. The battery can power a red light-emitting diode (LED) with sufficient lumen.

References

1. Jingling Ma et al., "Performance of Al-0.5 Mg-0.02Ga-0.1Sn-0.5Mn as Anode for Al-Air Battery", J. Electrochemical Soc., 161,3 (2014)

2. Zegao Sun et al., "Performance of Al-Air Batteries Based on Al-Ga, Al-In and Al-Sn Alloy Electrodes", J. Electrochemical Soc., 162,10 (2015)

3. Guha S, inventor; 2022 November 06. AQUEOUS COPPER-ALUMINIUM (Cu-Al) PRIMARY BATTERY. Indian Patent Application No. 202231063340 (Note-1).

Note-1: The author declares the following financial interests/personal relationships which may be considered as potential competing interests: Sirshendu Guha has a patent pending to Sirshendu Guha.

353b On the Good Practices of Electrochemical Reactive CO_2 Capture

Avishek Banerjee, Chudi Yue, Jounghwan Choi, Carlos Morales-Guio

Upgrading anthropogenic CO_2 from concentrated point sources or directly from the atmosphere is a valuable approach in closing the carbon cycle. Existing processes capture the CO_2, concentrate it into pure gas streams, transport it, and then convert it into fuels and chemicals in a separate process plant. This sequential approach results in higher energy and operating costs which can be reduced by integrating the capture and conversion steps to directly reduce the captured CO_2-bound adduct to value-added products. The capture and direct reduction of the captured CO_2-bound adduct is called Reactive CO_2 Capture (RCC). Understanding of RCC has been obscured by the higher intrinsic complexity of the system. The CO_2 capture media is a complex space of several buffer reactions that allow the co-existence of different carbon species in solution depending on CO_2 loading, temperature, pressure, and pH. In order to design improved capture agents and catalysts for integrated CO_2 capture and conversion, it is essential to identify the carbon source and the primary factors influencing product formation on a RCC catalyst. Using a gas-tight rotating cylinder electrode and leveraging a reaction-transport model developed in our group we characterize our cell to identify the active species undergoing reduction. In this talk, we discuss the reaction-transport model in detail. We also show ways to integrate the understanding of CO_2 loading, temperature, pressure, and pH to determine equilibrium speciation and further discuss the effect that these different conditions have on transport, thermodynamics and kinetics of RCC systems.

353c Free-Standing Porous Solid Electrolyte Layer for Efficient CO_2 Electrolysis

Sarah Adaryan, Rafael Verduzco, Haotian Wang

CO_2 reduction reactors can convert carbon dioxide (CO_2) into chemical fuels, enabling sustainable processes for CO_2 reduction and chemical production. However, conventional CO_2 reduction reactors use liquid-based electrolytes and as a result require costly post-reactor purification steps. Here, we report a free-standing and highly conductive PSE layer for CO_2 electrolysis produced by combining ion exchange particles with sulfonated polysulfone as a binder. We optimized the PSE by varying both the degree of sulfonation of the binder and the binder content in the free-standing PSE while characterizing the PSE morphology and electrochemical properties. The optimized PSE achieved an ionic conductivity as high as 10.4 mS/cm while maintaining excellent mechanical properties. When used in a CO_2 electrolyzer, formate selectivity consistently exceeded 90% at current densities up to 200 mA cm^{-2}, and the electrolyzer maintained a cell potential close to 3.5 V over 220 hours at 100 mA cm^{-2}. This composite PSE therefore addresses stability and performance shortcomings of previously reported PSEs and advances the development of large-scale, energy-efficient solid-state CO_2 electrolyzers.

353d What's inside the Black Box? Understanding Enzymatic Conversion of CO_2

Jenny Hancock, Siddharth Patwardhan

Carbon capture and conversion to methanol is an essential part of the global transition towards net-zero emissions. Rare-metal catalysis is the most developed and researched method; however, it carries cost and sustainability concerns.

Enzymes are becoming increasingly recognized as a powerful and sustainable alternative, beginning with carbonic anhydrase which can increase hydration of CO_2. This new method presents challenges as enzymes must be robust enough to work within adverse environments, such as exhaust outflows. One of the most effective ways to solve these issues is through immobilization. Unfortunately, many studies are limited by the complex methodologies used to immobilize enzymes, loss of activity upon immobilisation, and loss of enzymes from the matrix.

This study seeks to explore the use of enzymes immobilised on a novel bioinspired silica framework for improved enzyme functionality and retention. By including a novel application of multi-wavelength scanning using UV-vis spectroscopy, it also hopes to elucidate the dynamic chemical process behind CO_2 conversion to methanol.

Using different bioinspired silicas pore sizes effects on the carbonic anhydrases immobilisation efficiency, retention, and stability were characterised. Specifically, the different bioinspired silicas were synthesised using pentaethylenehexamine, diethylenetriamine, and polyethylenimine/L-Arginine amine additives to create these different pore sizes. Further work was also done on the impact of varying temperature, pH, leaching, and storage time on carbonic anhydrase.

With the goal of creating an enzyme cascade to make methanol more enzymes were tested on bioinspired silica. Products and intermediates formate, formaldehyde, methanol and the co-factor NADH were tested over time within the immobilized system. Data from other dependent variables will also be presented, such the absence of carbonic anhydrase and the inclusion of a co-factor regeneration system.

353e Systems Level Prediction of Best Product Conversion from CO_2 for the Novel Fast Growing Cyanobacterium *Synechococcus Elongatus* PCC 11801

Lokesh Venkatesh Babu, Pramod Wangikar

Greenhouse gases from chemical manufacturing industries contribute significantly to climate change and global warming, and their sequestration is imperative to reducing the carbon footprint. Cyanobacteria are photoautotrophic microorganisms capable of fixing CO2 from the environment via photosynthesis, and can be engineered to convert the CO_2 to value-added products. Despite progress in synthetic biology and strain engineering technologies, the commercial applicability of cyanobacteria has been limited due to the slow growth rate. However, the freshwater strain *Synechococcus elongatus* PCC 11801 recently isolated from our lab is one of the fastest growing strain with tolerance to different environmental stresses, making it a potential cell factory. To streamline and guide strain engineering efforts for PCC 11801, we generated a genome scale metabolic model for the strain that can be utilized with flux analysis to generate in silico predictions of hotspots for genetic perturbation that would maximize a specific value-added product in its metabolic network. The model was reconstructed using the standard protocol of draft reconstruction generation, manual curation, model debugging and validation. The genome scale model ('iJB785') of a phylogenetically close model strain *Synechococcus elongatus* PCC 7942 was used as the starting point. The draft reconstruction of PCC 11801 was generated by first mapping the homologous genes with iJB785 through bidirectional best hit analysis of the genomes. The model reactions were manually curated to reduce redundancy, and verified with the BiGG, KEGG and BRENDA database. 287 novel reactions identified pertinent to PCC 11801 were added de novo into the model with the corresponding gene-protein-reaction association. Model debugging was performed through flux balance analysis, flux variability analysis, single gene deletion analysis, and single reaction deletion analysis. Overall, the metabolic model consisted of 1052 genes and 1130 reactions. The model was validated by gene essentiality analysis with a sensitivity and specificity score of 0.9223 and 0.5425 respectively. The metabolic robustness of the model was estimated to be 39% from the model, which validates the amenability of the strain for genetic engineering. The model predicted that the strain would convert CO_2 to succinic acid at a higher production rate compared to other value-added products through the in silico predictions.

230a Efficient Thermal Storage: Unveiling a 2D Computationally Efficient Borehole Thermal Energy Storage (BTES) Model for Renewable Energy Integration.

Mohamed Bahr, Piyush Kumar Kumawat, Blake Billings, Palash Panja, Kody Powell

As we navigate towards an era dominated by the transition to entirely renewable energy sources, the search for viable, enduring energy storage methodologies becomes paramount. Within this evolving energy landscape, the technology of Borehole Thermal Energy Storage (BTES) emerges as a strategic asset for the smooth integration of renewable energies. However, the complex and resource-intensive nature of conventional 3D BTES modeling techniques poses significant barriers. This study introduces an innovative two-dimensional (2D) modeling technique designed to drastically reduce computational workload while maintaining good accuracy levels. Our approach rigorously models the intricate processes of thermal accumulation, retention, and release within a BTES framework across an extended seven-month timeline. A distinctive feature of our research is the deployment of the thermal mass average temperature parameter, an innovative concept that sheds new light on BTES system dynamics. The results of our investigation highlight the model's ability to improve BTES system efficiency, representing a substantial advance in the pursuit of enhancing renewable energy storage methods. Our findings reveal that this streamlined model operates with a speed that is seven times greater than conventional Computational Fluid Dynamics (CFD) models, showcasing an optimal balance of speed and accuracy. This investigation not only validates the practicality of this novel 2D approach but also sets the stage for future enhancements in BTES technologies, which are essential for propelling us towards a sustainable energy era. This endeavor seeks to spur ongoing innovation in the domain of energy storage solutions, further facilitating our global shift towards renewable energy adoption.

230b Thermodynamic Characterization of a Novel Compressed-Air Energy Storage Technology: Flooded-Mineshaft Compressed Air Energy Storage (FM-CAES)

Neil Stacey, Diane Hildebrandt, Peter Stacey, James Fox, David Glasser

With renewable energy sources representing a rapidly growing share of the global energy mix, their intermittent nature has led to growing interest in mechanisms of storing energy during periods of high availability for use during times of high demand. Several different types of compressed-air energy storage have been considered, including the conventional constant-volume approaches such as a salt dome or artificial vessel storage and, more recently, newer constant-pressure approaches such as submerged cavern storage. This manuscript proposes a novel approach where abandoned, flooded mineshafts are pressurized with air, displacing water downward in the shaft but upward in other hydraulically-linked shafts in the same geological complex. This approach is thermodynamically distinct from the existing compressed air energy storage methods because neither volume nor pressure are constant. A thermodynamic characterization of this storage method is presented, showing a relationship between depth and energy density, which is unique to this storage method. It is found that, in cases where shaft depth is sufficient, this approach offers energy density higher than an equivalent pressure constant volume scenario, albeit less than a constant-pressure scenario. In certain locations, geography may dictate that of all these various scenarios, the flooded-mineshaft approach may be the most readily achievable, and therefore, the technology merits further development

230c Optimal Scheduling of Pumped Thermal Energy Storage Plants for Multi-Vector Energy Provision

Alp Albay, Mehmet Mercangöz

Industrial decarbonization has major obstacles arising from the lack of high-grade heat supply from conventional renewable sources. Additionally, renewables have inherent intermittency issues. Advanced energy storage technologies are required to solve the latter problem. In this paper, we present a method to supply high-grade heat from renewable sources coupled with a storage technology to address these two problems. The underlying technology is a pumped-thermal energy storage (PTES) system, capable of charging and discharging with electrical power and discharging high temperature heat. To accomplish this, we propose an optimization-based state-of-charge (SoC) management solution. PTES systems have been the focus of several studies recently, such as [1] investigating the off-design operation of the system or [2] investigating the various layouts and conducting a technology review. The layout and design of the PTES system in this study was informed by these recent studies and built upon by the authors.

The PTES technology in question has a charging/discharging cycle based on a supercritical CO_2 recuperative Brayton cycle (SCBC) with molten salt as hot storage and water as cold storage [3]. The underlying technology was analyzed, and a model built in previous work [4]. In that study, in addition to the model building, an in-cycle SoC management algorithm was developed. This ensured that in either the charging or the discharging cycle, the two storage media would follow the same SoC trend. This allowed for consistent operation of the system as management of only one side of the cycle was adequate. The current study builds on this previous work by the authors to extract the model data as a function of the electrical power as well as the intended heat supply rate. A SoC balancing heat exchanger and a utility heat exchanger are integrated to the base design to facilitate high temperature heat supply. With these functions, an optimization problem is formulated for the cycle-to-cycle SoC management. Cycle-to-cycle SoC management involves controlling the stored energy of the system and balancing between the charging and the discharging cycles. Inherently, SoC management is also a scheduling problem, as it requires managing power for use in a given timeframe. To do this, the system is discretized into hourly segments to which the problem assigns charging/discharging power. An electricity and heat supply profiles are used in the objective function, maximizing the operating profit of the system.

We conducted three case studies with the optimization problem. First case study was a base case scenario, without heat supply. This was used as a benchmark for the other two studies. The second study was a flexible, variable heat supply scenario and the final one was a continuous heat supply scenario in which the system delivers constant heat to a coupled process. The economic and thermodynamic results of the study are presented in Table 1 and show that the proposed approach to the PTES heat supply can consistently be optimized while keeping within the SoC constraints under various scenarios. The resulting SoC profiles are shown in Figure 1 and the charge/discharge power levels are shown in Figure 2. While the heat supply scenarios outperformed the base case both economically and thermodynamically, there is a tradeoff between heat arbitrage and constant heat supply. The results highlight the potential of integrating PTES system with industrial processes and the importance of SoC management algorithms.

References

[1] F. Frate, L. Paternostro, L. Ferrari, U. Desideri, Off-design of a pumped thermal energy storage based on closed brayton cycles, Journal of engineering for gas turbines and power (09 2021). doi:10.1115/1.4052426.

[2] J.D. McTigue, P. Farres-Antunez, C. N. Markides, A. White, Pumped thermal energy storage with liquid storage, Elsevier eBooks (01 2021). doi:10.1016/b978-0-12-819723-3.00054-8.

[3] M. Marchionni, G. Bianchi, S. A. Tassou, Review of supercritical carbon dioxide (sco2) technologies for high-grade waste heat to power conversion, SN Applied Sciences 2 (03 2020). doi:10.1007/s42452-

020-2116-6.

[4] A. Albay, Z. Zhu, and M. Mercangöz, "State-Of-Charge (SOC) Management of PTES Coupled Industrial Cogeneration Systems," in *Turbo Expo 2024*, ASME, Jun. 2024.

Table 1: Objective function and round-trip efficiency of each scenario

	Operating Profit	η_{RTE}
Scenario 1	3.59e4	63.38
Scenario 2	4.49e4	74.45
Scenario 3	2.15e4	89.24

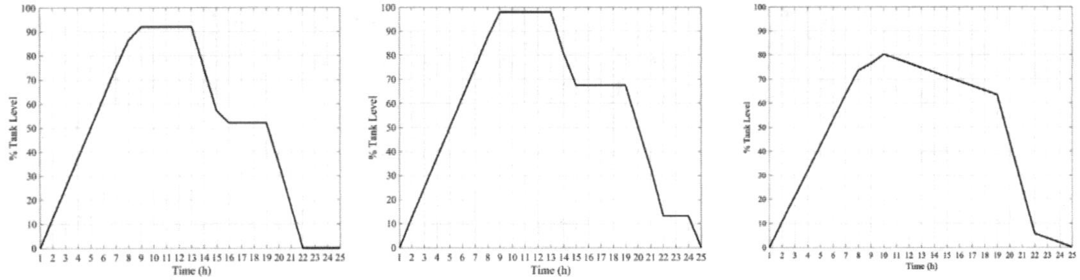

Figure 1: State-of-Charge profiles of each case. From left to right, no heat supply, variable heat supply and constant heat supply.

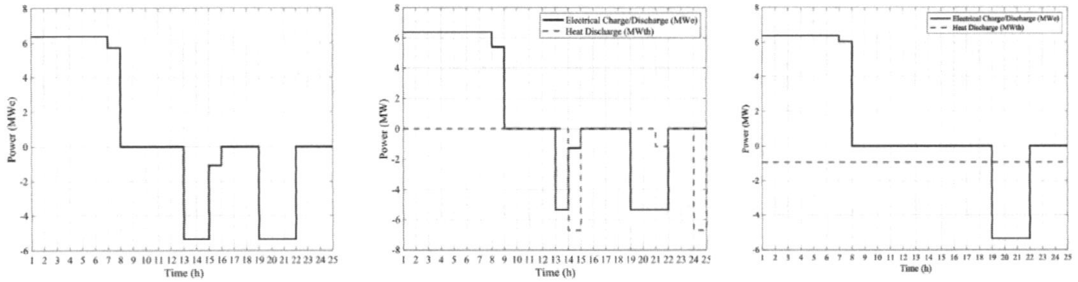

Figure 2: Optimal power profile for each case. From left to right, no heat supply, variable heat supply and constant heat supply.

230d Optimization of Liquid Air Energy Storage Systems in Future US Electricity Markets

Shaylin Cetegen, Paul I. Barton

Grid-scale energy storage demand is increasing due to the growing adoption of renewable energy sources, which are intermittent and necessitate reliable storage solutions to ensure a continuous and stable power supply [1]. Pumped hydro energy storage and compressed air energy storage currently supply over 96% of global grid-scale energy storage capacity; however, due to their location-constrained nature, they both face severely limited expansion opportunities [2]. Thus, alternative grid-scale energy storage solutions, such as liquid air energy storage (LAES), are sought to meet growing demand.

LAES is a clean and scalable long-duration energy storage technology capable of delivering multiple gigawatt-hours of energy storage. All LAES systems operate by following the same basic steps as outlined in Figure 1. First, air is taken from the atmosphere at off-peak times to charge the system. The influent air is cleaned and dried before it is compressed to high pressure and cooled to a liquid state in an air liquefaction unit. The liquid air is then stored at low temperature (~ 90 K) and atmospheric pressure in insulated tanks. Small energy losses of around 0.1-0.2% of the storage tank's total energy capacity per day enable long-duration storage for up to several weeks [3]. The system is discharged through the power recovery unit, within which liquid air is pumped from the storage tank to high pressure, evaporated, heated, and expanded to drive a generator to recover power. The round-trip efficiency of this process is enhanced through implementation of hot and cold thermal recovery systems throughout the process. Since the LAES process does not depend on the presence of any specific geography, and the components required to construct it (e.g., compressors, turbines, multi-stream heat exchangers, etc.) are readily available from existing industries, these systems can be constructed quickly in virtually any location where a small plot of land is available.

The inherent locatability of LAES systems unlocks nearly universal siting opportunities for grid-scale storage, which were previously unavailable with pumped hydro energy storage and compressed air energy storage. While the technical viability of LAES has been well established through several academic studies and industrial ventures reporting round-trip efficiencies above 60%, the economic viability of LAES has not yet been rigorously assessed across a broad range of electricity markets [2].

In this work, a mixed integer linear program was used to conduct a high-level economic analysis of LAES systems in eighteen US electricity markets under eight different decarbonization scenarios proposed by the National Renewable Energy Laboratory. By simultaneously optimizing the design and hourly operation of LAES systems to maximize their net present value over their projected lifespan, a yes/no indication of their economic viability can be obtained. The results of this analysis provide conservative estimates of the economic performance of LAES systems across the US and may serve as useful planning aids for LAES deployment. Levelized costs of storage are also calculated based on the optimal operating schedules obtained for all systems and compared to those of other long-duration energy storage technologies. Finally, sensitivity analyses are performed to elucidate how improved technical performance and economic incentives might affect the economic feasibility and, consequently, the adoption of LAES technology more broadly in the future.

References:

[1] Edenhofer, O., Pichs-Madruga, R., Sokona, Y., Seyboth, K., Kadner, S., Zwickel, T., Eickemeier, P., Hansen, G., Schlömer, S., von Stechow, C. & Matschoss, P. (Eds.). (2011). Renewable energy sources and climate change mitigation: Special report of the intergovernmental panel on climate change. Cambridge University Press.

[2] Vecchi, A., Li, Y., Ding, Y., Mancarella, P., & Sciacovelli, A. (2021). Liquid air energy storage (LAES): A review on technology state-of-the-art, integration pathways and future perspectives. Advances

in Applied Energy, 3, 100047.

[3] Borri, E., Tafone, A., Romagnoli, A., & Comodi, G. (2021). A review on liquid air energy storage: History, state of the art and recent developments. Renewable and Sustainable Energy Reviews, 137, 110572.

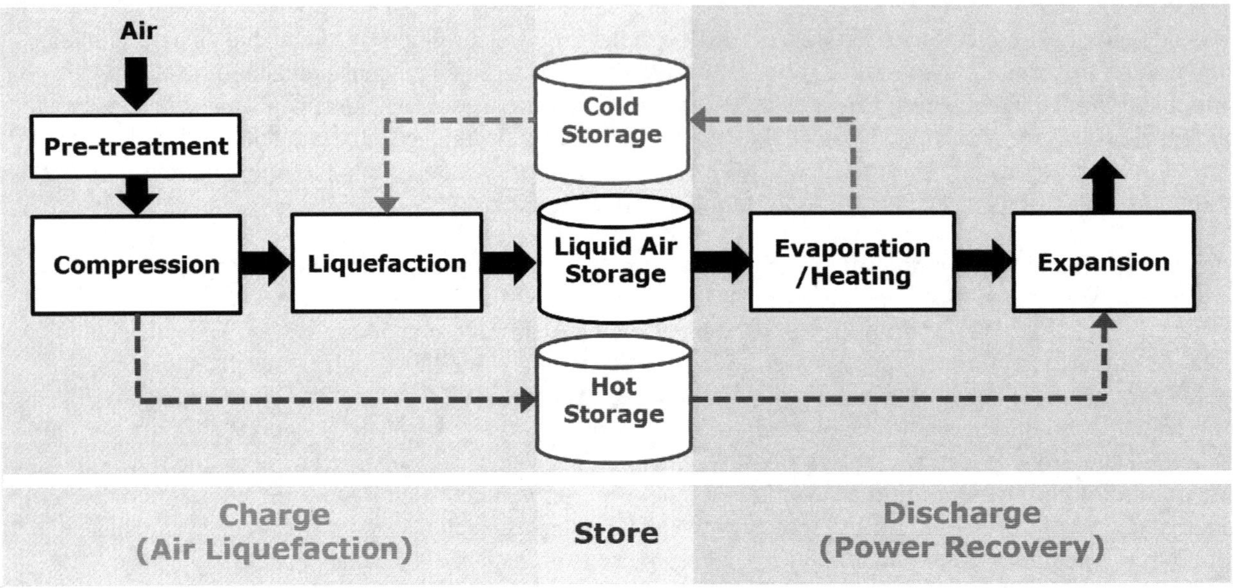

Figure 1: Standalone liquid air energy storage system with hot and cold thermal recovery.

230e Techno-Economic Analysis of Integrating Different Energy Storage Models in Highly Renewable Energy Systems

Ali Khosravani, Kody Powell

As the world pushes towards achieving 100% renewable energy future, the need for reliable energy storage systems emerges as a critical factor. Renewable energy sources, such as solar and wind, often produce electricity intermittently, creating gaps in supply. To bridge these gaps, energy storage systems play a vital role by storing excess electricity during times of abundance and releasing it when renewable energy alone is not sufficient to meet demand. Despite numerous valuable studies listing various energy storage technologies and assessing their technical capabilities, there remains a notable gap in evaluating these energy storage techniques within the context of a renewable energy system.

In this study, we developed a generic energy storage model including necessary parameters to describe an energy storage system such as capital cost, operating cost, storage duration, roundtrip efficiency, and lifetime. These parameters collectively characterize any energy storage system, providing a comprehensive framework for analysis. Additionally, utilizing Python programming, we constructed a renewable energy system model incorporating solar photovoltaics and wind turbines as primary renewable resources. This model integrates with our generic energy storage model. Furthermore, a backup natural gas power plant is incorporated into the system to supplement electricity generation during periods of insufficient renewable energy supply or when the storage system is depleted, as illustrated in the figure. The significance of our study lies in its ability to evaluate diverse (even new) energy storage systems described by the above-mentioned parameters within the context of renewable energy systems. We performed sensitivity analysis on the above-mentioned storage parameters. Our findings show that long-term energy storage technologies (with higher storage duration) will help us to have a more reliable and affordable energy system. This approach enables comprehensive performance assessments, offering valuable insights into the effectiveness and suitability of various energy storage solutions in real-world applications. It also enables decision-makers to anticipate potential challenges and opportunities, guiding long-term planning and investment decisions in the renewable energy sector.

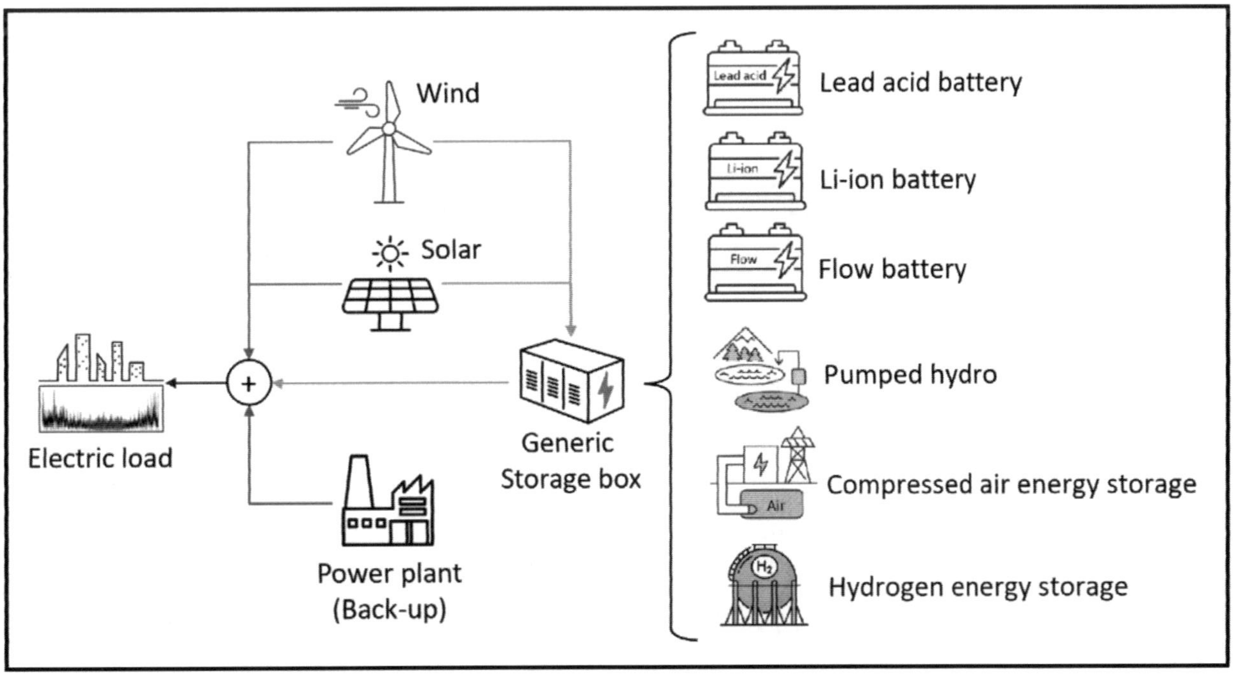

230f Optimal Design and Integration of a Green Hydrogen Production, Compression and Storage System for Solar Photovoltaic Power Smoothing

Yifan Wang, Paul Brooker, James Fenton

With the increasing penetration of solar photovoltaic (PV) power into the electric grid, intermittent solar energy can lead to frequent and steep ramping operations of conventional fossil fuel power plants [1,2]. Green hydrogen production via water electrolysis using renewable solar power can serve as a controllable load and provide a short/long duration energy storage system for addressing load fluctuations (i.e., PV smoothing) and improve grid resiliency. The generated green hydrogen will be stored under high pressure (HD) for further use (e.g., hydrogen fuel cell electric vehicle refueling).

Polymer electrolyte membrane (PEM) electrolyzers with high efficiency and quick dynamic response can be used for hydrogen production [3,4]. The capacities of the solar field and PEM electrolyzer affect the maximum load of hydrogen production and the corresponding performance of PV smoothing on cloudy days. In addition, energy consumption for the high-pressure compression and storage of hydrogen should be considered in the integrated hydrogen system. Both size and operating pressure of compressor and HD storage vessels need to be optimized based on the tradeoff between capital expenditures and effectiveness of PV smoothing.

An integrated hydrogen-based energy storage system (renewable solar field coupled with green hydrogen production, compression and storage) was developed based on dynamic modeling. A high-fidelity dynamic model of a PEM electrolyzer cell/stack was established with consideration of the mass/heat transfer coupled with electrochemical kinetics. Multi-stage hydrogen compression with a cascaded vessel storage configuration was designed and modeled. Reduced order models (ROM) of hydrogen production, compression and storage were developed based on their nonlinear standalone models for optimal design and operation of electrolyzer/compressor/storage vessels under fluctuating solar power.

The integrated energy storage system must minimize PV fluctuations to the grid while maintaining electrolyzer durability. To evaluate the transient performance of the PEM electrolyzer stack and integrated energy system, real-time PV data from Orlando Utilities Commission (OUC) solar farm is smoothed using the optimal power signal control algorithm. The optimal PV smoothing algorithm has the predictive capability using the rolling-horizon approach and shows the effectiveness of hydrogen-based energy storage system in smoothing the PV signal to improve the grid stability and flexibility.

[1] Wang, Y., Bhattacharyya, D., & Turton, R. (2021). Multi-objective Dynamic Optimization for Optimal Load-Following of Natural Gas Combined Cycle Power Plants under Stress Constraints. Industrial & Engineering Chemistry Research, 60(39), 14251-14270.

[2] Wang, Y., Bhattacharyya, D., and Turton, R. (2019). Evaluation of Novel Configurations of Natural Gas Combined Cycle (NGCC) Power Plants for Load-Following Operation using Dynamic Modeling and Optimization. Energy and Fuels, 34(1), 1053-1070.

[3] Kumar, S.S. and Himabindu, V., 2019. Hydrogen production by PEM water electrolysis–A review. Materials Science for Energy Technologies, 2(3), 442-454.

[4] Götz, M., Lefebvre, J., Mörs, F., Koch, A.M., Graf, F., Bajohr, S., Reimert, R. and Kolb, T., 2016. Renewable Power-to-Gas: A technological and economic review. Renewable energy, 85, 1371-1390.

230g Durational Testing of Lithium-Ion Battery Cells Using Grid-Informed Duty Cycles

Jaesung Kim, Dongwei Zhao, Audun Botterud, Fikile Brushett

Electrochemical energy storage is anticipated to play an important role in the decarbonization of by facilitating the deployment of variable renewable energy sources, and improving the efficiency, flexibility, and resilience of existing electrical infrastructure. Dominant in portable electronics and emerging in transportation, lithium-ion batteries (LIBs) represent the current state-of-the-art in energy storage due to their favorable combination of energy density, efficiency, and cost. Accordingly, LIBs are also being considered for stationary applications where they may support a range of grid needs. However, there is limited knowledge, at least in the open literature, of how duty cycles characteristic of different grid services impact the performance and lifetime of today's LIB systems. To this end, we are developing battery testing protocols informed by different end-uses (e.g., energy arbitrage, frequency regulation) and employing these routines to evaluate the durational performance of commercial LIB cells. We anticipate these studies will offer insights into the suitability of different battery chemistries for particular services, will facilitate improved representations of LIBs in grid models, and will aid in the identification and mitigation of degradation mechanisms within LIBs.

In this presentation, we will discuss our recent work on the performance and durability of different commercial 18650 LIBs as a function of both calendar aging at different states-of-charges (SOCs) and duty cycles characteristic of an energy arbitrage use-case. We find that battery performance is sensitive to cycling and to resting at high SOCs but, appears largely insensitive to charging rates (C/3 to C/20), at least for the chemistries considered (Figure 1).

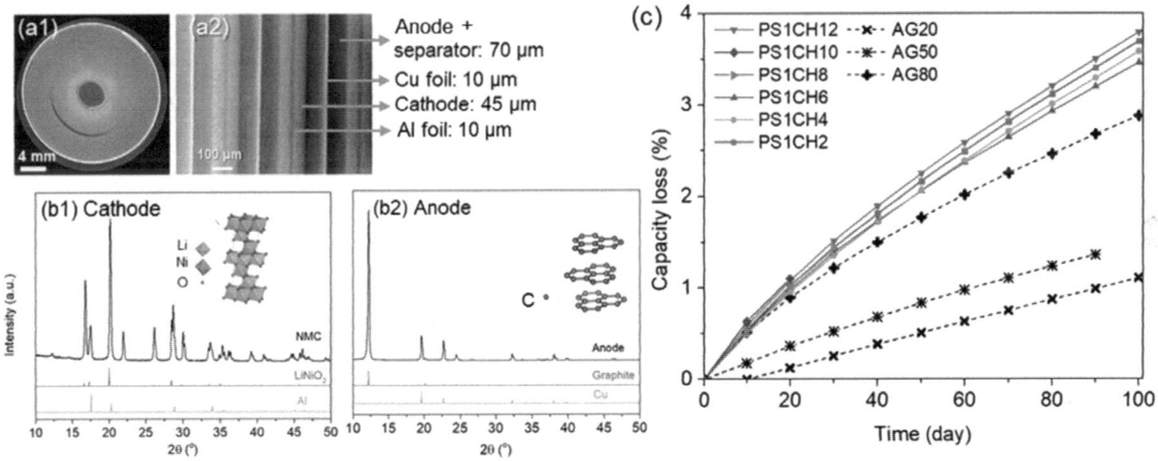

Figure 1 . (a1) X-ray tomography of the pristine commercial 18650 LIB with a $LiNi_xMn_yCo_zO_2$ (NMC) positive electrode (cathode) and a graphite + silicon negative electrode (anode) and (a2) vertical X-ray transmission image (inverted), XRD patterns of (b1) positive electrode and (b2) negative electrode with reference materials ($LiNiO_2$, Al, graphite, and Cu), (c) capacity loss over 100 days for cells cycled and aged at different SOCs in a simulated energy arbitrage application.

431a Integrating Carbon Capture Technologies in Hard-to-Abate Industries: A Techno-Economic Analysis

Bosong Lin, Haoxiang Lai, Lingyan Deng, Guiyan Zang

In this study, we develop process models to assess the performance and techno-economic viability of carbon capture solutions aimed at mitigating CO2 emissions in challenging sectors, notably the cement and steel/iron industries. These sectors are significant, contributing to roughly half of the U.S. industrial CO2 emissions (U.S. EPA, 2020). Our investigation covers a wide temperature range (300 to >1400°C) and CO_2 concentrations (4 to 25 mol%), characteristic of the emissions from these industries. The analysis extends beyond conventional amine-based carbon capture, our baseline technology, to explore promising alternatives like molten salt absorption, particularly for high-temperature flue gases.

The focus of our presentation is the integration of carbon capture technologies with heavy industries through detailed process modeling. This approach enables a comprehensive cost-performance analysis, evaluating energy consumption, process efficiency, capital and operating expenses, and scalability of various carbon capture and storage (CCS) systems tailored to specific industrial operations. We complement our study with sensitivity analyses on the cost of CO2 avoidance, identifying avenues for cost reduction and highlighting the potential for temperature-matched emission streams and sorbents. The ultimate goal is to map optimal carbon capture solutions to their most compatible industrial applications, providing a comprehensive assessment of their scalability potential in the targeted sectors.

431b Synergistic Conversion of Captured CO2 and Green H2 to Olefins and SAF By Microwave Catalytic Processing

Yujuan Chen, Brandon Robinson, Jianli Hu

The escalation of atmospheric carbon dioxide (CO_2) concentrations linked directly to human activities presents a formidable environmental problem, encompassing oceanic acidification, augmented sea levels, and the intensified occurrence of heat waves. Through the strategic utilization of CO_2, harmful emissions can be mitigated by transforming CO_2 into valuable chemicals. A distinguished methodology entails the conversion of CO_2 into C2+ hydrocarbons, notably light olefins such as ethylene, propylene, and butylene, which are pivotal for the chemical industry. The olefins can be oligomerized to from sustainable aviation fuel (SAF) to benefit aviation fuel industry. However, the utilization of CO_2 is impeded by its chemical inactivity and the elevated energy barriers associated with the formation of carbon-carbon bonds, rendering the efficient and selective transformation of CO_2 to C_2+ derivatives a challenging yet immensely promising venture. CO_2 is located at the bottom of the energetic ladder in thermodynamics, and the reaction of CO_2 requires considerable Gibbs energy input (the Gibbs formation energy, $\Delta G= -394.4$ kJ mol^{-1}). This research is pertinent to the processes of transforming CO_2 into methane and the non-oxidative coupling of methane to form ethylene. The primary challenges encountered include the formation of carbon deposits and inadequate selectivity favoring ethylene synthesis.

In this study, to overcome the activation energy barrier of CO_2 we have utilized microwave heating to energize the reaction. The application of electromagnetic energy, exemplified by microwave irradiation, can preferentially elevate the temperature of active sites on catalytic surfaces, thereby facilitating the activation of carbon-oxygen double bonds in CO_2, a crucial step in the hydrogenation of carbon dioxide. The incorporation of microwave energy has the potential to modulate reaction mechanisms, reduce the required thermal threshold for reaction initiation, and attenuate the kinetics of carbonaceous deposit formation. Consequently, these alterations can lead to a significant improvement in both the yield and selectivity of the desired chemical products.

In the present investigation, we have engineered a ruthenium-supported cerium dioxide (Ru/CeO_2) catalyst that demonstrates a notable selectivity for methane across a broad temperature spectrum, alongside a molybdenum catalyst supported on cerium dioxide (Mo/CeO_2), which catalyzes the conversion of methane into olefins. Our strategy encompasses the synthesis of lower olefins through a dual catalytic approach: firstly, the hydrogenation of CO_2 to methane facilitated by Ru/CeO_2, followed by the selective transformation of methane into lower olefins utilizing Mo/CeO_2. A single reactor consisted of two separate zones (thermal-microwave) was configured to allow methanation to take place in thermal zone followed by microwave-driven methane coupling zone. The resultant gaseous products are subsequently analyzed using an Inficon Fusion Micro-GC to elucidate the composition and efficiency of the catalytic processes. The catalysts used in either zone are characterized by XRD, XPS, TEM and in-situ Raman to correlate the catalytic structural changes with CO_2 conversion and selectivity to ethylene. Our approach achieved high CO_2 conversion of 70% , methane selectivity of 99.5%, and methane production up to 11445.8 umol CH_4/g.cat hr. This research shows that microwave-driven CO_2 transformation to ethylene can enable decarbonization of several industrial sectors by electrification. Efforts to mitigate CO_2 emissions in the chemical industry could catalyze a transition towards carbon-negative production, leveraging advanced technologies to transform environmental liabilities into sustainable assets.

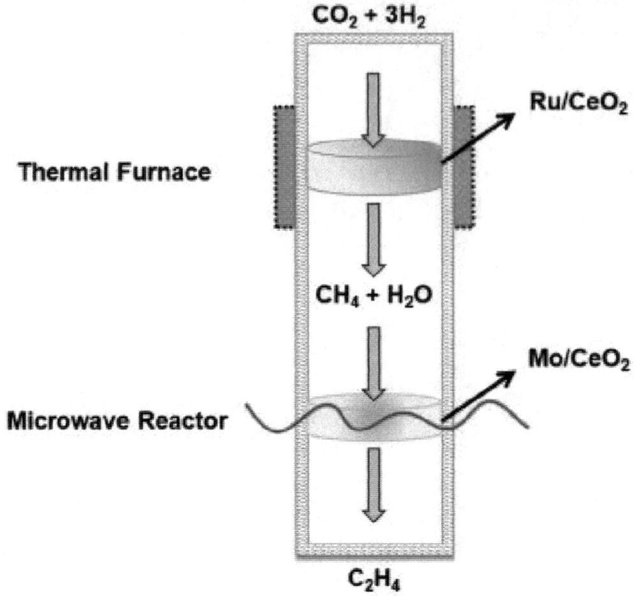

Figure 1. Reactor configurations for a hybrid heating reactor system

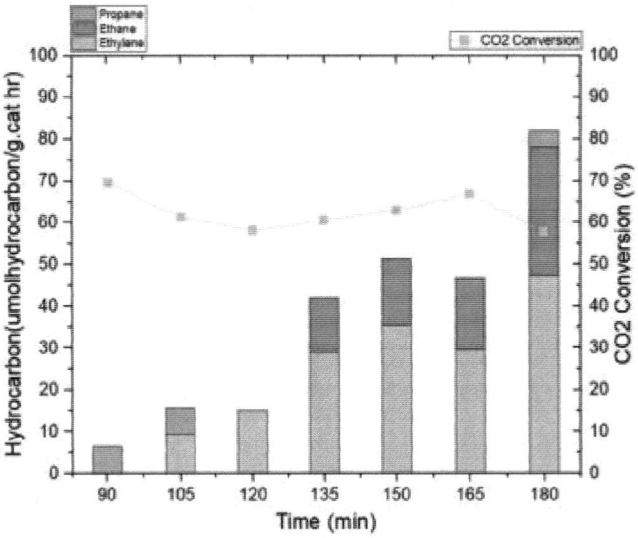

Figure 2. CO_2 conversion and C_2 hydrocarbon productivity under thermal microwave hybrid heating reactor (ambient pressure)

431c Comparative Techno-Economic and Life Cycle Assessment of Reactive CO_2 Capture to Synthetic Natural Gas

Alvina Aui, Hannah M. Goldstein, Wenqin Li, Nathan C. Ellebracht, Simon H. Pang

Reactive CO_2 Capture (RCC) is an emerging class of carbon management technologies that aim to efficiently produce low carbon fuels and chemicals. RCC is a process intensification approach integrating CO_2 capture and conversion to a value-added product in one system. RCC is particularly advantageous because i) it avoids intermediate CO_2 purification (reducing thermal energy demand) and ii) creates a marketable product in addition to CO_2 capture (additional revenue incentivizing CO_2 capture technologies), leading to potential energy and cost competitiveness. In this analysis, RCC is designed to capture and convert CO_2 to produce synthetic natural gas (SNG) as a method of long-duration energy storage. The production of natural gas is especially favorable because it has a wide range of uses, such as heat, electricity, transportation fuel, chemical production and an energy carrier in the power-to-gas framework. SNG is a promising energy carrier primarily because it has long-term storage capacity, and it can be used to overcome intermittency issues with renewable energy coupled grid systems. Additionally, unlike hydrogen, SNG can utilize readily available and existing natural gas infrastructures and therefore, has locational flexibility when deployed. SNG produced via RCC is considered a renewable and potentially carbon-neutral energy source that can be used to decarbonize multiple economic sectors.

In this contribution, I will discuss our comparative system analysis, which includes process modeling, techno-economic, carbon footprint and sensitivity analysis, to identify key metrics and evaluate the technical viability of RCC. Through RCC, CO_2 can be captured and directly converted into SNG in one system, with renewable hydrogen and a dual-function material. We compared RCC to a separated CO_2 to SNG process which involves two steps – i) CO_2 is first captured via Direct Air Capture, then ii) converted to SNG in a methanation facility. Our analysis suggests that RCC has 56-66% energy savings compared to a separated process. This resulted in a minimum SNG selling price of \$2.17/kg, 12-13% lower than the cost of a separated two-step process. The overall cost is dominated by sorbent and hydrogen operating costs (also evident in the sensitivity analysis). Compared to existing technologies such as conventional natural gas production offset with Direct Air Capture with Storage and anaerobic digestion of waste feedstocks, SNG via RCC costs 12-21% less. As an energy storage option, SNG from RCC is a low-cost option; it costs 6-79% less than other systems such as batteries, geothermal, hydropower, hydrogen, and others. The carbon footprint analysis resulted in around 8-10 g CO_{2e}/MJ when powered by renewable energy, but the carbon intensity is 15-17x greater if the process is powered by the US grid mix, suggesting that we must use renewable energy to drive these technologies. Our analysis suggests that RCC is a promising technology for energy generation, as it avoids the continued use of fossil fuels while offsetting carbon emissions, but it is not without its challenges.

431d Accelerated Deep Learning-Based Data Assimilation and CO2 Injection Optimization at the Illinois Basin-Decatur Carbon Sequestration Project

Masahiro Nagao, Takuto Sakai, Chin Hsiang Chan, Akhil Datta-Gupta

Carbon capture and storage (CCS) is widely considered to be one of the promising technologies for reducing greenhouse gas emission. Extensive research and field tests have been conducted to optimize the CO2 storage operation in different subsurface environments, such as depleted oil and gas fields or deep saline aquifers. It is essential to monitor the CO_2 plume effectively throughout the life cycle of a geologic CO_2 sequestration project to ensure safety and storage efficiency. However, there are several underlying challenges. First, field-scale high resolution numerical simulations of CO2 storage are computationally expensive involving complex physics associated with multi-component and non-isothermal fluid flow. Existing data assimilation methods entail solution of a non-linear inverse problem requiring numerous flow simulations, making it unscalable for field applications in many cases. Second, CO2 injection into the subsurface involves various risks, such as CO2 leakage and potential for induced seismicity. It is imperative to conduct a comprehensive evaluation considering multiple different factors, including pressure maintenance, seal integrity, and the potential for CO2 leakage from legacy wells, natural fractures, and faults. The optimization of CO2 injection project needs to involve multiple objectives, and these objectives often exhibit trade-offs, rendering the optimization of CO2 injection schedule challenging. We address all these challenges by leveraging the recent advancements in deep learning techniques, which significantly accelerate field-scale data assimilation and multi-objective optimization.

Our proposed approach is composed of two parts: data assimilation to calibrate the dynamic reservoir model by integrating field monitoring data, and then, optimization of the CO2 injection schedule. In the data assimilation framework, a deep neural network model is developed that takes available monitoring data such as well pressure and temperature measurements as inputs and predicts representative images of the subsurface flow field. The conventional approach to build a deep learning model involves a neural network architecture where the input data consists of the field monitoring data, and the output is pressure and CO2 saturation distribution at different timesteps. However, the high dimensionality of the spatiotemporal resolution of the pressure and saturation images makes the neural network training inefficient and impractical for field-scale applications. We improve the efficiency and scalability of the framework in three ways. First, instead of using multiple pressure and CO2 saturation maps for different timesteps, a single diffusive time of flight (DTOF) map is used for the output as a representative subsurface image. The DTOF represents the arrival time of pressure front propagation, which can be computed rapidly by the Fast Marching Method (FMM), an efficient front tracking approach that does not require expensive fluid flow simulations. Since the reservoir dynamics is compressed into a single DTOF image, the memory requirements and computational costs are reduced significantly, and also, the neural network architecture is simplified considerably. Second, we apply 'optimal' grid coarsening of geologic models to reduce the simulation time while preserving the simulation accuracy. The coarsening scheme is based on a 'bias-variance trade-off' that maximizes the computational time savings while minimizing the error in simulated monitoring data, such as well pressure and temperature data. Third, we apply a variational autoencoder-decoder (VAE) network to compress high dimensional DTOF images into lower dimensional latent variables. The use of image compression with VAE considerably simplifies the neural network architecture, enabling efficient training for large-scale reservoir applications. A regression model, consisting of feed forward and convolutional neural networks, is trained to predict latent variables of the VAE based on the available monitoring data. Subsequently, the predicted latent variables are fed into the trained decoder network to estimate the 3D DTOF image. The VAE can provide multiple DTOF image predictions to account for underlying uncertainties. Finally, the trained neural network can be used for reservoir model calibration and prediction of the CO2 plume migration using observed monitoring data. Calibrated reservoir models are identified by simply selecting multiple plausible realizations of the training data samples close to the predicted DTOF map based on the field monitoring data.

The calibrated reservoir models are used for optimizing the CO2 injection schedule. The proposed optimization workflow utilizes a recently developed architecture, Fourier Neural Operator (FNO), as a data-driven proxy model. The input of the FNO model includes permeability distribution, porosity distribution, and CO2 injection schedule, and it estimates the subsurface pressure and CO2 saturation images as output. The most time-consuming part of the data-driven framework is the training data generation, requiring numerous flow simulations. This challenge can be addressed by exploiting the super-resolution feature of the FNO. The FNO model can be trained using low resolution image data, and the trained model can predict high resolution image data by interpolating the solution within the Fourier domain. Therefore, training data generation can be conducted using coarsened reservoir models, reducing the computational cost significantly. Areal model coarsening is applied to the permeability and porosity fields, and training dataset is generated by running the numerical simulations with different CO2 injection schedules. The simulations provide coarse scale pressure and saturation distributions as training data. Although the FNO model is trained using the coarse scale data, the trained model can estimate fine scale pressure and CO2 saturation distributions by leveraging the super-resolution feature without significantly compromising the accuracy. For CO2 injection schedule optimization, a multi-objective genetic algorithm (MOGA) is utilized to properly account for the multiple and potentially conflicting objectives. The primary advantage of MOGA is its ability to optimize each objective function individually, unlike other optimization algorithms that tend to aggregate all objective functions into a single function. In the optimization framework, the FNO-based proxy model is used as a forward model and requires only a few seconds as opposed to several hours for pressure and CO2 saturation predictions. The objective of the optimization includes three different components: minimizing subsurface pressure increase, maximizing total CO2 injection amount, and maximizing storage efficiency. Since MOGA provides multiple optimal realizations on the Pareto front, users can select the most desirable realizations based on their specific requirements or constraints, such as maximum allowable pressure or target CO2 amount to be stored.

The power and efficacy of our approach is demonstrated by applying to the Illinois Basin-Decatur Project (IBDP), a large-scale CO2 storage test in saline aquifer. The data assimilation process is implemented with given field measurements including distributed pressure and distributed temperature sensing (DTS) data at an injection and a monitoring well. The data assimilation can be conducted in a few seconds after the neural network training is completed. CO2 plume evolution is predicted by running the simulation of the calibrated reservoir models identified based on the VAE latent space model selection. Subsequently, the calibrated reservoir model is utilized for the optimization framework. Training data for the FNO proxy model is generated at the coarse scale whereby an order of magnitude computational speed up is achieved compared to the original fine scale simulations. The accuracy of the trained proxy model is verified by comparing with a commercial reservoir simulator, and the prediction of the CO2 saturation and pressure distribution takes only about 10 seconds as opposed to several hours with the original fine scale simulation. Next, the MOGA is applied to optimize the CO2 injection schedule using the fast FNO proxy model. The multiple objectives considered in the optimization framework are improved considerably, providing optimized CO2 injection schedules. Orders of magnitude speed up is achieved compared to the traditional workflow that uses numerical simulation as a forward model.

Our proposed deep learning-based framework enables rapid data assimilation and injection schedule optimization for large-scale CO_2 storage applications. The data assimilation framework is significantly accelerated by exploiting the efficacy of deep learning, the concept of DTOF, and optimal coarsening of the geologic models. The evolution of the CO_2 plume images is predicted by running the calibrated reservoir models. In the optimization framework, the super-resolution feature in FNO reduces the computational cost associated with the training data generation significantly. Furthermore, the proxy model accelerates forward simulation by orders of magnitude, enabling examination of multiple optimization scenarios for large-scale field cases while considering the complex physic involved in the CO_2 storage applications such as non-isothermal and compositional fluid flow.

431e Evaluation of Thermophotovoltaics As a Decarbonization and Clean Energy Storage Technology: Techno-Economic Analysis and Uncertainty Assessment

Manish Mosalpuri, Mark Mba Wright, Fatima Toor

Thermophotovoltaics (TPV) is an emerging decarbonization and renewable energy technology that converts heat to electricity. **Figure 1** shows the TPV process schematic. It requires a heat source that can be any of these four- solar, chemical, nuclear, or electrical. The heat from these sources (in the range of 1300-2500 K) or waste heat sources reaches the thermal emitter, which converts it into radiation. The presence of a thermal emitter is the difference between a solar PV and a TPV system. The emitter emits a more suitable radiation spectrum (compared to the Sun in the PV system) that reaches the PV cell, which converts the part of incoming photons having higher energy than the PV cell bandgap into electricity. The photons having less energy than the bandgap of the PV cell (sub-bandgap photons) are not useful and do not convert to electricity but overheat the TPV cell, decrease the emitter temperature and heat-to-radiation efficiency, and thus reduce the overall efficiency.

TPV is increasingly becoming popular as an energy conversion pathway due to its numerous applications, some of which include energy storage, combined heat and power (CHP), waste energy recovery, and in-space programs. It can be used for CHP applications as it provides electricity and can be used to store heat due to its compatibility with storage applications. Due to its higher power density (>2.5W/cm^2) and reduced cost, there has been an increasing focus on energy storage using TPV, called thermal energy electrical storage (TEES) or thermal energy grid storage (TEGS), which is competitive with electrochemical batteries, and is suitable for long-duration energy storage applications.

In this study, we consider an energy storage model by Datas et al. (2019)[1] (**Figure 2**). It includes three energy sources: solar PV, grid, and natural gas; a boiler (running on natural gas); two heat storages: low-temperature energy storage (LTES) and high-temperature energy storage (HTES); a power generation unit (PGU, here, it is TPV); and a heat pump (either electric or thermal). It computes the energy going into each storage based on the interplay of electric and heat demand and generation.

We calculate and optimize the levelized cost of consumed energy (LCOE) and electricity ($LCOE_{el}$) for a TPV energy storage system for a building in Boone (Iowa) using an energy management algorithm (along with our modifications) in Python. For optimization, we vary the system size- nominal PV capacity, HTES maximum energy capacity and PGU maximum energy capacity. LCOE represents the cost of supplying heat and electricity; however, $LCOE_{el}$ only considers electricity costs. We perform a sensitivity analysis and Monte Carlo uncertainty assessment to better model the uncertainty in these input parameters- lifetime, nominal weighted average cost of capital (WACC), PGU/TPV efficiency, electricity price, PGU CAPEX (capital cost factor), PV CAPEX, HTES CAPEX, inflation rate, and natural gas price. The Monte Carlo method is an analysis tool that uses random numbers (generated using a fitted probability distribution for each parameter from literature data) to calculate the output for a modeled stochastic process.

The results indicate that after optimization (and using the energy management algorithm), LCOE reduces, and the self-consumption ratio (SCR, representing the amount of solar PV energy being utilized for energy storage and electricity supply) of solar PV increases from 57.2% to 61.7%. Less PV energy is lost, and the consumer utilizes more. For both LCOE and $LCOE_{el}$, the optimum system converges to the smallest system bounds: nominal PV capacity = 5 kW_{el}, HTES max. capacity = 15 kWh_{th}, PGU max. generation capacity = 0.5 kW_{el}. LCOE has a base-case value of \$0.08/kWh and optimizes to \$0.06/kWh. For our study, the base-case $LCOE_{el}$ is \$0.34/kWh, which is reduced by 35% to \$0.22/kWh on optimization.

The Monte Carlo uncertainty assessment (results shown in **Figure 3**) on an optimized system also produces the same mean LCOE of \$0.06/kWh (general extreme value distribution with a standard deviation of \$0.015/kWh) and $LCOE_{el}$ (general extreme value distribution with a standard deviation of

$0.048/kWh) of $0.23/kWh, respectively. These values are still higher than the average price of electricity = $0.124/kWh. However, the box plot (**Figure 4** shown for LCOE; $LCOE_{el}$ has a similar trend) shows that it is possible to reduce the cost further by improving these parameters- lifetime, PV CAPEX, inflation rate, natural gas price, and electricity price. These parameters have dominant effects on the cost, and thus, future research should be more focused on improving them to make the system economically feasible.

We thus observe that the LCOE and $LCOE_{el}$ reduce on optimization, and performing Monte Carlo uncertainty assessment reveals an enormous potential for cost reduction. Not many large-scale TPV prototypes exist, and our study provides a future research direction toward economic feasibility.

Our future work would involve performing a preliminary life cycle assessment (LCA) as the first step in quantifying potential decarbonization enabled by TPV.

Acknowledgements

We acknowledge the funding for the work by the Decarb 2040 Community Feasibility Grant Program.

References

(1) Datas, A.; Ramos, A.; del Cañizo, C. Techno-Economic Analysis of Solar PV Power-to-Heat-to-Power Storage and Trigeneration in the Residential Sector. *Applied Energy* **2019**, *256* (October), 113935. https://doi.org/10.1016/j.apenergy.2019.113935.

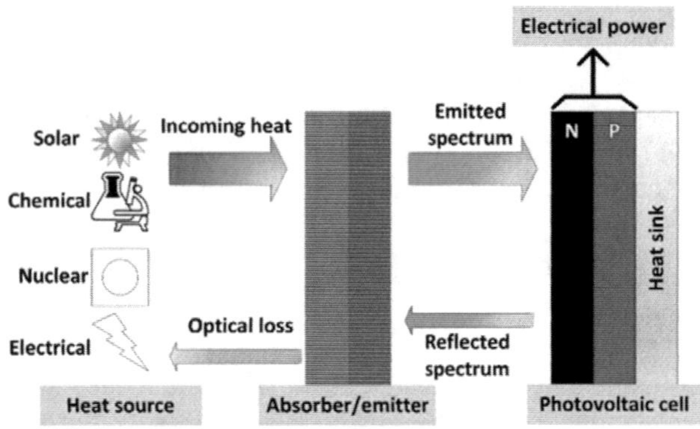

Figure 1. TPV process schematic showing heat conversion to electrical power by emitter and a PV cell

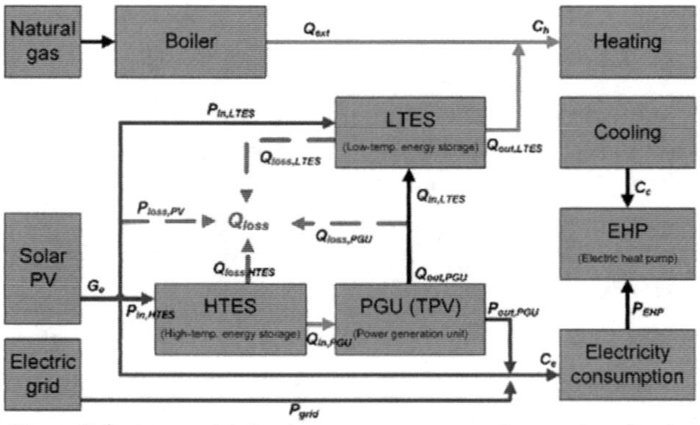

Figure 2. System model showing energy sources and energy transfers to satisfy electricity consumption and heating and cooling loads

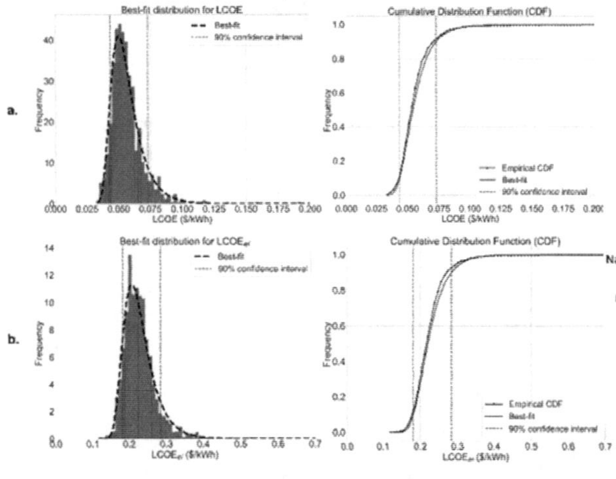

Figure 3. Monte-Carlo risk uncertainty assessment results (Probability density function, cumulative distribution function, 10%, and 90% confidence limits) for **a.** LCOE and **b.** LCOEel

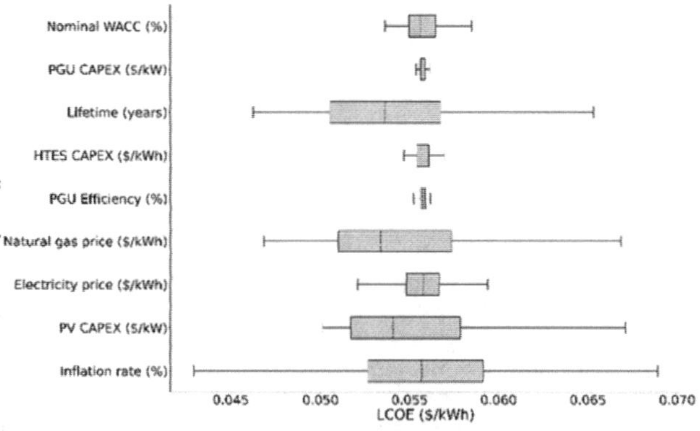

Figure 4. LCOE box-plot representing parameter uncertainty

431f Investigating the Design of Global-Scale Power-to-Methanol Production Systems: Large- or Small-Scale Chemical Plants?

Tibor Svitnic, Kai Sundmacher

Today, the chemical industry relies predominantly on fossil resources for feedstock and energy supply. The concentrated nature of these resources facilitates the operation of large-scale chemical plants and even complex production systems of various products, which benefit from the economies of scale and material/energy integration. However, with the goal to mitigate the negative consequences of climate change, the chemical industry faces a transition towards renewable sources of material and energy (e.g. wind and solar) that require larger land areas compared to fossil sources [1]. Consequently, the costs and losses associated with the energy/material supply to the production system increase, potentially negating the cost reductions achievable through the economies of scale by increasing the capacity. This would imply the existence of a threshold capacity beyond which a further increase ceases to be economically viable.

We investigate these scale dependencies within the context of methanol production, an important platform chemical with a global production capacity on the order of 100 Mt/a, which can be used for plastics production or directly as a shipping fuel [2]. Current fossil-based methanol synthesis processes can have a single-train capacity as high as 2,600,000 t/a with reports indicating installations of up to five such trains within a single production system [3]. In contrast, the newly emerging Power-to-Methanol processes using renewable sources of electrical energy, hydrogen and carbon dioxide, which are currently under construction, have capacities of up to 100,000 t/a [4]. Such disparity in scale leads to the following questions: To what extent would future solar and wind-based Power-to-Methanol processes be economically beneficial at scales approaching their fossil counterparts, given the aforementioned considerations? Alternatively, would a spatially distributed fleet of a greater number of small-scale plants be a more effective strategy to produce large quantities of methanol?

To offer quantitative insights into these aspects, we propose an optimization-based production system design method, in which the costs of energy/material supply from a given land area with solar/wind energy availability and the economies of scale of the chemical plant are modeled. Our study also includes the option to distribute the sub-processes away from the demand location of the system to benefit from reduced costs of energy/material transport through electricity transmission lines and pipelines. Importantly, heat integration/waste-heat utilization, which can lead to higher efficiencies for the Power-to-Methanol process as quantified in our previous modeling studies [5-7], are also considered. However, benefitting from such integration also constrains the sub-processes (e.g. water electrolysis and direct air capture of CO_2) to be installed in close proximity to the chemical plant, making the design problem even more interesting.

The proposed design approach is subsequently applied to assess case studies with different renewable energy conditions, representing promising, often remote [8], locations for greenfield Power-to-Methanol production systems. In this way, we contribute a modeling study, which can help in determining the economically optimal design, including the energy/material supply, sub-process localization for the production of methanol in large quantities, and is readily adjustable to other Power-to-X products.

[1] Nøland, J. K., Auxepaules, J., Rousset, A., Perney B., Falletti G. (2022) Spatial energy density of large-scale electricity generation from power sources worldwide. Sci Rep, 12, 21280, 10.1038/s41598-022-25341-9

[2] Tabibian S.S., Scharifzadeh M. (2023) Statistical and analytical investigation of methanol applications, production technologies, value-chain and economy with a special focus on renewable methanol. Renew Sustain Energy Rev, 179, 113281, 10.1016/j.rser.2023.113281

[3] Johnson Matthey (2020) World's largest single train methanol plants to use Johnson Matthey technology. URL: https://matthey.com/documents/161599/166306/Johnson-Matthey-Press-Release-Baofeng-IV-Methanol-Plantv07FINAL.pdf/7fbe5fa0-f118-832e-7d88-0b1668c9962a?t=1650968256207 (accessed: 03.04.2024)

[4] Methanol Institute (2024) Renewable Methanol: E-methanol and Biomethanol Plants. URL: https://www.methanol.org/renewable/ (accessed: 03.04.2024)

[5] Svitnič, T., Sundmacher K. (2022) Renewable methanol production: Optimization-based design, scheduling and waste-heat utilization with the FluxMax approach. Applied Energy, 326, 120017, 10.1016/j.apenergy.2022.120017

[6] Svitnič, T., Beer K., Sundmacher K., Böcher, M. (2024) Optimal design of a sector-coupled renewable methanol production amid political goals and expected conflicts: Costs vs. land use. Sustainable Production and Consumption, 44, 123-150, 10.1016/j.spc.2023.12.003

[7] Svitnič, T., Sundmacher K. (2024) Identifying standard and simple designs of Power-to-Methanol processes: The costs of complexity reduction. Energy Conversion and Management, 307, 118325, 10.1016/j.enconman.2024.118325

[8] Pfennig M., Böttger D., Häckner B., Geiger D., Zink C., Bisevic A., Jansen L. (2023) Global GIS-based potential analysis and cost assessment of Power-to-X fuels in 2050. Applied Energy, 347, 121289, 10.1016/j.apenergy.2023.121289

351a Experimental Studies on Enzymatic Ring-Opening Polymerization for Sustainable Polyester Synthesis

Victor Sharma, Peter V. Bonnesen, Yangyang Wang, Jong K. Keum, John F. Cahill, Yue Yuan

Sustainable engineering has become an integral part of the overall effort toward establishing a circular bioeconomy, including emerging topics such as plastic degradation and synthesis. In the example of polyesters, which are primarily produced by melt polycondensation from petroleum-based feedstocks, their production is energy-intensive processes with high carbon dioxide (CO_2) emission. Enzymatic polymerization provides a new platform that enables selectivity and mild reaction conditions. Since the '90s, exploratory work on polyester enzymatic synthesis has been conducted with hydrolases for polycondensation, transesterification, and ring-opening mechanisms. In the past decades, the aim of enzymatic polyester synthesis often targets high molecular weight, particularly in aliphatic polyester synthesis. Recently, research interests have shifted to bio-based substrates and to obtain novel molecular architecture, although the model enzyme, *Candida Antarctica* lipase (CALB), has never changed. The lack of understanding of the kinetics and thermodynamics of the reaction, and limited methods that can be used to correlate protein sequences to the polymerization performance, posed challenges to engineering enzymes toward effective and selective polymerization, instead, engineering targets are enzyme thermostability, hydrolysis activity, and solvent tolerance.

In this work, we used serine superfamily hydrolases as the model enzymes for ring-opening polymerization (ROP) and developed a workflow to compare the enzyme performance. In addition to the physical-chemical properties of obtained polymer and oligomers, such as molecular weight and crystallinity, we conducted time-resolved measurement using nuclear magnetic resonance (NMR) spectroscopy and small-angle x-ray scattering (SAXS) to study the reaction kinetics. We also used open-port sampling interface mass spectrometry (OPSI-MS) to monomer the depletion and to realize the limits of different characterization techniques. The developed workflow can evaluate a panel of enzymes and lysates, thus providing a versatile platform for rapid enzyme engineering. By combining enzymatic method and polymer characterization methods, the study will lay down the groundwork for sustainable polyester synthesis and supplement enzyme engineering efforts towards engineering effective enzymes suitable for large-scale application and accessing diverse macromolecular architectures.

351b Alternative Feedstock Utilization By the *Rhodococcus Jostii* Strain PET for Biomanufacturing Beyond Earth

Hakyung Lee, Eunseo Lee, Yuxin Tian, Richa Guleria, Alexandra Smith, Luke Roberson, Mark Blenner, Daniel H. Yeh, Yinjie Tang, Jinjin Diao, Tae Seok Moon

Biomanufacturing at alternative habitats is drawing attention in space research. Using productive microbes as laborers presents a strategy to overcome cost limitations and waste generation of space exploration. This study investigated the potential of three different alternative feedstocks sourced or found beyond Earth: Regolith for micronutrients, Deconstructed polyethylene terephthalate (PET) for carbon, and Permeate from an anaerobic membrane bioreactor (AnMBR) treating fecal waste for macronutrients. This study aims to develop and optimize the chemical process for using *Rhodococcus jostii* strain PET (RPET) as a promising chassis to upcycle post-consumer PET into value-added chemicals (i.e., lycopene) in an alternative medium. For a partial alternative medium, regolith simulant was added into a reduced medium containing only carbon (PET monomers), nitrogen, and phosphorus. For a completely alternative medium, the nitrogen and phosphorus sources were replaced by AnMBR permeate. In these experimental settings, the successful utilization of alternative resources by RPET for valuable product generation was demonstrated.

First, regolith simulant supported RPET growth as a micronutrient source through bioleaching from ore particles and utilizing acidified regolith leachate. With 1 g/L dosage of lunar and Martian regolith simulant particles, growth achieved 66 % and 117 %, respectively, compared to the conventional medium. The active biosorption by the organic acid formation facilitated nutrient uptake, including calcium, magnesium, and iron. From the tolerance test to the acidified regolith leachate, RPET exhibited robustness against metal ions such as aluminum, titanium, and chromium present in regolith simulants that are supposed to be toxic to the strain. Second, regolith simulant enhanced the lycopene production, with Martian regolith simulant particles resulting in a 1.6 times higher yield compared to the conventional medium. Lunar regolith simulant achieved similar levels despite lower growth and carbon source consumption, indicating effective simulant utilization. The efficiency may be possibly due to enhanced activation of iron-related carotenoid enzymes. Regolith simulants are rich in iron, particularly the red-colored Martian regolith simulants. Third, in the completely alternative medium, where 40% AnMBR permeate was diluted in the water, growth levels were comparable to those in the conventional medium, even without the regolith simulant. The nutrient diversity of the wastewater other than nitrogen and phosphorus provided effective growth support. Lycopene production increased by 1.4 times in the presence of regolith simulant, which supplemented the trace elements lacking in the permeate that can support the lycopene producing pathway. Furthermore, the microbial growth and lycopene production have been tested in microgravity simulating high aspect ratio vessels to verify the feasibility of both partial and completely alternative feedstock utilizing medium in lower gravity conditions than Earth.

In summary, the viability of utilizing alternative feedstock was affirmed, and the potential recipe for RPET cultivation in space was proposed. This approach will contribute to space biomanufacturing research by significantly reducing space mission costs and advancing sustainable habitat construction in extraterrestrial environments.

351c Enhancing Biological CO_2 Conversion to Ethanol By *Clostridium Ragsdalei*: Insights from Batch and Continuous Operations

Rahul Thunuguntla, Hasan Atiyeh, Mari S. Chinn, Ralph S. Tanner

Biological conversion of CO_2 into biofuels offers a promising avenue for both greenhouse gas mitigation and resource utilization. This study investigates the conversion of CO_2 into ethanol by *Clostridium ragsdalei* P11 in 3-L stirred tank reactors operated in both batch and continuous modes using a gas mix H_2:CO_2:N_2 (60:20:20). Enzyme activities in the acetyl-CoA pathway, including carbon monoxide dehydrogenase (CODH), hydrogenase (H_2ase), formate dehydrogenase (FDH) and alcohol dehydrogenase (ADH), were evaluated. Results demonstrate that strain P11 produces 7.0 g/L ethanol from CO_2 in batch fermentations. During batch fermentation, H_2 and CO_2 conversion efficiencies were 60% and 40%, respectively. The supplementation of CO to CO_2 improved ethanol production. Continuous CO_2 fermentation with CO supplementation yielded 22 g/L ethanol. Increased activity of CODH and H_2ase was observed during the initial growth phase, while the rise in ADH activity closely correlated with ethanol production. Enzyme activities exhibited a 3.6-7 fold increase during continuous compared to batch fermentations. These findings highlight the robust potential of strain P11 for CO_2 conversion into ethanol and provide a promising groundwork for further optimizing CO_2 fermentation processes.

383g Separation of 2,3-Butanediol from Fermentation Broth Via Cyclic and SMB Adsorption over Nano-MFI Zeolites

Jianpei Lao, Qiang Fu, Marco Avendano, Yadong Chiang, Matthew Realff, Sankar Nair

The biomass-based platform molecule 2,3-butanediol (2,3-BDO) has a wide range of applications in production of sustainable fuels, chemicals, synthetic rubber, and others. However, the selective separation of 2,3-BDO from multicomponent fermentation broths presents challenges due to its low concentration, high solubility in water, high boiling point, and presence of many other species. Here we demonstrate remarkably selective enrichment and recovery of 2,3-BDO from a corn stover hydrolysate fermentation broth by a pure-silica nano-MFI-type zeolite adsorbent. By means of cyclic and simulated moving bed (SMB) adsorption processes, we obtained concentrated aqueous 2,3-BDO streams from the fermentation process stream with ~ 93% purity and 3-fold enrichment, and > 98% purity and 8-fold enrichment, respectively. These findings provide strong support for large-scale adsorptive separation for biobased 2,3-BDO production.

351e Continuous Production of Natural Blue Phycocyanin in Pilot-Scale Bioreactor

Mohammad Redwanur Rahman, Md Tabish Noori, Klaus Hellgardt

Natural blue pigment has numerous health benefits over synthetic phthalocyanine pigments. One of the most potential sources of natural blue pigments is the microalgae and cyanobacteria that have high concentrations of phycocyanin protein. Natural phycocyanin produced by cyanobacteria attracted a lot of interest, which boosted market demand because of its brilliant color and antioxidant qualities. One effective way to fulfill the growing demand for phycocyanin while reducing environmental impact is through the scalability of cyanobacterial production in bioreactors. Bioreactors are used to produce food- and pharmaceutical-grade biomass and biomolecules from microorganisms. Photosynthetic microalgae and cyanobacteria are currently cultured in bioreactors, either in batch or semi-continuous modes. Those approaches are less efficient, labor-intensive, and have high downtime and operating costs due to regular sterilization and cleaning in between cycles. To address those challenges, researchers are working on the development of more efficient production techniques. In this study, we have optimized the growth kinetics of the cyanobacteria *Synechocystis* spp. 6803, aiming for the maximum production of phycocyanin through the design of experiments (DoE). A series of batch and continuous-design experiments have been conducted in small lab-scale bioreactors to optimize light illuminations and types. Significant differences in dry cell weight, dry biomass, and specific growth rates (SGR) at different light illuminations have been observed. The growth rate of the cells highly depends on the initial cell density, the intensity of light, and the type of LED illumination. The maximum biomass and phycocyanin concentration in batch mode was observed in a lower and narrow white LED illumination of 200–300 μmol.m^{-2}s^{-1}. Based on those results, a factorial-designed continuous approach has been applied in lab-scale bioreactors. The results of the continuous experiments showed some promising outcomes toward a steady state of growth with continuous nutrient dilution and regular harvest from the late exponential phase. Higher biomass and phycocyanin production were achieved through increasing light illumination from 200 μmol.m^{-2}s^{-1} to 550 μmol.m^{-2}s^{-1} over the exponential phase. Finally, those results were transferred to a 120-liter pilot-scale bioreactor, where the steady state of the growth was maintained with a 0.15 to 0.18 dilution rate and an average of 10 liters per day of wet biomass harvest. Phycocyanin concentration was observed at 30–50 mg/g dry biomass at a 500 μmol.m^{-2}s^{-1} white LED light illumination. This continuous technique has the potential to increase profitability to one-quarter of the current approaches practiced in the industrial phycocyanin production process. However, further research is needed to prolong the steady state and develop the automated bioreactor technology, aiming at the highest possible natural phycocyanin pigment production.

429a Kinetic Modeling of the Epoxidation of High Oleic Palm Oil Using Amberlite IR-120 As Catalyst

Steven Solarte Sr., Alvaro Orjuela, Laura R. Conde

Kinetic modeling of high oleic palm oil epoxidation using IR-120 acid ion exchange resin as catalyst

Steven A. Solarte, Alvaro Orjuela, Laura R. Conde.

Department of Chemical and Environmental Engineering, Universidad Nacional de Colombia, 111321, Bogotá D.C., Colombia

In recent years, the chemical industry has been actively integrating renewable raw materials as substitutes for conventional fossil feedstocks. Among these, vegetable oils have emerged as promising alternatives, particularly for sensitive applications like cosmetics, food products, pharmaceuticals, and biocompatible polymers. Their inherent properties, environmentally friendly composition, renewability, low toxicity levels, high lubricity, low boiling points, and notable biodegradability make vegetable oils invaluable sources of various chemical compounds. Oleochemical derivatives can effectively serve as direct or functional substitutes for many fossil-based products, either partially or entirely. This is of particular importance for products with significant negative environmental impacts such as fossil-based polymers.

Polymers play a crucial role in human development. However, the environmental pollution resulting from mismanagement at the end of their use, the health hazards associated with certain additives (such as phthalate-based plasticizers), and the high costs and complexity of recycling have fostered the development of renewable and biodegradable materials and additives for the polymer industry. Consequently, the biopolymer market reached a valuation of USD 11.5 billion in 2022. Within this market, biobased epoxides constitute a large portion, serving as fundamental components in the production of resins, polyols, polyurethanes, polyesters, polyethers, and as additives for different applications (including plasticizers, lubricants, thermal stabilizers, surface coatings, hydrogen chloride scavengers in PVC). Currently, most biobased epoxides correspond to epoxidized vegetable oils, primarily utilized as plasticizers for PVC, monomers in epoxy resins, and feedstock for polyols used in polyurethanes.

Epoxidized vegetable oils with high oxygen oxirane content, low remaining iodine values, and minimal acidity are considered superior raw materials within the biopolymer industry. Consequently, they are primarily derived from highly unsaturated oils such as linseed, soybean, sunflower, and canola. Nonetheless, such raw materials have higher costs compared to the predominant oil within the global market, namely palm oil. Despite its lower cost and widespread availability, palm oil is rarely used as a raw material for epoxides due to its low unsaturation content, which limits its theoretical maximum oxirane index in-between 3-3.5 g-O/100 g-oil. Recently, due to phytosanitary issues in palm cultivation, hybrid species have been extensively cultivated. Surprisingly, the oil extracted from the fruits of these hybrid palms exhibits a higher content of oleic acid chains in the triglycerides. This has facilitated the large-scale production of high oleic palm oil (HOPO), characterized by iodine values ≥75 g-I_2/100 g-oil. Using HOPO as feedstock, it would be possible to obtain epoxides with theoretical oxirane oxygen content ≥4% wt., making them suitable to produce plasticizers and polyols (2).

The quality of epoxidized oils and their oxirane oxygen content are significantly influenced by the synthesis method. Industrially, epoxidized vegetable oils are typically synthesized via the Prileschajew reaction using H_2O_2 as the oxidant, a percarboxylic acid as the oxygen carrier, and H_2SO_4 as the catalyst. Despite achieving high conversion, this process exhibits limited selectivity, resulting in productivity below 80%. This was confirmed in a recent study that experimentally optimized HOPO epoxidation using H_2SO_4 (1). The study revealed productivities up to 80% and epoxidized products with oxirane indexes ~3.5 g-O/100 g-oil. However, challenges arose due to the corrosiveness of sulfuric acid and the required downstream neutralization, leading to waste generation. Consequently, there is a need for more selective, less problematic, and more environmentally conscious processes and catalysts. Various studies have proposed the use of heterogeneous catalysts to overcome the limitations of traditional ones. Among these, ion exchange resins have shown promising results for large-scale implementation. Several ion exchange resins, including Amberlite IR-120, Amberlyst 39, Amberlyst 36, Indion 225, Seralite SRC-120, Dowex 50WX2, Zerolit 325, and Aquivon PW79S, have been explored. Particularly, gel-type Amberlite IR-120 stands out as a suitable catalyst for epoxidation due to its activity and stability. However, as HOPO only recently became commercially available, heterogeneously catalyzed epoxidation has not been explored, and there is a lack of available kinetic models, which are necessary for further process design and scale-up.

In this respect, Amberlite IR-120 was assessed as a catalyst in the epoxidation of HOPO with H_2O_2 via the in-situ peracetic acid process. A Box-Behnken experimental design was employed to investigate the effect of catalyst loading, acetic acid loading, temperature, and molar ratio of reactants on the selectivity and yield of the epoxidation reaction. The progress of the reaction was monitored by measuring the oxirane oxygen and iodine content using a greener and safer spectroscopic method (near-infrared spectroscopy). The obtained experimental data were used to correlate a kinetic model using a pseudo-homogeneous approach and considering ring opening as a side reaction. Based on literature reports, different kinetic models were evaluated (i.e. Power law, Eley-Rideal, LHHW), and their corresponding kinetic parameters were determined by regression. All models exhibited a relative error below 10%, indicating their suitability in describing the reaction kinetics. These models can be further employed for process design, kinetic optimization, and scale-up. Additionally, further experiments were conducted to assess the reusability of the catalyst. It was observed that Amberlite IR-120 enhanced both selectivity and productivity compared to traditional catalysts, and it could be reused for multiple cycles without significant loss of activity.

1.Bohórquez, W. F., Orjuela, A., Narváez Rincón, P. C., Cadavid, J. G., & García-Nunez, J. A. (2022). Experimental optimization during epoxidation of a high-oleic palm oil using a simplex algorithm. *Industrial Crops and Products, 187* (Part A), 115321. https://doi.org/10.1016/j.indcrop.2022.115321.

2.Orjuela, Á., Bohórquez, W., Díaz, M. A., García, J. A., Narváez, P. C., Cadavid, J. G. (2022). *Polioles grasos: producción, retos y oportunidades para el sector del aceite de palma.* Bogotá. https://doi.org/10.56866/9789588360935

429b The Effects of Solvent on the Mechanisms and Kinetics of Levulinic Acid Production from Biomass Using a Polymeric Acid Catalyst

Zhexi Zhu, Ranil Wickramasinghe, Xianghong Qian

Levulinic acid is one of the critical platform chemicals for the conversion of lignocellulosic biomass to biofuels and bio-based products. However, current conversion technology is not yet commercially viable due to the high production cost and low yield with generally less than 70%. We have developed a patented technology for converting lignocellulosic biomass into levulinic acid in one step at mild conditions with high conversion and high yield reaching over 90%. Our unique catalyst consists of two polymer chains acting synergistically to hydrolyze the biomass and convert the hydrolyzed sugars to levulinic acid. However, the product yield is strongly dependent on the solvent and solvent mixtures. Here the effect of solvent on mechanisms of levulinic acid production is elucidated using cellulose substrate. MALDI-MS combined with GC-MS was used to determine all the intermediate- and side- products. Moreover, the kinetics of the levulinic acid production and activation energies for the conversion are determined at different solvent and solvent mixtures.

ID. 690730. CATALYTIC PYROLYSIS OF LIGNOCELLULOSIC BIOMASSES: A SUGAR CANE BAGASSE AND SALICORNIA STUDY

Jeimy Judith Bautista-Martinez, *Sandro González-Arias, Ricardo García-Morales, Octavio Elizalde-Solís and Edgar Ramírez-Jiménez, Chemical Petroleum Engineering Department, Superior School of Chemical Engineering and Extractive Industries, Instituto Politécnico Nacional, Mexico City, Mexico.*

Abstract

The transformation of lignocellulosic biomass into high-value energy compounds has become a topic of great interest due to the problems we currently face such as the energy crisis, climate change and the greenhouse effect. Biofuels (bioethanol or biobutanol), biochemicals, synthesis gas and biochar have become a valuable replacement for hydrocarbons obtained from crude oil.

Slow catalytic pyrolysis experiments were carried out on lignocellulosic biomass using sugarcane bagasse (SCB) of Mexican origin (*Saccharum officinarum L.*) and salicornia (*Salicornia bigelovii*) as feedstock, using H-ZSM-5 zeolites as catalyst, the objective of this experimentation was to determine the optimal conditions for slow catalytic pyrolysis. The experiments were carried out in a laboratory scale batch reactor, in temperature ranges of 400°C, 450°C, 500°C, 550°C and 600°C, with different catalyst-to-biomass ratios, 1:1, 1:2 and 1:3, particle sizes of 850 μm - 297μm and 0.0070 μm -0. 0029 μm for sugar cane bagasse and salicornia, respectively, with a heating rate of 10°C/min and a residence time of 30 minutes once the reaction temperature was reached, a feed flow of 85 mL/s of Nitrogen (N_2) was used to guarantee an inert atmosphere. Three products were obtained, a solid phase (charcoal), a liquid phase (pyrolysis liquid), and a gaseous phase (non-condensable pyrolysis compounds), for each of these products were used quantitative and qualitative analysis techniques of the organic compounds present in them, Fourier Transfer Infrared Spectroscopy (FTIR) was used to determine the compounds present in both the solid and liquid phases, gas chromatography for the liquid and gas phases, and HPLC-mass for the liquid phase only. The highest yield is obtained at temperatures of 500 and 550°C at a 1:3 ratio for both raw materials, it was observed that the production of aromatics at these conditions was very favorable. Regarding the product of the solid phase, the SCB is the one that generated less charcoal (6.7%) this at 500 °C and 1:2 catalyst-to-biomass ratio, in other words, about 90% of the raw material was transformed into gas and pyrolysis liquids; on the other hand, the salicornia has a higher production of non-condensable. During the experiments, the accumulation of gases was observed once the reaction reached 300°C; this is attributed to the halophytic properties of the plant (salinity). Based on the data obtained from chromatography and FTIR, the presence of oxygenated compounds is observed, mainly alcohols, aromatic groups, and aldehydes, the presence of the phenol family and some alkanes is also evident in the liquid fraction.

Keywords:

Catalytic pyrolysis, sugarcane bagasse, salicornia, biofuel, biochemicals, hydrogen, biomass.

Introduction

Currently the creation and improvement of technologies for obtaining biofuels is of great interest in the world, technologies that can guarantee an efficient use of natural resources and the preservation of the environment [1], one of these technologies is the pyrolysis of biomass, a renewable energy source that in addition to obtaining biofuels and chemical products reduces organic waste from agro-industrial sectors. Pyrolysis offers a fast route to decompose the polymeric structure of biomass, which generates high-value liquid products, such as mainly aromatic hydrocarbons, i.e., benzene, toluene and xylene, as well as phenols, acids, carbonyls, esters, ethers and alcohols, where ethanol stands out as an additive in gasoline to improve octane and quality [2].

One of the objectives of pyrolysis and its variants (thermal or catalytic pyrolysis) is to optimize the process conditions to maximize the production of pyrolysis liquids, charcoal or gases [3]. These variables are usually temperature (from 300 to 600°C, or 400 to 1000°C depending on the type of pyrolysis), particle size, heating range, type of catalyst, catalyst-to-biomass ratio and residence time, which directly impact the yields of the product in addition to its composition [3,4,5].

In recent years, the development of biofuels has been promoted around the world, as well as their consumption and production, seeking a sustainable path. During the decade from 2010 to 2020, global biofuels production experienced a growth of 36%, reaching around 163,000,000 m^3 in 2019, showing an evident growing trend. The top five producers of liquid biofuels are the United States (42%), Brazil (24%), Indonesia (6%), China (3%) and Germany (3%). The remaining 22% is distributed among the other countries of the world [6].

Although Mexico is not among the countries with the highest production of biofuels worldwide, it is among the most relevant emerging bioeconomies, thanks to the evident potential of renewable sources in the future production of bioenergy, within the framework of the goals and targets of the 2030 Agenda for Sustainable Development (Objective 7: Ensure access to accessible, reliable, sustainable and modern energy for all) [7]; these energies range from solar, hydroelectric, geothermal, wind biomass.

Sugarcane bagasse is a potential waste for obtaining a wide range of products, however, in this work it is analyzed exclusively the production of added-value products, that is, the obtaining of ethanol and aromatics. Mexico is a sugar country par excellence, in the last 10 years, the area corresponding to the sugarcane harvest increased by 16 thousand hectares [8, 9].

The commitment to salicornia (*bigelovii*) as lignocellulosic biomass to obtain biofuels has gained great relevance in Mexico, mainly in the northern states of the country where the environment benefits the cultivation of this halophytic plant, since its irrigation requires seawater, and it is favored in dry climates. According to research carried out by the University of Sonora, salicornia has multiple uses, one of them being the generation of biofuels, especially to obtain biodiesel; another sector that benefits from the advances in obtaining biofuels from salicornia is air transport, in a few years the aviation sector will grow at a remarkable rate and consequently the demand for fuels will also increase, in this way the incorporation of renewable fuels is essential, the implementation of different projects for the production of bio jet fuel is planned[10].

Experimental Methodology

Raw material

Agro-industrial residues from the Tierra Blanca Sugar Mill, Veracruz (sugarcane bagasse) and from crops in the state of Baja California Sur (salicornia) were used. The preparation of the raw material consists of 4 main stages: drying, grinding or crushing, sieving and characterization, which aim to take advantage of most of the raw material from the biomass by homogeneously integrating stems and leaves. A particle size of 850-297 μm was defined for sugarcane bagasse (see Fig. 1) and 0.007-0.0029 μm for salicornia (see Fig. 2), since the smaller particle size, the greater the heat transfer, which benefits the conversion of biomass into biofuels or recoverable products [11].

Fig. 1. Sugarcane bagasse

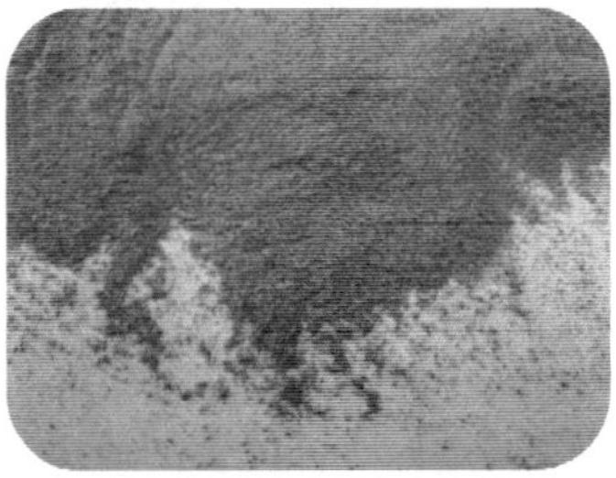

Fig. 2. Salicornia

Methods

The catalytic pyrolysis reactions were carried out in an experimental system (see Fig. 3), which has a batch titanium tubular reactor, loaded with 1 gram of biomass in the catalyst-to-biomass ratios (1:1, 1:2, 1:3). Nitrogen (N_2) was used to maintain an inert atmosphere, which also acts as a drag gas and diffuser, regulated by a globe valve and a flowmeter (SHINAGAWA) to deliver 85 mL/s (±5 mL/s). The internal temperature was monitored with a K-type thermocouple in a concentric thermowell inside the reactor. A heating ramp of 10°C/min was applied until the selected reaction temperatures were reached: 400°C, 450°C, 500°C, 550°C and 600°C. Each reaction had a residence time of 30 minutes. At the end of the reaction, the reactor is removed from the furnace and cooled. At the reactor outlet, a system captures condensable and non-condensable gaseous products for further analysis.

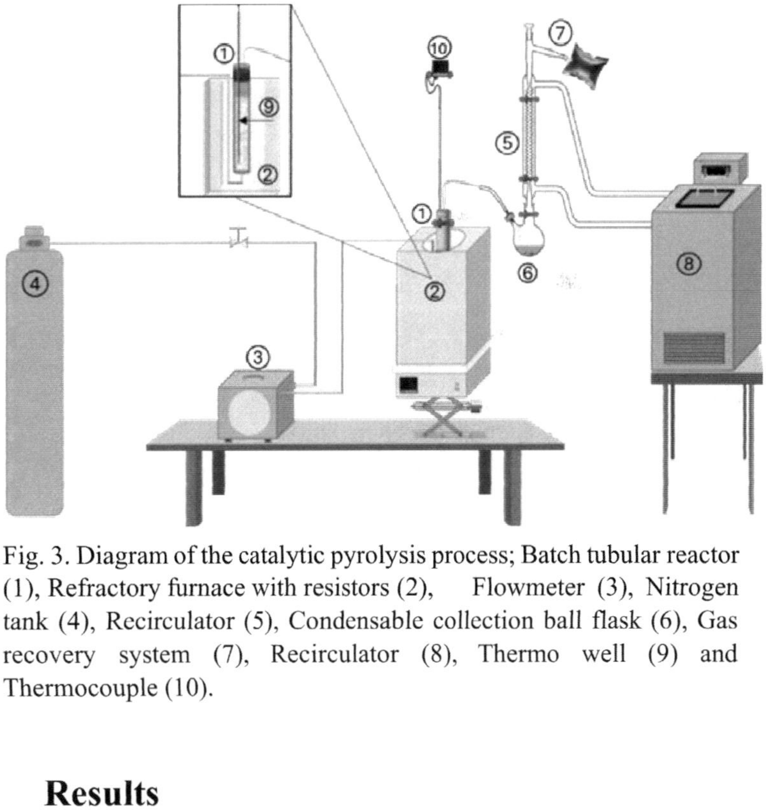

Fig. 3. Diagram of the catalytic pyrolysis process; Batch tubular reactor (1), Refractory furnace with resistors (2), Flowmeter (3), Nitrogen tank (4), Recirculator (5), Condensable collection ball flask (6), Gas recovery system (7), Recirculator (8), Thermo well (9) and Thermocouple (10).

Results

Figures 4 and 5 show the yields of pyrolytic products as a function of reaction temperatures for sugarcane bagasse (see Fig. 4) and salicornia (see Fig. 5), as shown in Fig. 4a and 5a, the gas phase yield increases as the reaction temperature increased, in the case of sugarcane bagasse, the highest gas yield occurs at 600 °C, as a result of decarboxylation and decarbonylation reactions. Salicornia peaks at 67% at 550°C with a ratio of 1:3, due to solid phase decomposition. As can be seen in Fig. 4b and 5b, with respect to the liquid phase, there was an increase in yield when the reaction temperature was increased from 500°C to 550°C and from 450°C to 500°C for sugarcane bagasse and salicornia, respectively, in Fig. 4b a maximum of 550 °C with a biomass ratio of 1:3 is observed. This is attributed to the conversion of aromatic rings, stabilizing the compounds formed, while in Fig. 5b, the maximum point is seen between 450 °C and 500 °C with a ratio of 1:1, due to the conversion of the alkyl chains, however, a further increase in the reaction temperature decreases the yield of liquid products. The performance of the solid phase shows an opposite trend, salicornia generates more solids, which is attributed to its high salt and mineral content. Finally, the lower the presence of catalysts, the higher the performance of the liquid and gas phase.

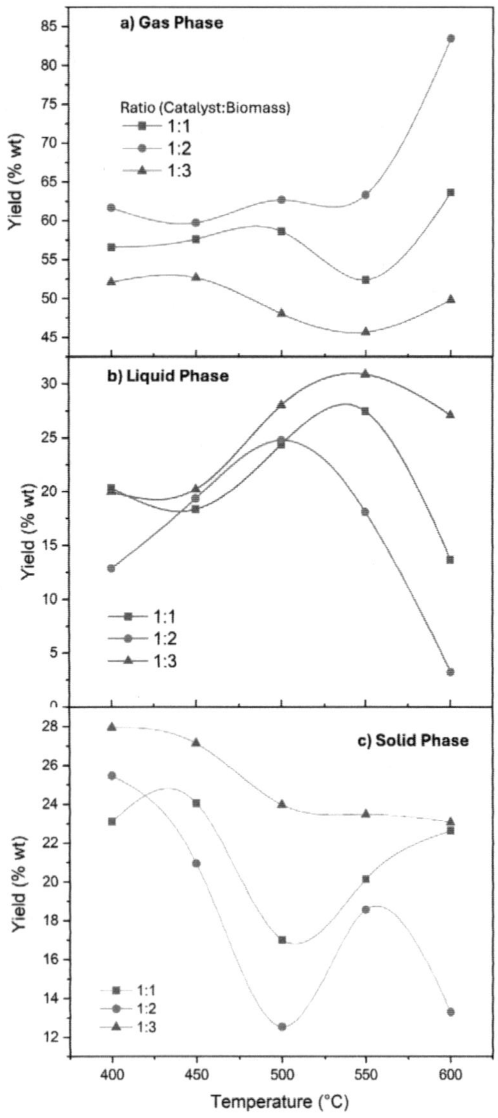

Fig. 4. Yield of products from catalytic pyrolysis of sugarcane bagasse.

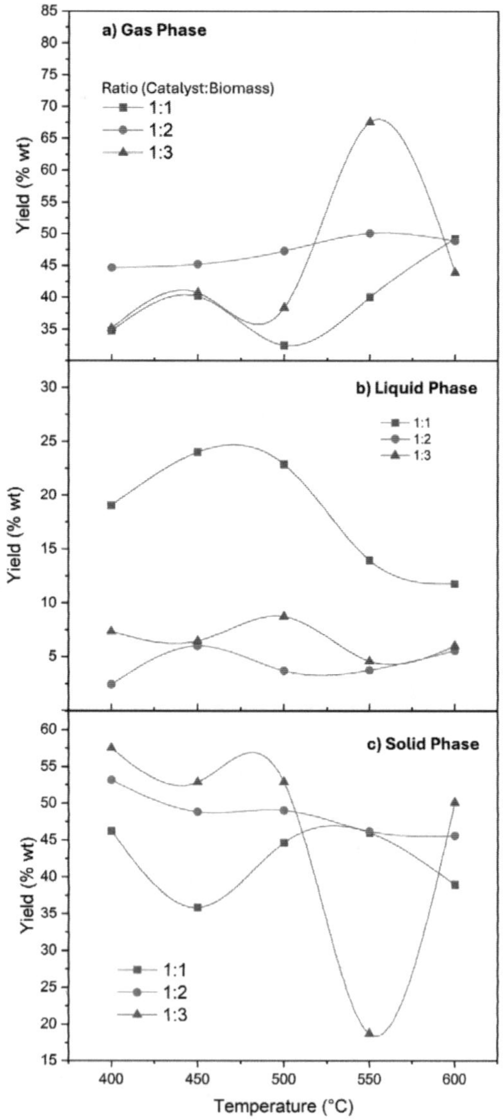

Fig. 5. Yield of Salicornia catalytic pyrolysis products.

Liquid Phase: Condensable Volatile Products

The liquid phase products were analyzed by means of FTIR (Fourier Transform Infrared Spectroscopy) using a catalyst-to-biomass ratio of 1:3 at different reaction temperatures.

The functional groups in the liquid phase, identified by FTIR, reveal the chemical composition and quality of the oil. Fig. 6 for sugarcane bagasse shows that as the temperature increases, the oxygenated groups (OH, C-O) decrease, favoring the formation of more stable phenolic compounds and some hydrocarbons from 500°C.

Fig. 7, referring to salicornia, indicates that the increase in temperature also eliminates the oxygenated groups (C-O). At 400-450°C, a high intensity of these groups is observed, indicating a partial breakdown of cellulose and hemicellulose. From 500°C, these groups disappear, favoring the formation of more stable compounds, such as aromatic hydrocarbons.

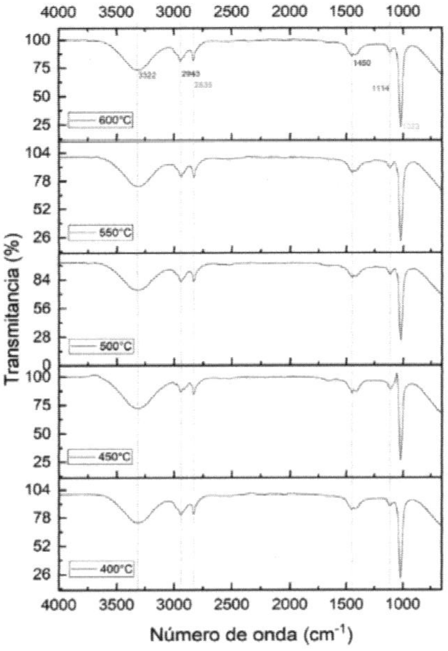

Fig. 6. Results of the FTIR analysis, sugarcane bagasse.

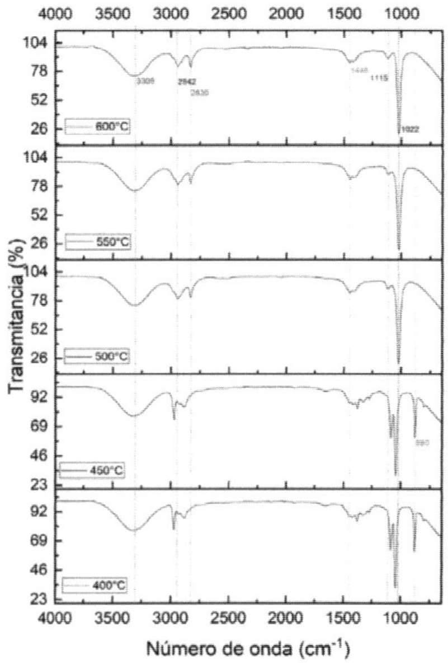

Fig. 7. Results of the FTIR analysis, Salicornia.

Liquid chromatography

Figures 8 and 9 show a profile of liquid compounds, mainly aromatics, alcohols and ketones.

The Figure 8 shows the distribution of products for sugar cane bagasse, it can be seen that at a 1:1 ratio a higher phenol production is observed at 500 (35%) and 550°C (20%) due to lignin conversion. At the 1:2 ratio, phenols predominate by 35% at 500°C and 44% at 550°C, benefiting the deoxygenation of the biomass. Regarding the 1:3 ratio, a maximum conversion of the biomass with 44% phenol is observed at 500 °C, improving the quality of the bio-oil at 550 °C due to the decomposition of aromatic lignin. Figure 9 shows the product distribution for salicornia, at the 1:1 ratio, shows a higher production of acetone (59%) at 500°C, and isopropanol (60%) at 550°C from the decomposition of cellulose and hemicellulose. Regarding the 1:2 ratio, phenol formation increases, as well as butanol formation (50%) at 550°C by dehydration and decarboxylation reactions. At a ratio 1:3, the improvement of the quality of the bio-oil is appreciated, since it favors the production of aromatics, the mixture of added-value products is varied to this reaction condition, the salts and minerals present in salicornia naturally catalyze the decomposition of lignin, a tendency to produce more gases at higher temperatures is observed.

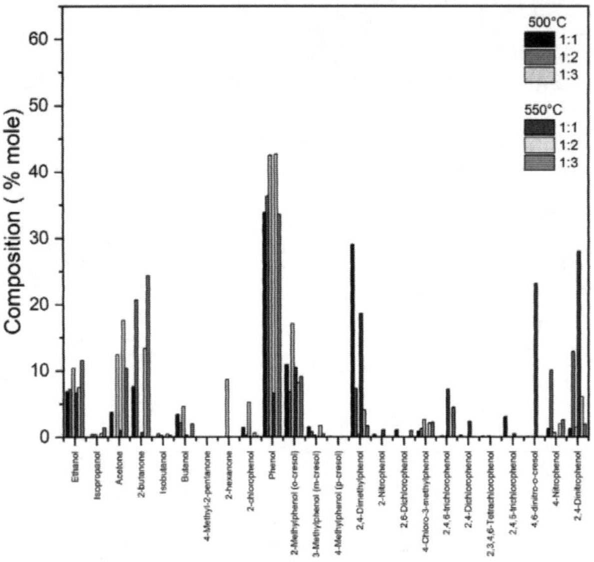

Fig. 8. Gas chromatography (liquid phase), for sugarcane bagasse

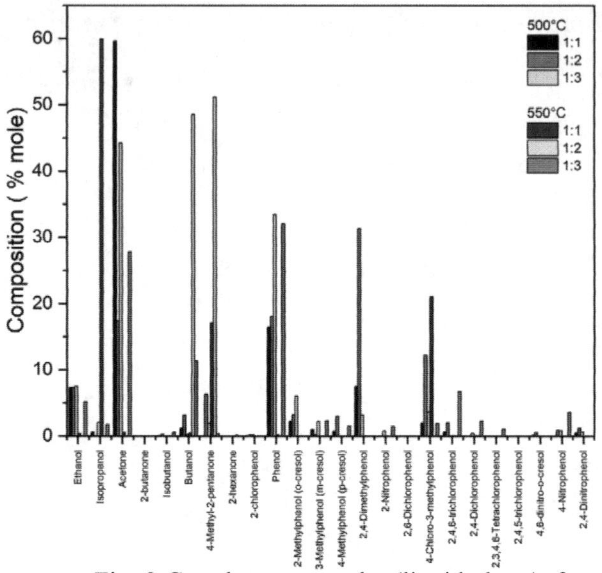

Fig. 9 Gas chromatography (liquid phase), for salicornia.

Gas Phase: Non-Condensable Volatile Products

The 1:3 ratio was selected because it presented compounds with higher added-value compared to the 1:1 and 1:2 ratios that mainly produce CO_2 and CO.

Figure 10 corresponding to the gas phase of sugarcane bagasse shows that at low temperatures the CO_2 and CO fractions are found in greater proportion due to decarboxylation and decarbonylation, which reduces oxygen in solid and liquid products and increases gas production. By increasing the temperature, H_2 production peaks at 500 °C (65%) and decreases at 550 °C (60%) and 600 °C (45%), this reduction is linked to the breakdown of cellulose and hemicellulose, breaking C-H bonds. Regarding the Lower Heating Value (LHV), the highest value of the Heating Value is obtained at 500 °C, driven by the increase in combustible gases (H_2, CO, and CH_4) and the reduction of oxygenated compounds and non-combustible gases.

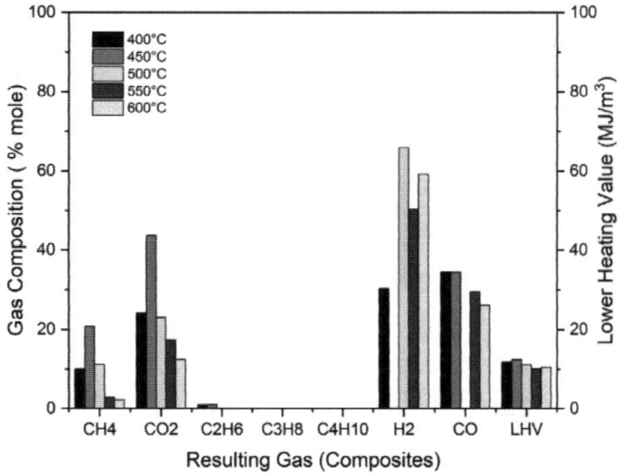

Fig. 10. Gas chromatography, sugarcane bagasse

Thanks to Fig. 11, corresponding to the salicornia gas phase, at low temperatures there is a greater production of CO_2 and that it decreases significantly with increasing temperatures. Higher concentrations of CO, H_2, CH_4, C_2H_6, C_3H_8 and C_4H_{10} are observed at higher temperatures, which increases the calorific value of the gas. The highest calorific value is obtained at 500 °C due to a higher concentration of high calorific value

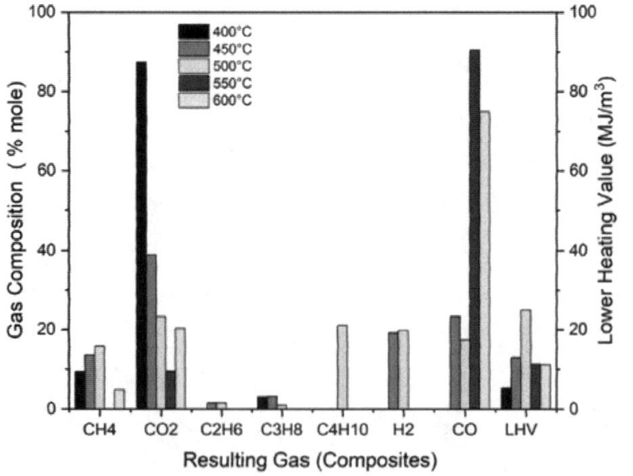

Fig. 11. Gas chromatography, Salicornia

compounds. At 550 °C, an increase in CO yield is observed, corresponding to the maximum gas yields of the decarboxylation and decarbonylation reactions.

Solid Phase: FTIR Analysis

Results of the FTIR analysis in solid phase of catalytic pyrolysis, catalyst-to-biomass ratio (1:3), at different reaction temperatures.

To analyze the solid phase of both raw materials, the FTIR technique was used. Figure 12 represents biochar from sugarcane bagasse; as the temperature increases, there is a reduction in the amount of oxygenated functional groups and greater carbon stability, forming a more stable biochar with a more aromatic structure. Figure 13 shows the transformation of salicornia biochar into a material with a higher carbon content. The concentration of salts and minerals is stabilized, contributing to the properties of the material. With the increase in temperature, the presence of oxygenated groups decreases and the formation of aromatic structures benefits.

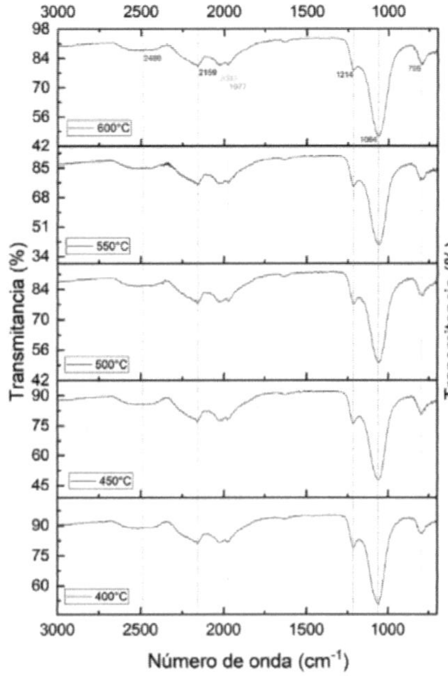

Fig. 12. Results of the FTIR analysis for sugarcane bagasse.

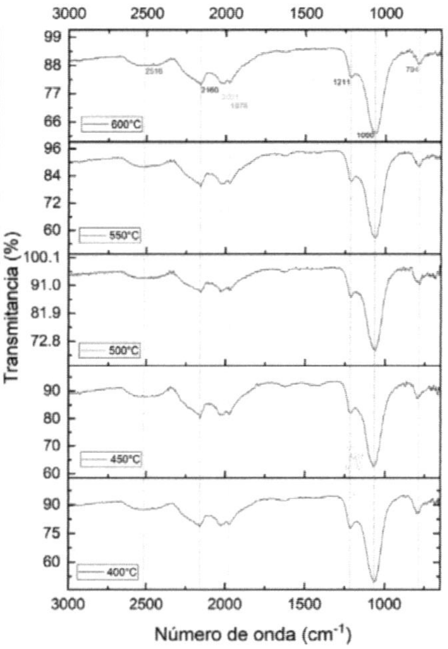

Fig. 13. Results of the FTIR analysis for Salicornia.

Comparison between Thermal and Catalytic Pyrolysis (Catalyst-to-Biomass Ratio 1:3)

Liquid phase

To make the comparison between both pyrolysis (thermal vs catalytic), the catalyst to biomass ratio 1:3 was selected, which presents the highest formation of added-value products, improving efficiency in obtaining condensable and gases, since the 1:1 and 1:2 ratios presented a profile of liquid compounds very similar to those obtained by thermal pyrolysis (oxygenated compounds that affect the quality of the liquid phase). Figure 14 corresponds to the liquid phase yields obtained from sugarcane bagasse from both thermal and catalytic pyrolysis, similar to Fig. 15 which corresponds to the yields obtained from salicornia: Fig. 14 and Fig. 15 show that thermal pyrolysis produces a greater quantity of condensable products for both raw materials. However, this condensable product blend contains more oxygenated compounds. In the case of catalytic pyrolysis, although the yield is lower, the products obtained have a better quality, the use of catalyst transforms the oxygenated compounds resulting in products of greater value for the chemical industry.

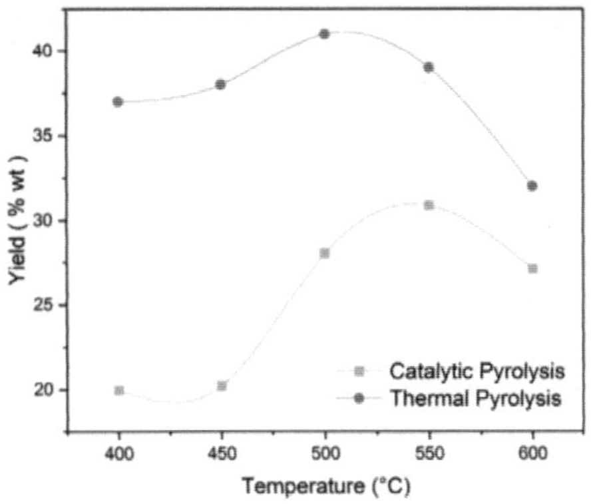

Fig. 14. Comparison of Liquid Performance in Thermal and Catalytic Pyrolysis of Sugarcane Bagasse

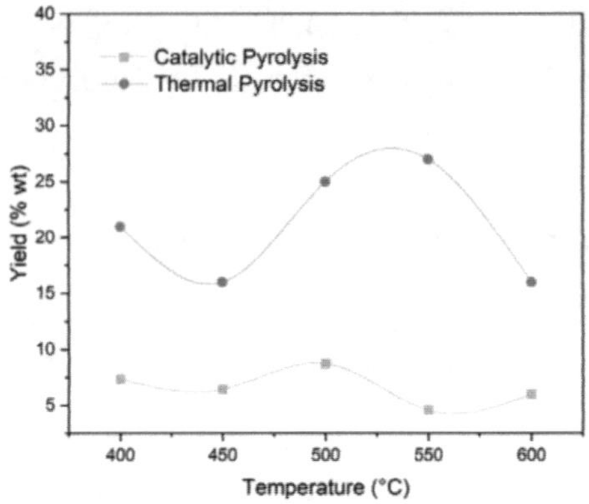

Fig. 15. Comparison of liquid performance in thermal and catalytic pyrolysis of salicornia

Gas phase

Figures 16 and 17 correspond to gas chromatography of sugarcane bagasse, Fig. 16 corresponds to catalytic pyrolysis, presents a more selective behavior towards valuable gases such as H_2 and CH_4, resulting in higher quality gas. Figure 17 shows the distribution of gases resulting from thermal pyrolysis, it is observed that CO_2 and CO are the predominant gases due to the decomposition of the oxygenated groups, generating a greater volume of these gases for all reaction temperatures.

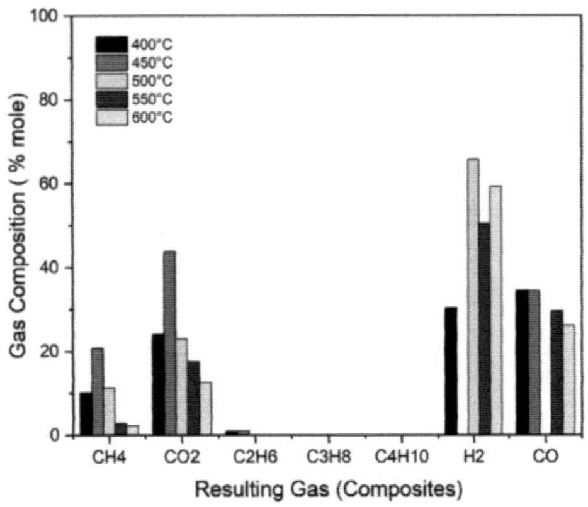

Fig. 16. Gas chromatography resulting from catalytic pyrolysis of sugarcane bagasse

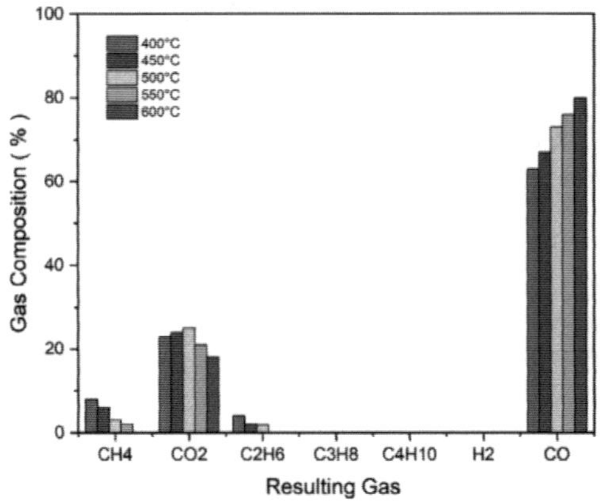

Fig. 17. Gas chromatography resulting from thermal pyrolysis of sugarcane bagasse

Figures 18 and 19 correspond to salicornia gas chromatography, Fig. 18 shows the distribution of pyrolysis gases at the different reaction temperatures, a reduction in oxygenated gases and an increase in the production of H_2, as well as CH_4, C_2H_6, C_3H_8 and C_4H_{10} with the increase in reaction temperature, the presence of H-ZSM-5 improves the conversion of biomass for industrial use. The salt- and mineral-rich composition of salicornia contributes significantly to the production of gases, since these elements in the presence of zeolite accelerate the reactions of decomposition or transformation of biomass, there are studies in which zeolite is supported by magnesium to improve catalytic performance, in this study, magnesium is naturally found in the raw material, suggesting a similar catalytic effect without the need for additional support[12]. Figure 19 shows the composition of the gases resulting from thermal pyrolysis, with higher amounts of oxygenated gases observed for all temperatures.

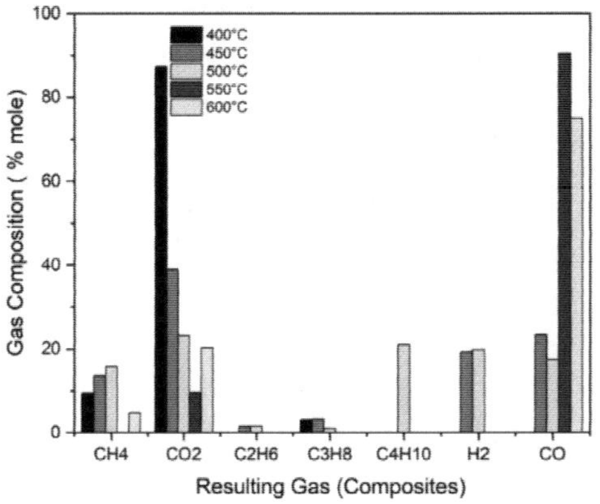

Fig. 18. Gas chromatography resulting from catalytic pyrolysis of salicornia

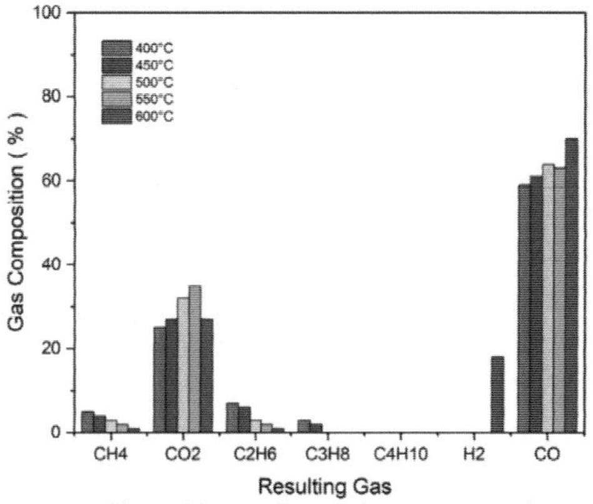

Fig. 19. Gas chromatography resulting from thermal pyrolysis of salicornia

As part of the analysis, it is important to highlight that there was a difference between the heating furnace and the reactor approximately of 100 °C. This allows to thermally understand the conversions that biomass polymers are going through

On the other hand, it was observed that the Salicornia tended to generate more non-condensable gases and vapors, while sugarcane bagasse was more prone to production of condensable liquids.

Conclusions

The presence of the H-ZSM-5 catalyst significantly improved the production of alcohols, ketones, and aromatics for both raw materials. The catalytic improvement of pyrolysis gases promotes a gas rich in H_2, which impacts the efficiency of the process resulting in greater selectivity towards this desired product.

The results obtained allowed to define the optimal operating conditions for the conversion of sugarcane bagasse and salicornia by catalytic pyrolysis, these were: catalyst-to-biomass ratio 1:3, and temperature range of 500 to 550°C. These conditions improved the efficiency of the catalytic conversion process, favoring the production of compounds of greater interest (alcohols, ketones, aromatics, and H_2) in addition to improving selectivity towards added-value products. The selection of these variables and parameters are a key point to maximize the performance of catalytic pyrolysis and enhance its application at an industrial level, sustainably taking advantage of agricultural waste in the generation of biofuels and chemical products.

Compared to thermal pyrolysis, in catalytic pyrolysis bio-oils are associated with better quality and greater production of added-value compounds, in terms of gas production, these gases (H_2, CH_4, C_2H_6, C_3H_8) have a higher calorific value, this shows that the addition of a catalyst, in this case a zeolite, optimizes the selectivity and efficiency of pyrolysis, which is advantageous for applications where gases of high purity and energy value are needed or liquid precursors in the petrochemical industry.

Salicornia is a plant with great potential for future research. It is hypothesized that, due to its properties such as its elemental composition, the conditions of the pyrolysis system (reaction temperatures and N_2 atmosphere), and the use of the H-ZSM-5 catalyst, it is possible to obtain higher-value products in both the liquid and gas phases. This can be attributed to a potential synergy between the salts present in the biomass and the H-ZSM-5 support, as the reaction conditions are like those used in the synthesis of Mg, Ca, K, and P/H-ZSM-5 catalysts. Previous studies on the production of aromatics through catalytic pyrolysis have shown that modifying the catalyst with salts or metals enhances biomass conversion, and some of these salts (Mg, Ca, K, and P) are naturally part of the plant's chemical composition, yielding results similar to those in this study [12].

References

1. Manals-Cutiño, E., Penedo-Medina, M., & Giralt-Ortega, G. (2011). Análisis termogravimetrico y térmico diferencial de diferentes biomasas vegetales. *Tecnología química*, *31*(2), 180-190.
2. Onay, O., & Kockar, O. M. (2003). Slow, fast and flash pyrolysis of rapeseed. Renewable energy, 28(15), 2417-2433.
3. Sandoval Jabalera, R., Delgado Vigil, M. D., & Collins Martinez, V. H. (2010). PIRÓLISISA BAJA TEMPERATURA DEL BAGAZO DE CAÑA DE AZÚCAR PARA LA PRODUCCIÓN DE COMBUSTIBLES LÍQUIDOS. *Repositorio Nacional*, 14. Retrieved on 12/04/2023: http://repositorionacionalcti.mx/recurso/oai:cimav.repositorioinstitucional.mx:1004/1291
4. Maschio, G., Koufopanos, C., & Lucchesi, A. (1992). Pyrolysis, a promising route for biomass utilization.
5. Encinar, J. M., Gonzalez, J. F., & Gonzalez, J. (2000). Fixed-bed pyrolysis of Cynara cardunculus L. Product yields and compositions. Fuel Processing Technology, 68(3), 209-222.
6. Torroba, A., Orozco Montoya, R. A., & Productivo, D. (2023). Atlas de los biocombustibles líquidos 2022-2023. Atlas de los biocombustibles líquidos.
7. Gamez, M. J. (2022, 24 mayo). Objetivos y metas de desarrollo sostenible - Desarrollo Sostenible. Desarrollo Sostenible. https://www.un.org/sustainabledevelopment/es/objetivos-de-desarrollo-sostenible/
8. CONADESUCA., (2023). 10mo Informe Estadístico del Sector Agroindustrial de la Caña de Azúcar en México. https://www.siiba.conadesuca.gob.mx/siiaca/docext/10mo_Informe_Estadistico.pdf
9. CONADESUCA., (2023). Listado de ingenios que realizaron zafra 2022/2023. https://www.gob.mx/cms/uploads/attachment/file/863949/Ingenios_zafra_22-23.pdf
10. Rios, D. M. M. (2019). Optimización de la cadena de suministro para la producción de bioturbosina. http://ri-ng.uaq.mx/handle/123456789/1847
11. Mutsengerere, S., Chihobo, C. H., Musademba, D., & Nhapi, I. (2019). A review of operating parameters affecting bio-oil yield in microwave pyrolysis of lignocellulosic biomass. Renewable and Sustainable Energy Reviews, 104, 328-336.
12. Zheng, Y., Wang, F., Yang, X., Huang, Y., Liu, C., Zheng, Z., & Gu, J. (2017). Study on aromatics production via the catalytic pyrolysis vapor upgrading of biomass using metal-loaded modified H-ZSM-5. Journal of Analytical and Applied Pyrolysis, 126, 169-179.

429d Thermodynamic Evaluation of Mixed Metal Redox Carriers for Biomass Utilization Using Chemical Looping

Sudeshna Gun, Tanay Jawdekar, Sonu Kumar, Falguni Akulwar, Liang-Shih Fan

Anthropogenic climate change poses a severe threat to the environment and human well-being, necessitating a transition to a sustainable, carbon-neutral future. Biomass, a renewable resource with abundant availability, offers a promising alternative to fossil fuels for producing electricity, hydrogen, and liquid fuels. Due to its carbon-neutral nature, biomass has become an attractive source for producing hydrogen and liquid fuels with a minimal carbon footprint. However, the relatively low energy density and wide geographical distribution of biomass necessitate effective strategies to ensure the techno-economic viability of biomass conversion technologies.

Chemical looping technologies have emerged as a promising approach for biomass conversion into high-purity syngas and hydrogen utilizing iron-based redox carriers in a multi-reactor system. This technology generates high-purity products, negating post-reactor separation processing before utilization in various chemical industries, such as liquid fuel or ammonia synthesis. The scheme involves reducing the redox carrier in the first reactor (reducer) with biomass, operating in either co-current or counter-current mode to generate syngas or carbon dioxide, respectively. In counter-current mode, the reduced carrier undergoes steam-mediated partial oxidation in the second reactor (oxidizer) to produce hydrogen. While the exothermic oxidizer compensates for the endothermic reducer, the latter's heat demand depends heavily on the thermodynamics of the redox carrier. Consequently, optimizing process efficiency requires developing redox carriers with favorable thermodynamic properties that can significantly reduce the endothermic heat demand of the process. Such a reduction in heat requirement can significantly reduce the oxygen carrier requirement of the entire process.

This study conducts comprehensive process simulations using biomass as the feedstock to investigate the design of chemical looping carriers for syngas and hydrogen production, aiming to minimize the endothermic heat requirement of the reduction process. Thermodynamic evaluations are performed for various metal oxides, and based on their endothermic heat requirements in the reducers, two metal oxides are selected and used to create bimetallic metal oxides with varying compositions. Thermodynamic evaluations for hydrogen and syngas generation are performed for the mixed metal oxide composites, and exergy efficiencies are compared across different compositions. Product yields are also compared across all compositions to assess the efficiency gains achieved using bimetallic oxygen carriers. The findings from this study contribute to the development of more efficient and cost-effective biomass conversion technologies, paving the way for a more sustainable energy future.

661b Understanding the Effect of Particle Characteristics on Flowability of Biomass Slurries

Glenn Matamoros, Sampath Gunukula, Clayton Wheeler, C. Luke Williams

Feeding wet biomass slurries into a high-pressure, high-temperature reactors presents significant challenge such as feed pump plugging issues. To address this, we have explored methods to improve slurry flowability and minimize plugging through adding flow aids. However, the addition of flow aids can increase the overall cost of the biomass conversion process. In this study, we investigate alternative approaches to enhance biomass particle flowability without relying on flow aids. Specifically, we examine the effects of biomass particle size distribution, the presence of small-particle-size (<1 mm) materials (byproducts of size reduction), and biomass densification. To conduct our analysis, we designed and constructed a continuous cold flow loop with a progressive cavity pump and transparent piping. The biomass slurry flow rate was 190 L per hour for the cold flow experiments, encompassing various biomass compositions including softwood, hardwood, clean softwood with 10% bark, and clean hardwood with 10% bark. The biomass solids concentrations in the slurries was increased from 1 wt% at the onset of the experiment, until pump plugging occurred. Throughout the experiments, parameters such as slurry viscosity, volume concentration of biomass solids, biomass particle porosity, particle size distribution, as well as settling and deposition velocities of biomass particles were measured. These comprehensive measurements provide valuable insights for enhancing biomass slurry flowability in industrial scale biorefineries.

661d Characterization of Sulfur Profile and Fate Study of Loblolly Pine Using Pyrolysis-GC/MS-FPD

Gary Lopez, Jian Shi

The sulfur compounds produced during thermochemical conversion are significant factors when selecting a biomass as a biofuel feedstock due to the deactivation of catalysts and equipment wear. Loblolly pine samples representing levels of high, medium, and low sulfur content were selected for analysis to determine the sulfur profile and fate. The biomass samples were hand fractionated into five categories: chips, bark, limbs, needles, and particulates. Analysis was conducted using a micro-pyrolysis unit attached to a gas chromatography mass spectrometer equipped with a flame photometric detector. Results provided both qualitative and quantitative information about the sulfur compounds produced during the pyrolysis of the loblolly pine samples at 550 °C and 900 °C. Additionally, model sulfur compounds, including both inorganic (sodium sulfide nonahydrate, and potassium bisulfate) and organic (L-methionine, cystine, 2-aminoethyl hydrogen sulfate) species were tested at the same pyrolysis conditions. This work provides insights and feedback for sulfur reduction strategies and the design of biorefinery systems.

661e How Do Feedstock Variabilities Impact the Convertibility of Lignocellulosic Biomass? a Data-Oriented Case Study on Corn Stover in the United States

Xihui Kang, Chang Dou, Ling Ding, Yining Zeng, Bryon Donohoe, Kenneth Sale, Ning Sun

The valorization of low-cost lignocellulosic wastes into biofuels and bioproducts contributes to a low carbon emission bioeconomy. Although it is widely acknowledged that the inherent variability of lignocellulosic biomass affects its conversion, no study thoroughly investigated the relationship between feedstock properties and conversion efficiency. Using corn stover as an example, this study employed a multiscale analytical characterization approach coupled with data analytic methods to understand some key critical material attributes of feedstock quality and elucidate their impacts on the subsequent conversion.

We observed a significant decrease in moisture and ash content for corn stover after one year of indoor storage, likely due to water evaporation and soil contamination reduction. No significant difference was found in volatile solids, fixed carbon content, elemental components, and cellulose crystallinity index across corn stover samples. The ash content, largely originating from the soil, significantly impacted the mineral composition of corn stover, with the highest SiO_2 content (12.2%) observed in high ash groups (ash content > 15%). Moisture content was closely related to biomass degradation during storage, with higher moisture content leading to more severe degradation, thus altering corn stover characteristics. Interestingly, self-degradation improved the sugar yields (glucose: from 21% to 45%), indicating the potential benefits of self-degradation. Different anatomic fractions (i.e. stalk, leaf, husk, and cob) demonstrated varied convertibility; pretreatment and self-degradation enhanced sugar yields. The correlation analysis highlighted the negative impact of lignin on sugar yields among corn stover samples with different ash and moisture content. However, no correlation was found between the variability (such as lignin content and cellulose crystallinity) and sugar yield, likely due to the deconstruction of the cell wall after self-degradation. This study emphasized the importance of optimizing pretreatment methods to enhance sugar yield and considering the balance between promoted sugar yields and mass loss during self-degradation.

661c Air Classification of Forestry Residues for Fast Pyrolysis

Nepu Saha, Jordan Klinger, Tiasha Bhattacharjee, Yidong Xia, Vicki Thompson, Oluwafemi Oyedeji, James E. Parks, Mehrdad Shahnam, Yupeng Xu

Understanding critical biomass attributes through efficient fractionation is crucial for advancing sustainable pyrolysis for renewable energy and chemical production. This study investigates the intricate relationship between biomass preprocessing and pyrolysis product yields, employing the air classification technique for the treatment of loblolly pine residues with varying moisture content. A comprehensive exploration of the physicochemical properties of air-classified loblolly pine informs a sophisticated pyrolysis simulation model. Given the complex and multifaceted nature of biomass pyrolysis, operating across diverse temporal and spatial scales, a pyrolysis kinetics-based CFD–DEM simulation method is employed to predict product yields. Results showed that the elevated moisture content amplifies particle adhesiveness, necessitating augmented air velocities for effective separation, thereby influencing the efficiency of the separation process. While carbon and hydrogen contents exhibit relative stability across diverse moisture contents and blower frequencies, the oxygen content undergoes noticeable changes. For example, the oxygen contents were measured as 29.2 and 38.6 wt% in the light fraction of 30% moisture content sample at blower frequencies of 10 and 20 Hz, respectively. An intriguing finding emerges from pyrolysis simulation, indicating that a lower blower frequency in air classification moderately enhances bio-oil yield and significantly improves its quality, particularly in terms of water content. For instance, the water content in the bio-oil was about 1.5% and 10% in the heavy and light fractions, respectively from 10% moisture sample under 15 Hz blower frequency.

99g Agile Systems Analysis Framework to Assess the Sustainability Implications of Cellulosic Biorefineries Using Consolidated Bioprocessing

Kay Glass, Muhammad Hafi Wadgama, Rui Shi

The current aviation industry contributes substantially to transportation-related emissions, making it a critical target for decarbonization efforts. Sustainable Aviation Fuel (SAF) is a renewable drop-in biofuel with the potential to significantly reduce the carbon footprint compared to conventional jet fuel. SAF can be produced through several ASTM-approved pathways, each with defined blending limits. Among these pathways, the alcohol-to-jet (ATJ) route offers distinct advantages, utilizing cellulosic ethanol to produce SAF blendstock, with renewable naphtha as a co-product. Previous studies found sustainable supply of ethanol is the key the feasibility of SAF. However, cellulosic ethanol biorefineries have historically experienced mixed economic outcomes, often resulting in plant closures.

A new approach has been developed by Center for Bioenergy Innovation (CBI) researchers, where cellulose biomass is configured to be directly fermented into either ethanol or isobutanol using a consolidated bioprocessing (CBP) approach. Following this approach, cellulose biomass is directly fermented into either ethanol or isobutanol using a consolidated bioprocessing (CBP) method. *Clostridium thermocellum*, a thermotolerant cellulolytic bacterium, is utilized to simultaneously produce enzymes, break down cellulose, and carry out fermentation within a single vessel. This integration simplifies the process while still allowing for separate processing of other biomass streams.

Past studies have assessed the economic performance of biorefineries and employed global sensitivity analysis to identify key cost drivers. Yet, there has been limited exploration of SAF-focused designs that account for strain design changes under uncertainty across all system parameters. This presentation provides an overview of the latest advancements in agile techno-economic analysis (TEA) and life cycle assessment (LCA) for CBP to produce cellulosic ethanol biorefineries. It features a case study using an agile TEA/LCA framework to simulate a consolidated bioprocessing system, with integrated global sensitivity analysis. Key factors such as inflation, commodity prices, reactant conversions, and production capacity are evaluated in TEA in relation to past biorefinery performance, underscoring the need for further model refinement in this field.

99b Feasibility of Photovoltaic Energy: Life Cycle Assessment, Circular Economy, and Criticality Indicators

Natalia Cano, Laura Franco-García, Rebecca Saive Sr.

Introduction:

The feasibility of photovoltaic (PV) energy as a sustainable energy source is crucial for addressing contemporary environmental challenges. This research explores the integration of methodologies such as Life Cycle Assessment (LCA), Social Life Cycle Assessment (S-LCA), and Costing Life Cycle Assessment (C-LCA) in evaluating the viability of PV energy production systems. Additionally, it emphasizes the importance of incorporating Circular Economy indicators and Criticality evaluations to assess sustainability dimensions comprehensively.

Methodology:

Life Cycle Assessment (LCA), Social Life Cycle Assessment (S-LCA), and Costing Life Cycle Assessment (C-LCA) serve as foundational methodologies for estimating environmental, social, and cost indicators in PV energy production. However, Circular Economy indicators and Criticality evaluations must be conducted in parallel to ensure a holistic evaluation. This integrated approach is essential as separate assessments may overlook interconnected sustainability dimensions.

Key Evaluations:

a) End-of-life Routes: Assessing different end-of-life routes after the solar panel's lifespan is crucial for sustainable management. Circular economy indicators are essential for optimizing resource utilization and minimizing waste.

b) Social Impacts: Social Life Cycle Assessment enables the evaluation of social risks and impacts throughout the PV energy supply chain, considering both positive and negative effects on communities and stakeholders.

c) Economic Analysis: Costing Life Cycle Assessment evaluates economic indicators such as Capital Expenditure (CAPEX), Operational Expenditure (OPEX), and Net Present Value (NPV) under diverse scenarios, providing insights into the financial viability of PV energy systems.

d) Futuristic Scenarios: Considering critical raw materials parameters and end-of-life routes is imperative for envisioning future PV energy scenarios. Understanding supply risks and economic importance informs sustainable decision-making.

e) Sustainability Hotspots: Identifying sustainability hotspots across the technology's life cycle supports informed decision-making and guides further technology development. This proactive approach enhances the overall sustainability of PV energy systems.

Outlook:

Innovative solutions such as agri-PV systems can be implemented to overcome gaps and limitations in PV energy. Agri-PV systems integrate agricultural activities with PV energy generation, maximizing land use efficiency and promoting sustainable practices. By combining agriculture and renewable energy production, agri-PV systems offer multifaceted benefits, including enhanced food security, reduced land competition, and increased renewable energy generation capacity.

Conclusion:

The feasibility of photovoltaic energy hinges on a comprehensive evaluation framework that integrates LCA, S-LCA, C-LCA, Circular Economy indicators, and Criticality evaluations. By considering environmental, social, and economic dimensions in tandem, stakeholders can make informed decisions to advance the sustainability of PV energy systems. Embracing innovative solutions such as agri-PV systems further strengthens the case for transitioning towards renewable energy sources and mitigating climate change impacts.

99c Life Cycle Sustainability Assessment (LCA) of Nature-Inspired Bio-Syngas Technologies for Olefin Synthesis

Semra Bakkaloglu, Aurora Cavaliere, Paola Lettieri, Andrea Paulillo, Massimiliano Materazzi, Nilay Shah

Olefins serve as fundamental building blocks for numerous chemicals and materials. However, conventional methods of olefin production rely heavily on fossil fuels, thus raising significant environmental concerns. Nature-inspired bio-syngas technologies present a promising sustainable alternative by harnessing renewable biomass to generate syngas, which can then be converted into olefins.

This study constitutes a follow-up investigation into the life cycle environmental and cost assessment presented in 'Techno-environmental analysis of olefins production from syngas via sorption-enhanced synthesis' by Cavaliere et al. in 2024. Our approach employs a dynamic Life Cycle Assessment (LCA) to evaluate the environmental impact of these bio-syngas technologies, which involve the integration of methanol synthesis and methanol-to-olefin generation within the same reactor, alongside traditional syngas-to-olefins processes throughout their cradle-to-grave life cycle.

The results of our analysis predict the climate change impacts over a standard 100-year time frame, considering factors such as the average shelf-life and end-of-life treatment of the final olefin-based product. Furthermore, we assess other environmental impacts, including resource consumption, acidification, land use, and toxicity. In addition to environmental considerations, our study conducts a life cycle cost (LCC) analysis from cradle to gate, identifying hotspots in the overall production cost. We then perform a comparative analysis to investigate the economic competitiveness of the proposed process. To enhance efficiency and both economic and environmental performance, we incorporate exergy analysis as a complementary tool.

Our findings shed light on key environmental and economic aspects associated with various stages of the technology, spanning from biomass feedstock acquisition to olefin production. Ultimately, the study concludes that bio-syngas technologies hold the potential to offer a more sustainable route for olefin synthesis compared to traditional methods.

99g Biogas Utilization in Refuse Power Plants (BURP)

Ahmed Essam

Problem: As the energy industry shifts towards a heavier priority on decreasing the environmental impact caused by power plants, an innovation that repurposes existing coal processing facilities to use renewables has been in high demand. However, limited availability and processing demands of renewables necessitates integrating non-renewable resources to facilitate this transition. One promising solution lies in co-firing waste coal with biogas while utilizing carbon capture and storage (CCS). This approach provides a remediation pathway for reducing waste coal quantities and providing greenhouse gas mitigation strategies.

Methods: The study included generating multiple different cases, both as retrofit on a waste coal power plant in Grant Town, West Virginia and greenfield cases. Two assessments were conducted to evaluate the viability of the cases, A Techno-Economic Analysis (TEA) to investigate the economic viability of the proposed cases and a Life Cycle Assessment (LCA) to investigate the environmental impacts of co-firing biogas from various sources including landfills as well as anaerobic digesters fed by animal manure, municipal solid waste, and wastewater sludge. The TEA utilized ASPEN Plus V14 simulations to aid the process estimation for the proposed retrofit. The Levelized Cost of Electricity (LCOE), and Total As-Spent Costs (TASC) were then used as the basis for justifying feasibility. An additional greenfield case has been analyzed in an alternate location. The LCA assessment followed a cradle-to-grave approach and covered impacts from the biogas production, burdens from waste coal extraction, transportation, combustion and CO2 emissions, including avoided emissions. A combustion model was constructed for the power plant, with the mass and energy balances utilized to compile the LCA inventory.

Results: The viability of implementing a new powerplant was determined to be strongly influenced by the accessibility to biogas and the cost associated with pipeline construction. Increased amounts of cofired biogas resulted in a lower greenhouse gas abatement incentive, necessary to become feasible against a non-retrofit process. The findings indicate that by replacing approximately 30 wt.% of the waste coal with biogas from landfills and capturing 90% of the CO2, the power plant's emissions could be lowered from a global warming potential (GWP) of 1400 to -1 kgCO2eq/MWh achieving negative emissions. Cofiring with (23-29) wt.% biogas from the anaerobic digestion processes also achieved negative CO2 emissions. Though, the other environmental impact categories like acidification and ozone depletion potentials increased with the increased biogas feed ratios and using CCS, the large GWP reduction provided significant offsets, underscoring the advantages of the co-firing approach.

99e Carbon Footprint Assessment of Production of Ethylene from U.S. Corn Ethanol

Kirti Richa, Pahola Thathiana Benavides, Ulises Gracida Alvarez, Jennifer Port, Troy Hawkins

Ethylene is one of the most common feedstocks used in production of plastics such as polyethylene (PE). Currently, ethylene and PE are produced almost exclusively from fossil resources.

The purpose of this study was to evaluate the potential carbon footprint estimate of producing polymer-grade ethylene by using U.S. corn ethanol as feedstock. The functional unit for this assessment was 1 kg of ethylene and the system boundary was cradle-to-ethylene manufacturing output gate.

The system boundary of corn ethanol derived ethylene included ethanol production and conversion to ethylene. The ethanol production from corn is divided into corn farming, ethanol manufacturing, and ethanol transportation stages. Previous studies from Argonne National Laboratory (ANL) and the Greenhouse gases, Regulated Emissions, and Energy use in Technologies model version 2022 (GREET 2022) were leveraged for data for these steps in the ethanol value chain. Sensitivity analyses were conducted to understand the impact of variability in the carbon intensity (CI) of ethanol due to Life Cycle Assessment (LCA) allocation methods, sustainable farming, and fuel switching, and its influence on ethylene carbon footprint.

The carbon footprint analysis covered bio-based ethylene production pathways via corn ethanol dehydration from the following routes:

1) Stand-alone ethanol-to-ethylene processes

2) A co-processing route via fluid catalytic cracking (FCC) process wherein corn ethanol was co-processed with vacuum gas oil (VGO)

Process inventory data for stand-alone processes was derived from published literature sources (secondary data). The inventory for co-processing ethanol to produce ethylene was generated based on primary data, modeled/simulation data (benchmarking and internal calculations) and secondary data. Additionally, background global warming potential (GWP) values for process inputs were derived from ANL GREET 2022. One challenge associated with co-processing is the estimation of the carbon footprint of products with bio-based content due to the combination of the feedstocks utilized (bio- and fossil-based). For this study, laboratory data based on carbon-14 (C-14) analysis was used to generate process yields for corn ethanol feed. In the case of products containing both bio-derived and fossil components, C-14 content is used to calculate how much of the product is derived from bio- vs. fossil-derived feedstocks. Additionally, modeled yields based on co-processing yields of ethanol and VGO and base FCC unit yields was used to generate bio-ethylene yields for comparison purposes (Modeled Yields Case).

The conventional, fossil-based ethylene production pathway was included as reference case for comparison with corn-ethanol derived, bio-based ethylene. For this analysis, the steam cracking LCA developed by ANL for GREET 2022 was leveraged to represent a carbon footprint estimated range for fossil-based ethylene. The range was based on feedstock choices for steam cracking such as naphtha and ethane.

This study demonstrated that corn ethanol-based ethylene showed lower carbon footprint estimate when compared to fossil-based ethylene production (103% - 127% reduction in the base case scenarios). This reduction is due to the addition of biogenic CO_2 sequestration credits to the bio-based product. The study also evaluated the impact of low CI ethanol on the carbon footprint of bio-ethylene. For example, substituting natural gas used in ethanol production with renewable natural gas (RNG) from dairy and swine manure and utilizing heat and power from corn stover collected during farming reduced the carbon footprint estimate of bio-ethylene compared to the baseline corn ethanol production. This reduction is due to the addition of biogenic CO_2 sequestration credits and avoided emissions of current waste management practices by use of RNG.

99f Soil Carbon: An Overlook Component of Biofuel Life Cycle Assessment

Yuan Yao, Kai Lan, Bingquan Zhang

Converting renewable biomass to biofuel has great potential to enhance energy security and decarbonize the transportation sector. Forest residues are abundant biomass sources in the United States. As overstocked forest residues are often linked to wildfire risks, they emerge as viable feedstocks for biofuel production, offering benefits of energy generation and fire hazard reduction. While numerous life cycle assessments (LCA) have explored the environmental and decarbonization benefits of biofuels from forest residues, many overlook or minimize the impacts of residue removal on the soil carbon pool, assuming such effects to be negligible when considering the overall life cycle greenhouse gas (GHG) emissions of biofuels.

This study addresses a critical question: Does the alteration in soil organic carbon (SOC) due to the removal of forest residues significantly affect the life cycle GHG emissions of biofuels? To answer this question, we utilize an innovative LCA framework that dynamically links process-based simulations, soil carbon modeling, and forest growth modeling. Different climate conditions and forest productivity are simulated to understand whether variations in environmental conditions would affect the conclusion. The analysis encompasses current practices, including the natural decay of forest residues, and innovative biofuel production scenarios, such as those incorporating carbon capture and storage (BECCS) and the co-production of biofuel and biochar, given the strong interest in using BECCS and biochar for carbon dioxide removal.

Our results show the considerable impact that residue removal has on SOC levels, which in turn contributes to significant GHG emissions in the life cycle carbon footprint of forest residue-derived biofuels. The impact of SOC change is particularly larger for emerging systems like BECCS and the co-production of biofuels and biochar. These insights underscore the necessity of incorporating soil carbon assessments into future LCA studies, biofuel policies, and forest management strategies, even when utilizing waste biomass without land-use change. Our integrated LCA framework also offers a more holistic approach to estimating the carbon footprint of biofuels derived from residue biomass to inform decision-making towards sustainable energy and forest management practices.

561a Pulp and Paper Sludge Valorization: Techno- Economic and Life Cycle Assessment for Enhanced Methane Production

Erfan Hosseini, Selen Cremaschi, Zhihua Jiang

The pulp and paper industry (PPI) generates a significant volume of wastewater that undergoes physical and biological treatment [1,2], resulting in a considerable quantity of pulp and paper sludge (PPS) [3], which is one of the major waste streams of PPI [4]. As global energy demand escalates and concerns over energy security and climate change intensify, anaerobic digestion (AD), which transforms organic materials into methane and carbon dioxide without oxygen [5], has emerged as a versatile technology for renewable energy production. This study undertakes a life-cycle assessment (LCA) and techno-economic analysis (TEA) to explore the valorization of PPS. The focus is on utilizing AD of PPS to generate combined heat and electricity.

The LCA aims to estimate the life cycle environmental impacts of electricity and heat co-generated in a combined heat and power (CHP) unit using biogas produced from AD of PPS. This assessment assumes the plant is in the United States and employs the OpenLCA software. Concurrently, the TEA employs a process simulation of PPS AD built utilizing Aspen Plus. Economic parameters such as net present value (NPV), cash flow rate of return (CFRR), and payback period are estimated and compared across three scenarios.

Scenario A, the Base Case, replicates the conventional operational conditions of AD, serving as a benchmark for assessing the economic and environmental viability and methane production efficiency of a standard AD that utilizes PPS as the feedstock. Scenario B, PPS is pretreated via alkaline pretreatment before AD. Alkaline pretreatment is a promising method to enhance enzymatic hydrolysis for various lignocellulosic materials [6,7]. Studies indicate that alkaline pretreatment effectively disrupts the floc structure of PPS, leading to improved biodegradation efficiency [6]. Research has shown a significant increase in soluble chemical oxygen demand (SCOD) removal efficiency, ranging from 83% to 93%, compared to 70% for untreated sludge [6]. Alkali pretreatment, typically involving hydroxides like NaOH or KOH [8], could be substituted with the by-product of the kraft pulping process, Green Liquor Dregs (GLD), offering a cost-effective alternative [9]. Scenario C explores the co-digestion of food waste (FW) with PPS, leveraging FW's nitrogen-rich composition [10]. Previous studies have identified nitrogen deficiency as a challenge in PPS AD [11,12]. The optimal carbon-to-nitrogen ratio for AD feedstock typically falls between 20-30% [13]. Pulp and paper sludge, with its high carbon-to-nitrogen ratio, accelerates nitrogen consumption by methanogens, hindering microbial population growth and prolonging digestion. Conversely, a low ratio increases ammonia release, inhibiting the AD process.

In this study, for the sake of simplicity in comparison, the management of 500 tons of sludge is chosen as the functional unit. All emissions, materials, wastewater treatment, energy consumption, and product generation are standardized based on this functional unit.

For the AD-CHP process of sludge valorization, as depicted in Fig. 1, the system boundary encompasses biogas production in the AD unit, electricity, heat generation in the CHP unit, and the storage and disposal of digestate. While some generated heat and electricity are utilized within the AD plant, excess electricity is exported to the national grid, and excess heat is utilized either onsite or offsite of the pulp and paper plant. This study does not consider sludge collection and transportation in the comparison and assumes that sludge is generated on-site. This assumption is consistent across all scenarios; thus, it does not affect the comparison.

The life cycle inventory (LCI) involves compiling and quantifying all inputs and outputs that pass through the system boundaries, including input mass/energy and output mass/work done, as depicted in Figure 2. The LCI analysis aggregates emissions and extractions related to the consumption of raw materials, water, energy, wastewater, and transportation.

The impact assessment phase follows inventory analysis in the LCA process. This phase evaluates potential environmental impacts based on inventory flow results. Life Cycle Impact Assessment (LCIA) aims to translate LCI results into environmental impacts, such as effects on natural resources, the environment, and human health [14]. In the final step, results from the inventory and impact phases are interpreted to meet the study objectives. This involves checking for completeness, sensitivity, and consistency and deriving conclusions, limitations, and recommendations.

We developed an Aspen Plus simulation that integrates the AD stoichiometric model (Buswell), CHP production, and recirculation system (Figure 3). This study focuses on utilizing gas turbine technology for CHP due to its reliability, low emissions, abundant high-grade heat, and lack of cooling requirement [15,16]. The gas turbine engine system, based on the Brayton cycle, is depicted in Figure 3. Biogas from the digester is compressed, then combusted, and the resulting high-temperature gases expand through a turbine, generating power. Exhaust gases are directed to a Heat Recovery Steam Generating Unit (HRSG) for further energy recovery. The simulation calculates the rate of biogas generation and biogas composition. Furthermore, the simulation results, including heat and power consumption required for plant operation and power generated in the gas turbine, are used to precisely determine the appropriate CHP unit sizing, and estimate energy consumption within the AD-CHP system.

Under typical operational conditions (Scenario A), before entering the anaerobic process, the PPS is heated to 35 °C (mesophilic temperature). Within the digester, the feedstock undergoes agitation with a recirculated stream of biogas in the absence of oxygen, producing 10,444 m^3 of methane per day. The daily consumption of 500 tonnes of feedstock with 12% total solids (TS) and a 30% conversion rate equates to a methane yield of 179.46 ml CH_4/g volatile solids (VS). The biogas is conveyed from the top of the AD to the CHP unit. The CHP unit generates 1.65 GW of electricity and 1.56 GW of heat, totaling 3.21 GW of energy output. The CHP plant operates with a total efficiency of 83%, comprising an electricity generation efficiency of 43% and a heat generation efficiency of 40%. Of the generated heat, 24% (366 kW) is allocated for heating the sludge and compensating for heat loss from the digester walls. Approximately 174 kW of the electricity generated is dedicated to compressing the recirculated biogas stream (RECIR1) to match the pressure at the bottom of the digester (3.5 atm), with the surplus electricity being exported to the grid.

This talk will discuss the detailed results of the TEA and LCA for all three scenarios. The study findings highlight the potential for commercially producing biogas from PPS, presenting promising solutions to enhance sustainability and resource efficiency.

References

1. van Oel PR, Hoekstra AY. Towards Quantification of the Water Footprint of Paper: A First Estimate of its Consumptive Component. *Water Resources Management.* 2012;26(3):733-749. doi:10.1007/s11269-011-9942-7
2. Pokhrel D, Viraraghavan T. Treatment of pulp and paper mill wastewater—a review. *Science of The Total Environment.* 2004;333(1-3):37-58. doi:10.1016/j.scitotenv.2004.05.017
3. Priadi C, Wulandari D, Rahmatika I, Moersidik SS. Biogas Production in the Anaerobic Digestion of Paper Sludge. *APCBEE Procedia.* 2014;9:65-69. doi:10.1016/j.apcbee.2014.01.012
4. Gottumukkala LD, Haigh K, Collard FX, van Rensburg E, Görgens J. Opportunities and prospects of biorefinery-based valorisation of pulp and paper sludge. *Bioresour Technol.* 2016;215:37-49. doi:10.1016/j.biortech.2016.04.015
5. Kamali M, Gameiro T, Costa ME V., Capela I. Anaerobic digestion of pulp and paper mill wastes – An overview of the developments and improvement opportunities. *Chemical Engineering Journal.* 2016;298:162-182. doi:10.1016/j.cej.2016.03.119
6. Lin Y, Wang D, Wu S, Wang C. Alkali pretreatment enhances biogas production in the anaerobic digestion of pulp and paper sludge. *J Hazard Mater.* 2009;170(1):366-373. doi:10.1016/j.jhazmat.2009.04.086
7. Navia R, Soto M, Vidal G, Bornhardt C, Diez MC. Alkaline Pretreatment of Kraft Mill Sludge to Improve Its Anaerobic Digestion. *Bull Environ Contam Toxicol.* 2002;69(6):869-876.

doi:10.1007/s00128-002-0140-4
8. Loow YL, Wu TY, Md. Jahim J, Mohammad AW, Teoh WH. Typical conversion of lignocellulosic biomass into reducing sugars using dilute acid hydrolysis and alkaline pretreatment. *Cellulose*. 2016;23(3):1491-1520. doi:10.1007/s10570-016-0936-8
9. Sewsynker-Sukai Y, Naomi David A, Gueguim Kana EB. Recent developments in the application of kraft pulping alkaline chemicals for lignocellulosic pretreatment: Potential beneficiation of green liquor dregs waste. *Bioresour Technol*. 2020;306:123225. doi:10.1016/j.biortech.2020.123225
10. Lin Y, Wang D, Liang J, Li G. Mesophilic anaerobic co-digestion of pulp and paper sludge and food waste for methane production in a fed-batch basis. *Environ Technol*. 2012;33(23):2627-2633. doi:10.1080/09593330.2012.673012
11. Bayr S, Rintala J. Thermophilic anaerobic digestion of pulp and paper mill primary sludge and co-digestion of primary and secondary sludge. *Water Res*. 2012;46(15):4713-4720. doi:10.1016/j.watres.2012.06.033
12. Meyer T, Edwards EA. Anaerobic digestion of pulp and paper mill wastewater and sludge. *Water Res*. 2014;65:321-349. doi:10.1016/j.watres.2014.07.022
13. Veluchamy C, Kalamdhad AS. Influence of pretreatment techniques on anaerobic digestion of pulp and paper mill sludge: A review. *Bioresour Technol*. 2017;245:1206-1219. doi:10.1016/j.biortech.2017.08.179
14. Hijazi O, Munro S, Zerhusen B, Effenberger M. Review of life cycle assessment for biogas production in Europe. *Renewable and Sustainable Energy Reviews*. 2016;54:1291-1300. doi:10.1016/j.rser.2015.10.013
15. Spellman FR. *Water & Wastewater Infrastructure*. CRC Press; 2013. doi:10.1201/b13968
16. Breeze P. Gas turbines and combined cycle power plants. In: *Power Generation Technologies*. Elsevier; 2005:43-61. doi:10.1016/B978-075066313-7/50005-X

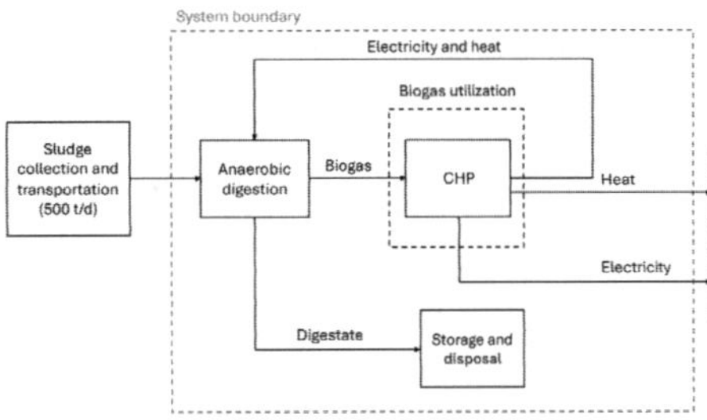

Figure 1- system boundary for AD-CHP system

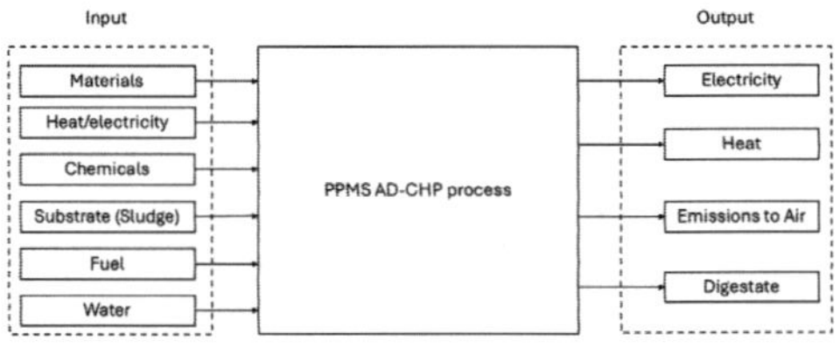

Figure 2- Overview of life cycle inventory analysis of AD-CHP system

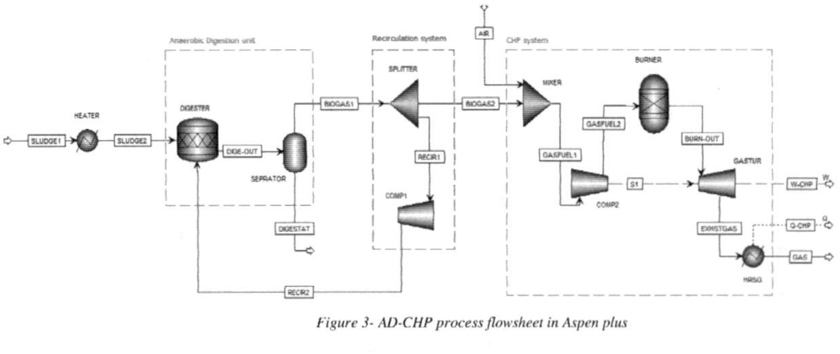

Figure 3- AD-CHP process flowsheet in Aspen plus

561b Integration of Biogas Model (ADM1) for Economic and Environmental Analysis of Rng Production

Katherine Wild, Juan Carlos Dominguez, Lisa Schulte, Mark Mba Wright

Renewable natural gas (RNG) from farm anaerobic digestion (AD) can provide clean energy and additional products to farmers. Novel co-digestion substrates, such as prairie biomass, can provide synergistic qualities for AD that stabilize the reactor and increase methane (CH_4) production. Prairie biomass often requires less material input than other biomass and can impact the environment beneficially (such as the reduction of soil erosion).

This study focuses on a techno-economic analysis (TEA) and life cycle assessment (LCA) for the co-digestion of cattle manure and prairie biomass utilizing a biogas model, Anaerobic Digestion Model no. 1 (ADM1). While the full synergistic relationship of cattle manure and prairie biomass was not entirely captured, the addition of manure to prairie biomass increased CH_4 production for all scenarios analyzed.

The scenario with the largest CH_4 production from AD had the lowest estimated minimum fuel selling price (MFSP). This scenario was a 1:9 volatile solids (VS) ratio of prairie biomass to manure, with a MFSP decrease of 19.4% when compared to the mono-digestion of prairie biomass. While the feedstock price for prairie biomass was almost three times greater than that of cattle manure, the feedstock price did not significantly impact the MFSP. All impact categories analyzed had avoided burdens from the use of manure. The cultivation of prairie biomass had burdening effects due to the use of resources such as fertilizer.

561c Environmental Sustainability of Biofuels for Maritime Decarbonization on a Planetary Scale

Eric Tan

Marine biofuels are a promising solution for reducing the environmental impact of the shipping industry and reducing greenhouse gas emissions. They are produced from renewable sources and can be used as a direct substitute for traditional fossil fuels. By using marine biofuels, we can not only reduce greenhouse gas emissions but also improve air quality and diversify fuel sources, leading to greater energy security. However, there are also challenges associated with the use of marine biofuels, such as the need for sustainable feedstocks and the potential competition with food production. Therefore, it is crucial to carefully consider the environmental and social impacts of biofuel production and use to ensure that sustainability criteria are met.

Additionally, to our knowledge, the implications of displacing biofuels for fossil marine fuels, such as commercial very low sulfur fuel oil (VLSFO), have yet to be examined on a planetary scale. We have attempted to link biofuel production to planetary boundaries to determine if sustainability thresholds are met. This study links a bottom-up model covering the production of up to 80% of marine biofuels to the planetary boundaries framework. We will determine the extent to which sustainability thresholds are transgressed or exceeded by marine fuels derived from biomass. This presentation will discuss the opportunities for achieving absolute sustainability of marine biofuels for marine propulsion.

561e Life Cycle Assessment of Functionalized Ultraporous Superactivated Hydrochars for Carbon Capture

Swarna Saha, Toufiq Reza

Carbon dioxide accounts for 77% of annual greenhouse gas emissions, contributing significantly to environmental crises such as global warming and climate change. Consequently, research into carbon capture technologies is vital to mitigate these environmental risks stemming from CO2 emissions in power plants and energy-intensive industries. Adsorption of CO2 in porous carbon-based adsorbents is a well-established method due to its higher capacity, renewability, cost-effectiveness, and faster kinetics, especially when the adsorbent surface is functionalized with nitrogen and oxygen functionalities. However, different functionalization methods for carbon capture have various environmental impacts, spanning from the production stage of system infrastructure to the end of their lifecycle. Therefore, conducting comprehensive and appropriately designed life cycle assessment (LCA) studies is crucial to provide clear evidence on the comparative sustainability of different functionalization methods. This research aims to compare the environmental impacts of functionalizing super activated hydrochars derived from loblolly pine using zeolite and CO2 gas during hydrothermal carbonization (HTC) for enhanced carbon capture. Environmental impacts were evaluated using a combined approach for the life cycle inventory, physical and economic input data. Simulation results indicate that utilizing CO2 during functionalization for carbon capture applications yields a positive impact compared to the zeolite functionalization method.

561f Process Design and Economics for Production of Sustainable Aviation Fuel (SAF) from Biomass Derived Intermediate

Kevin Dyke, Devin Fan, Michael Anthony Lucci III, Grant Zeszutek, Dhruv Patel, Muhammad Hafi Wadgama, Rui Shi

Sustainable aviation fuel (SAF) feedstocks and production technologies play an important role in decarbonizing aviation. Depending on the feedstock and technologies used to produce it, SAF can reduce life cycle GHG emissions significantly compared to conventional jet fuel. In effort to achieve United States' goal to have net zero emissions by 2050, it is vitally important to advance the SAF production in the coming decades. On the other hand, while the demand for high performance jet fuel in the aviation industry is currently stands at 6.95 million barrels a day and is expected to increase with the demand for flights continually growing, in 2022, only 15.8 million gallons of SAF were produced. One of the key reasons for the low yield is due to the current high cost of manufacturing SAF. To this end, it is important to understand the feasibility and key economic drivers in order to develop a comprehensive strategy for scaling up technologies to produce SAF on a commercial scale across various regions in the U.S.

In this study, we investigated the alcohol-to-jet (ATJ) manufacturing processes, developed conceptual design, and perform techno-economic analysis (TEA) and sensitivity analysis. For the base case, 95% ethanol first undergoes dehydration to produce ethylene and water. The ethylene stream then undergoes oligomerization to produce longer chain olefins while the other stream is integrated with various heat exchangers further downstream as a coolant. The resulting product stream from the oligomerization is further separated. Propene and butene can be hydrogenated to produce propane and butane for sale and ethylene can be recycled back into the oligomerization feed. The higher-molecular-weight olefins are sent to a distillation column where the distillate and bottoms streams are also hydrogenated to produce natural gasoline and jet fuel, respectively. This process was scaled to produce 11 million gallons annually assuming continuous, year-round plant operation based on the current scale of mid-size refinery, Fulcrum Bioenergy. The minimum fuel selling price and impact of feedstock cost were determined following established approaches for cost analysis using n^{th}-plant project costing and financing. This study also explored various strategies to further reduce operating costs. These included scenarios involving the optimization of recycling streams and oligomerization reagents, minimizing the cost of feed streams, on-site hydrogen production and storage, as well as reducing the overall cost of feed streams. Finally, this study identified opportunities to advance the SAF production by developing an improved and more efficient refining process through integrated biorefinery design.

730a From Waste to Sweetness: Catalytic Strategies and Reaction Engineering for Upcycling Greek Yogurt Acid Whey

Wenjia Wang, Owen J Dziedzic, Claire Lesnjak, Xiaolei Shi, Jarryd Featherman, James Miller, Scott A Rankin, George W Huber

In the last decade, the dramatic rise in Greek yogurt's market share in the US has resulted in the production of about 1.8 million metric tons of byproducts Greek yogurt acid whey (GAW) annually. Managing GAW is challenging for the dairy industry due to the uneconomical processing costs stemming from its low protein and high mineral content. GAW is often disposed of in wastewater plants where the GAW's high biological (BOD_5) and chemical oxygen demands (COD) can overwhelm the system. There is a crucial need for innovative, sustainable methods to utilize GAW, reducing environmental impacts and improving agricultural and industrial use.

Our research team is developing a process to transform GAW into a glucose-galactose syrup (GGS). This process uses H2SO4-catalyzed hydrolysis of lactose into GGS. GGS has a similar sweetness to high fructose corn syrup. Undesired humins and 5-Hydroxymethylfurfural (HMF) are formed in this process. We develop a kinetic model that predicts the desired GGS production and undesired product formation. The kinetic model indicated that both glucose and galactose degrade into humins first and then humins can further be converted into HMF. Galactose has a higher rate of humin formation than glucose. We have designed and operated a pilot-scale reactor (227 L/hr) to achieve up to a 90% yield and selectivity of GGS with minimal byproduct formation based on the kinetic model. This pilot scale reactor allows us to make larger amounts of GGS that we are testing in a variety of products.

The GGS can be further converted into tagatose, a low-calorie sweetener that is 90% as sweet as sucrose but with 60% fewer calories. We developed an effective tagatose production process from GGS and described the chemistry of tagatose production.

730b Ru/CNT Catalyst for Supercritical Water Gasification of Biomass

Yukihiko Matsumura, Mizuki Kodama, Mohammed Ahmed Mohammed Ali, Takashi Noguchi, Mengli Zhang, Shuhei Inoue

Supercritical water gasification is a promising technology to gasify wet biomass for energy recovery. It has been studied by many researchers and its commercialization is sought. However, due to the high price for the process, it is not easy to commercialize this process. This high price is due to the fact that for complete gasification, high temperature or effective catalyst is needed. When high temperature is employed, the reactor has to endure high pressure under high reaction temperature, and the cost of the reactor is high. When effective catalyst such as ruthenium is employed, the cost of catalyst is very high. To circumvent this problem, our group developed Ru/CNT catalyst, where ruthenium is impregnated onto carbon nanotube. First, stability of carbon nanotube in supercritical water was studied. Carbon nanotube reacted with supercritical water, but its rate was more than one order of magnitude lower than that of activated carbon, which was employed as catalyst for supercritical water gasification by Antal's group, University of Hawaii. Then, ruthenium was impregnated onto the carbon nanotube, and the catalyst was packed into the reactor. When glucose solution was fed, complete gasification was achieved with small amount of catalyst. The flow rate of the feedstock as well as catalyst amount in the packed bed was changed, and carbon gasification efficiency was determined experimentally. The result was well expressed with the first order reaction model for both homogeneous and heterogeneous reactions. The amount of ruthenium needed is more than one order of magnitude lower than previous study, showing the high possibility of achieving low-cost commercial supercritical water gasification process.

730c Overall Transesterification Rate of Palm Oil with Methanol Catalyzed By Acid in Oil-Methanol Two Phase: A Novel Two-Phase Overall Reaction Model

Yoshimitsu Uemura, Yuki Mitsutake, Fon Yee Han, Thanh Hoai Trinh, Katsuki Kusakabe

Transesterification of triglyceride (TG) with methanol in two-phase batch reactor to produce fatty acid methyl ester (FAME) is one of a few successful industrial processes for producing biofuels. We have proposed a novel overall reaction model of TG transesterification, in which (1) TG in continuous oil phase passed the interface from oil-side to methanol-side, and diffuses through methanol droplets, and simultaneously undergoes transesterification with methanol during diffusion and (2) methanol in methanol droplets passed the interface from methanol-side to oil-side, and reacts with TG in oil phase. Based on this model, the overall reaction rate was expressed by a pseudo-first order reaction kinetics for TG concentration in oil phase when the oil-side mass transfer is negligible. This oil-side mass transfer free condition was proved empirically under the experimental condition used in this study. The proposed model successfully reproduced experimental results of overall rate of palm oil transesterification with methanol catalyzed by sulfuric acid at 298, 308 and 323K in a batch reactor.

730d Characterization of Torrefied Woody Biomass

Yoshinobu Yokogawa, Hajime Kida, Kenichiro Tanoue

To improve the biomass co-firing rate in co-firing power generation, woody biomass was semi-carbonized to improve fuel properties. After semi-carbonization, the biomass samples were crushed using a pulverizer and the change in bulk density was investigated. Evaluation of crushability by bulk density was examined.

161a Techno-Economic Analysis and Life Cycle Assessment of Non-Thermal Plasma Co-Upcycling of Waste Plastics and CO_2 to High-Value Chemicals

Samirah Gnangbe, Harish Radhakrishnan, Xianglan Bai, Mark Mba Wright

Single-use plastic waste is accumulating rapidly in landfills and oceans, causing significant harm to the environment and human health. In the United States, over 80% of plastic waste ends up in landfills, while around 30% of it is dumped into the oceans, posing a severe threat to the ecosystem[1,2]. Single-use plastics are mainly composed of polyolefins, polyethylene (PE) and polypropylene (PP), manufactured from fossil fuels. Therefore, plastic waste contains a significant amount of carbon, which has the potential to be recovered and converted into valuable products at the end of its life cycle. However, this potential is often limited due to inefficient waste management practices and a lack of cost-effective recycling methods.

Plastic upcycling into value-added products can establish a circular economy for plastics and mitigate the environmental burden caused by their end-use and disposal. Non-thermal plasma has gained recognition as an effective chemical upcycling technology to convert plastic and CO_2 waste to valuable chemical commodities[3,4]. However, the technical and economic feasibility of employing this technology for plastic waste treatment has not been thoroughly investigated yet.

In this study, we conducted the techno-economic analysis (TEA) and life cycle assessment (LCA) of a non-thermal plasma technology that upcycles plastic waste and waste CO_2 to oleochemicals and hydrocarbons. A conceptual facility with a daily processing capacity of 200 metric tons (MT) of waste PE with a 20-year lifetime was designed. The process was modeled based on the experimental results at the laboratory scale. The process simulations, mass and energy balance, and TEA analysis were performed with the open-source platform BioSteam 2.38.6.

The study evaluated two plasma carrier gas compositions: CO_2 plasma (scenario 1) and CO_2/O_2 plasma (scenario 2). The decomposition of waste PE through CO_2 and CO_2/O_2 plasma processes results in the production of both liquid and gaseous products. The liquid product comprises fatty alcohols, fatty acids, fatty aldehydes, olefins, and paraffins while the gas yield includes carbon monoxide, hydrogen, and C_1-C_5 hydrocarbons. All the liquid compounds are recovered in the process, and the gases are combusted to generate energy for the plant operations.

All the scenarios were economically feasible, with an internal rate of return (IRR) ranging from 42.4% to 43.7% in favor of scenario 2. The TEA showed that the total installed cost of CO_2 plasma and CO_2/O_2 plasma were respectively \$52.7 million and \$56.9 million. The annual revenue of scenarios 1 and 2 were \$145 and \$163.7 million respectively. The net present value (NPV) was \$744.4 million and \$857 million, respectively.

The LCA was performed in OpenLCA 1.11.0 and the EcoInvent database. The global warming potential ranged from -1.9 to -0.73 kgCO2e/kg PE. Negative emissions were achieved in all scenarios, as the credits earned from the products surpassed the carbon emissions generated in the refinery. Sensitivity analysis showed that capital cost and fatty alcohol price have a significant impact on the profitability of the facility, while utilities and the production of fatty alcohols and olefins have the most significant impact on the environmental factors.

These findings indicate that non-thermal plasma is a viable alternative for the upcycling of plastic waste and CO_2 waste. This technology has great potential to produce carbon-negative chemicals, reducing our dependence on petrochemicals.

Acknowledgments: This research is supported by the Department of Energy Office of Energy Efficiency and Renewable Energy under contract no. DE-EE0009943. We thank our collaborators, Dr. Xianglan Bai and her student Harish Radhakrishnan, for performing the experimental work for the project.

References

1. United States Environmental Protection Agency (U.S. EPA). Plastics: Material-Specific Data; U.S. EPA: Washington, D.C., 2023; https://www.epa.gov/facts-and-figures-about-materials-waste-andrecycling/plastics-material-specific-data (accessed Oct 20,2023).
2. Yadav B, Pandey A, Kumar LR, Tyagi RD. Bioconversion of waste (water)/residues to bioplastics- A circular bioeconomy approach. *Bioresour Technol*. 2020;298:122584. doi:10.1016/j.biortech.2019.122584
3. Nguyen HM, Carreon ML. Non-thermal Plasma-Assisted Deconstruction of High-Density Polyethylene to Hydrogen and Light Hydrocarbons over Hollow ZSM-5 Microspheres. *ACS Sustain Chem Eng*. 2022;10(29):9480-9491. doi:10.1021/acssuschemeng.2c01959
4. Bogaerts A, Neyts EC. Plasma Technology: An Emerging Technology for Energy Storage. *ACS Energy Lett*. 2018;3(4):1013-1027. doi:10.1021/acsenergylett.8b00184

161b Techno Economic Analysis and Life Cycle Assessment for Synthetic Spider Silk Production

Melika Tajipour, Bojing Jiang, Pallavi Dubey, Fuzhong Zhang, Marcus Foston, Mark Mba Wright

Synthetic spider silk has emerged as an innovative biomaterial, notable for its exceptional strength and thermal resistance. Its applications range from textile production to medical instruments. The advancement of synthetic biology has facilitated the generation of recombinant spider silk proteins, utilizing Escherichia coli (E. coli) as a production host, and introducing a sustainable alternative to traditional materials. This paper explores the economic viability and environmental implications of synthetic spider silk manufacturing on a large scale. To assess the feasibility of synthetic spider silk production, a techno-economic analysis (TEA) was conducted to examine the economic aspects of the manufacturing process.

This analysis considered various factors, such as production costs, market demand, and potential profitability, to evaluate the economic feasibility of large-scale production. In addition, a life cycle assessment (LCA) was performed to analyze the environmental impact of the production process, covering the complete life cycle of synthetic spider silk, from raw material extraction to end-of-life disposal. Based on laboratory data, we developed and validated a comprehensive process model that encapsulates the essential phases of synthetic fiber production, including Bacterial Cell Culture, Protein Separation, and Fiber Processing, within the BioSTEAM environment.

A detailed techno-economic assessment reveals that the cost of producing spider silk can vary significantly, depending on the operational efficiency and technological maturity of the production facilities. The results indicate a minimum sale price of 29 USD per kilogram and associated greenhouse gas (GHG) emissions of 38 kg CO2-equivalent per kilogram of silk produced. These outcomes are attributed to enhanced protein yields, process optimizations, and economies of scale inherent in mature production operations.

Compared to natural silk fiber, which has a reported environmental impact of 58.73085 kg CO2 equivalent, our synthetic spider silk production demonstrates a reduced global warming potential of 43.43 kg CO2 equivalent, underscoring the environmental benefits of our approach. Furthermore, sensitivity analysis conducted in this study identifies the key techno-economic analysis parameters affecting the minimum sale price of the silk. Notably, fiber yield, measured in kg/kg of glucose, emerged as the most significant factor, followed by the cost of raw materials. This analysis underscores the importance of optimizing yield and managing raw material costs to enhance the economic and environmental sustainability of synthetic spider silk production.

In this study, various fiber yield percentages were analyzed to determine their impact on the minimum selling price (MSP) of synthetic spider silk. The findings reveal that at a 16% fiber yield, the MSP is 29 USD per kilogram, whereas a lower yield of 4.5% results in a significantly higher MSP exceeding 100 USD per kilogram. Conversely, achieving a higher fiber yield of 27% reduces the MSP to 18 USD per kilogram, highlighting the crucial influence of fiber yield on the economic viability of synthetic spider silk production and more research is needed to understand how other fiber qualities might affect the value and benefits of the synthetic fibers.

161c Techno-Economic Analysis of a Modular Bioprocess System for Producing Polyhydroxyalkanoates (PHAs) from Food Waste Via Heterogeneous Fermentation.

MST Sharifa Khatun, Arna Ganguly, Zhiwu Wang, Mark Mba Wright

Plastic pollution is among the most significant challenges the world faces. Annually, 350 million tons of petroleum-based plastic waste are produced, and estimated to reach 500 million tons by 2050. The packaging industry is the largest and growing consumer of synthetic plastics derived from fossil fuels, accounting for 26% of the total volume of plastic used. Plastic pollution detrimentally impacts environmental and human health by contaminating ecosystems, including soil and groundwater, due to its non-biodegradable nature. PHA, a Bioplastics, can be produced from organic waste, such as food waste, via fermentation by using different types of cultures and possesses physicochemical properties resembling or superior to common petroleum-based plastics. However, the variability of food waste properties poses technical challenges for bioplastic production efforts. Heterogeneous fermentation of food waste, utilizing advanced bioprocessing techniques and specialized microbial cultures, can handle diverse food waste compositions. This study evaluates the techno-economic feasibility of a modular bioprocess system that converts food waste into PHAs via heterogenous fermentation using specialized microbial cultures.

A process model of the PHA production targeting specific components of food waste (FW) was developed using BioSTEAM. A simulation was performed to estimate the mass and energy balance of the plant, assuming that 450MT/day of FW is processed in a day. The mass balance showed conversion yields of 104MT of PHA from 450MT food waste (FW) per day, of which 300MT regular food waste,100MT high salinity glycerol waste, and 50MT FOG. The recovered PHA biopolymer will be further used for both flexible and rigid packaging products. We estimated capital costs and the PHA minimum selling price (MSP) from the preliminary economic analysis. The fixed capital investment is estimated at about $77 million, and the minimum Selling Price (MSP) is $3.67/kg, which falls within the market price range of ($2-$7)/kg. The result showed that capital expenses are mainly attributed to equipment, especially the anaerobic digester, and fermenter, totaling $28 million. For Operating Costs, utilities and raw materials constitute the significant expenses. Sensitivity analysis indicates that PHA yield is the most sensitive parameter, followed by the costs associated with fermentation culture and solvent used in the extraction process. Power consumption and electricity prices have a significant impact on the MSP. These results indicate that producing PHA through the heterogeneous fermentation of food waste offers a promising solution for food waste management. Future work will investigate the lifecycle impacts of the technology.

161d Design and Analysis of a Direct Pathway Process for the Enantioselective Production of Biomass-Derived (*R*)-γ-Valerolactone

Seungwon Lee, Jaehyun Cha, Byeongchan Ahn, Joungho Park, Kyoungseon Min, Wangyun Won

Introduction

γ-valerolactone (GVL) is considered a key chemical in biorefineries. This is because GVL can be used for a variety of purposes, including fuel, solvent, and precursor to other products. However, the high cost of GVL has been a barrier to its commercialization. This high cost is due to the numerous reaction steps and complex separation processes required in conventional biorefinery processes. Therefore, we propose an improved approach that addresses the limitations inherent in conventional biorefineries to lower production costs.

Traditional biorefinery process typically involve the hydrolysis of biomass to sugars, followed by their fermentation or further processing into desired products. However, these multi-step paths suffer from inefficiencies due to significant product loss at multiple stages. To overcome these issues, we explored an improved method. These include one-step oxidation of biomass under harsh conditions, followed by an enzymatic reaction to produce GVL. Additionally, the enzymes used to produce GVL can produce GVL with high optical purity without the need for additional separation processes. Therefore, our strategy is to produce optically active GVL that has the potential to be utilized as pharmaceuticals (antileukemic steganacin or antihypertensive drugs) and sold at high prices.

The GVL produced from the proposed process can be utilized in various sectors of the chemical industry, contributing to the industry's progress toward the goal of carbon neutrality. Therefore, through this strategy, we aim to improve the economic and environmental viability of GVL to promote its widespread use.

Methods

In this study, we designed and evaluated a process to synthesize (*R*)-GVL from biomass composed of various components, following a pretreatment step and utilizing enzymatic reactions. Initially, lignocellulosic biomass underwent pretreatment using sulfuric acid under high temperature and pressure conditions (170°C, 7.8 atm), resulting in the production of levulinic acid (LA) and furfural (FF). At this stage, both cellulose and hemicellulose components were converted into LA in a one-step process and remained in the aqueous phase, leaving lignin as a solid residue.

Following this, lime was used to adjust the pH before separating the solid biomass residue from the liquid phase. The solid biomass residue was then separated into lignin and other components through a separation process. The lignin was utilized to make electrode materials for additional income. A portion of the liquid solvent was recycled back to the pretreatment reactor, with the remainder directed to the enzymatic reactor. Due to the enzymatic reactor's operation at a lower temperature (30°C) compared to the pretreatment reactor, significant energy was required to reduce the temperature of the stream. This necessitated the introduction of a separation process to decrease energy consumption and to efficiently utilize the thermal energy of the stream.

The stream entering the enzymatic reactor contained LA, which was converted into (*R*)-4-hydroxyvaleric acid ((*R*)-HA), and subsequently converted into (*R*)-GVL under acidic conditions. Following this, the components of the stream, including water, unreacted LA, FF, (*R*)-HA, and (*R*)-GVL, were efficiently separated. FF was converted into LA, and unreacted LA was recycled to produce more (*R*)-GVL. Unreacted HA was also separated to increase the efficiency of (*R*)-GVL production. The separated water was recycled back to the pretreatment reactor to fulfill the solvent requirements for the pretreatment reaction. Furthermore, to minimize energy consumption, heat integration was implemented within the process. A comparison was then made between our strategy and conventional (*R*)-GVL production strategies through a techno-economic analysis (TEA). Additionally, a cradle-to-gate life cycle assessment (LCA) was conducted to analyze the environmental impact factors from the raw material production stage to the (*R*)-GVL production process.

Results

The approach in this study efficiently utilized various components of the biomass with only a one-step pretreatment and simplified the reaction by converting the biomass to (*R*)-GVL via a highly optically selective enzymatic reaction. However, simplifying the reaction steps was not enough to reduce the cost of the process. Typically, integrating biological reactions operating at lower temperatures and pressures with chemical reactions operating at higher temperatures and pressures into a single integrated process required a significant amount of energy due to the large temperature difference. Therefore, we analyzed energy-intensive steps in the process and introduced appropriate separation processes to reduce energy consumption. Furthermore, by performing heat integration, we further reduced the amount of energy required, thus lowering the costs while simplifying reaction steps and optimizing the process.

To verify the economic feasibility of our process, we conducted a TEA. First, due to the small size and immaturity of the (*R*)-GVL market, we estimated the market price range of (*R*)-GVL using the market prices of products that can be made with (*R*)-GVL. As a result, we found that the market price range of (*R*)-GVL was higher than that of optically inactive GVL. The minimum selling price (MSP) of (*R*)-GVL produced in our process was 1517.3 $/ton and it was lower than the estimated market price range for (*R*)-GVL and within the market price range for optically inactive GVL, making it competitive even in the market for optically inactive GVL. Considering the current immaturity of the (*R*)-GVL market and the necessity for the produced (*R*)-GVL to be sold in the optically inactive GVL market, this process demonstrated economic feasibility not just for future scenarios but also at present. Moreover, compared to the MSP of the conventional process, (*R*)-GVL produced in our process showed a lower MSP, as the conventional process incurred high costs due to multiple reaction stages and purification steps, such as chiral chromatography, leading to a higher MSP.

Moreover, a cradle-to-gate life cycle analysis (LCA) was performed to assess the environmental impact of the proposed process. The results showed a global warming potential (GWP) of 2.33 kg CO_2 eq./kg GVL, which is lower than that of the conventional process. This is because the proposed process used less energy compared to the conventional process. It is especially notable that the factor with the most impact on GWP was lime (1.02 kg CO_2 eq./kg GVL), not heating requirement energy (0.65 kg CO_2 eq./kg GVL). This is due to the large pH difference between the pretreatment reactor and the enzyme reactor, which resulted in the use of a large amount of lime. The large pH difference is a limitation of this process and cannot be further improved in its current state. To improve the process from an environmental perspective, future efforts should focus on creating low-pH enzymes and researching high-pH pretreatment methods to reduce this difference.

Conclusion

In this study, we propose an innovative approach within the biorefinery sector, suggesting a method to enhance both economic viability and environmental sustainability. Unlike traditional approaches, this research develops a streamlined process for the production of optically active GVL, paving the way for a cost-effective alternative. As a result of the process design, we were able to reduce the MSP of (*R*)-GVL to 1517.3 $/ton, which is below the market price range of (*R*)-GVL and within the market price range of optically inactive GVL. Given the current state of the global GVL market, where the market size for optically active GVL is not large, a significant portion of the production must be sold in the market for optically inactive GVL. Therefore, this method, which reduces the MSP of (*R*)-GVL, presents commercial feasibility even at the current stage of the immature (*R*)-GVL market. Considering the potential market expansion for high-value products derived from (*R*)-GVL in the future, the commercialization prospects for (*R*)-GVL are expected to increase.

Furthermore, this approach allows for the reduction of the environmental impact. The findings suggest that the global warming potential of our process was 2.33 kg CO_2 eq./kg GVL, which is lower than the conventional process. Considering GVL's utility not only in its own right but also as an important platform chemical for producing other products, the given process signifies a substantial role in advancing toward carbon neutrality in the chemical industry. Through these analyses of economic and environmental benefits, this research underscores the necessity of an innovative approach that can improve both economic viability and environmental sustainability. Adopting various methodologies in biorefinery can contribute to a greener chemical industry.

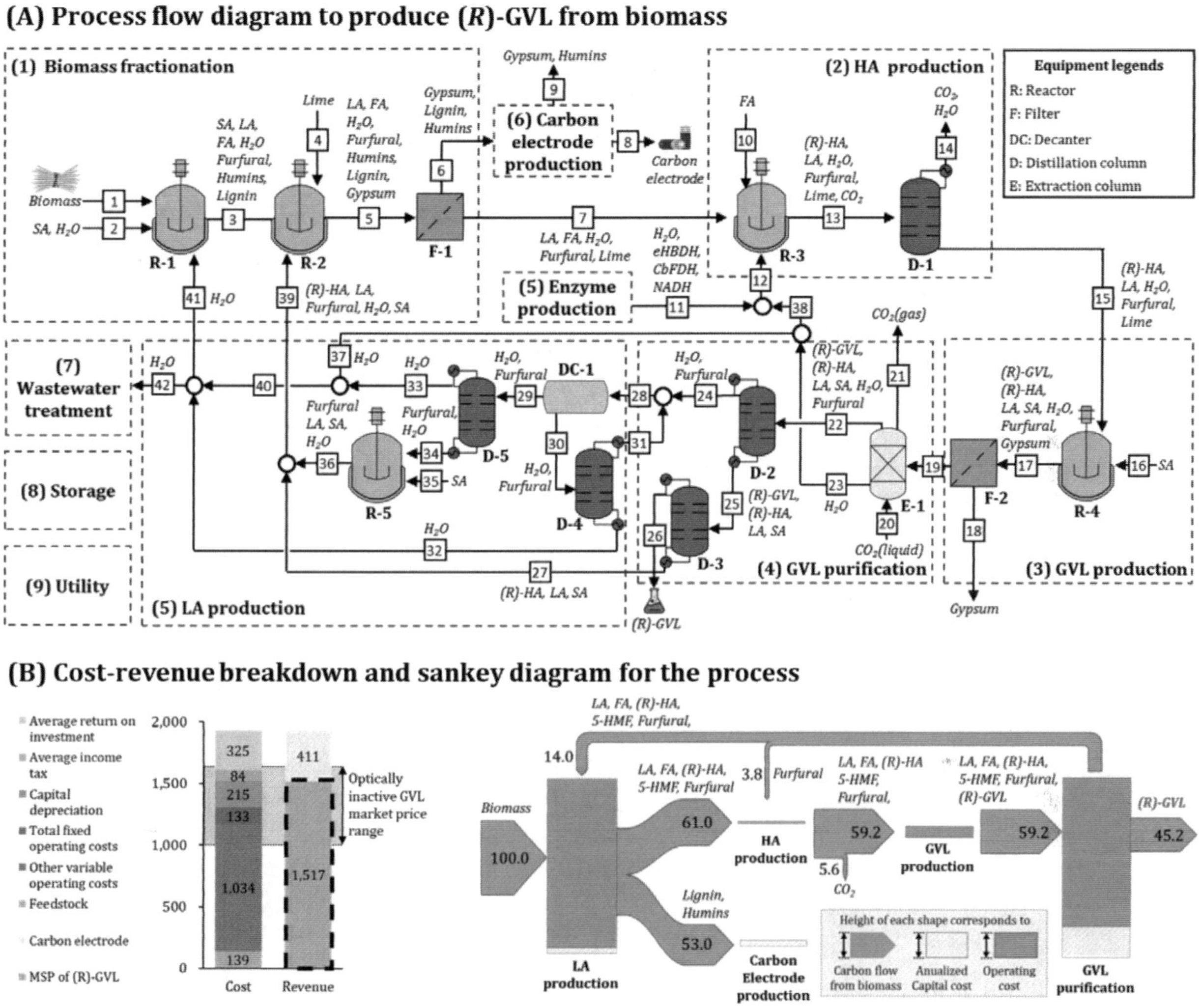

161e Life Cycle Assessment and Sensitivity Analysis of a Plant-Based Meat Alternative Containing Heme Protein

Kay Glass, Hariteja Nandimandalam, Rui Shi

According to the United States Environmental Protection Agency, beef production accounts for 38% of US agricultural greenhouse gas emissions, and 3.5% of total United States greenhouse gas emissions. To mitigate the impact of food protein production, the United States aims to advance food protein production methods with greenhouse gas emissions that are 50% or less of current methods by 2028. Meat alternatives have been developed to mitigate beef greenhouse gas emissions. One nascent additive to some meat alternatives is heme protein produced via fermentation, added to enhance flavor and verisimilitude to meat. While previous life cycle assessments of meat alternatives of ground beef have included products containing heme protein as an additive, these studies have not quantified the impact of the heme additive alone, nor its contributions to the product's environmental impact in depth. To study the environmental impact of heme protein production, and its contribution to the impact of a meat alternative, two cradle-to-gate life cycle assessments (LCA) were conducted using material and energy inputs obtained from information provided by Motif Foodworks™. The LCA of meat alternative production compared the impact of producing the meat alternative with ground beef, using a functional unit of 1 kg of meat alternative patty following the ReCiPe 2016 impact assessment method. Heme protein production resulted in an impact of 117 kg CO_2 eq. per kilogram of heme protein produced, while the plant-based meat alternative patty had an impact of 3.45 kg CO_2 eq. per kilogram, 90% lower than the same quantity of ground beef. Among all environmental impact categories, the impact of the heme-containing meat alternative was lower than that of ground beef, except for freshwater ecotoxicity, which was 9% greater than ground beef. The environmental impact of the meat alternative showed sensitivity to the proportion of renewable energy in the electricity used in the process, and the agricultural practices used to grow ingredients. Key drivers of greenhouse gas emissions were determined via sensitivity analysis. Greenhouse gas emissions were most sensitive to the heme protein production rate. Electricity, ingredient procurement, and heme production parameters each affect product environmental impact and are targets for future improvements to sustainability. These results can be used to illustrate bottle necks in system sustainability and inform the public of the environmental impact of meat alternative and plant-based protein production.

731a Multiple Criteria Productivity Analysis Applied to Perovskite Solar Devices from Lab to Factory Production

Nathan Dodd, Rachael Rothman

The process of understanding the consequences and impacts of industrial scale manufacturing decisions is highly time consuming, requires multiple areas of expertise and a variety of different, and occasionally conflicting, philosophies. Multiple Criteria Productivity Analysis consolidates a variety of methodologies to create a single holistic method for assessing the time, cost and environmental impact of manufacturing decisions. MCPA is an approach which uses life cycle inventory data in combination with activity-based analysis to understand the relationship between output productivity as a measure of output per unit time and environmental productivity as a measure of output per unit environmental impact. A tool has been created in conjunction with the development of the MCPA methodology. Significant advances have been made in understanding the requirements of and relationship to each of the distinctive data subsets which exist, creating a rich data flow from the environmental impact of raw materials to the capital and operational costs of machines and tools through to the time taken to undertake each individual activity. The MCPA method is herein demonstrated for the production of perovskite devices in a university lab environment and a full scale production scenario with detailed results on the trade-offs between increased output productivity on environmental productivity within the film deposition process based industry, providing the evidence for holistic industrial decision making alongside product design focussed outcomes.

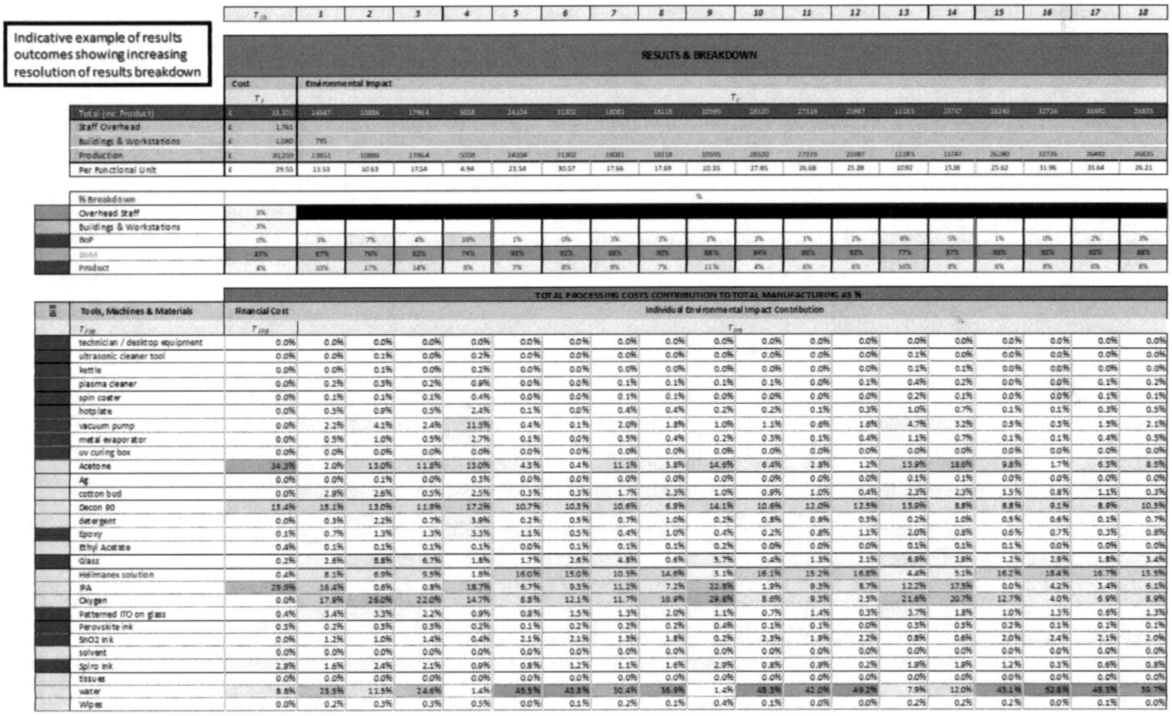

731b Integrated Systems for Continuous CO_2 Separations from Industrial Flue Gas Sources.

Rohan Sartape, Vamsi Vikram Gande, Meenesh Singh

Managing the constant accretion of greenhouse gases such as CO_2 in the earth's troposphere is the challenge of this century. Out of the available technologies for CO_2 capture, water-driven CO_2 capture is attractive since it incurs a low energy penalty. The Migration Assisted Moisture Gradient (MAMG) technology is a water-based CO2 capture system that involves CO_2 capture and release based on the bicarbonate-carbonate equilibria. The CO_2 capture occurs in an organic phase, and the captured CO_2 migrates to the aqueous phase, which is separated by the membrane under an external bias. Herein, we present the results of improving the energy efficiency of the MAMG process by integrating the hollow fiber membrane for gas contacting. Moreover, we also demonstrate the results from the large-scale operation of the MAMG process in a Bipolar Membrane Electro Dialysis unit (BPMED) unit for CO_2 capture from flue gas containing impurities. Parameters such as the residence time of the gas in the hollow fiber membrane, surface area of the membrane, effects of CO_2 concentration, impurities such as SO_x, and relative humidity of the flue gas on the rate of CO_2 capture will also be presented.

731c Critical Technoeconomic Appraisal of Product-Specific Sequential and Integrated CO2 Capture and Electrochemical Utilization

Ahmed Badreldin, Ying Li

The swift advancement of low-temperature electrochemical CO_2 reduction (eCO2R) technologies offers a promising avenue for harnessing anthropogenic CO_2 to produce value-added renewable chemicals and fuels. The current economic value proposition of eCO2R for production of C1 and C2 hydrocarbon and oxygenate type products typically entails technoeconomic analyses (TEA) on conventional sequential carbon capture routes followed by the rapidly evolving eCO2R step. However, these models are seemingly coupled with impractical separation and purification steps, which skews the economic viability of certain eCO2R processes. Further, with the advent clearer catalyst and electrolyte based fundamental insights, integrated reactive capture technologies with amine, hydroxide, or carbonate as capture medium and electrolyte holistically offer a more cost-competitive pathway towards realizing eCO2R processes. To that end, sequential-eCO2R, reactive capture with amine, reactive capture with hydroxide, and sorption enhanced-eCO2R process systems are examined in this work. TEA on the aforementioned four process systems is undertaken on several key high market size chemical classes including carbon monoxide, carboxylic acids and their salts, alcohols, and hydrocarbons (i.e., ethylene) based on contemporary literature for each system-product vector. Parametric variations in terms of required voltage biases, current density, faradaic efficiency of products and byproducts, single-pass conversion efficiency, carbonate formation ratio, levelized cost of electricity, and electrolyzer cost are considered in sensitivity analyses under the conventional scenario where a single target product is purified and sold, and in the case where byproducts are also separated and purified to commercial levels prior to being factored as revenue streams in net present value (NPV) models. Therefore, cost breakdowns, sensitivity analysis, and NPV models are performed on each system-product vector to identify key bottlenecks and offer targeted performance perspectives for future development of integrated capture-eCO2R systems. This work provides insight into future research tailored towards integrated carbon capture and reduction.

731d Effective CO_2 Utilization Using Coke Oven in Iron Making Process for Sustainability

Gihun Kwon, Chang Houn Rhee, Younghun Kim, Sangwook Joo, Beom-Sik Kim, Hai Woong Park

[Introduction]

The CO_2 recycling from exhaust gases in chemical industry has been received attention due to its potential for saving power energy and reducing CO_2 emissions. Among various industries, the iron making industry generates a significant amount of CO_2, contributing to the formation of greenhouse gases. Iron is produced by the reaction of iron oxide and coke, as shown infollowing equation below;

Fe_2O_3, Fe_3O_4 + Coke → Fe + CO_2 (1)

In order to reduce the amount of greenhouse gas in iron making industry, the development of CCU (CO_2 Capture and Utilization) technology specialized for the steel and iron making process is necessary due to difficulty of change in process or raw materials. Here, we demonstrate CO_2 utilization technology using heat and unused carbon at coke oven in iron making industry.

[Methods]

CO_2-rich gas is injected into the hot coke oven. The Boudouard reaction takes place between CO_2 and unused carbon, resulting in the production of CO, which contains heat energy. The increased production of CO gas is utilized as a fuel source to generate power or heat. A conceptual scheme for the Boudouard reaction with CO_2 in the coke oven is shown in upload Image. The Boudouard reaction can be represented by the following equation below;

CO_2 + C(unused carbon) → 2CO (2)

[Results and discussion]

A Coke oven is used to convert coal into coke by heating it to high temperatures (~ 1,200 °C) in an oxygen-free atmosphere. The main products are coke, coke oven gas (COG), and tar, etc. Coke is used to produce iron by reducing iron oxide. COG mainly contains CH_4, CO and H_2 gas and is used as fuel gases to produce power or heat. The tar is usually sticks to the oven wall and then carbonized, turning it into unused carbon. The unused carbon sources are currently released by air-combustion.

To utilize CO_2 in the coke oven, CO_2-rich gas is injected. The Boudouard reaction occurs between heat and CO_2 and unused carbon, generating CO. The produced CO is mixed with the original COG, resulting in increased COG emissions.

As a result of the analysis, the overall COG flow has increased by approximately 10% and the heat capacity has increased by about 7%. This indicates that the amount of COG that can be used as an energy source has increased due to the conversion of injected CO_2 to CO. Ultimately, by reducing the external LNG supply, it was possible to provide energy supply within the steelworks, resulting in cost savings from LNG usage and a reduction in CO_2 emissions. In 2023, the completion of the demo plant demonstrated the effectiveness of utilizing CO_2 through injection into coke.

[Conclusions]

CO_2 utilization technology, specifically designed for the iron making industry, gas been developed using coke oven. The heat within the coke oven serves as the reaction energy for CO_2 reduction, while the unused carbon attached to the coke oven wall acts as the source of CO_2 reduction. The CO_2 injected into the coke oven was converted into CO, a heat gas, through the Boudouard reaction. The completion of demo plant confirmed the result of increasing the total heat capacity of COG by 7%. This can lead to economic benefits by reducing reliance on energy sources such as LNG in the steel industry, as well as environmental benefits by reducing CO_2 emissions.

731e Proposal for an Upcycling Process to Commercialize All Components of By-Product Oil in Edible Oil Refinery

Naomi Shibasaki-Kitakawa, Kousuke Hiromori, Makiko Kato

Vegetable oils are important for us, not just for food but also for industrial uses. Every year, the world produces about 215 million tons of these oils, and the amount grows by 5% each year. Out of this total, 60% is used for food, 15% goes into making chemicals known oleochemicals, and the remaining 25% is turned into biodiesel fuel. In addition to these primary uses, about 20% of the oils processed are inedible and become by-products during the refining process. Currently, these by-product oils, rich in fatty acids and containing a few functional ingredients, are incinerated for heat recovery due to the absence of economically viable technologies for their efficient utilization.

Elemental technologies have been developed to convert fatty acids, their primary component, into biodiesel via esterification, or to separate and recover trace functional ingredients. However, conditions of high temperature and reduced pressure are required to suppress the reverse reaction in esterification and to utilize multi-stage molecular distillation in the recovery process. As a result, these requirements lead to high energy consumption and low overall efficiency, creating significant challenges for the viability and sustainability of these technologies.

Table 1 details the content and generation volumes of each component in the distillation by-products obtained from the deodorization process of four types of plants. It also presents the potential value of these components as functional ingredients, along with their market prices and growth rates (Compound Annual Growth Rate, CAGR). For fatty acids and triglycerides, the table shows the prices when converted into fatty acid esters, which can be used as biodiesel.

We have therefore developed an integrated reaction separation process aimed at commercializing all components found in these by-products. The key technology behind this innovation is a continuous reaction separation process that utilizes ion exchange resins as both catalysts and adsorbents. As previously reported[1)], this method enables the simultaneous complete conversion of fatty acids and triglycerides into esters and the separation of phenolic functional ingredients such as vitamin E with high purity. Remarkably, this process operates under mild conditions at 50°C and atmospheric pressure.

Our findings reveal that when paraffins are present in the distillate oil, they precipitate at room temperature due to their low solubility in the fatty acid ester-alcohol solution. By washing with alcohol, we have successfully recovered the paraffins purity to more than 95%. Notably, with a melting point of about 60°C, this substance proves to be an excellent base material for lipstick, benefiting from its card house structure. As this is the world's first biomass-derived product of its kind, we have also registered CAS numbers for them. Furthermore, when phytosterols are contained in the distillate oil, they automatically precipitate out after the distillation for alcohol recovery due to their low solubility in fatty acid esters, enabling their recovery with over 95% purity. The remainder is mainly fatty acid esters, which are used as biofuels for vehicles and power generation.

We have already established a practical process for rice bran oil distillate, confirming that its potential value can be realized as planned. Currently, we are advancing the demonstration of this process for palm distillate oil.

1)Hiromori et al., J.Clean.Prod.,189, 223-230(2018)

Table 1 Generation and composition of each by-product oil in edible oil refinery and its potential value

items		market price (CAGR[%])	by-product oils (deodorizer distillate)			
			rice bran	palm	rapeseed	soybean
amount generated [t/y]		Japan	545	—	5,035	2,310
		Global	8,250	**11,391,900**	135,595	297,320
composition [wt%]	**tocotrienol**	**$833/kg(6.6)**	**2.4**	**0.38**	—	—
	tocopherol	**$66/kg (4.3)**	**2.3**	**0.12**	**4.0**	**9.1**
	fatty acid	$1/kg (10.0)	44.0	82.00	45.0	41.0
	triglyceride	$1/kg (10.0)	14.0	14.00	10.0	9.0
	phytosterol	**$13/kg (7.4)**	**5.0**	**0.40**	**10.0**	**17.0**
	bio-paraffin	**$66/kg (5.3)**	**1.5**	—	**3.0**	—
	hydrocarbon	$0.3/kg	30.8	3.10	28.0	23.9
potential value of each waste oil [$/kg]			**23.8**	**4.3**	**6.6**	**8.8**
overall potential value of amount generated [million dollars]		Japan	13	—	33	20
		global	**196**	**48,985**	**895**	**2,616**

731f Continuous Active Pharmaceutical Ingredient (API) Manufacturing from Biomass-Derived Building Blocks: An Integrated Process to Enable Sustainable Manufacturing

Yuma Miyai, Hung Lin Lee, Aibolat Koishybay, Hubert Hellwig, Loic Bovy, Tianjie Zhu, Rodolfo Romañach, Muhammad Noor Alam, Cornelis Vlaar, Jean-Christophe Monbaliu, Allan Myerson, Torsten Stelzer

The pharmaceutical industry aims to advance its technological capability in process development and manufacturing to achieve improved process efficiency, manufacturing safety, product quality, and sustainability. In this regard, research activities in continuous manufacturing, process analytical technologies (PAT), and process modeling have increased significantly in the last decade. However, sustainability in pharmaceutical manufacturing and its environmental impact still poses challenges that need to be addressed. The pharmaceutical manufacturing is frequently linked with the generation of significant wastes, primarily stemming from the extensive use of petroleum-derived chemicals as well as numerous work-up, isolation, and purification processes inherent in batch manufacturing. To that end, continuous manufacturing, in which several reaction steps can be telescoped, has the potential to reduce the number of processing steps needed, thereby improving the overall environmental friendliness of the process. In addition, there is an opportunity to achieve sustainable manufacturing of APIs through the use of biomass-derived building blocks and bio-based solvents, replacing conventional petroleum-based chemicals.

In this work, the multi-institutional project aims to demonstrate the continuous manufacturing of the antibiotic API Nitrofurantoin from a biomass-derived building block and utilizing bio-based solvents, while ensuring compliance with the product quality attributes outlined by the U.S. Pharmacopeia (USP). Specifically, the process integration for the multi-step continuous synthesis developed by CiTOS (University of Liege) and the crystallization process developed by CDI (University of Puerto Rico) and MIT including continuous reactive crystallization is demonstrated on a computer-controlled continuous manufacturing platform. The integrated system is monitored by real-time quality control via PAT supported by advanced analytics where process data is captured in real-time and analyzed for process optimization and control. This work aims to advance the current science by demonstrating reliable process operation with minimal human interventions while addressing the need for sustainable pharmaceutical manufacturing.

731g Evaluating Glycerol Carbonate Production from CO_2 and Glycerol: An Integrated Process Analysis

Octavio Del-Mazo-Alvarado, Carlos Prieto, Antonio Sánchez, Cesar Ramírez, Adrián Bonilla-Petriciolet, Mariano Martin

The decarbonization of the chemical industry stands as one of the most important challenges in the transition towards a renewable-based system. Currently, fossil resources serve a dual role: as energy supplier (accounting for around 42% of the total usage) and as feedstock for the different processes (representing the remaining 58%) (Kätelhön et al., 2019). Among the alternatives proposed for establishing a sustainable chemical industry, CO_2 utilization has attracted great attention in the last years as feedstock. The aim extends beyond a mere substitution of current fossil resources, encompassing the creation of novel chemical products in accordance with the principles of green chemistry and engineering. In this context, Glycerol Carbonate (GC) emerges as a key green chemical that is gaining increasing attention (Pirzadi & Meshkani, 2022). GC offers two main benefits: its wide reactivity involving an extended range of applications including the synthesis of green polymers or surfactants and the possibility of valorising glycerol, a by-product of the biomass-based industry. Different pathways have been explored in the production of GC. The most straightforward approach (at least theoretically) is the direct combination of glycerol and CO_2 (Luo et al., 2022). However, this route is highly constrained by thermodynamics, offering low performance results. Presently, the most promising alternative lies in transcarbonation, involving the reaction between an alcohol (glycerol) with a carbonate (highlighting dimethyl carbonate) (Christy et al., 2018).

In this work, an integrated process assessment for the synthesis of GC is performed employing CO_2 and glycerol as the primary raw materials. The selected synthesis follows the transcarbonation reaction between glycerol and Dimethyl Carbonate (DMC). The overall process consists of three main stages. The first one involves the production of urea via the combination of CO_2 and ammonia. Subsequently, in the second step, the alcoholysis reaction between the produced urea and methanol synthesizing DMC. Finally, the DMC is combined with glycerol for the final synthesis of GC. The entire process is modelled using ASPEN Plus V14 including some experimental equations for this particular synthesis using Fortran statements. A comprehensive techno-economic evaluation is performed for the proposed process obtaining the final technical and economic results toward the implementation of this green route to produce GC.

The techno-economic evaluation reveals promising results. The two main raw materials of the GC production are glycerol and CO_2 with a global yield in the synthesis of 0.78 kg glycerol/kg GC. Some small amounts of ammonia and methanol are required to mitigate process losses. The capital cost, for a facility with an operating capacity of 166,870 t of GC/year, is estimated at 171.5 M$. The total operating cost for GC produced by this integrated green process amounts to 1558 US$/t of GC where the utilities and the raw materials constitute the two most important contributions. These results demonstrate competitiveness with other alternative processes for the synthesis of GC showing the potential of process integration in designing a new efficient green chemical industry.

References

Christy, S., Noschese, A., Lomeli-Rodriguez, M., Greeves, N., Lopez-Sanchez, J.A. (2018). Recent progress in the synthesis and applications of glycerol carbonate. Current Opinion in Green and Sustainable Chemistry, 14, 99–107

Kätelhön, A., Meys, R., Deutz, S., Suh, S., Bardow, A. (2019). Climate change mitigation potential of carbon capture and utilization in the chemical industry. Proceedings of the National Academy of Sciences, 116, 11187–11194.

Luo, C., Wang, J., Lu, H., Wu, K., Liu, Y., Zhu, Y., Wang, B., Liang, B. (2022). Atmospheric-pressure synthesis of glycerol carbonate from CO2 and glycerol catalyzed by protic ionic liquids. Green Chemistry, 24, 8292–8301.

Pirzadi, Z., Meshkani, F. (2022). From glycerol production to its value-added uses: A critical review. Fuel, 329, 125044.

Acknowledgements

The authors are grateful to the PSEM3 research group for its support in the development of this work, the funding received from the European Union's Horizon 2020 research and innovation program under the Marie Sklodowska-Curie grant agreement no. 778168 (IProPBio proyect), and also acknowledge the funding received from the CONAHCYT and the FPU, Spain grant (FPU21 /02413) to C.P.

573a Techno-Economic Analysis and Life Cycle Assessment of Hydroxylamine Eco-Manufacturing Via Wastewater Electrochemical Reduction

Manish Mosalpuri, Wenzhen Li, Mark Mba Wright

The manufacturing of nitrogen-based chemicals is dependent mainly on the Haber-Bosch process, which produces cheap ammonia (NH_3) from hydrogen (H_2) and atmospheric nitrogen (N_2) gas. However, this process has high energy requirements due to the high process temperature (500^0 C) and pressure (100-150 bar). It is also responsible for significant CO_2 emissions (400 metric tons annually) and is energy-intensive- it consumes around 2% of the world's primary energy. Hydroxylamine (NH_2OH) is an important chemical intermediate currently produced from NH_3 formed by the Haber-Bosch process, followed by the Ostwald process (which generates nitric oxide, NO from NH_3) and NO reduction by H_2. We propose an alternative reaction mechanism to the conventional method described above, which involves the electroreduction of nitrate ions (NO_3^-) to NH_2OH, as depicted in **Figure 1**. Wastewater from fertilizer run-offs, nuclear power plants, and slaughterhouses is a rich source of NO_3^-, which harms the ecosystem.

Many researchers are now focusing on manufacturing chemicals via electrochemical processes. Electrochemical processes are typically executed at ambient temperature and pressure, thus requiring mild operating conditions and low process energy.

We conducted a techno-economic analysis (TEA) and a life cycle assessment (LCA) of this electrochemical system converting NO_3^- to NH_2OH. We conducted TEA using a general model by Orella et al., 2020 [1]. We performed a sensitivity analysis to calculate the maximum possible reduction in NH_2OH production cost and discover its influencing parameters. The most dominant parameter influencing the cost was the separation factor, followed by lifetime, product Faradaic Efficiency, electricity price, and catalyst loading. We obtained the optimum projected cost of NH_2OH as \$2.06/kg, very close to its market price estimated at \$1.72/kg. We obtained the base cost, or the current production cost, as \$6.14/kg. Our analysis suggests an immense potential for cost reduction, which motivates further research.

We also conducted an LCA of the process system depicted in **Figure 1**. The goal of this LCA study is to calculate the environmental impacts of the proposed eco-manufacturing pathway to produce NH_2OH in a wastewater treatment plant in Iowa. We determined if the proposed alternate method of producing NH_2OH is a sustainable option compared to the most prevalent method of producing almost every N-based chemical, the Haber-Bosch process. This study's life cycle inventory (LCI) data was obtained from balanced chemical reactions, the preliminary TEA model, and literature, scaled to the functional unit: 1 kg of NH_2OH produced. We performed a 'Cradle to Gate' LCA study, focusing on the system from resource extraction to transport and not including the use and waste management phases of NH_2OH. We used openLCA 1.11.0 to perform LCA using Tool for Reduction and Assessment of Chemicals and Other Environmental Impacts (TRACI) 2.1 as the impact assessment method. It evaluates these impact categories- acidification, carcinogenics, ecotoxicity, eutrophication, fossil fuel depletion, global warming, non-carcinogens, ozone depletion, respiratory effects, and smog. We modeled the NO_3^- reduction unit process in openLCA based on Dominguez-Ramos et al. (2013) and Orella et al. (2020). [1,2]

We compared our results for the proposed pathway with the conventional method of NH_2OH production. We found the proposed method is environmentally friendly for all impact categories, even using grid electricity to operate the electrical components. We also observed that the proposed pathway further reduces emissions using solar photovoltaic (PV) energy instead of grid electricity.

Figures 2 and 3 show the life cycle impacts comparison of producing 1 kg NH_2OH produced using the conventional and proposed route in Iowa using grid and solar PV electricity, respectively.

Our analysis shows that the proposed electrochemical pathway has less environmental impact than the conventional pathway. The economic feasibility is expected to improve based on future technological considerations. The proposed pathway has massive potential and could be used as an alternative to conventional processes, including the Haber-Bosch process, followed by the Ostwald process and hydrogenation to manufacture NH_2OH.

Acknowledgments: The authors gratefully acknowledge the funding for the project from the National Science Foundation (NSF) with the NSF Award number 2036944. We thank our collaborators- Dr. Shuang Gu, Dr. Wenzhen Li, and his students (Hengzhou and Yifu), for performing the experimental work for the project.

References

[1] Orella, M. J., Brown, S. M., Leonard, M. E., Román-Leshkov, Y., & Brushett, F. R. (2020). A General Technoeconomic Model for Evaluating Emerging Electrolytic Processes. Energy Technology, 8(11).

[2] Dominguez-Ramos, A., Singh, B., Zhang, X., Hertwich, E. G., & Irabien, A. (2015). Global warming footprint of the electrochemical reduction of carbon dioxide to formate. Journal of Cleaner Production, 104, 148–155.

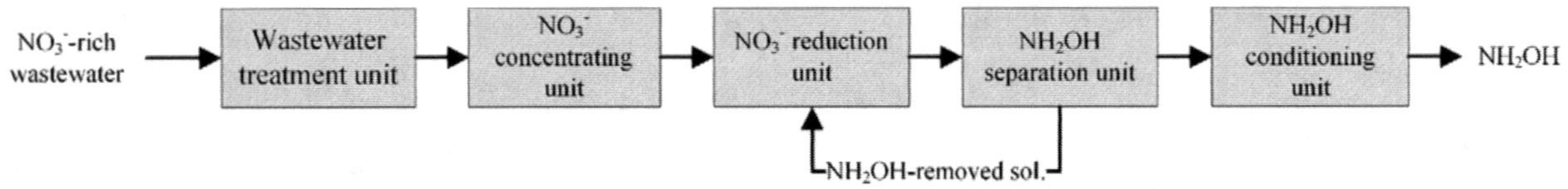

Fig. 1 Entire NO3- to NH2OH conversion system, showing individual processes.

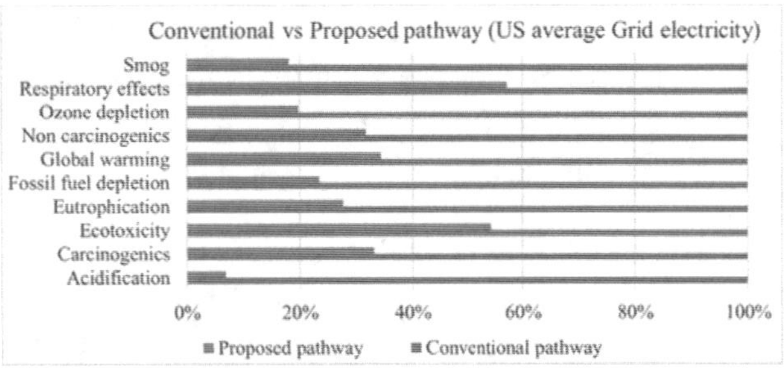

Fig. 2 Life cycle impacts for producing 1 kg NH2OH via the conventional and proposed pathway using grid electricity

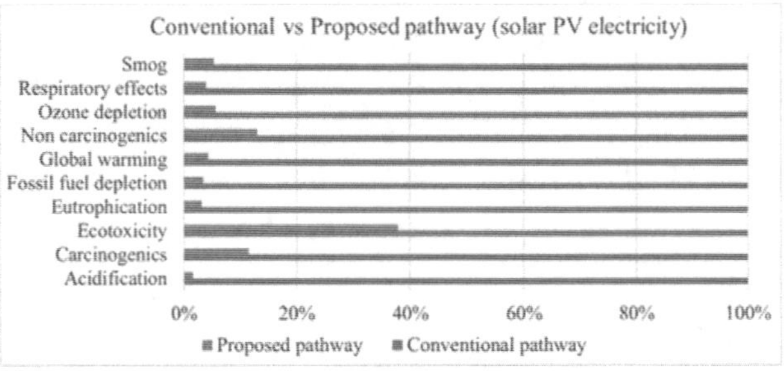

Fig. 3 Life cycle impacts for producing 1 kg NH2OH via the conventional and proposed pathway using solar PV electricity

573c Sustainable Resources for the Synthesis of Aromatic Feedstock

Marvin Wittek, Andreas Widjaja, Guido Waidmann, Alfons Drochner, Bastian Etzold

Introduction

Today's chemical industry is based mostly on commodity chemicals that are produced on a large scale. They make up the raw materials that are used to manufacture many of the industrial products such as plastics, electronics, fertilizers and more. Aromatic compounds, such as benzene, toluene and xylene (BTX) make up a significant portion of the commodity chemicals. In 2022 the production of benzene amounted to roughly 60 million metric tons worldwide with projections showing further increase in production capacities.[1] The application sectors of the BTX fraction are multifaceted. While benzene is typically used in gasoline blending to improve knocking characteristics, it also serves as the chemical building block for many other chemical compounds such as ethylbenzene or aniline. Another important field stems from the oxidation of p-xylene to produce terephthalic acid, an important monomer for the production of various polymers, such as polyethylene terephthalate (PET). Currently BTX compounds are almost exclusively produced from fossil resources. Roughly 70% of the total world supply of BTX stems from reformed naphtha. The remainder originates from cracked gasoline byproduct from ethylene plants and coal liquids that are produced in coke ovens.[2] With respect to the transition of the chemical industry towards more sustainable production processes and away from the use of fossil resources, it can be expected that the aforementioned raw material streams for the production of many industrial products will decrease in availability.

Thus, a review of the open literature was conducted to identify possible renewable resources and thermochemical conversion processes to obtain aromatic feedstock. Furthermore, means of increasing the aromatic content of respective material streams were investigated. Other concepts with the goal to transform renewable resources that are known to literature, such as Bio refineries or hydrothermal conversion processes and more were not considered, as aromatic compounds typically only make up small fractions of respective product streams.

Results

Figure 1 shows possible resource streams and respective upgrading processes to obtain aromatic feedstock that were considered. Renewable biomass is mainly comprised of wood, fats and oils. In terms of quantity the most relevant chemical compounds in this category are Cellulose, Lignin and other polysaccharides as they make up the main constituents of wood. In comparison fatty acids have limited availability and typically production pathways of industrial products that are structurally related have already been established. Furthermore, the competition with biofuels and food production makes the use of oils and fats to produce aromatics increasingly unattractive. In contrast, the challenge with the utilization of the constituents of wood lies in the extraction of pure chemical compounds from respective resources due to chemical linkage and oxygen content present in lignocellulose. Similarly wastes such as plastics or scrap tires require pretreatment processes to break chemical bonds and generate material streams that can be utilized.

Pyrolysis and Gasification processes are actively being investigated to transform renewable biomass and wastes into liquid oils and syngas respectively. In both cases subsequent possibilities for upgrading are greatly dependent on the quality of the obtained oils and gas mixtures. Sulfur-, nitrogen- and oxygen compounds can negatively impact performance and stability of heterogeneous processes that employ catalysts to facilitate for example methanol synthesis or aromatization reactions. Thus, prior to upgrading of resource streams costly purification is often required and further research in this area is required to find suitable methods to decrease the content of impurities in pyrolysis oils as well as gasification product mixtures.

Alternatively, capturing carbon dioxide from the atmosphere can present a pathway that is not reliant on purification processes. Using clean hydrogen, that originated for example from water electrolysis, methanol can be produced, which can act as a platform chemical to produce aromatic compounds. However, currently methanol obtained in this way is more expensive than the conventional fossil resource-based production pathways and further research is required to lower the costs of renewable methanol.

Depending on the degree of impurity atoms in the form of nitrogen, oxygen, sulfur and others, present for example in pyrolysis oils, different routes for aromatization are considered. Commonly used catalysts for the upgrading of hydrocarbon mixtures as well as methanol are based on zeolite materials. Here, the MFI-type zeolite ZSM-5 is the most prevalent used support- as well as active-material. Generally the proton form of the zeolite HZSM-5 is actively being investigated in upgrading processes like pyrolysis due to its desulfurization and deoxygenation properties to obtain higher quality oils. Other applications of ZSM-5 include bifunctional catalysts combining acidic and metal species. Additional doping with metal atoms such as zinc, gallium or silver increase the selectivity towards aromatic hydrocarbons in the product mixture of upgrading processes. However, the use of an active metal species introduces the vulnerability towards impurities like sulfur and nitrogen containing compounds. Current topics of interest in the research field are the underlying mechanisms of aromatization reactions as well as limited catalyst stability and respective mechanisms of deactivation.

[1] Statista Search Department, "Market volume of benzene worldwide from 2015 to 2022, with a forecast for 2023 to 2030", can be found under https://www.statista.com/statistics/1245172/benzene-market-volume-worldwide/, 2023.
[2] R. A. Meyers, Handbook of Petroleum Refining Processes, McGraw-Hill Education, New York, N.Y., 2016.

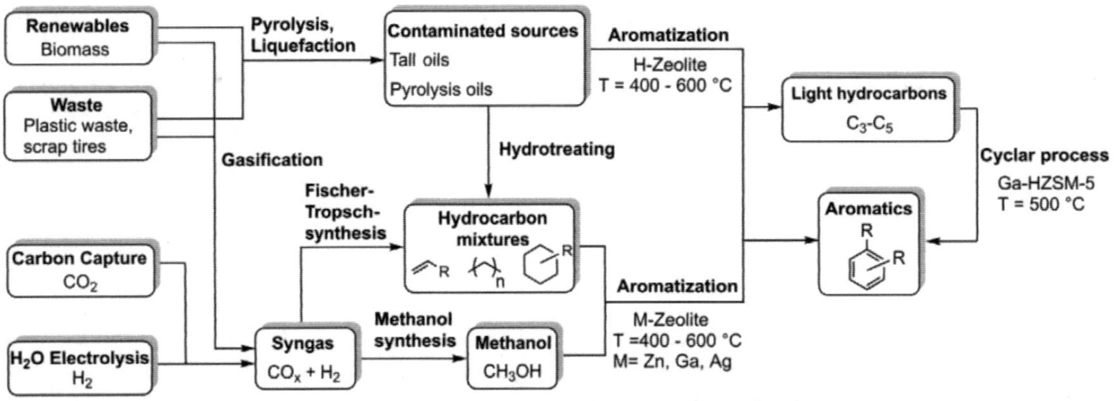

Figure 1.: Considered renewable or cyclable resources and respective synthesis routes towards renewable aromatic hydrocarbons. Depending on the degree of contamination of the raw material, additional metal modification (indicated as „M-zeolite") can be used to further increase the aromatic yield in the upgrading process.

573d Using Electrochemical CO_2 Reduction to Recover Lithium Carbonate from Spent Lithium-Ion Battery Compounds

Shelby Atherton

Recycling lithium-ion batteries is becoming more important than ever before due to the global push towards a more electric economy. The objective of my research is to facilitate the recovery of Lithium Carbonate (Li_2CO_3) from Lithium Cobalt Oxide ($LiCoO_2$) in a high-pressure electrochemical CO_2 environment. $LiCoO_2$ is commonly used as the cathode material of lithium-ion batteries and current recycling technologies have a variety of disadvantages including the requirement for high energy inputs and the production of environmental pollutants. We hypothesize that lithium can be recovered from spent $LiCoO_2$ cathodes by forming insoluble Li_2CO_3 through electrochemical redox reactions performed in CO_2 gas expanded electrolytes.CO_2 expanded electrolytes present a novel reaction media for the oxidation of $LiCoO_2$ and reduction of Li^+ with CO_2, therefore, this work will focus on discovering fundamental knowledge about this system including thermodynamics, reaction mechanisms and selectivity.

573e Techno-Economic Analysis of Itaconic Acid Production Using a Biorefinery Approach

Nkechi Miriam Nwaogaraku, Mark Mba Wright

Introduction

The pursuit of eco-friendly industrial methods and a reduction in our reliance on fossil fuels has prompted research into green or biobased chemicals. These biochemicals are usually produced from renewable biomass, and several researchers have proposed increased product titres using affordable and alternative lignocellulosic feedstocks as viable strategies for effective IA production and can replace their fossil-based equivalents [1]. One such biochemical is itaconic acid. The year 2004 saw the introduction of twelve bio-based building blocks, including itaconic acid (IA), as Top Value-Added Chemicals from Biomass by the US Department of Energy. Only IA offers options for replacing methacrylic and acrylic acids among the others [2]. Itaconic acid (IA) has numerous uses in the pharmaceutical, medicinal, and agricultural industries, making it a promising commodity biochemical. Currently, the industrial biomanufacturing of IA is done using *Aspergillus terreus*, microbial biomanufacturing is a viable method for producing high-value chemicals with minimal environmental impact and significant economic benefits. We propose an alternative production mechanism to the conventional method described above, which involves utilizing a different microbial fermentation process as depicted in Fig 1.

The BIOSTEAM simulation software was used to design the Itaconic Acid biorefinery. According to Nieder et al [1], It was assumed that findings reported at the laboratory scale could be applied to an industrial setting. Therefore, the findings are sufficient for a conceptual level of study and could be validated and optimized using a pilot plant prior to implementation. The combined feedstock is made up of 70% sugarcane bagasse and 30% trash (by weight), comprising of 40.7% cellulose, 27.1% hemicellulose, 21.9% lignin, 6.7% extractives, and 3.5% ash. The block flow diagram (BFD) of the biorefinery with the major operating units is shown in Fig. 2, the simulation results were validated with experimental data from the published studies. Given that the sugars found in lignocellulosic biomass are in the form of complex carbohydrates, pretreatment, and enzymatic hydrolysis are necessary to hydrolyze the carbohydrates to simpler forms. Stoichiometric reactions were used for the pretreatment reactor, enzymatic hydrolysis reactor, and fermentation tanks [4]. The simple sugars from the reactor are mixed with an aqueous stream containing water and nutrients for microbial fermentation. Following fermentation, the bio-based chemicals are recovered through the downstream process.

The downstream process is based on an industrial process, using two evaporation and crystallization steps, followed by discoloration, final crystallization, and drying [5]. The volume of the stream is reduced by 75% using the first evaporator, which has a threefold impact. Following the initial phase of crystallization, the crystals are taken out of the liquid and put through a basket filter before moving on to the decolorization stage. Another phase of single-effect evaporation and crystallization is applied to the permeate fraction. With 2% (w/v) activated carbon, the crystals are decolorized [1]. The overall recovery of IA from the fermentation product stream to the dried crystals is 69.14 % as shown in Figure 3. The economic evaluation of the biorefinery would be carried out by estimating the capital cost, operating costs, and profitability. Equipment purchase costs are estimated using the BIOSTEAM software. The estimation of the capital cost considered direct and indirect costs would be calculated based on literature data. The economic assumptions are based on work from previous researchers, using a 25-year operation period, a minimum return rate of 10%, income tax of 5%, and operating at 310 days. The goal of this TEA study is to obtain a competitive minimum product selling price comparable to the current market price of 1800 US$/t [3].

Acknowledgments: The authors gratefully acknowledge the funding for the project from the National Science Foundation (NSF) with NSF Award number 2242763. We thank our collaborators- Dr. Zengyi Shao, and her students, for performing the experimental work for the project.

References

1. Mieke Nieder-Heitmann, Kathleen F. Haigh, Johann F. Görgens. 2018. "Process design and economic analysis of a biorefinery co-producing itaconic acid and electricity from sugarcane bagasse and trash lignocelluloses.

2. Nisha Devi, Shubhangi Singh, Shivakumar Manickam, Natália Cruz-Martins, Vinod Kumar, Rachna Verma, and Dinesh Kumar, October 2022. "Itaconic Acid and Its Applications for Textile, Pharma and Agro-Industrial Purposes".

3. Richa Bafana and R. A. Pandey 2017. "New approaches for itaconic acid production: bottlenecks and possible remedies".

4. Andrés Suazo, Fidel Tapia 1, Germán Aroca 1 and Julián Quintero. 2023. "Techno-Economic and Life Cycle Assessment of a Small-Scale Integrated Biorefinery for Butyric-Acid Production in Chile".

5. V. F. PFEIFER, CHARLES VOJNOVICH, AND E. N. HEGER. 1952. "Itaconic Acid by Fermentation with Aspergillus Terreus".

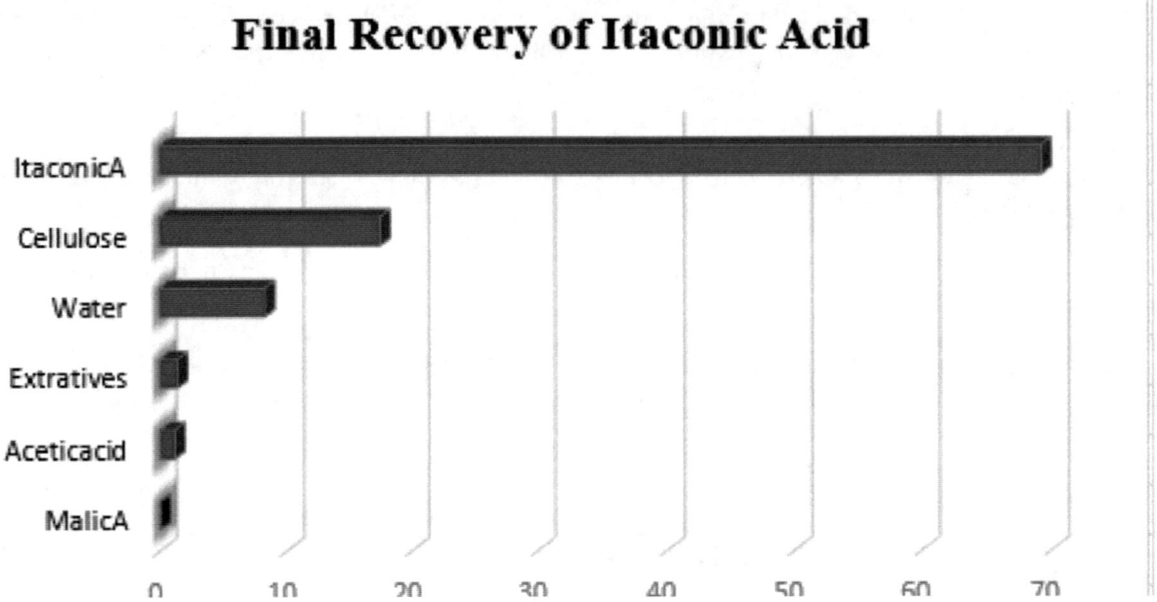
Final Recovery of Itaconic Acid
ItaconicA
Cellulose
Water
Extratives
Aceticacid
MalicA
0
10
20
30
40
50
60
70

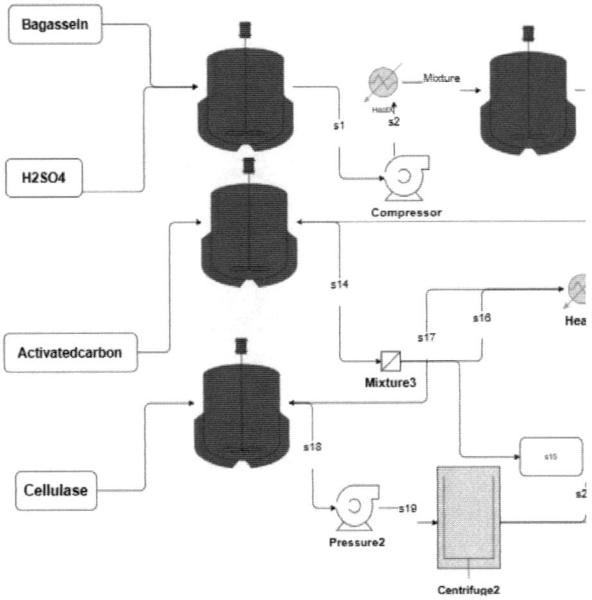
BagasseIn
H2SO4
Activatedcarbon
Cellulase
Mixture
s1
s2
Compressor
s14
s16
s17
Mixture3
s18
s19
Pressure2
Centrifuge2

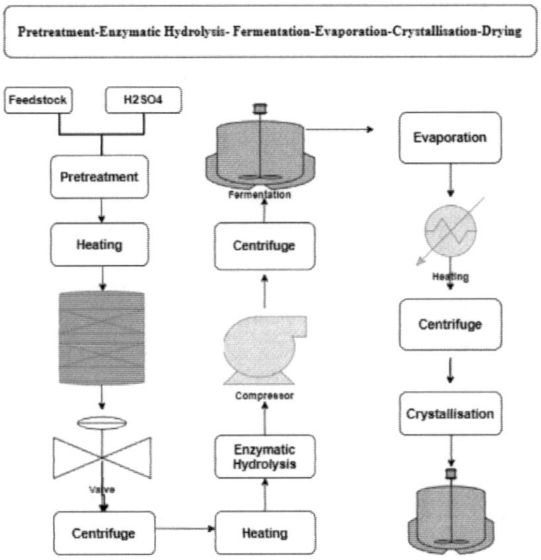
Pretreatment-Enzymatic Hydrolysis- Fermentation-Evaporation-Crystallisation-Drying
Feedstock
H2SO4
Pretreatment
Heating
Valve
Centrifuge
Heating
Enzymatic Hydrolysis
Compressor
Centrifuge
Fermentation
Evaporation
Heating
Centrifuge
Crystallisation

573f Optimizing Building Envelope Design for Sustainable Energy Performance

Sahar Fereidoni, Leila Fereidooni, Alibakhsh Kasaeian

This study investigates the impact of various building envelope design parameters on the embodied and operational energy consumption of a residential building. The key parameters examined include the glazing ratio of the southern and northern walls, glazing type, and wall assembly composition. The wall assembly variables encompass the thickness and material of the interior and exterior insulation layers, as well as the outermost layer material. Simulation results indicate that increasing the glazing ratio in the southern and northern walls from 0 to 80% leads to a 12.7% and 15.7% increase in total energy consumption, respectively. Additionally, using triple-pane windows instead of double-pane windows increases the embodied energy by 27.5%, but results in a 4.44% reduction in total energy consumption due to their superior thermal insulation properties. The simultaneous variation of glazing ratio and wall assembly revealed that the minimum possible total energy consumption over the building's life cycle can be achieved by using aerogel insulation with a thickness of 100 mm in the interior layer and 60 mm in the exterior layer, fiber cement panels as the outermost layer, and triple-pane windows. These findings highlight the importance of optimizing building envelope design to balance embodied and operational energy for sustainable building performance.

573g Redefining Lignin Valorization By Integrating Chemical Depolymerization and Microbial Funneling to Produce Specialty Bioproducts

Sarada Sripada, Samuel Scholz, Anna Qualls, Maximiliano A. Garcia, Bumkyu Kim, Timothy J. Donohue, Daniel R. Noguera, Steven D. Karlen, Canan Sener

To improve the sustainability and economics of lignocellulosic biorefineries, there has been a growing interest in upgrading lignin into bioproducts and specialty biofuels through integrated chemical and biological processes. We have developed a biomass-to-bioproduct pipeline that involves chemical fractionation and depolymerization of biomass by catalytic hydrogenolysis followed by microbial funneling (using an engineered microbe *Novosphingobium aromaticivorans*) of depolymerized lignin to produce 2-pyrone-4,6-dicarboxylic acid (PDC), a precursor for bioplastics.

This pipeline used reductive catalytic fractionation (RCF) to fractionate lignocellulosic biomass and depolymerize lignin into a mixture of monomeric and oligomeric phenolic compounds. The RCF process used methanol as a solvent to fractionate lignin from cellulose and hemicellulose and a Pd/C catalyst with hydrogen to reductively depolymerize the lignin. A solid carbohydrate pulp was obtained, which can be further upgraded into biofuel. The solvent used in RCF plays a critical role in fractionating and solubilizing the lignin and may significantly impact phenolic monomer/oligomer yields. While methanol is conventionally employed in RCF, the safety concerns and cost issues surrounding the use of methanol indicate the need for green alternatives with lower boiling points to lower the reactor pressure and, thereby, the reactor cost. In addition, compatibility with biological systems is a prerequisite for integrated chemical and biological processes. RCF was explored with ethanol, ethylene glycol, isobutanol, and aqueous mixtures of these solvents, and the lignin monomer/oligomer product yields per kg of biomass and product distributions were analyzed and compared. This research aimed to develop more efficient solvent systems for the RCF of lignocellulosic biomass to maximize the yield of lignin monomers while minimizing process economics and environmental impact.

573h Techno-Economic Analysis of Green Hydrogen Production in South Korea and Saudi Arabia, Using Monte Carlo Simulation Based on Meteorological Data

Yewon Kim, Sanghun Lee

Because of the ongoing climate crisis, hydrogen is increasingly recognized as a viable, sustainable energy solution. In particular, green hydrogen, produced using renewable energy sources, is considered a promising alternative to fossil fuels. Nonetheless, the inherent regional and seasonal variability of renewable energy sources directly impacts the production of green hydrogen, affecting its availability and cost-effectiveness on a regional and monthly basis. This study analyzed the levelized cost of green hydrogen (LCOH) in South Korea and Saudi Arabia, employing Monte Carlo simulations incorporating monthly meteorological data. This approach enabled an evaluation of how the variability and regional disparities in LCOH influenced the economic viability of hydrogen trade.

This research employed Monte Carlo simulations to assess the variability of the levelized cost of hydrogen (LCOH) due to solar energy fluctuations. Regionally averaged monthly data on global horizontal irradiance(GHI) and temperature from the National Solar Radiation Database NREL are used to model weather variability. The simulations estimated monthly solar power generation capacities and hydrogen production volumes. An economic analysis formula is then applied to these estimates, calculating the LCOH and presenting it in a probability distribution to highlight monthly cost variability.

The analysis showed that South Korea exhibited higher LCOH values than Saudi Arabia, with significant monthly differences in hydrogen production quantity and LCOH due to seasonal weather changes. In contrast, Saudi Arabia maintained relatively stable hydrogen production quantity and LCOH values. This suggested that the variability of weather conditions could significantly impact the production quantity and costs of green hydrogen.

This research forecasted the variability of green hydrogen production volumes and production costs in the two countries, offering suitable data to evaluate the economic feasibility of international green hydrogen trade and the transition to a green hydrogen economy. Additionally, this study found the importance of considering monthly weather variability for stable green hydrogen production and economic viability improvement. It was also expected to contribute to establishing variability management strategies for hydrogen production and import.

573i Advancing Green Ammonia Synthesis in Dynamic Offshore Wind Power-to-X Systems

Philipp Rentschler, Niklas Brinker, Christoph Klahn, Roland Dittmeyer

In this study, the feasibility and dynamic operability of a Power-to-Ammonia process chain, in particular its integration into offshore wind farms, was analyzed in detail. Using the dynamic process simulation software AVEVA Process Simulation® (APS), simulation-based analyses of the process were carried out under different conditions, such as full and partial load, in order to gain a deeper understanding of the transient behavior of the synthesis and the entire offshore process chain [1].

In the context of global warming, this approach also represents an important step towards reducing greenhouse gas emissions, particularly in chemical production. The central role of ammonia since its synthetic production in 1913 by Fritz Haber and its lasting impact on agriculture and human civilization underline the importance of this study [2]. Since 72% of hydrogen in the conventional process chain is produced by steam methane reforming and 26% by coal-fired power plants, ammonia synthesis contributes about 1-2% of global carbon dioxide emissions [2], [3]. Despite the century-long dominance of the Haber-Bosch process in global ammonia production, the need for alternative, emission-free synthesis routes are becoming apparent, especially in view of the growing world population and the associated increase in demand for food and ammonia-based fertilizers. Coupling ammonia synthesis with electrolyzers to produce green hydrogen therefore offers a major lever for reducing emissions in this important industrial sector.

The implementation of an offshore production facility is an innovative solution that avoids the need for conventional infrastructure to transport electrical energy to shore and can use existing material transport routes such as pipelines or tankers. In the German hydrogen flagship project H2Mare, the environmentally friendly offshore production of hydrogen and hydrogen derivatives is being investigated by an interdisciplinary consortium [4]. Despite its many advantages, offshore integration poses a number of challenges, particularly in the development of chemical processes that are confronted with unstable energy and raw material sources, as well as unusual weather and site conditions [5]. A major hurdle is the costly connection of wind farms to the mainland via submarine cables and offshore substations. An island solution that avoids direct grid connection therefore offers potential cost advantages and the opportunity to develop remote offshore sites. This option is particularly attractive due to the above-average and consistent wind speeds at sea, allowing wind farms to operate for longer periods compared to onshore alternatives [5].

The design process was as follows: the individual sub-systems consisting of the wind turbine, electrolyzer and synthesis plant were modelled and simulated individually, optimized and then the overall performance was evaluated in a coupled manner (see figure). The system was designed for a nominal output of 100 MW electrolysis nominal load.

The initial focus of the design process was on the development and modelling of the sub-system of the dynamic ammonia synthesis process without coupling with the wind turbine and electrolyzer, and the subsequent comparison with conventional stationary plant concepts. This design includes two reactor bed arrangements with an intermediate heat exchanger to efficiently model an intercooled reactor. The adiabatic reactors were modelled using the empirical Temkin-Pyzhev rate equation for a wustite iron catalyst [6]. A key aspect was the separation of the gaseous product from the unreacted reactants by two-phase separation using cooling and condensation. A recycle stream was provided to increase the overall efficiency. The design includes control loops and buffer tanks to stabilize operation under fluctuating conditions. A dual compressor pressure stabilization strategy was also designed and implemented. The compressors and control valves were validated using commercially available components. In addition to controller tuning, the ammonia synthesis sub-system was then optimized under steady-state conditions to maximize overall efficiency.

The next step was to integrate the previously developed model into the dynamic Power-to-Ammonia process chain, which simulates the direct coupling of wind turbines with electrolyzers and ammonia synthesis plants in off-grid systems. The APS software was used to enable comprehensive simulation of the entire periphery, including pumps, compressors, heat exchangers, kinetic reactions, and electrical components. This detailed analysis provided insight into the transient behavior of the overall system. In particular, the adaptability of the system to volatile H_2 injection flows and the ability to operate over a 24-hour cycle under real wind conditions were analyzed. The hydrogen flow was reduced from 18,000 Nm^3/h to 4,000 Nm^3/h, corresponding to a reduction in ammonia production from approximately 235 t/d to 47 t/d. The nitrogen flow was matched to the volatile hydrogen flow to ensure a constant feed ratio of 3:1. In a further step, a hydrogen storage tank was implemented and evaluated as a function of the H_2 storage pressure to counteract fluctuations in the volatile feed flow and ensure stable operation. For the isolated ammonia synthesis sub-system, an efficiency of 53% at full load and 60% at partial load (50% load) was calculated, as less cooling capacity is required. In coupled operation of the entire power-to-ammonia process chain with the electrolyzer, the efficiency of the entire process chain is 37%, as the peripheral energy consumption and efficiencies of the electrolyzer also play a role.

The final step was an optimization loop in which a bi-directional heat exchanger was incorporated into the reactor configuration to increase efficiency. This heat exchanger uses the hot gas flow from the first reactor, heated by the exothermic reaction, to preheat the inlet flow of the same reactor. This strategy allows efficient heat recovery and contributes to the energy efficiency of the overall process. An electric heat exchanger was also used to precisely control and regulate the reactor temperature. In addition to ensuring the desired operating parameters, this also allows temperature control during the start-up process after a plant shutdown. The comparative analysis of this arrangement with the originally proposed design showed a significant improvement in efficiency. Efficiencies of approximately 43% at full load and 35% at partial load were achieved for the overall process.

In summary, this study investigated the integration of a Power-to-Ammonia process chain with offshore wind parks and electrolyzers. Using dynamic flow sheet simulations, the individual sub-systems were modelled and optimized for efficient and dynamic operation. The overall efficiency of the power-to-ammonia process chain was 43% at full load and 35% at partial load. This emphasizes the transient operation of ammonia synthesis under fluctuating conditions and demonstrates an innovative approach to future-proof, sustainable and dynamic energy and chemical production that adapts to the dynamic requirements of renewable energy sources.

[1] "AVEVA Process Simulation." Accessed: Apr. 01, 2023. [Online]. Available: https://www.aveva.com/en/products/process-simulation/

[2] S. Ghavam, M. Vahdati, I. A. G. Wilson, and P. Styring, "Sustainable Ammonia Production Processes," *Front. Energy Res.*, vol. 9, p. 580808, Mar. 2021, doi: 10.3389/fenrg.2021.580808.

[3] J. Humphreys, R. Lan, and S. Tao, "Development and Recent Progress on Ammonia Synthesis Catalysts for Haber–Bosch Process," *Adv Energy and Sustain Res*, vol. 2, no. 1, p. 2000043, Jan. 2021, doi: 10.1002/aesr.202000043.

[4] Federal Ministry of Education and Research, "How partners in the H2Mare flagship project intend to produce hydrogen on the high seas." [Online]. Available: https://www.wasserstoff-leitprojekte.de/projects/h2mare

[5] P. Rentschler, C. Klahn, and R. Dittmeyer, "The Need for Dynamic Process Simulation: A Review of Offshore Power-to-X Systems," *Chemie Ingenieur Technik*, vol. 96, no. 1–2, pp. 114–125, Jan. 2024, doi: 10.1002/cite.202300156.

[6] M. I. Temkin and V. Pyzhev, "Kinetics of Ammonia Synthesis on Promoted Iron Catalysts," *Acta Physicochimica U.R.S.S.*, vol. 12, pp. 327–356, 1940.

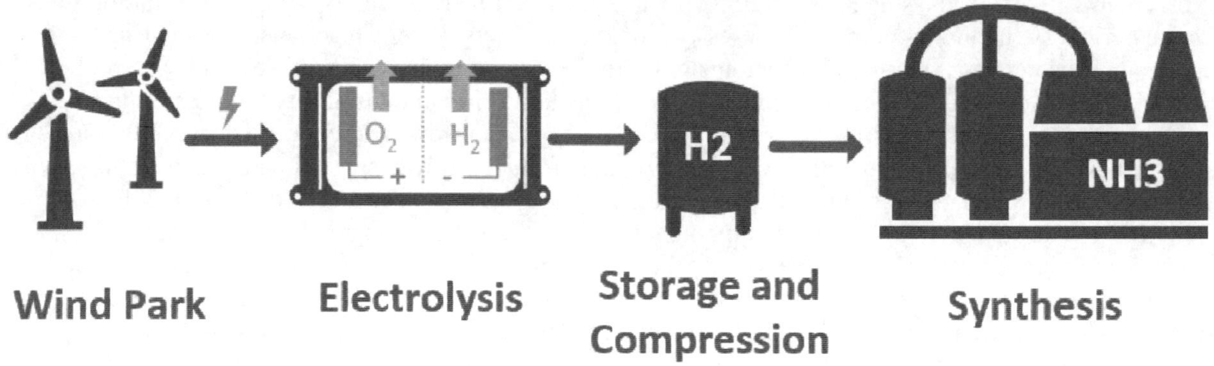
O2
H2
+
-
H2
NH3
Wind Park
Electrolysis
Storage and Compression
Synthesis

573j A Case Study of Valorizing Landfill Gas to Renewable Energy in an Industrial Zone

Jian Fang, Helen Lou, Andrew W Kolp, Daniel Chen

Landfills contribute to climate change primarily through the release of greenhouse gases, particularly CH_4. The U.S. EPA actively works to minimize landfills' carbon footprint and promote renewable energy projects utilizing landfill gas (LFG). This study investigates four LFG conversion strategies, flaring, electricity generation, renewable natural gas (RNG), and methanol production in Southeast Texas. The research assesses capital investments, operational costs, carbon footprint reduction, and potential profits for these strategies. The electricity generation strategy stands out, achieving profitability from the first year, projecting a peak profit after 10 years. In contrast, the RNG strategy, while eventually profitable, takes longer to break even due to a higher initial investment, becoming profitable in its tenth year without carbon credit trading or in its fifth year with trading ($40/ton). The RNG strategy leads to carbon footprint reduction, cutting 1,073,195 tons of CO_2 equivalent by 89% in the 10th year. Additionally, this study explores methanol production strategy and highlights the potential synergy between landfill facilities and the hydrocarbon industry.

573k Assessing the Sustainability of Recovering Rare Earth Elements from a Toxic Wastewater Slurry: A Combined Life Cycle Assessment and Technoeconomic Analysis Study

Adam Smerigan, Rui Shi

Phosphogypsum (PG) is a waste generated from fertilizer production consisting primarily as a slurry of gypsum, acidic water, and various heavy metals (e.g., U, Th, rare earth elements). PG is stored indefinitely in 'stacks', which slowly release these toxic and radioactive chemicals into the environment. Since PG is generated at hundreds of millions of tonnes per year, indefinite storage in stacks is quickly becoming untenable necessitating new treatments. Some potential new treatments generate useful products for use in road construction and agriculture, but are not economically feasible. A new treatment that recovers high-value rare earth elements (REEs) from PG could make PG remediation profitable.

REEs (consisting of lanthanides, scandium, and yttrium) are considered critical materials and are used in a variety of modern technologies including wind turbines, lasers, consumer devices, and electric vehicles. The current production of REEs is achieved by an energy and chemically intensive process (beneficiation, leaching, separation by solvent extraction, and refinement) from the mining of REE ores in primarily one geopolitical region. Investigations into a more sustainable supply of REEs from abundant secondary sources, such as toxic PG waste, is vital to securing a stable global REE supply chain. With a minimum REE content of ~0.02 wt% rare earth elements (REEs), PG has the potential to meet the annual rare earth element consumption within the United States (~9,000 tons REE/year) while also making PG remediation profitable. Developing a sustainable source of REEs from PG will enable clean and efficient transition to green energy technologies.

Currently, it is unclear how to best recover the dilute REEs from the PG waste (conventional solvent extraction is inefficient and has high environmental impact). In this work, we propose a treatment train (technological readiness level < 3) for the recovery of REEs from PG which includes a bio-inspired adsorptive separation to generate a stream of pure REEs. We assessed its financial viability and life cycle environmental impacts via life cycle assessment (LCA), techno-economic analysis (TEA), and identify targeted improvement opportunities through global uncertainty/sensitivity analysis and scenario analysis. We determined that this system can be profitable (net present value of $200 million, internal rate of return of 17%, and minimum selling price (MSP) of $48/kg REO) and shows reduced environmental impact in several ReCiPe 2016 environmental impact categories (land use, ecotoxicity, eutrophication, ionizing radiation, material resource depletion, and human toxicity) when compared to conventional REO production and PG stacking. However, the system underperforms in other impact categories (climate change, terrestrial acidification, ozone depletion, photochemical oxidant formation, particulate matter formation, and fossil resource depletion). The endpoint analysis shows that the system outperforms conventional REO production in ecosystem quality (93.0% of the impact) and resource depletion (96.2% of the impact) but underperforms in human health (213% of the impact). The life cycle environmental impacts and financial viability are primarily driven by chemical consumption in the leaching, concentration, and wastewater treatment process sections ($15/kg REO). In addition to high chemical costs, the large capital cost of the selective adsorption resin ($18/kg REO) limits profitability. Scenario analysis shows that the system is profitable at capacities larger than 100,000 kg/hr PG for PG with REE content above 0.5 wt%. The most dilute PG sources (0.02-0.1 wt% REE) are inaccessible using the current process scheme (limited by the cost of acid and subsequent neutralization) requiring further examination of new process schemes and improvements in technological performance. Overall, this study evaluates the sustainability of a first-of-its-kind REE recovery process from PG and uses these results to provide clear direction for advancing sustainable REE recovery from secondary sources.

573l Advancing Automotive Plastics Recycling through Molecular Recycling Technologies: Innovations for a Circular Economy and Sustainability

Mahshid Mokhtarnejad, Gary Hawkins, Hendrik Mainka

The urgent need to address the global plastic waste crisis and its impact on climate change has created a demand for a materials revolution. Volkswagen Group of America Inc. (VWGoA) and Eastman Chemical Company are addressing the global plastic waste crisis by advancing molecular-level chemical recycling technologies. This innovative method allows for the conversion of hard-to-recycle automotive plastics into raw high-quality materials for new specialty plastics and fibers without compromising quality or performance. We focus on various plastic waste streams, including mixed plastics, colored materials, car carpets, and textiles. Our collaboration addresses recycling challenges and provides environmental metrics, contributing to global waste management efforts and reduced carbon footprint. This process offers a sustainable alternative to virgin fossil-based plastics, crucial for reducing carbon footprints and diverting automotive waste from landfills. By implementing this technology, VWGoA has reduced greenhouse gas emissions by 25-50%. Examples like the Atlas Cross Sport Spoiler Assembly and Passat R-Line Diffuser demonstrate successful mixed plastic recycling, showcasing its effectiveness and sustainability through a circular economy in automotive applications. This study underscores the technical feasibility and commitment to circular economy principles within the automotive industry.

573m Exploring the Effect of Facility Construction in Biofuel Life Cycle Analyses through Greenhouse Gas Emissions

Huy Nguyen, Rui Shi

Life cycle analysis (LCA) has been widely used as a crucial tool to determine the environmental benefits of biorefineries. While extensive LCA research has been conducted on the environmental footprints of various biofuels and bioproducts, the environmental impact of facility construction has not been well understood and quantified in the field of biorefinery LCA.

Most LCAs on biofuels and bioproducts to date neglect the effect of facility construction in their assessment. In the regulatory domain, the Environmental Protection Agency (EPA) implemented the Renewable Fuel Standard (RFS) to promote biofuel use. The guideline for biorefinery LCA for the fuel production and distribution stage under this standard particularly specified that emissions associated with physical infrastructure (i.e., facility construction) are not included in the EPA's assessment[1]. On the other hand, no quantitative studies have been done to verify or support such assumptions. The goal of this study is to address this gap by quantifying the Greenhouse gas (GHG) emissions associated with the construction of equipment used in biorefineries to validate such assumptions.

This study investigates two bioprocess pathways based on technical reports from the National Renewable Energy Laboratory (NREL) and examines the potential emissions from biorefinery infrastructure. The equipment sizing for pressure vessels and storage tanks is performed following the ASME Boiler and Pressure Vessel Code and the API 650, respectively. GHG emissions associated with the production of pressure vessels and storage tanks are estimated based on equipment weight and emission factors of construction materials from Norgate et al. (2007) and the Ecoinvent database. Emissions related to pump and conveyor belt production are determined on OpenLCA with the Ecoinvent database. Four different methods are used to determine the weight and material consumption of pressure vessel manufacturing, thus further characterizing the GHG emissions from biorefinery equipment construction. This study also conducts a scenario analysis on emissions calculated with three different head types for pressure vessels. Additional scenario analysis is performed with emission factors for carbon steel and stainless steel production from Norgate et al (2007), the Ecoinvent database, Jing et al. (2019), Burchart-Korol (2013), Renzulli et al. (2016), and the Greenhouse gases, Regulated Emissions, and Energy use in Technologies (GREET) model.

Preliminary results indicate equipment construction contributes a small fraction to the baseline biorefinery GHG emissions. This fraction is higher as more environmentally friendly approaches are utilized to produce biofuel, but it is still low. Furthermore, the GHG emissions from equipment construction vary depending on the method used to calculate pressure vessel weight. The variation primarily stems from the undefined design temperature, design pressure, and dimensions of the pressure vessels and the heuristics each method employs to find these unspecified parameters. Also, there is a significant deviation in GHG emissions from equipment construction between using the Flanged and Dished head and the other two head types (Hemispherical and 2:1 Semi-Elliptical) for all the pressure vessels in the plant. This is because the Flanged and Dished wall thickness is higher than the thickness of the other two head types which leads to higher head weight. Additionally, emissions also vary with different data sources for stainless steel and carbon steel emission factors, which is due largely to deviations in emission factors and differences in carbon steel and stainless steel weight fractions.

The results of our study suggest the EPA's assumption to exclude facility construction from its biorefinery LCA requirement under the RFS is justifiable. Future studies should expand the scope of the LCA and incorporate a landscape of biorefinery design scenarios to have a holistic understanding of the environmental impact of biorefinery facility construction.

1. US EPA. *Lifecycle Analysis of Greenhouse Gas Emissions under the Renewable FuelStandard.* https://www.epa.gov/renewable-fuel-standard-program/lifecycle-analysis-greenhouse-gas-emissions-under-renewable-fuel (accessed 2023-05-08).

573n Electrically Enhanced Thermochemical CO_2 Splitting for Fuel Production

Jayni Hashimoto, Jordan Monroe, Alicia Bayon, Alonzo Mendez, Olivia Tamburro, Christopher L. Muhich

Thermochemical gas splitting has been identified as a promising method of harnessing solar energy to produce solar fuels. In thermochemical CO_2 splitting metal oxides, MO_x are thermally reduced (T > 1200 °C) to $MO_{x-\delta}$ using concentrated solar energy, followed by an oxidation step with CO_2 to regenerate MO_x and form CO, which can be mixed with H_2 to form liquid fuels. However, solar thermochemical processes are plagued by high solar field costs due to the extraordinarily high reduction temperatures required. Ceria has been identified as the state-of-the-art material due to high crystal and cyclic stability, and fast oxidation kinetics. Methods of lowering the reduction temperature of ceria have been well-studied, but can have detrimental effects on the CO production rates. Previous studies have demonstrated the ability to reduce ceria under only the application of a strong electric field. The E-field induced reduction of ceria changes the chemical potential of lattice oxygen, making the formation of O vacancies more favorable. As the formation of O vacancies is tunable with the applied potential, removal of the potential from the system, allows ceria to re-establish its original thermodynamics and retain oxidation favorability. One major challenge to the usage of an electric field in thermochemical systems is the method by which a sufficiently large field is applied on a large-scale. In this work we use molten salts to generate an electronic double layer (EDL) at the electrolyte-ceria interface. The EDL acts as a parallel plate capacitor, allowing for large specific capacitances to amplify low applied voltages and produce large electric fields to drive the reduction of ceria. We demonstrate an electrically enhanced two step thermochemical CO_2 splitting process, where a low potential is applied to the ionic liquid to generate an electric field via EDL to reduce ceria, then the field is removed and ceria is oxidized using CO_2, producing CO. Specifically, we show molten salt compositions compatible with generating the electronic double layer with CeO_2 and their ability to generate high electric fields. We also demonstrate the ability of these EDL electric fields to reduce ceria at temperatures below 1000 °C and perform isothermal CO_2 splitting cycles. By allowing for significant reductions in the needed temperatures, the E-field enhanced CO_2 splitting process significantly drives down the CO production costs as the number of heliostats needed decreases, the radiative losses are lowered, and the constraints on materials of construction are loosened.

573o Tribological Properties of Bio-Based Lubricants Obtained from Castor Oil Fatty Acids and Isopentanol

Marliete Melo Neta, Igor Figueredo, Paulo Roberto C. F. Ribeiro Filho, Celio Cavalcante Jr., Francisco Murilo Tavares de Luna

Chemically modified vegetable oils may be used for several applications such as lubrication. Some bio-based lubricants may reduce friction and prevent direct metal-to-metal contact through a thin film, with polar groups attached to long carbon chains that enable the oil molecules to cling to metal surfaces while sliding over themselves. In this study, a bio-based lubricant synthesized from castor oil using isopentanol was evaluated for thermal stability and tribological properties. The bio-lubricant sample was obtained from an esterification reaction between Castor oil fatty acids and isopentanol, followed by epoxidation and epoxide ring-opening reactions using the same alcohol as the nucleophilic agent. A Thermal Gravimetric Analyzer (TGA) Model DTG-60 (Shimadzu, Japan) was used to monitor the weight changes of the samples, while the derivative of a TG curve (DTG) evaluated the rate of weight loss during heating versus temperature, from 30 °C to 700 °C at a constant heating rate of 10 °C/min. For tribological behavior, four-ball experiments were carried out using a DHR-3 rheometer (TA Instruments, New Castle, USA), with experimental conditions modified from the ASTM D4172 method within the limits of the rheometer. The tribological tests were conducted at a sliding velocity of 0.46 m/s, a temperature of 75 °C, a load force of 55 N and for 1 h (Figure 2). The balls are made of a chrome steel alloy (AISI52100, 64 HRC), with diameter of 12.7 mm and surface roughness of 0.015 μm. An optical microscope (Zeiss, Germany) with 100× zoom was used to study the morphology and measure the wear scar diameter (WSD). Samples of Castor oil fatty acids and Mineral Oil were used as reference for tribological comparisons with our synthesized sample.

573p Studying Precipitation Rates Under Semicontinuous Operation of a CO_2 Mineralization Reactor Utilizing Flue Gas and Produced Water

Quinn Bennett

Flue gas, a high concentration point CO_2 waste stream created across many industries, offers an opportunity for capturing CO_2 at rates much greater than direct air capture. In this work, flue gas is contacted with produced water, a brine waste produced from hydraulic fracturing (fracking) that contains high concentrations of Ca^{2+} ions, to mineralize CO_2, forming calcium carbonate ($CaCO_3$).[1] This mineralization route simultaneously sequesters CO_2 into a valuable commodity product, mitigates two common waste streams, and can utilize "off-the-shelf" technologies including common crystallization reactors for mineralization and a tandem chlor-alkali cell to provide hydroxide. Herein the design and operational conditions of the crystallizer are studied, and the chlor-alkali cell is simulated with a concentrated NaOH stream.

Crystallizer parameters including CO_2 concentration in flue gas, temperature, and reaction pH were studied in a two-level design of experiments. The $CaCO_3$ precipitation rate and the material properties of the collected product were characterized. Rates as high as 16.59 ± 0.37 $\mu mol.L^{-1}.s^{-1}$ were sustained in a bench scale (1 L) crystallizer allowing full recovery (~100% yield) of Ca^{2+} as $CaCO_3$. The reactor design also influences the crystal structure of the collected precipitate. Examples of designs that will produce a majority aragonite (orthorhombic) structure as opposed to the more stable and valuable calcite (trigonal) structure will be shown. Effects of reactor parameters on particle size and the influence of other divalent cations (Mg^{2+}, Sr^{2+}, Ba^{2+}) were also studied. Overall, this study of CO_2 mineralization utilizing flue gas and produced water is aimed at providing the necessary information to scale this process to a pilot scale (200 L crystallizer) capable of mineralizing >10 kg CO_2/day as well as informing techno-economic analyses of this promising process.

1. Zhu, B.; Wilson, S.; Zeyen, N.; Raudsepp, M. J.; Zolfaghari, A.; Wang, B.; Rostron, B. J.; Snihur, K. N.; von Gunten, K.; Harrison, A. L.; Alessi, D. S. Unlocking the Potential of Hydraulic Fracturing Flowback and Produced Water for CO2 Removal via Mineral Carbonation. *Applied Geochemistry* **2022**, *142*, 105345. https://doi.org/10.1016/j.apgeochem.2022.105345.

573q "a Review of Existing Data and ASTM E681 Testing for Flammability Limits of Hydrofluorocarbons and Their Mixtures: Towards a Predictive Model"

Kevin Turner

Flammability limits of gases and their mixtures are critical parameters for ensuring the safety and efficiency of various industrial processes, particularly in refrigeration systems. The ASTM E681 standard provides a reliable methodology for discerning these limits experimentally, yet there is a pressing demand for predictive models capable of quickly and accurately estimating flammability limits. With the advent of next generation refrigerants, characterized by their lower global warming potential (GWP), a thorough understanding of the flammability characteristics of both current and low-GWP refrigerants is essential for the transition to novel refrigeration systems.

This research focuses on collecting reliable experimental data regarding the flammability limits of refrigerants currently in circulation: pure hydrofluorocarbons (HFCs) and their mixtures. Through meticulous literature review, we investigate the consistency and reliability of existing flammability data. Moreover, we adopt ASTM E681 testing guidelines in the design of an experimental apparatus, considering the effect of temperature, pressure, humidity, and composition on the flammability limits of HFC mixtures.

In this study, we aim to develop a model for predicting the flammability limits of pure refrigerant gases and their mixtures under varying conditions of temperature, pressure and composition. Our approach involves a comprehensive literature review, data collection of experimental flammability limits, and a rigorous statistical analysis to identify correlations between flammability limits and thermodynamic properties.

Through this multi-faceted approach, we seek to establish a comprehensive understanding of the flammability limits of refrigerant gases and mixtures. Ultimately, the goal is to provide engineers and safety professionals with an efficient tool for assessing flammability hazards, thereby facilitating the design and operation of safe and sustainable refrigeration systems.

573r Excess Absorption of Type II Porous Liquids for CO_2/H_2 Separation

Lu Lu, Ryan Lively

Porous liquids are believed to have potential in gas separations and can combine the benefits of the fluidity of solvents and the high capacity of porous solids. Type II porous liquids are rigid, discrete, porous host molecules dissolved in solvents that are too bulky to enter the pores. Porosity can be maintained in the liquid state, and the gas solubility in these porous liquids can be drastically increased over that of pure solvents. While Type II porous liquids have been primarily investigated for their synthesis and the existence of porosity, few isotherms have been reported to show the gas absorption ability of existing Type II porous liquids. Therefore, this work focuses on studying the CO_2 sorption behavior of Type II porous liquid and utilizes solution phase and adsorption thermodynamics to understand the role of porosity in Type II porous liquid. We measure the sorption isotherms of Type II porous liquid based on a specific porous organic cages CC13 and three solvents, namely 2'-hydroxyacetophenone (2HAP), 2-isopropylphenol (2IPP), and 2-chlorophenol (2CP) with different weight loadings. The excess absorption of all PLs is calculated to elaborate the absorption deviation from the additive combination of the cage and the solvent. A non-porous control molecule and CC13 with different crystalline structures are used to probe the mechanism of the positive deviation from volume additivity. Another POC, CC19, is first used to form Type II porous liquids and excess absorption can also be observed in the new PLs. The results of this study provide valuable insights into the potential gas separation applications of Type II porous liquids.

573s Perovskite Oxides As New Family of Tunable CO_2 Sorbents

Seyedamin Razavi, Vahid Rahmanian, Mahe Rukh, Runxia Cai, Mohammadreza Kosari, Saad A. Khan, Fanxing Li

CO_2 adsorption by solid sorbents represents an attractive option for carbon capture due to its potential for low energy consumption, ease of operation, and minimal corrosion or sorbent loss. This study introduces perovskite oxides as a new family of highly tunable solid sorbents. Despite a relatively limited sorption capacity (~up to 0.68 wt.%) which ties to their surface area, perovskite oxides offer remarkable structural and compositional flexibility to tailor their sorption thermodynamics and kinetics. Using $Sr_xLa_{1-x}FeO_3$ (x=0, 0.2, 0.5, and 0.7) as a model system, we demonstrated that varying the A-site composition of the perovskite oxide leads to substantial change in the CO_2 adsorption and desorption behavior, allowing a tunable CO_2 release within a temperature range of 180 to 500+°C. A strong correlation between the oxide surface area and sorption capacity was also established. Despite the low surface area inherent to perovskite oxides, we managed to enhance their surface area from 3 to 5 m^2/g (prepared by a salt assisted reactive grinding method) to ~30 m^2/g using a electrospinning method. $Sr_{0.2}La_{0.8}FeO_3$ prepared via electrospinning exhibited a relatively moderate desorption temperature (~240°C) and a CO_2 sorption capacity of 0.68 wt.%. Based on analysis of the adsorption isotherms, the $Sr_{0.2}La_{0.8}FeO_3$ sorbent adsorbs CO_2 via chemisorption at low pressures (0-1 kPa) and it switches to physisorption at higher CO_2 partial pressures (10-120 kPa). TEM characterizations revealed that perovskite synthesis via the electrospinning method results in the formation of perovskite nanorods, which exhibit high surface area and unique surface properties compared to oxides prepared with conventional methods. This study demonstrates the tunability of perovskite oxides as a new family of CO_2 sorbent materials and the potential to further enhance their surface area towards practical applications in CO_2 capture and utilization.

573t Assessing and Unlocking the Potential for Carbon Mineralization in Peridotite Rock for Carbon Dioxide Removal

Soyoung Choi, Ah-Hyung Alissa Park

Climate change has reached a critical level, necessitating urgent actions in the form of carbon dioxide removal (CDR) technologies from the air. Among these, carbon mineralization process, especially in-situ carbon mineralization, is a critical technology for CDR due to its potential for large-scale storage capacity and security at relatively low cost. Peridotite rock, known for its high potential and reactivity, remains unexplored for engineered in-situ carbon mineralization site. Additionally, peridotite rock typically contains Ni at around 2000 ppm, making it a potentially valuable unconventional source for energy-relevant elements. In this regard, this study aims to investigate the parameters influencing the reactivity of peridotite and how these parameters alter the morphological and chemical properties of solid, as well as the aqueous chemistry due to the reactions at rock-H_2O interfaces. While these parameters are inherently interdependent, this study focuses on elucidating the effect of each parameter, such as pH, temperature, and water composition using spectroscopic and electron microscopic analysis. Moreover, we seek to assess the impact of each parameter to gain insights into the feasibility of in-situ carbon mineralization of peridotite rock-host reservoir and the sensitivities of the related parameters during geological CO_2 storage. By shedding light on these factors, our study aims to contribute valuable knowledge to the development of effective strategies for geological CO_2 storage with comprehensive consideration of aqueous chemistry, thereby aiding in climate change mitigation.

573u Surface Wettability and Pore Size Effect on CO_2 Diffusivity in Water-Saturated Kaolinite Nanopores

Minjunshi Xie, Zhehui Jin

Geological CO_2 sequestration (GCS) is a promising method to alleviate CO_2 emission and induced dire consequences. Evaluating CO_2 potential leakage would be of great importance as the ultimate goal of GCS is to permanently store CO_2 in geological locations. Diffusive loss through caprock is identified to be one of the possible ways for CO_2 leakage, which is closely associated with CO_2 diffusivity. Caprocks typically consist of clay minerals and contain a considerable number of nano-scale pores. Besides, caprocks are saturated with formation water. Even though CO_2 self-diffusivity and diffusivity in water/brine are widely studied, its diffusivity in water-saturated nanopores remains less understood.

In this work, we use molecular dynamics (MD) simulations to study CO_2 diffusivity in water-saturated nanopores under *in-situ* GCS condition. Kaolinite, a common type of clay minerals in caprocks, is used as substrate. It has surfaces of different wettability and pore sizes are also considered in our simulation set-up. We find that CO_2 diffusivity is higher in water-saturated hydrophobic confinements than in hydrophilic ones. It is further reduced by stronger surface hydrophilicity due to increasing content of H_2O. Moreover, smaller pore size can inhibit diffusion of CO_2 as the result of stronger surface-CO_2 interactions. But there is a sudden drop for CO_2 diffusivity in all types of confinements when the pore width is smaller than 1.0 nm as all the molecules are adsorbed on the surface. Our study should provide fundamental understanding into the effect of surface wettability and pore size on CO_2 diffusion in caprock, which may help evaluate the CO_2 leakage in GCS projects.

573v Hydrogen Storage in Silica-Kaolinite Shale and Its Impact on the Integrity: Molecular Modeling and Experimental Investigation

Alankaa Al-Harbi, Mohammed Jaber Almarri, Safwat Abdel-Azeim, Mohammed Ali Saleh, Ibnelwaleed Hussein

Hydrogen has emerged as an attractive alternative to traditional fossil fuels, offering potential as both a primary energy source and a versatile industrial raw material. Its attractiveness stems from its high specific energy capacity of 120-142 MJ/kg, widespread availability, and clean combustion product of only water vapor. However, challenges such as its lower volumetric and energy densities at room temperature, requirements for high-pressure vessel storage conditions (typically operated at 700 bar), and susceptibility to hydrogen embrittlement have complicated efforts to manage seasonal fluctuations in hydrogen supply and ensure long-term energy security. In response to these challenges, researchers have increasingly turned their attention to underground storage as a potential candidate. Previous investigations have explored hydrogen storage within pores of various minerals, revealing differences in uptake rates. Notably, studies have observed that hydrogen uptake tends to follow the order of illite > kerogen > kaolinite > quartz > calcite. However, much of the existing research has focused on the processes of adsorption and injection, often examining specific scenarios such as storage within calcite-rich carbonate or interactions with shale organic phases in the presence of shale gas.

Therefore, this research explores the potential to store hydrogen in a depleted shale gas reservoir which is silica (quartz (SiO_2)) -kaolinite($Al_2Si_2O_{54}$) interface as it is closer to the natural shale reservoir. This investigation aimed to study the hydrogen adsorption affinity, H_2 reactivity in various conditions (temperature, pressure, gas mixtures, wet/dry environments) both theoretically and experimentally. The work is divided into several stages. Initially, Density Functional Theory (DFT) is employed to explore hydrogen adsorption, surface chemistry, and active adsorption sites on the interface. Subsequently, Reactive Force Field molecular dynamics (ReaxFF MD) is used to get an insight into hydrogen reactivity and replacement within the interface, considering a pure hydrogen environment as well as the presence of gas mixtures (CH_4/CO_2) and water vapor under typical of regional reservoir conditions (450 K and 500 bar). Finally, molecular modeling coupled with the experimental investigation to assess the impact of hydrogen adsorption/desorption on the shale integrity. This investigation spanned a range of temperatures from 50 to 100°C and pressures from 20 to 100 bar for understanding hydrogen behavior in this context. Then the shale samples were characterized through methods including Fourier transform infrared (FTIR) spectroscopy, X-ray diffraction (XRD), scanning electron microscope (SEM), and X-ray photoelectron spectroscopy (XPS) to a comprehensive understanding of how hydrogen interacts with shale under different conditions. The initial results revealed the physisorption of hydrogen on the shale surface, supported by low adsorption energies (0.04 to -0.21 ev) and by Bader charge analysis indicating no charge transfer. This work has successfully shown the potential of depleted shale gas reservoirs to store hydrogen as less hydrogen will penetrate through the shale surface so hydrogen loss risk will be reduced, and it can be stored to supply the energy demand.

573w Optimizing the Reverse Supply Chain Network for Photovoltaic Panels Under Uncertainty

Funda Iseri, Halil Iseri, Natasha Chrisandina, Shivam Vedant, Eleftherios Iakovou, Efstratios Pistikopoulos

Within the global energy transition, there is a remarkable surge in the investment in photovoltaic (PV) systems. This exponential growth has elevated solar PV as a favorite among renewable technologies, surpassing even wind power [1]. However, this rapid expansion brings to light the urgent need for innovative end-of-life (EoL) management strategies tailored for solar panels [2]. Hence, designing a robust reverse supply chain for PV panels is crucial for efficient waste management, recovery of valuable resources, cost reduction, sustainability alignment, and regulatory compliance, all of which contribute to long-term environmental and economic goals. This calls for the creation of a system that can efficiently manage the flow of materials, processes, information, and finances in the reverse direction from the traditional forward supply chain.

In this research, we present a framework for modeling and optimizing decisions for recovering valuable materials from EoL PV panels and reintroducing them into the production cycle, thereby promoting a circular economy. A mixed integer optimization model is proposed to i) configure the optimal network of collection centers, recycling facilities and manufacturing sites along with the selection of various recycling technologies and transportation routes[3], and ii) address the uncertainties inherent within the system, which may arise from various factors such as the volume of PV waste generated, future raw material prices, demand for new PV panels, and regulatory changes, or technological advancements in recycling technologies. The proposed holistic framework incorporates adaptive strategies for planning under various scenarios ensuring resilience and agility of the network.

References

[1] IEA. Renewables 2022. Technical report, IEA, 2022

[2] S. Weckend, A. Wade, and G. Heath. End-of-life management: solar photo-voltaic panels. Technical report, International Energy Agency and International Renewable Energy Agency, 2016.

[3] E. Iakovou, E. N. Pistikopoulos, J. Walzberg, F. Iseri, H. Iseri, N. J. Chrisandina, S. Vedant, and C. Nkoutche. Next-generation reverse logistics networks of photovoltaic recycling: Perspectives and challenges. Solar Energy, 271:112329, 2024. ISSN 0038-092X. doi: https://doi.org/10.1016/j.solener.2024.112329.

573x Directing Climate Mitigation throughout Energy Transformation: Feasibility of Synthetic Methane Integrated with Direct Air Capture

Changgwon Choe, Hankwon Lim

With an increase in environmental awareness, energy transformation toward a sustainable energy economy has received much interest. The hydrogen (H_2) economy with green H_2 is under consideration for the transformation, but there are several challenges to be overcome. In this context, further conversion of green hydrogen into carbon-based transportation fuel based on the carbon hydrogenation process has been explored. The representative fuel is methane which is a prime element of natural gas. Natural gas is a primary energy resource widely consumed in households and industries around the world, and it has extensive networks for distribution to end-users. In line with this trend, this research aims to investigate the feasibility of methane synthesis integrated with water electrolysis and direct air capture for negative carbon emission. The synthesized methane may become an alternative to conventional fossil-based natural gas; then, it also utilizes the existing infrastructures for supply. The case study objecting several countries is performed to identify whether it can have the economic and environmental potential for climate mitigation. Further, this study aims to provide comprehensive guidelines in the aspects of economic and environmental competitiveness, global challenges, and future directions.

573y Optimal Compressor Operational Policies for LNG Storage Tanks Under Unloading Mode Operation: A Numerical Approach

Suraj Prakash Singh, Rajagopalan Srinivasan, Iftekar Karimi

The increasing energy demand forces our dependency on fossil fuels, leading to excess CO_2 emissions. These emissions are major greenhouse gas contributors, responsible for climate change, ozone layer depletion, and increasing earth's surface temperature[1]. Therefore, the government is forcing industries to shift from fossil fuels to renewable energy resources, and natural gas (NG) acts as a transition fuel[2]. Over larger distances, NG is transported under cryogenic conditions for ease of handling, safer operation and higher energy density. NG under cryogenic conditions is called 'Liquefied Natural gas' (LNG). Despite heavy insulation, a continuous heat ingress in the tank results LNG evaporation, known as 'boil-off gas' (BOG). BOG generation pressurizes the tank and needs to be managed appropriately for the plant's safety. La Spezia (Italy, 1971) and Partington (UK, 1993) are the two major incidents that occurred due to over-pressurization of LNG tank[3]. In plants, compressors regulate tank pressure and are operated heuristically at varying capacities (0%, 50%, 75%, and 100%) to maintain safe tank pressure and optimize the compressor energy requirement. During LNG unloading, compressors are operated continuously at 100% capacity due to the lack of well-known tank dynamics. Thus, it is necessary to understand the dynamics of the storage tank under unloading operation, and optimize compressor energy.

Optimizing the compressor's energy consumption has been studied by many researchers over the years. Shin et al.[4] observed that a compressor operated at higher capacity is safe but increases the terminal's operating cost. They developed a dynamic model to estimate tank pressure and provided an algorithm to regulate pressure through the compressor under minimum capacity. Further, Jang et al.[5] improved the model by providing a new algorithm to distribute the compressor load and optimizing the energy consumption. They developed a dynamic model to estimate tank pressure using the Peng-Robinson equation of state for better accuracy. Ye et al.[6] considered the compressor and low-pressure (LP) pump to optimize energy requirements. They considered pump running or on standby mode, number of compressors and compressor capacity in the analysis. The optimization decreases the number of compressors and LP pumps operation. Li and Li[7] estimated the operating condition that generated minimum BOG by varying influencing factors and managed it using a single compressor operation. Effendy et al.[8] provide a dynamic model to estimate BOG generation by considering the storage tank and recirculation line during holding mode operation. As mentioned earlier, all the studies focus on BOG generation or compressor operation during holding mode operation, but none focus on unloading operation. Our work seeks to fill these gaps by considering the Effendy et al.[8] model as a base for estimating BOG generation and developing a new unloading operation model. The model considers the additional feature of unloading the nozzle and compressor to manage generated BOG under a safety range. The developed model helps to formulate new compressor operation policies to optimize energy consumption.

This paper develops a dynamic model of an industrial LNG storage tank. The tank has a diameter of 80m and a height of 40m, and it has a nozzle for unloading cargo. The tank can fill a maximum of 30m of LNG and drain a minimum of 2m from the tank. Initially, the tank contains some LNG inventory and is unloaded to the maximum level. Unloading at high pressure generates a large amount of BOG due to flashing and increases the tank pressure abruptly. The dynamics of the unloading operation are not well known, so the compressor is operated at 100% capacity to regulate safe tank pressure. In this work, we estimate tank pressure using Peng-Robinson equation of state and compare it with plant data to calculate the critical parameters for better model accuracy. Further, tank pressure is regulated by a compressor operated at variable capacities and performs several scenarios by varying initial LNG levels, namely 20% (6m), 40% (12m), 60% (18m), 80% (24m), and 100% (30m). Our model recommends new policies for safe operation and optimal compressor energy requirements in the receiving terminal.

References:

(1) *Climate Change: Atmospheric Carbon Dioxide | NOAA Climate.gov.* http://www.climate.gov/news-features/understanding-climate/climate-change-atmospheric-carbon-dioxide (accessed 2024-03-27).

(2) *EI Key Findings global gas 2021 DT.* http://ceros.mckinsey.com/energy-insights-global-gas-key-findings-2021 (accessed 2023-02-19).

(3) *Rollover in LNG Storage Tanks 2012-2015* -. https://giignl.org/document/rollover-in-lng-storage-tanks/ (accessed 2024-03-26).

(4) Shin, M. W.; Shin, D.; Choi, S. H.; Yoon, E. S. Optimal Operation of the Boil-off Gas Compression Process Using a Boil-off Rate Model for LNG Storage Tanks. *Korean J. Chem. Eng.* **2008**, *25* (1), 7–12. https://doi.org/10.1007/s11814-008-0002-9.

(5) Jang, N.; Shin, M. W.; Choi, S. H.; Yoon, E. S. Dynamic Simulation and Optimization of the Operation of Boil-off Gas Compressors in a Liquefied Natural Gas Gasification Plant. *Korean J. Chem. Eng.* **2011**, *28* (5), 1166–1171. https://doi.org/10.1007/s11814-010-0487-x.

(6) Ye, Z.; Mo, X.; Zhao, L. MINLP Model for Operational Optimization of LNG Terminals. *Processes* **2021**, *9* (4). https://doi.org/10.3390/pr9040599.

(7) Li, Y.; Li, Y. Dynamic Optimization of the Boil-Off Gas (BOG) Fluctuations at an LNG Receiving Terminal. *Journal of Natural Gas Science and Engineering* **2016**, *30*, 322–330. https://doi.org/10.1016/j.jngse.2016.02.041.

(8) Effendy, S.; Khan, M. S.; Farooq, S.; Karimi, I. A. Dynamic Modelling and Optimization of an LNG Storage Tank in a Regasification Terminal with Semi-Analytical Solutions for N2-Free LNG. *Computers & Chemical Engineering* **2017**, *99*, 40–50. https://doi.org/10.1016/j.compchemeng.2017.01.012.

573z Understanding Microbial Processes of Algal-Fixed Carbon Using a Co-Culture Porous Microplate

Hyungseok Kim, Vanessa L. Brisson, John R. Casey, Jeffrey A. Kimbrel, Kristina A. Rolison, Christopher A. Vaiana, Peter K. Weber, Rhona K. Stuart, Xavier Mayali, Cullen R. Buie

Photosynthetic microalgae have potential as a renewable bioenergy source, but the role of their microbiome in outdoor ponds is poorly understood. While the bacteria can consume up to 50% of the fixed carbon, they are unavoidable and are known to impact the algal productivity. In this work we present a novel co-culture method ("porous microplate") to study how the associated bacteria spatiotemporally respond to algal exometabolites *in vivo*. Mechanistic design of the porous microplate allowed us to identify taxonomically distinct responses in or between bacterial isolates. The results were consistent with their individual consumption patterns of the algal exometabolites, identified using untargeted metabolomics, suggesting a competitive interaction in our co-culture system. Our systems biology approach exemplifies how the microbes spatiotemporally interact while competing for limited resources. The presented method can also be useful in understanding more advanced manufacturing systems examining interactions in, and between, other biological domains such as bacteria, fungi and yeast.

573ab Fabricating an Economical and Sustainable Fermentation Medium to Produce Drop-in-Fuel and Value-Added Bioproducts from Alternative Feedstocks

Shraddha Maitra, Bruce S. Dien, Kristen Eilts, Nurzhan Kuanyshev, Yoel Cortes-Pena, Jeremy Guest, Yong-Su Jin, Vijay Singh

Utilization of abundant renewable carbon present in lignocellulosic biomass can reduce the environmental concerns raised due to the use of fossil fuels. This can also be a driving force for scientific and technological advancement of biorefining facilities to efficiently produce drop-in fuels. Previously, a wide variety of non-transgenic bioenergy crops have shown promising results for the production of cellulosic fermentable sugars that can be conveniently converted to bioethanol and high value bioproducts. In the last few years, the energy density of bioenergy plants such as energycane, sugarcane, and sorghum, has been further increased by genetic engineering. These high biomass bioenergy crops are being metabolically engineered to sequestrate carbon in their vegetative tissues and direct the carbon flux for the synthesis and accumulation of energy-rich lipid molecules like triacylglyceride (TAG) molecules (Vanhercke et al. 2019; Parajuli et al. 2020; Luo et al. 2022). Therefore, transgenic bioenergy crops cater to both cellulosic sugars and vegetative lipids to produce drop-in fuel and value-added biochemicals (Maitra et al. 2022a; Maitra et al. 2022b). However, because the lipids in the vegetative tissues of transgenic crops are present in a complex form and lower amounts (Parajuli et al. 2020; Maitra et al. 2021), application of appropriate bioprocessing technologies is needed to efficiently deconstruct the recalcitrant lignocellulosic structure to recover vegetative lipids along with cellulosic sugars from the feedstock without degenerating lipids. Chemical-free hydrothermal pretreatment has been successfully established for the recovery of both cellulosic sugars and vegetative lipids that maintain the lipid profile during processing. During the processing, the biomass residues get enriched with lipids that were recovered at the end of the process (Maitra et al. 2022a; Maitra et al. 2022b).

Although the recovery of cellulosic sugars from bioenergy crops has been successfully established, the hydrolysate that is produced is not suitable for most microorganisms for growth and fermentation. The utilization of hydrolysate for fermentation still needs either an external supply of nutrients, additional detoxification steps, or both. There are no reports on the utilization of hydrolysate for fermentation without additional nutrients, salts, or detoxification. Detoxification steps and the addition of nutrients such as vitamins, yeast nitrogen base (YNB), yeast extract, and peptone (YP) increase the cost of the process. To alleviate the issue, the study presents the fabrication of a new sustainable and economic fermentation media using hydrolysate and juice derived from transgenic oil producing sugarcane.

Successful development of noble transgenic biomass has been achieved. Development of scalable processing technology is needed for sustainable biomanufacturing. The hydrolysate produced, especially from transgenic oilcane, had a detrimental effect on growth and fermentation. Unfortunately, both the engineered *S. cerevisiae* and commercial strains capable of utilizing both C6 and C5 carbons were unable to grow on 100% hydrolysate without any external nutrient. Hydrolysate was analyzed for the presence of compounds responsible for the inhibition. The hydrolysate was found to be rich in carbohydrates (~ 120 g/l of total fermentable cellulosic sugars), lipids (TAGs, DAGs, and phospholipids), and phenolics (various metabolites and compounds generated due to pretreatment) which could have an inhibitory effect on microorganisms. The amount of sugar degradation products such as HMF and furfurals were negligible. The first step was to add growth-supporting nutrients without any additional cost. To this end, juice from transgenic oilcane was not edible, rich in nitrogen, and had all the vitamins (LC-MS analysis) needed for growth. The media was designed with different ratios of hydrolysate to juice. All the combinations of hydrolysate to juice worked except ratios < 1:1. A 1:1 ratio of undetoxified hydrolysate and juice without any external nutrients was the optimized growth medium for both the lab and commercial strains. Once, the growth medium was optimized for the strains, operating conditions were optimized to improve the fermentation efficiencies.

It was demonstrated that pH control, microaerophilic conditions, and adaptation of strains in the same medium prior to inoculation enhance the fermentation efficiency. The specific and volumetric ethanol productivity of adapted engineered *S. cerevisiae* were 0.17 ± 0.001 and 1.04 ± 0.00, respectively, while that of commercial strain was 0.19 ± 0.02 and 1.06 ± 0.004, respectively, in the new sustainable media i.e., hydrolysate: juice (1:1). The adapted lab and commercial strains exhibited 1.7 times and 3.1 times higher specific ethanol productivity and 5.4 times and 5.0 times higher volumetric ethanol productivity, respectively. Adapted lab and commercial strains also demonstrated 1.1 times and 3.1 times higher specific consumption rates, respectively. The growth and fermentation kinetics of the strains suggest that uptake of substrates by the cells improved on adaptation and the improved carbon flux was directed towards the fermentation pathway. The production of byproducts such as glycerol and acetic acids were negligible by the strains. The growth and fermentation kinetics on the constructed media demonstrated that the proposed combination of hydrolysate and juice can be used as a noble economic and sustainable fermentation medium. The requirement of expensive nutrients such as vitamins, minerals, nitrogen, and buffer salts which are crucial for cell growth, metabolism and fermentation can be furnished without extra cost. Moreover, since the hydrolysate and juice would be derived from transgenic bioenergy crops which would be grown in marginal land there would be no competition for limited cultivable land.

References

Luo G, Cao VD, Kannan B, Liu H, Shanklin J, Altpeter F (2022) Metabolic engineering of energycane to hyperaccumulate lipids in vegetative biomass. BMC Biotechnol 22:24 . https://doi.org/10.1186/s12896-022-00753-7

Maitra S, Dien B, Long SP, Singh V (2021) Development and validation of time-domain 1 H-NMR relaxometry correlation for high-throughput phenotyping method for lipid contents of lignocellulosic feedstocks. GCB Bioenergy. https://doi.org/10.1111/gcbb.12841

Maitra S, Long SP, Singh V (2022a) Optimizing Chemical-Free Pretreatment for Maximizing Oil/Lipid Recovery From Transgenic Bioenergy Crops and Its Rapid Analysis Using Time Domain-NMR. Front Energy Res 10:1–13 . https://doi.org/10.3389/fenrg.2022.840418

Maitra S, Viswanathan MB, Park K, Kannan B, Alfanar SC, McCoy SM, Cahoon EB, Altpeter F, Leakey ADB, Singh V (2022b) Bioprocessing, Recovery, and Mass Balance of Vegetative Lipids from Metabolically Engineered "Oilcane" Demonstrates Its Potential as an Alternative Feedstock for Drop-In Fuel Production. ACS Sustain Chem Eng 10:16833–16844 . https://doi.org/10.1021/acssuschemeng.2c05327

Parajuli S, Kannan B, Karan R, Sanahuja G, Liu H, Garcia-Ruiz E, Kumar D, Singh V, Zhao H, Long S, Shanklin J, Altpeter F (2020) Towards oilcane: Engineering hyperaccumulation of triacylglycerol into sugarcane stems. GCB Bioenergy 12:476–490 . https://doi.org/10.1111/gcbb.12684

Vanhercke T, Belide S, Taylor MC, El Tahchy A, Okada S, Rolland V, Liu Q, Mitchell M, Shrestha P, Venables I, Ma L, Blundell C, Mathew A, Ziolkowski L, Niesner N, Hussain D, Dong B, Liu G, Godwin ID, Lee J, Rug M, Zhou X-R, Singh SP, Petrie JR (2019) Up-regulation of lipid biosynthesis increases the oil content in leaves of Sorghum bicolor. Plant Biotechnol J 17:220–232 . https://doi.org/10.1111/pbi.12959

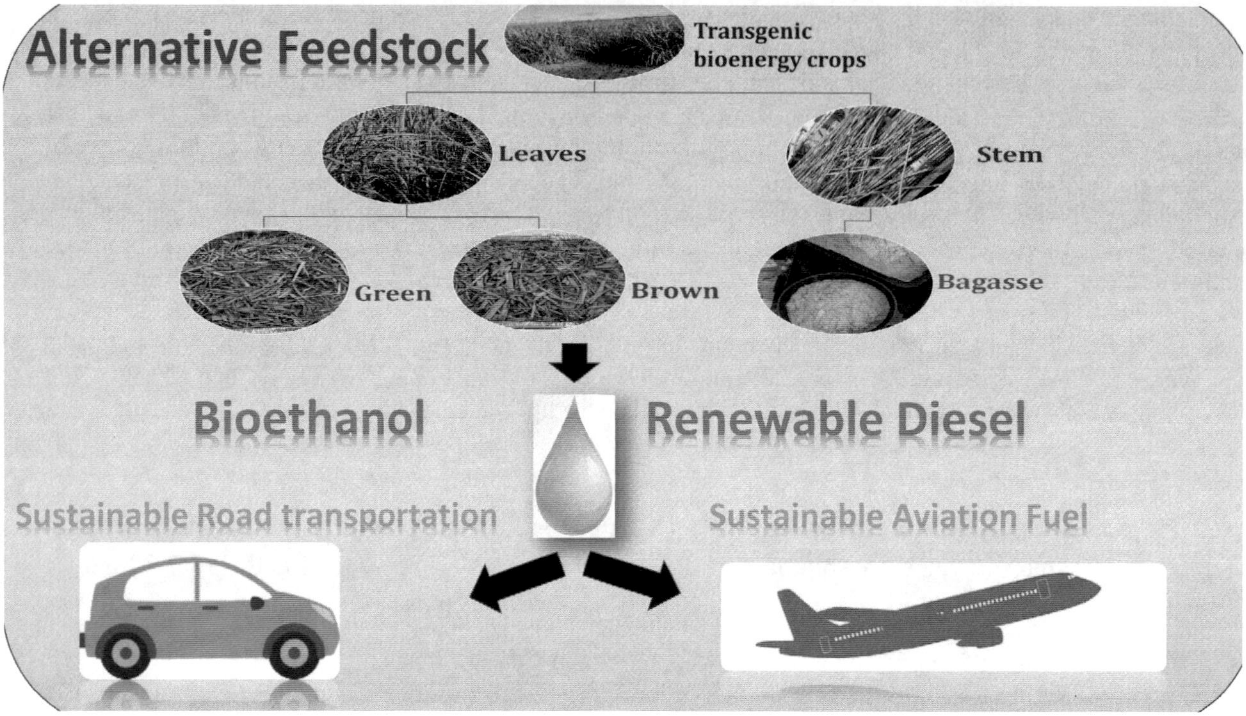
Alternative Feedstock
Transgenic
bioenergy crops
Leaves
Stem
Green
Brown
Bagasse
Bioethanol
Renewable Diesel
Sustainable Road transportation
Sustainable Aviation Fuel

573ad Systematic analysis on the energy transition pathways for sustainable natural gas production

Meire Ellen Gorete Ribeiro Domingos, Daniel Florez Orrego, François Maréchal

Abstract

According to the International Energy Agency, in the present context of market uncertainty, the natural gas prices are foreseen to remain extremely volatile [1], which reinforces the need of alternative pathways to overcome its limited availability and guarantee the supply of this important commodity. The combination of anaerobic digestion and thermal gasification ensures efficient use of biomass, since the unconverted biomass from digesters can be used to produce syngas through thermal gasification. On the other hand, hydrogen production from water electrolysis has recently drawn attention as a versatile solution for balancing intermittent renewable electricity generation, particularly from sources like wind and solar. In addition, reversible solid oxide cells (rSOCs) are an electrochemical energy conversion technology that can produce both electricity from fuel (gas-to-power) and fuel from electricity (power-to-gas), depending on resource availability and demand. Hydrogen is produced in an rSOC operated as an electrolyzer, using excess power from renewable resources. Next, it is combined with syngas from biowaste and residue gasification to produce methane. The rSOC system can also be operated in fuel cell mode by oxidizing methane to produce electricity and balance power requirements of the integrated system if renewable electricity is not available from the grid.

This work investigates the integration of biodigestion, gasification and rSOCs aiming to increase the sustainable natural gas yield, while evaluating the tradeoffs of minimizing both the CO_2 emissions and the total cost of the different configurations. In order to assess the impact of different scenarios for the energy transition in the sustainable natural gas production, the seasonal excess and deficit of electricity is considered, either in the current energy mix scenario and also for a future energy mix scenario, in which a full renewable grid is modeled based on the generation (taking into account GIS-based land-restriction, geo-spatial wind speed and irradiation data, and the maximum electricity production from renewable sources was derived considering EU-wide low restrictions) and demand (full electrification of the residential and mobility sectors is assumed) [2].

Given the fluctuating costs of seasonal energy and the intermittent availability of renewable energy sources, the integration of other technologies, such as CO_2 injection and storage systems (e.g., SNG, CO_2), are also assessed. The OSMOSE Lua platform is used to address the optimization problem, seeking to minimize energy consumption, the overall cost of the chemical plant, and also the CO_2 emissions. A mixed integer linear programming (MILP) method is employed to identify optimal system configurations under different economic scenarios. An incremental financial analysis that incorporates the uncertainty related to the acquisition and selling costs of the feedstock and fuels produced, when embedded in a volatile market, by using the Monte Carlo method is performed. This methodology helps elucidating the feasibility of the proposed processes.

The proposed superstructure is shown in Fig 1. The biodigestion process is modelled considering a biomethane potential of 300 Nm^3 CH_4 per t of volatile solids using organic wastes [3]. A water scrubbing technology is considered for the biogas upgrading, as it can obtain methane concentration of > 99% CH_4 consuming around 0.20 kWh/Nm^3. The CO_2 is also recovered in the upgrading system with a purity above 99.5% [4]. Upgraded biomethane is marketed and the CO_2 rich stream follows to the biomethane production. The CO_2 from the anaerobic digestion unit could be liquefied and stored in a tank at -50 °C and 7 bar (1,155 kg/m^3). Liquefied CO_2 can be later regasified and fed to a methanation system, in which the hydrogen necessary is provided by the rSOC system under the electrolyzer mode, operating at 1 bar, 800 °C with steam conversion rate of 81%. The rSOC system can also be operated in fuel cell mode by oxidizing methane to produce electricity. The rSOC system is modelled considering the concentration, ohmic and activation overpotentials [5].

The gasification of the digestate coming from the anaerobic digestion unit occurs at atmospheric pressure and uses steam as gasification agent. After leaving the gasifier, the syngas is treated to remove tars and impurities. A fraction of the char produced in the pyrolysis step is combusted to supply the heat required by the endothermic drying, pyrolysis and reduction reactions. Hydrogen coming from SOEC is used to adjust syngas composition for the methanation reaction ($(n_{H2}-n_{CO2})/(n_{CO}+n_{CO2}) = 3.1$).

The methanation system is based on the TREMP® process [6], in which a series of methanation beds are intercooled by recycling or indirect inter-cooling in order to achieve higher conversion. Simulations are performed in Aspen Plus® software, using Peng-Robinson EoS with Boston-Mathias modifications.

A systematic approach combines different energy solutions and time-varying energy demands to identify operating conditions and arrangements for maximizing the sustainable natural gas production. The storage systems enabled the seasonal storage of the CO_2 streams, which were utilized to produce CH_4 when renewable electricity is available (see Fig 2). CH_4 is stored and used to generate power to balance the system demands via SOFC. This strategy enables harvesting excess renewable electricity and producing value added fuels from biogenic emissions. Availability of renewable electricity impacts not only overall generation and demand balance, but also extent of use of the CO_2 and the performance of its management system to produce value-added products and handle the intermittency of renewable sources. The uncertainty of the market prices for the commodities has been also considered, revealing that the use of constant inputs costs may lead to misleading results regarding the likelihood of loss of risky projects. The implementation of more rigorous carbon taxes and the technologies maturation may reduce the financial risks and give support to decision-makers in implementing biorefineries and exploring new business opportunities to trigger the decarbonization of important commodities.

References

1. IEA. Russian supplies to global energy markets. International Energy Agency; 2022 Available from: https://www.iea.org/reports/russian-supplies-to-global-energy-markets Accessed 17 March 2024
2. Santecchia A, Castro-Amoedo R, Nguyen TV, Kantor I, Stadler P, Maréchal F. The critical role of electricity storage for a clean and renewable European economy. Energy & Environmental Science. 2023;16(11):5350–70.
3. Wellinger A, Murphy JD, Baxter D. The biogas handbook: science, production and applications. Elsevier; 2013.
4. Lems R, Langerak J, Dirkse E. “Next generation biogas upgrading using highly selective gas separation membranes” Showcasing the Poundbury project. DMT Environmental Technology: Joure, The Netherlands. 2008;
5. Ni M, Leung MK, Leung DY. A modeling study on concentration overpotentials of a reversible

solid oxide fuel cell. Journal of Power Sources. 2006;163(1):460–6.
6. Topsoe, H. From solid fuels to substitute natural gas (SNG) using TREMP Topsøe Recycle Energy-efficient Methanation Process. 2009; Available from: https://www.netl.doe.gov/sites/default/files/netl-file/tremp-2009.pdf Accessed 17 March 2024

573ae LATP-incorporated Cellulose Membrane for Selective Lithium Extraction from Salt Lake Brine under an Electrochemical System

Jiwoo Lee, Dongju Seo, Shik Rou Kong, Gyudae Sim, Youngjune Park

The demand for lithium is expected to continue rising due to its critical role in the clean energy transition. However, conventional lithium extraction methods from salt lake brines have limitations. These methods rely on solar evaporation, which takes 10 to 24 months and requires substantial water consumption in dry regions such as the lithium triangle in Chile, Bolivia, and Argentina.

Recently, lithium-selective membranes have emerged as a promising alternative to traditional evaporation methods. In this study, we developed an LATP/CA membrane for selective lithium extraction. The NASICON-type lithium-ion conductor LATP ($Li_{1.3}Al_{0.3}Ti_{1.7}(PO_4)_3$) was utilized for its high lithium selectivity, while cellulose acetate (CA) was used to enhance the mechanical properties of the membrane.

To assess its practical applicability, two types of lithium separation tests were conducted. First, tests were performed under binary conditions with two mixed cations to evaluate lithium selectivity. Second, lithium separation tests were conducted using simulated brine solutions. In these simulated brine tests, different applied currents were used to observe performance variations. Under the 200 μA condition, lithium selectivity was found to be Li/Mg: 467.3, Li/Na: 44.8, and Li/K: 22.7. This indicates that the LATP/CA membrane exhibits high lithium selectivity compared to other cations, which can be attributed to the crystal structure of LATP. Consequently, the application of the LATP/CA membrane could be a promising approach to address the environmental and supply challenges associated with conventional evaporation methods, contributing to sustainable lithium mining.

729f Optimal Design of the Bio-Jet Fuel Supply Chain Under Uncertainty

Shunmei Wu, Congqin Ge, Zhihong Yuan

Abstract

Motivated by grand carbon neutrality, China is going through a systematic and in-depth energy revolution to achieve environmental protection and sustainable development. Needless to say, biomass is expected to play a key role[1]. China is the second largest aviation (jet fuel) market around the world. The production of jet fuels from renewable bioresources is a promising way for aviation industry to reduce the dependence on fossil and achieve the carbon emission reduction targets[2]. The development of bio-jet fuels is of great significance for achieving carbon neutrality.

The high level of uncertainty in feedstock availability, market fluctuations, and other factors make the bio-jet fuel industries face great challenges[3]. In order to improve the utilization efficiency of biorenewable energy and promote the rational allocation of biological resources, this work carries out a multi-period stochastic optimization framework for the bio-jet fuel supply chain network under multiple uncertainties relevant to feedstock supply, customer demand and price fluctuations with the maximum profit expectation. Via the tailored benders decomposition algorithm for the formulated large-scale mixed integer linear programming, the overall supply chain network structure and related operation decisions such as the location/scale, the type of production technology, the distribution relationship and distribution volume among each mode in the network and so on are revealed. Meanwhile, the result can provide the decision makers with appropriate balanced economic performance and robustness to uncertainties. Finally, effectiveness of the proposed framework is further investigated by sensitivity analysis of the key parameters in the optimization model.

References

[1] Yue, D., F. You and S. W. Snyder. Biomass-to-bioenergy and biofuel supply chain optimization: Overview, key issues and challenges. COMPUTERS & CHEMICAL ENGINEERING 66: 36-56. doi:10.1016/j.compchemeng.2013.11.016

[2] Tiwari, R., R. Mishra, A. Choubey, S. Kumar, A. E. Atabani, I. A. Badruddin and T. M. Y. Khan. Environmental and economic issues for renewable production of bio-jet fuel: A global prospective. FUEL 332. doi:10.1016/j.fuel.2022.125978

[3] Pannok, M., M. Finkbeiner, H. Fasel, J. Riese and S. Lier. Transformable Decentral Production for Local Economies with Minimized Carbon Footprint. CHEMBIOENG REVIEWS 7(6): 216-228. doi:10.1002/cben.202000008

290a Atmospheric Methane Removal: a National Academies Research Agenda

Christopher W. Jones

Methane is the second most important greenhouse gas behind carbon dioxide contributing to human-driven warming. Large reductions in methane emissions are needed to limit end-of-century warming; however, global methane emissions continue to rise.

The urgency to limit warming in the near- and long-term, and potential barriers in technologies, policies, and investments needed to reduce methane emissions at the scale required, have motivated researchers to begin exploring the nascent concept of atmospheric methane removal. There are a small number of peer reviewed publications on the topic, reflecting limited investments in research funding to date. At this critical moment for resource investment and decision making around climate action, an authoritative study on the state of knowledge of atmospheric methane removal was requested from the National Academies.

Atmospheric methane removal encompasses a range of technological approaches to either physically remove and store atmospheric methane or accelerate the conversion of methane in the atmosphere to a less radiatively potent form. Five key technological approaches for atmospheric methane removal include: methane reactors and methane concentrators (both partially closed systems), and surface treatments, ecosystem uptake enhancement, and atmospheric oxidation enhancement (open systems lacking a physical boundary). The Committee assessed these five technology categories against a set of criteria developed as meaningful indicators of each technology's advancement towards a pathway for methane removal at atmospheric concentrations and to assess key research needs. This committee also considers the state of foundational knowledge along social and policy dimensions which affects the development and acceptance of research on atmospheric methane removal among different publics.

This presentation provides an overview of the research agenda on atmospheric created by the National Academies, highlighting priority areas for future research and action particularly relevant to the AIChE community.

290f Carbon Utilization Markets, Infrastructure, Research, and Development: A National Academies Assessment

Ah-Hyung Alissa Park

Carbon utilization addresses two underexamined needs in a net-zero GHG emissions future: durably storing CO_2 in products and approaches to circular-carbon systems of short-lived carbon-based products. This presentation describes the National Academies research agenda on carbon dioxide utilization, highlighting the key research needs for mineralization, chemical, and biological processes and cross-cutting research themes.

The committee identified many needs for essential or high-value carbon-containing products in a net-zero GHG emissions future, such as cement, concrete, and aggregates; aviation fuel; agrochemicals; and platform chemicals. It identified where CO_2 utilization is favored as a sustainable carbon source, such as to form oxygenates, single-carbon products, and products that durably store carbon, and addressed the differing sustainability of products based on both the carbon source and the lifetime of the product.

The committee's report describes four pathways for CO_2 conversion: (1) mineralization of CO_2 into inorganic carbonates, (2) conversion of CO_2 into elemental carbon materials, and (3) chemical and (4) biological pathways for CO_2 conversion into organic chemicals and fuels. For each pathway, research needs were identified and described by their research status, research area, and product class. Recommendations were directed at federal agencies or other actors to address those needs. Themes in research needs were developed at the reaction-, systems-, and demonstration and deployment-levels. This presentation will describe the committee's findings and recommendations relevant to the chemical engineering and chemistry research community.

In addition to describing the research agenda, the presentation will also draw in some of the study's major conclusions on markets, policy, technology assessment, and infrastructure needs for CO_2 utilization.

290c Convergent Research and Innovation at the NYS Center for Plastics Recycling Research and Innovation

Amit Goyal

This talk will provide an overview of plastics recycling research being conducted at the NYS Center on Plastics Recycling Research & Innovation at the University at Buffalo. The Center focuses on addressing a range of contemporary issues related to plastics recycling, with the goal of advancing plastics recycling in NY state and beyond. The Center is based on ***multidisciplinary, convergent research*** and involves faculty and students from several schools and colleges spanning the disciplines of *operations research*, *environmental engineering*, *chemistry*, *chemical engineering*, *materials science*, *economics, physics, data science* and *communications*. Ongoing Tasks include: **1:** Mapping the reverse supply for plastics in NYS and deep dive into the structure of the collection, disposal, and recycling industries; **2:** Improving the efficiency of businesses involved in collecting and processing plastics; **3:** Plastics in NYS agriculture/food and medical industry; **4:** Further understand contemporary attitudes toward recycling and current recycling behaviors and to design and test campaigns to increase recycling knowledge and minimize contamination in recycling; **5:**Microplastics in Natural Environments; **6:** Advanced High-Speed Plastic Sorting Technique by Molecular Contrast Infrared Imaging for advancing plastics recycling; **7:** Determining percentage of recycled plastic content in plastic products via transformative new technologies. The talk will provide a summary of results obtained from these projects.

Acknowledgements: The NYS Center on Plastics Recycling Research & Innovation at the University at Buffalo acknowledges funding and support from NYDEC.

290d Systems Engineering for Renewable Energy: Novel Approaches for New Problems

Christos Maravelias

We discuss how process systems engineering (PSE) studies can facilitate the development of novel strategies for the production of renewable fuels and chemicals. Importantly, we show how PSE can be used to identify technological and economic drivers, and, ultimately, guide future research efforts. After an overview of the methods used for these studies, we discuss the synthesis and analysis of two broad classes of strategies: (1) biomass-to-fuels/chemicals systems and (2) systems that employ solar energy for power and fuels generation. Finally, we discuss how the synthesis of integrated renewable energy systems leads, naturally, to new types of problems for which traditional approaches are insufficient; and outline progress towards the development of novel methods to address these challenges.

290e Chemical-Free UV Unit That Degrades PFAS in Landfill Leachate Using Non-Toxicboron Nitride

Abraham Nelson

Despite being a well-known chemical contaminant, per/polyfluoroalkyl substances (PFAS) are used extensively due to their commercial, industrial, and household applications in nonstick cookware, water-resistant fabrics, firefighting foam, fast food packaging, and more. Once these products are no longer in use, they are disposed of and processed by hazardous waste facilities. PFAS contained in these products can potentially leak into landfill leachate and subsequently seep into the local water supply. The contamination of the water supply near hazardous waste facilities poses a tremendous health concern for the communities that live nearby and consume this as their sole source of water. According to the National Academies of Sciences, Engineering, and Medicine, PFAS consumption in humans is directly linked with higher rates of kidney and testicular cancers, impaired endocrine systems, fertility issues, and more. Additionally, landfills are disproportionately located in underserved areas, resulting in low-income and minority communities being unequally vulnerable to these health concerns. Due to the pressing need for effective and affordable water treatment solutions, our team aimed to explore the ability of BN/UVC photocatalysis to break down PFAS in landfill leachate. Our team also examined the location of known and suspected PFAS occurrences, socially vulnerable communities (i.e., communities with increased susceptibility to the adverse impacts of natural hazards due to systemic barriers and structural inequalities), and landfill sites in the Harris County area to better understand the relationships between these variables. Providing an energy-efficient, non-toxic method to treat PFAS-containing wastewater emanating from landfills will reduce the amount of PFAS and other organic pollutants entering the water supplies of nearby communities, giving many people the chance to enjoy the prosperity of cleaner water.

352a Overview of Doe-Fecm's Carbon Storage Infrastructure Projects

William Aljoe, Darin Damiani

The Department of Energy's Office of Fossil Energy Carbon Management (DOE-FECM) has been supporting R&D on geologic storage of captured carbon dioxide (CO_2) though its Carbon Transport and Storage Program for more than two decades. A component of the program is the Carbon Storage Infrastructure (CSI) portfolio, which includes a variety of engineering design studies, field projects, and technical assistance partnerships. The CSI projects are helping to facilitate and enable commercial CO_2 storage project development; testing diverse and emerging storage techniques; and providing technology transfer and expertise to geologic carbon storage project developers and key stakeholders, including affected communities. The federal funding for CSI projects is strategically invested to help support the safe, equitable, and environmentally responsible development and deployment of large-scale carbon transport and storage facilities and regional storage hubs.

As of April 2024, the CSI project portfolio includes 41 active projects with a total planned value of approximately $684 million. DOE funding for these projects is approximately $480 million, with the remainder coming from the project performers and their partners in industry, academia, State government, and other organizations. By the end of October 2024, the CSI portfolio is expected to increase by more than 25 projects, with an expected total DOE funding level of around $600 million. The rapid expansion of the portfolio has been enabled through funding from several different provisions of the Bipartisan Infrastructure Law (BIL) enacted in November 2021.

Projects in the CSI portfolio fall into three broad categories:

- **CarbonSAFE** – Since its inception in 2016, projects in the multi-year, multi-phase CarbonSAFE Initiative have focused on identifying (Phase I), investigating, characterizing, designing, and permitting (Phases II and III), and constructing (Phase IV) large-scale storage facilities, each designed to hold 50+ million metric tons of CO_2. Bolstered by funding from the Bipartisan Infrastructure Law's (BIL) $2.5 billion allocation for Carbon Storage Validation and Testing (BIL Section 40305), 36 new projects have been added to the CarbonSAFE portfolio over the past two years, with additional projects to be selected via DOE Funding Opportunity Announcements through 2026. The overall goal of the CarbonSAFE projects is to provide access, by 2030, to storage capacity of 2 billion metric tons (MT), with 5.5 billion MT identified contingent storage resources nationwide, and enabling the injection of at least 65 million MT per year. By October 2024, it is expected that 13 Phase I and 6 Phase II projects will have been completed, with 20 Phase II and 21 Phase III projects either underway or in negotiation for award.
- **CO_2 Transport Infrastructure** – Projects include three stages of transport infrastructure development:
    - **Pre**-Front-end Engineering Design (Pre-FEED) studies of multi-modal CO_2 transfer facilities which are designed to: accept input from one or more modes of transport (pipeline, truck, rail, maritime); process, compress, and temporarily store (if necessary) the CO_2 to meet downstream transport requirements; and offload the CO_2 onto a different transport mode.
    - Funding from BIL Provision 40303 is supporting front-end engineering and design (FEED) studies for large-scale CO_2 transport from anthropogenic sources to CO_2 conversion or secure geologic storage locations. All modes of CO_2 transport (pipeline,

truck, rail, barge, and ship), including any combination of transport modes, may be included in the projects.
  - Future Growth Grants from the Carbon Dioxide Transportation Infrastructure Finance and Innovation (CIFIA) program (BIL Provision 40304) will provide financial assistance for developing and building extra CO_2 transport capacity up front (in advance of contracted demand) that will then be available for future carbon capture, conversion, and storage facilities as they are developed and come into operation. The Future Growth Grants effort is coordinated with DOE's Loan Programs Office, which offers loans for design and construction of large-scale CO_2 transport infrastructure upon which the projects funded via Future Growth Grants can be based.
- **Technical Assistance & Collaboration** – Over the past 5 years, DOE has built on the success of its Regional Carbon Sequestration Partnerships (RCSP) Initiative (which ran from 2003 through 2019) by awarding four "Regional Initiative" (RI) projects to teams whose experience in the RCSP Initiative enabled them to identify and address the full range of challenges facing commercial deployment of carbon capture, utilization, and storage. Each of the four RI projects is charged with providing technical, procedural, and outreach assistance to carbon management efforts across a broad region of the U.S. (East-Midwest, Southeast, Great Plains, and West-Southwest). To help address the rapidly increasing demand for technical assistance, six projects were recently selected to augment the efforts of the Regional Initiative projects by providing technical and community engagement assistance to developers of specific, well-defined, large-scale storage facilities or regional carbon management hubs. Ten additional projects are led by State Geologic Surveys or organizations with similar responsibilities, as a means of assembling, organizing, and analyzing geologic data needed to facilitate the development of future commercial carbon storage projects within their respective States. DOE is currently evaluating another set of new projects with the objective to provide assistance in addressing the unique challenges that arise when multiple storage-based projects are sited within a single geologic basin or sub-basin. Challenges may include detecting pressure interference among neighboring storage facilities, establishing legal frameworks for pore space and mineral rights, delineating transportation rights-of way, and navigating diverse regulatory jurisdictions.

Future CSI activities include Carbon Storage Technology and Operations REsearch Facilities (CarbonSTORE), an initiative to develop field laboratories at CarbonSAFE project sites or similar locations at which experiments can be conducted to:

- Compare performance of advanced vs. existing technologies for monitoring carbon storage
- Obtain data to improve storage facility performance
- Reduce uncertainty regarding the potential long-term impacts of CO_2 injection

The Carbon Basin Assessment and Storage Evaluation (CarbonBASE) initiative, to be launched in late 2024, may include research on the following topics:

- Models and monitoring systems designed for assessing/ forecasting/predicting basin-scale storage resource performance and impacts
- Tools, data and technologies that stakeholders (e.g., industry, regulators, investors, insurers, federal/state land managers) can use to assess/confirm isolation of influences between independent projects or to detect interferences
- Studies aimed at enhancing the current understanding of basin-scale geomechanical influences and impacts
- Best practices for efficient and safe CO_2 storage resource management at the basin scale

Regional Resource Assessment for CO_2 Storage in New Mexico and Surrounding Areas

Dr. Jiyue Wu, Petroleum Recovery Research Center, New Mexico Institute of Mining and Technology, Socorro, NM

Dr. Sai Wang, Petroleum Recovery Research Center, New Mexico Institute of Mining and Technology, Socorro, NM

Dr. Na Yuan, Petroleum Recovery Research Center, New Mexico Institute of Mining and Technology, Socorro, NM

Dr. Xiaojing Ge, Petroleum Recovery Research Center, New Mexico Institute of Mining and Technology, Socorro, NM

1. Introduction

The "Regional Resource Assessment for CO_2 Storage in New Mexico and Surrounding Areas" project aims to explore and assess statewide resources in New Mexico for potential CO_2 storage through mineralization. This initiative, funded by the U.S. Department of Energy (DOE), focuses on CO_2 storage via mineralization within basalt formations, stratigraphic units, and mining waste sites. The New Mexico Institute of Mining and Technology leads this effort, investigating the mineralization potential to provide a long-term storage solution as part of broader carbon management strategies.

This DOE project seeks to mitigate climate change impacts by capturing CO_2 emissions and securely storing them. Mineralization, particularly within basalt-rich areas, offers an effective way to transform CO_2 into stable minerals that lock carbon in solid form. New Mexico's diverse geological structures, including extensive basalt formations, make it an ideal location for evaluating CO_2 storage using mineralization.

---

2. Literature Review

2.1 CO_2 Mineralization Processes

CO_2 mineralization, also known as carbon mineralization or carbonation, involves the chemical binding of CO_2 with minerals that contain divalent cations like Mg^{2+} or Ca^{2+}, forming stable carbonates. This approach provides a long-term, secure method of storing CO_2, minimizing leakage risks. Basalt formations, due to their mineralogy, are particularly suitable for such processes. Studies highlight that mineralization rates and stability depend on mineral composition, reaction conditions, and geological characteristics of the site. Key studies include those by Schaef and McGrail (2009), who investigated basalt dissolution kinetics under varying temperatures and pH conditions, providing valuable insights into the reactivity of basalt minerals with CO_2 under conditions relevant to storageesearch, such as Nielsen and Fisk (2008), examined the specific surface area of basalts and its impact on dissolution kinetics, which is critical for understanding mineralization rates. These studies underscore that factors like grain size, surface area, and mineral composition play essential roles in CO_2 sequestration efficiency.

2.2 Global Examples of Basalt-Based CO_2 Storage

International projects, like the CarbFix initiative in Iceland, demonstrate effective and safe CO_2 storage in basalt formations. CarbFix has shown that CO_2 can quickly mineralize within basalt, stabilizing as solid carbonates with minimal risk of re-release. In New Mexico, the Regional Resource Assessment Project aims to replicate the CarbFix success by applying similar methods under local geological conditions. Furthermore, Ramos et al. (2019) and Dungan et al. (1986) provided comprehensive chemical analyses of New Mexico basalt formations, such as the Sevilletta and Capulin basalts, identifying favorable chemistries for mineralization .

The project alss studies on basalt composition in regions like the Taos Plateau, which have shown high concentrations of divalent cations, promoting efficient CO_2 reactions. By analyzing these studies, the project builds upon a foundation of successful mineralization efforts worldwide, adapting methods for the distinct conditions found in New Mexico.

---

3. Project Goals and Objectives

The primary objectives of this project focus on identifying and assessing suitable sites within New Mexico for CO_2 storage, specifically through mineralization processes in basalt and other geologically appropriate formations. This effort aims to contribute significantly to the development of long-term, stable carbon storage solutions within the region. Achieving these objectives requires a combination of geological, hydrological, and socio-economic studies, supported by both experimental and simulation-based approaches.

To begin, the project prioritizes screening and site identification, a process that involves a detailed analysis of New Mexico's geological and hydrological characteristics. By evaluating these factors, researchers can pinpoint the most promising sites for CO_2 storage, focusing on areas where natural rock formations are well-suited to facilitate the mineralization process. This stage is essential for narrowing down locations with optimal conditions for safe and effective CO_2 storage.

Following site selection, the project moves into detailed characterization of the target formations. This involves a comprehensive study of the petrological, mineralogical, geophysical, and geomechanical properties of the identified rock formations. Understanding these properties allows the team to evaluate how these rocks will interact with injected CO_2, particularly with regard to their ability to form stable carbonate minerals over time. This step is crucial for confirming the suitability of the rock formations for large-scale mineralization and ensuring that CO_2 will remain securely stored.

Another core objective is the reaction dynamics study, which involves laboratory experiments to analyze how CO_2 interacts with the selected rock types under controlled conditions. By studying reaction rates, the project seeks to identify the conditions that optimize mineralization, including factors like temperature, pressure, and fluid composition. These laboratory studies provide valuable data that help refine the mineralization models, ensuring that CO_2 injection is both efficient and stable when scaled up to field applications.

The project also includes a storage capacity assessment using simulation models to estimate how much CO_2 each site can realistically hold. These simulations not only provide insights into the potential storage capacities but also analyze the economic viability of scaling up CO_2 storage efforts. By assessing costs and operational feasibility, this objective supports a realistic approach to implementing CO_2 storage solutions across New Mexico, balancing scientific rigor with economic practicality.

Finally, the project emphasizes stakeholder engagement to build awareness and foster support within local communities. This involves actively engaging with stakeholders through outreach activities, educating the public about the benefits and safety of CO_2 storage. By involving communities early in the process, the project aims to build trust and create a foundation of support that will be essential for the long-term success and acceptance of CO_2 storage initiatives.

Together, these objectives form a comprehensive strategy for assessing New Mexico's potential for large-scale CO_2 mineralization. Each step contributes to the groundwork needed to make CO_2 storage a viable component of the region's carbon management strategy, ultimately paving the way for future deployment of these methods on a larger scale.

4. Current Status of CO_2 Storage via Mineralization in New Mexico

4.1 Basalt Formations

The project team has identified multiple basaltic regions in New Mexico with substantial potential for CO_2 mineralization, focusing on locations where geological characteristics favor the stable and long-term storage of CO_2 through mineralization processes. These basaltic regions, rich in minerals that readily react with CO_2 to form stable carbonates, offer an ideal natural setting for testing and implementing mineralization strategies. Two primary sites have emerged as significant candidates for further exploration: the Taos Plateau Volcanic Field (TPVF) and the Raton-Clayton Volcanic Field.

The Taos Plateau Volcanic Field (TPVF) is particularly promising due to its unique geological composition and structure. This basalt field encompasses a broad range of basalt compositions, with layered structures that are advantageous for CO_2 injection and mineralization. Notably, the TPVF contains high levels of olivine, a magnesium iron silicate mineral that reacts effectively with CO_2 to form stable carbonate compounds. Olivine's high reactivity supports efficient CO_2 mineralization, potentially accelerating the carbonation process and enhancing storage stability. Additionally, the TPVF's structure, with its distinct layers, enables strategic injection points and minimizes the likelihood of CO_2 escape, making it a valuable natural laboratory for testing various mineralization scenarios.

The Raton-Clayton Volcanic Field is another key site, distinguished by its extensive geological diversity and accessibility. With over 125 known volcanic vents, this field presents a wide variation in basalt types across its landscape. This diversity allows researchers to test CO_2 storage and mineralization across multiple basalt compositions, providing insights into how different rock chemistries influence CO_2 reaction rates and carbonate stability. The presence of both fine-grained and coarse-grained basalts within the field offers opportunities to study how rock texture impacts mineralization processes, as well as the long-term containment of CO_2. The field's accessibility further supports ongoing sampling, monitoring, and experimental injection trials, positioning it as a versatile and resource-rich environment for mineralization research.

By focusing on the TPVF and the Raton-Clayton Volcanic Field, the project leverages New Mexico's unique geological resources to pioneer effective CO_2 storage solutions. Both fields provide not only the necessary mineral compositions for successful CO_2 mineralization but also a range of structural and compositional variations that contribute to a deeper understanding of basalt-based carbon storage. These insights will be crucial for optimizing storage techniques, not only within New Mexico but potentially across other basalt-rich regions globally.

4.2 Mining Waste Sites

In collaboration with the mining sector, this project explores the feasibility of using mining waste as an alternative medium for CO_2 storage. Mining waste often contains minerals with similar reactivity to basalt, making it capable of supporting mineralization processes that stabilize CO_2 as solid carbonates. The use of mining waste is particularly attractive as it not only increases the overall storage capacity but also repurposes existing industrial byproducts. This approach aligns well with goals to minimize additional land use and environmental impact by utilizing already-disturbed sites. Additionally, using mining waste reduces the need for pristine natural sites, which can help mitigate land access and environmental concerns often associated with CO_2 storage projects. The project team is working closely with mining companies to characterize waste materials, assess their mineral composition, and determine their CO_2 uptake potential through a series of laboratory tests and field studies.

---

5. Methodology

5.1 Site Screening and Characterization

To identify and evaluate potential CO_2 storage sites, this project uses a structured, multi-step screening process informed by geological and hydrological data. The team begins by gathering extensive regional data on geological formations, focusing on factors like mineral composition, rock porosity, and overall site accessibility. Field mapping and sampling are key components of this phase, allowing researchers to directly analyze the mineralogical, petrological, and geomechanical properties of target rock formations. This in-depth characterization provides essential baseline data that informs the assessment of each site's suitability for CO_2 injection and mineralization. Sites with high concentrations of reactive minerals, stable geological structures, and minimal risk of leakage are prioritized. Furthermore, hydrological data is analyzed to ensure that the potential injection of CO_2 will not adversely impact local groundwater systems.**5.2 Laboratory and Simulation Studies**

The project incorporates both laboratory experiments and simulation models to gain a detailed understanding of CO_2 storage dynamics within basalt and mining waste. Laboratory tests are conducted using batch and flow-through reactors that simulate in-situ conditions, allowing researchers to study how CO_2 dissolves, reacts, and precipitates under various temperature and pressure settings. By recreating natural formation waters in these reactors, the project team can observe the specific reaction pathways that lead to stable carbonate formation. This experimental data provides insight into optimal storage scenarios, helping to refine mineralization techniques for both basalt and mining waste materials. Following laboratory tests, the team uses simulation software to model storage capacities and CO_2 distribution within the formations. These models assess how injected CO_2 will behave over time, predicting its spread and reactivity, which aids in determining the scalability of storage solutions. The combination of laboratory results and simulation models offers a comprehensive approach to understanding CO_2 behavior, enabling the team to make informed decisions about field deployment.

---

6. Global Context and Comparisons

6.1 International Basalt-Based CO_2 Storage Projects

Research from various global projects highlights the effectiveness of basalt formations as a medium for CO_2 storage. In addition to the CarbFix project in Iceland, which has become a leading example of CO_2 mineralization, projects in regions such as the United States and Asia are investigating basalt as a storage solution due to its widespread availability and high mineral reactivity. The CarbFix project, for instance, has demonstrated that CO_2 can be rapidly mineralized in basalt formations, transforming into stable carbonate minerals and effectively eliminating leakage risks. Lessons from CarbFix have been instrumental in refining the New Mexico project's approach, as the project seeks to adapt these techniques to fit the unique geological context of the southwestern United States. This international comparison underscores the potential for basalt-based CO_2 storage and reinforces the viability of mineralization as a sustainable and secure carbon management strategy.

6.2 Regional Collaborations

To strengthen its regional approach, the project collaborates with neighboring states and organizations, including the Arizona Geological Survey. This partnership enhances regional outreach and research efforts, creating a more cohesive carbon management strategy for the southwestern United States. By sharing data, methodologies, and best practices, this collaboration fosters a unified approach to addressing CO_2 emissions and storage needs. The joint efforts also extend to coordinated outreach across multiple communities, enabling the project to educate a broader audience on the environmental and economic benefits of CO_2 storage. By aligning goals and resources across states, this regional network amplifies the project's impact, contributing to a more comprehensive strategy for carbon management in the region.

---

7. Identification and Characterization of Potential Storage Sites

7.1 Taos Plateau Volcanic Field

The TPVF's layered volcanic structure provides a range of CO_2 storage horizons, making it a prime candidate for mineralization studies. The project team has gathered detailed structural data on the TPVF's fractures, vesicles, and mineral textures to assess injectivity and stability for CO_2 storage.

7.2 Albuquerque Basin

Located in the Rio Grande Rift, the Albuquerque Basin has subsurface basalt zones ideal for CO_2 storage. These formations, intersected by wells at various depths, provide a rich dataset to evaluate scalability and injectivity.

7.3 Raton-Clayton Volcanic Field

This field's thick basalt layers are well-suited for CO_2 storage. The varied basalt compositions present an opportunity to study mineralization across different lava flows, enabling comparative studies within the same field.

---

8. Storage Capacity and Method

The project employs simulations to estimate CO_2 storage capacity in New Mexico's basalt formations. Laboratory tests have shown that CO_2 reacts favorably with basalt under controlled conditions, forming stable carbonates. The simulations confirm that New Mexico's basalt formations offer significant storage potential, and initial economic analyses suggest that these formations could be suitable for large-scale storage if deployed effectively.

The combination of lab data and simulation results provides a comprehensive understanding of the mineralization capacity and long-term stability of CO_2 in basalt formations. These findings support the economic feasibility of upscaling storage solutions in New Mexico, setting the groundwork for future commercial deployment.

---

9. Community and Stakeholder Engagement

Engagement with local communities is a key component of this project. Outreach efforts include workshops, educational programs, and collaborations with institutions such as the University of New Mexico. The project's Community Benefits Plan emphasizes the importance of educating local communities on the environmental and economic benefits of CO_2 storage.

By involving stakeholders in the planning process, the project has successfully raised awareness and gathered input from communities near potential storage sites. Specific outreach efforts include collaboration with mining companies, landowners, and local governments. Engaging these groups early in the project fosters trust and helps address concerns about environmental impact, ensuring broader community support for the project.

---

10. Summary and Next Steps

The Regional Resource Assessment Project demonstrates the viability of New Mexico's basalt formations and mining wastes as effective mediums for CO_2 storage via mineralization. The results thus far indicate that large-scale implementation is feasible, given the region's unique geology and the project's thorough site assessment methodologies.

Moving forward, the project will focus on advanced modeling, expanded laboratory experiments, and detailed site assessments to refine storage strategies. Collaborations with additional stakeholders and communities are planned to strengthen regional support. This project positions New Mexico as a leader in mineralization-based CO_2 storage, contributing valuable insights to the global carbon management community.

References

CarbFix Project. "CarbFix: Turning CO_2 into Stone." Accessed from https://www.carbfix.com/.

Dungan, M.A., Wulff, A., and Widom, E. (1986). "Petrology and Geochemistry of Basaltic Rocks from the Taos Plateau, New Mexico." Journal of Volcanology and Geothermal Research, 30(1), pp. 29-54.

Nielsen, H., and Fisk, M.R. (2008). "Basalt Weathering Processes and CO_2 Sequestration Efficiency." Geochemistry, Geophysics, Geosystems, 9(3), pp. 225–240.

Ramos, F., Bradshaw, M., and Singleton, M. (2019). "Mineralogy and Geochemistry of New Mexico Basalt Formations for Carbon Sequestration." Environmental Earth Sciences, 78(2), pp. 120-128.

Schaef, H.T., and McGrail, B.P. (2009). "Dissolution Kinetics of Basalt in CO_2-Saturated Water." Environmental Science & Technology, 43(8), pp. 2538–2544.

352c CO_2 Storage Site Selection: A Case Study from Oklahoma, US

Priyank Jaiswal, Aaditya Khanal

We present a site-selection case study for sequestering CO_2 in Oklahoma, US. The selection is made at two levels. First, a state-wide search is conducted hierarchically where we try to avoid areas with major faults and recent seismicity (2013 – 2018) as much as possible while remaining close to emission points and existing CO_2 pipelines (Figure 1). This narrowed our selection to two counties near the Oklahoma City metro area, Kingfisher and Canadian. The two counties are known for their oil and production from STACK play, which comprises a set of stacked reservoirs of various ages and depositional styles. Next, using a 3D seismic survey from Canadian County, we try to gain a deeper understanding of the structures and stratigraphy in the context of the STACK's CO_2 storage potential. We use data attributes such as impedance and coherence to compare the different formations and realize that although over 3000 ft of Cambrian – Permian Age sediments are available for injection in theory, in practice they all have unique risks. For example, the deeper formations have high continuity but are dominated by faults and fractures while the shallower formations are stratigraphically compartmentalized but largely free of faults. In the ongoing work, we rank the formations by evaluating the pros and cons of injecting into various formations. Various criteria such as formation depth, reservoir pressure, composition, legacy wells, and stratigraphic architecture determine the fate of the injected plume.

In this article, we demonstrate our ranking strategy through two representative sets of dynamic injection simulation examples. The simulator is physics-based and uses realistic relative permeability and capillary pressure curves from experiments on the closest analogs. The simulations account for the variation in relative permeability endpoints and entry pressure, as well as other uncertainties such as the expected changes in pore architecture and the impact of various injection rates on near-wellbore phenomena such as halite precipitation. The first set of simulations is done for the Pennsylvanian section comprising formations of Desmoinesean, Missourian, and Virgilian ages, that are dominantly siliciclastic. In the study area, the Pennsylvanian section drapes over the deeper structures and is overlain by an anhydrite-rich Permian regional seal. Although the Pennsylvanian sections can sustain good fluid flow due to the high porosity (5% – 15%) and high permeability (100s of mD), their complex internal stratigraphic architecture leads to highly uneven plume dispersion. The other set of simulations is done for the Meramec Age formation, a self-sealed and mixed siliciclastic-carbonate "tight" formation with primary storage in the pre-existing fractures. The primary trapping mechanism in the Pennsylvanian is capillary storage while in the Meramec is mineralization.

The 3D seismic volume used in this study is from El Reno, a city to the west of the Oklahoma City metro. It is hereafter referred to by the same name, the El Reno survey. Data were acquired using IVI type vibroseis source that simultaneously performed 4 linear 10 – 100 Hz sweeps each 12s long with a start and end taper of .3s and a peak ground force of 70% of the weight of the vibroseis. In all the surveys, the receiver type used was a SM24 10Hz resonance frequency geophone. Data were acquired with a 2ms sample interval and a bin spacing of 25 m (82.5 feet). The survey area is ~ 125 square miles comprising 1428 inlines (north-south) and 564 crosslines (east-west). The data were processed to zero-phase reverse polarity and imaged using an advanced orthorhombic pre-stack time migration (PSTM) to account for both the vertical transverse anisotropy (VTI) and horizontal transverse anisotropy (HTI). Data have a bandwidth of 15–65 Hz with a 35–40 Hz dominant frequency.

The study is broken into three parts – seismic interpretation, model building, and plume simulation. Seismic interpretation follows the standard practice of well-to-seismic-tie, which allows the interpreter to relate the log tops to seismic wiggles, followed by fault and horizon picking. Although the study area is densely covered by wellbores of various vintages, wells containing sonic and density logs, which are needed for simulating synthetic traces are scarce. This limited our well dataset to two boreholes only, hereafter referred to as A and B. Both are vertical wells that mainly sample the Meramec, which has been the recent focus of the operator's exploration efforts. Formation depths in both wells were provided by the operator driven by the subsurface knowledge accumulated from decades of exploration in the region. The well-tying is done in the time domain with the understanding that if the depth-to-time conversion of the well log (e.g., using a check shot or sonic log) is accurate, the waveform character of the synthetic trace will align with the corresponding waveform character of the seismic data. When the correlation is not apparent, a match is achieved in an interpretive manner by "stretching" and "squeezing" the individual wavelets in the synthetic seismic trace and/or bulk shifting it which implies changing the interval velocities. Theoretically, such adjustments are justified because of the inherent conversion inconsistencies between stacking and interval velocities. Here, we focused on correlating the wavelet characters corresponding to the tops of the Hunton, Osage, and Chester formations, obtaining an overall match of 79% in Well A and 85% in Well B.

Following the seismic interpretation, a geo-cellular model was constructed. These models, essential for flow simulations, are a collection of various petrophysical properties such as porosity, permeability, water or hydrocarbon saturation, and lithofacies within a regularized grid pattern. Geocellular modeling requires various data types, including geophysical, geologic, petrophysical, and engineering, to be integrated at the reservoir scale. The process is challenging for two reasons – not all the modeling parameters can be determined at all locations, and the size of the simulation grid is often much larger than the size of the samples used for lab measurements. As a result, the flow parameters such as porosity and permeability, must be significantly upscaled depending on the size of the cell chosen for simulation. For example, to achieve the simulation completion within a reasonable time, the cell size is often kept at 10m X 10m X 10m. In comparison, samples used for lab measurements are cm scale. In this case study, we addressed the conundrum of upscaling by populating the cells statistically following a facies-based expected value.

Preliminary results from the numerical experiments show that CO_2 trapped by dissolution, capillary trapping, and mineralization is significantly affected by the various uncertain reservoir properties and decision parameters. The vertical permeability anisotropy significantly impacted the plume migration throughout the reservoir. We observed significant dissolution in the higher permeability region, whereas capillary trapping was dominant in the low permeability zones. As a result, the mobile phase CO_2 disappeared in the reservoir faster in high permeability zones compared to low permeability zones. The faults and fractures resulted in significant channeling of the injected CO_2 within the formation. We observed a substantial reduction in the pH of the aquifer brine from the initial value of around 7. The pH near the well bore rapidly dropped to 4.45, gradually reducing to 4.13 during the active injection period. The pH change was only seen close to the wellbore for the injection layer, and there was no pH change far away from the injection site. However, the pH in the upper layer close to the caprock was significantly depressed beyond the wellbore due to the buoyant CO_2 spreading throughout the formation. This reduction in pH had significant implications for Meramec formation but a limited impact on the Pennsylvanian formation. For each case, the capillary pressure enhanced the mineralization, which can be attributed to the occlusion of $ScCO_2$ in pores due to capillary forces rather than being displaced as a continuous plume. Furthermore, the rate of dissolution and the transport of dissolved CO2 is governed by the pore-scale configuration of different phases, which is controlled by the capillary pressure. Our simulation results also showed a significant impact of divalent ions (Ca^{++} and Mg^{++})on the mineralization and dissolution of minerals such as calcite and dolomite. We also observed that the highest injection rate led to the least fraction of mobile CO_2 in a reservoir. This is mainly because relative permeability hysteresis and capillary pressure trap a more significant fraction of the injected CO_2 at a high injection rate. We also identify locations on the reservoir that are suitable for safely containing injected CO_2 based on the reservoir porosity and permeability. Finally, we present preliminary results for the loss of injected CO_2 from neighboring leaky abandoned wells. Our results showed that injecting in the sweet spot resulted in a rapid dissolution of CO_2, leading to a reduced probability of diffusion through the existing wells. The result from the study provides critical insights for the proper selection of CO_2 storage sites for secure long-term storage of the injected CO_2, accounting for various reservoir and operation uncertainties.

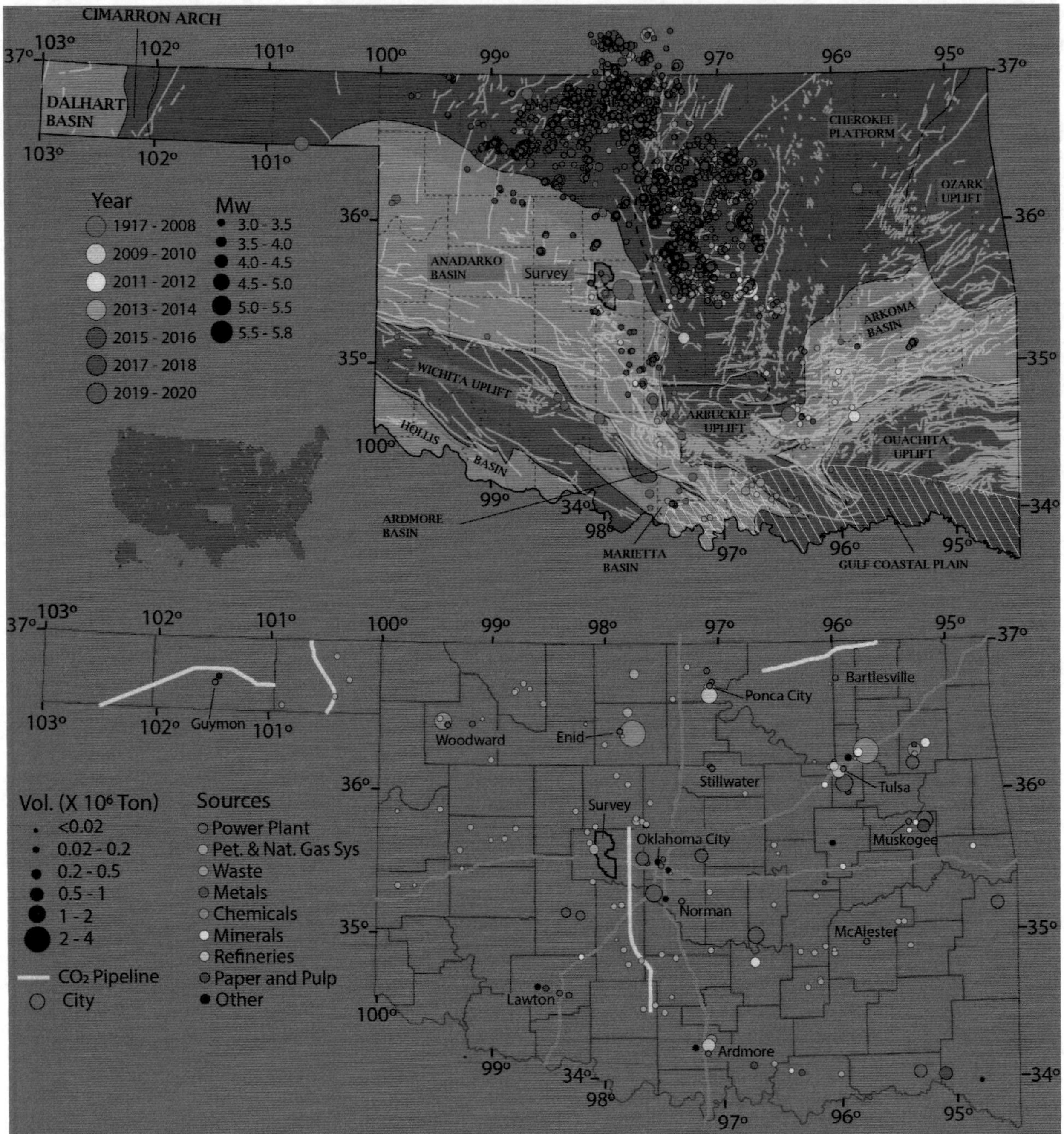

Figure 1. Base map. (A) State map of Oklahoma showing counties in dotted lines, major faults in semi-transparent solid blue lines (source: Oklahoma Geological Survey), and > 3 Mw epicenters color coded by year-range and size coded by magnitude (source: USGS). Key tectonic provinces are labeled. Inset shows location of Oklahoma within USA. (B) State map of Oklahoma showing counties in solid lines, major cities and freeways, and reported emissions for 2019 color coded by source -type and size coded by CO2-equivalent volume (source: OGS). Also shown are the current CO2 pipelines in blue. In (A) and (B), solid black polygon is the seismic survey used in this paper. It is located in Canadian County.

352d Implications for Developing Multiple CCS Projects in Same Geological Setting – US Gulf Coast Case Study

Seyyed Hosseini, Zhichang Wang

CO_2 injection into geological formations will increase the local and regional pressure. Currently many projects have been planned and developed across USA and all those projects looking for best geological formations close to major sources of CO_2 trying to reduce the pipeline length and cost. Many of the proposed projects end up being in very close proximity to each other. Texas and Louisiana Gulf Coast is one of those major targets for CCS that has this specific feature. We have developed analytical models that closely model two-phase flow behavior of the CO_2 injection into porous media and allows for fast analysis of interference among multiple CCS projects. Our results show that interreference among projects is likely and it can potentially hinder project implementations both from technical point of view and legal challenges for the CCS community.

352e Can Fault Activation Cause Long-Term Caprock Integrity Issues? Findings from a Self-Sealing Field Experiment

Jens Birkholzer, Yves Guglielmi, Christophe Nussbaum, Frederic Cappa, Tanner Shadoan, Jonathan Ajo-Franklin, Florian Soom, G W Lanyon, Paul Cook, Chet Hopp, Verónica Rodríguez Tribaldos, Michelle Robertson, Todd Wood, Senecio Schefer

The widespread pressure increases expected from large-scale deployment of CO_2 storage in the deep subsurface may cause unwanted geomechanical effects such as caprock failure per fault activation. As previously reported (e.g., Guglielmi et al., 2021; Cappa et al., 2022), we conducted a series tens-of-meter scale experiments of fault activation within the Main Fault intersecting the Opalinus Clay formation in the Mont Terri underground research laboratory in Switzerland. These tests involved short-term fluid injections into the fault to explore how aseismic and seismic events could jeopardize the integrity of a sealing caprock overlying a CO_2 sequestration reservoir. The experiments demonstrated that significant fluid leakage can occur along the initially low-permeability shale faults when rupture is activated. However, an apparent mechanical closure of the fault was observed as soon as injection ceased and fluid pressure dropped. Unfortunately, it was not clear at the time whether the apparent mechanical closing would lead to complete hydraulic sealing of the fault. There were some indications of a small residual opening remaining that could allow for small long-term leakage.

Here we report on findings from long-term monitoring of the time-dependent sealing of the activated patch of the Mont Terri fault. The monitoring study started in 2020 right after the highly permeable flow path of about 10 m length had been created by injecting water inside the fault zone. The hydraulic transmissivity of the flow path increased by several orders of magnitude from initially very small (10^{-12} m²/s) to about 3.2 10^{-7} m²/s. In order to observe whether and/or when the fault transmissivity would return to its initial low transmissive state, pressure pulse tests were repeated every two weeks during the year following the fault activation. Local fault transmissivity was estimated by matching the analytical Cooper-Papadopoulos solution to the observed pulse pressure decay vs time. Fault transmissivity variations were then compared to fault movements monitored during and after activation by a network of distributed optical fibers (Distributed Strain Sensing, DSS), local three-dimensional borehole displacement sensors, and changes in p-waves velocities measured with a Continuous Active Seismic Source Monitoring system (CASSM).

At the injection point, the strong transmissivity increase observed during the high-pressure fault activation was initially associated with dilatant slip caused by normal faulting, with a strike slip component in good accordance with the local state of stress, and subsequent further opening due to a decrease in the effective normal stress (Cappa et al. 2022). The peak dilation value of about 400 micro-strains measured during the injection experiment decreased with time until a permanent irreversible fault zone deformation of about 200 micro-strains was established about 70 to 200 days after activation, mainly related to irreversible fault shear displacement. During the first 10 days after activation, the pressure pulse tests showed a fast drop in transmissivity from 3.2 10^{-7} m^2/s during the high-pressure activation phase to 1.8 10^{-9} m^2/s (Figure 1). This period was dominated by normal closure of the fault due to the injection pressure drop and subsequent effective stress increase. After 10 days and until 70 days, there was a slower transmissivity decrease related to an approximately ~1.6 10^{-11} s^{-1} compaction creep of the fault zone. After 70 days, fault deformations progressively followed an extensional regime while transmissivities showed a slightly accelerated decrease. We hypothesize that factors such as clay mineral swelling may have influenced this long-term behavior.

We discuss this complex evolution of fault self-sealing processes using different long-term fault transmissivity relationships by drawing parallels with generic creep laws used to describe long term deformation of clay rocks at the laboratory scale (Figure 1). Due to the multiple competing processes involved, describing the fault sealing process with a single relationship proves challenging. By extrapolating these models, we estimate complete fault sealing may take at least 50 years. However, it is important to note that our estimations are based on experiments conducted at shallow stress levels, involving a limited volume of fluid leakage and a brief fault activation duration. While our data align reasonably well with laboratory-scale studies under similar stress and rock conditions, the variations in clay content within faults spanning multiple mudrock layers and the nature of fluids may influence fault sealing timescales differently than observed in our experiment. This underscores the necessity for additional observations at relevant basin-scale depths to enhance our understanding of these processes.

Figure Caption

Figure 1. Evolution of predicted fault transmissivity over time compared to experimental data estimated from the series of pulse tests (blue dots with error bars taken at 20% to take into account possible uncertainty of measurements). Gray lines are obtained for 100 inversions with a first transmissivity law, while the orange lines correspond to 100 inversions with a second transmissivity law. The red dashed line represents the mean solution among the 100 best-fit solutions for each law.

References

Guglielmi, Y., Nussbaum, C. Cappa, F., DeBarros, L., Rutqvist, J. and J. Birkholzer [2021] Field-scale Fault Reactivation Experiments Highlight Aseismic Leakage in Caprock Analogs: Implications for CO_2 Sequestration, International Journal of Greenhouse Gas Control, 111(103471).

Cappa, F., Guglielmi, Y., Nussbaum, C., De Barros, L. and J. Birkholzer [2022] Fluid Migration in Low-Permeability Faults Driven by Decoupling of Fault Slip and Opening. Nature Geosci 15(9), pp 1-5.

Shadoan, T.A., Ajo-Franklin, J.B., Guglielmi, Y., Wood, T., Robertson, M., Cook, P., Soom, F., Daley, T.M., Hopp, C., Tribaldos, V.R. and P. Marchesini [2023] Active-Source Seismic Imaging of Fault Re-Activation and Leakage: An Injection Experiment at the Mt Terri Rock Laboratory, Switzerland. Geophysical Research Letters, 50(23), p.e2023GL104080.

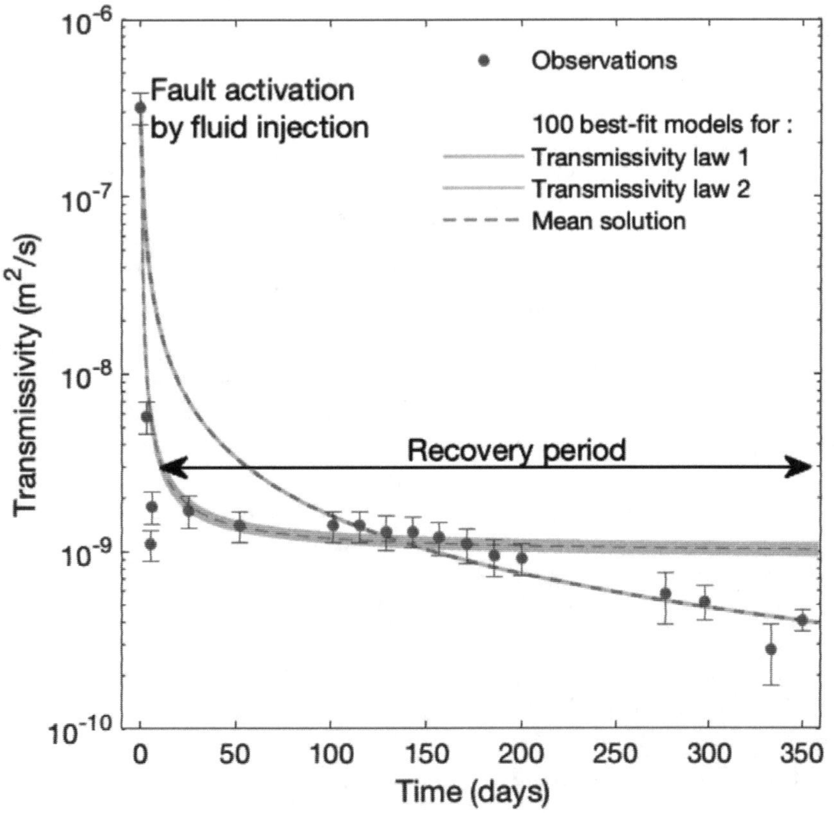

352f Olivine Dissolution in CO_2 Saturated Water Under Reservoir Conditions

Qiaoyun Chen, Martin Trusler

Geological carbon storage (GCS) is one of the most promising and reliable techniques to mitigate the atmospheric CO_2 concentration. Among the four trapping mechanisms of GCS, CO_2 mineralization provides the most permanent way for CO_2 fixation by converting CO_2 into carbonate rocks. Olivine is one of the most promising minerals for CO_2 mineralization because it is abundant in Earth's upper mantle. The Olivine dissolution kinetics under reservoir conditions are important to predict the long-term fate of the injected CO_2. In this work, experimental studies on Olivine dissolution in CO_2 saturated water at pressures from 76 to 155 bar, and at temperatures from 373.15 to 473.15 K, were conducted.

Experiments were performed with a batch reactor system, and liquid samples were withdrawn from a reactor with an auto-sampling system. The schematic of the experimental system is shown in Fig. 1. The two batch reactors were equipped with a heater jacket, a thermocouple, a pressure sensor, and a stirrer shaft assembly with impellers. The liquid from reactor 1 was transferred to reactor 2 via a transfer tubing between them. The CO_2 was pressurized by a syringe pump. The auto-sampling system was used to collect the filtered liquid samples from reactor 2. Two six-port sampling valves were used: one for collecting liquid samples, the other for collecting in-line solvent. N_2 was used to purge the in-line solvent together with the liquid sample to one of the centrifuge tubes containing some pre-added additional solvent, of the fraction collector.

Ultra-pure water and gem-quality Olivine particles were added to reactors 1 and 2, respectively, before injecting CO_2 and heating up the batch reactor system. Once CO_2 and water reached equilibrium in reactor 1, the CO_2 loaded water was transferred from reactor 1 to reactor 2 to provoke the Olivine-CO_2-water reaction under stirring. Liquid samples that were withdrawn from reactor 2 periodically were analyzed by ICP-MS, so the concentrations of Mg, Si and Fe were obtained. The recovered Olivine particles were analyzed by XPS. The initial dissolution rate of Olivine was determined by the following equation:

$r = (\mathrm{d}c/\mathrm{d}t)(V/A)$. (1)

Here, r denotes the dissolution rate, V denotes the volume of the solution, A denotes the mineral exposure area to the solution, c denotes ion concentration, and t denotes time.

The XPS analysis results of the particles showed that during the Olivine dissolution, secondary phase was formed on the particle surface along with the Olivine dissolution. Silica was mainly formed at low temperature, while Fe-rich secondary phase was mainly formed at higher temperatures. The formation of the Fe-rich secondary phase can be first easily identified by the change of the colour of the particle, and can also be revealed by the decreasing trend of the concentration of Fe during the reaction. The Olivine initial dissolution rate declined first and then inclined with the increase of temperature, when the pressure was fixed. This non-monotonic trend showed that the dissolution rate did not only depend on temperature, but the Fe-rich secondary phase on the particle surface has a stronger inhibition of Olivine dissolution. Fig. 2 shows some results of initial dissolution rates against temperature. Here the dissolution rates of Mg and Si were first calculated, after which the stoichiometric Olivine dissolution rate was converted based on the chemical formular, finally an average of the rates calculated by Mg and Si was taken. Fig. 2 also shows that the influence of pressure on olivine dissolution was slight until the temperature was increased to 473.15 K.

The results from this study extend the database of Olivine dissolution rates up to a temperature of 473.15 K. This study helps understand the long-term fate of the injected CO_2 and the geochemical reaction under reservoir conditions, and helps optimize the design of a GCS project.

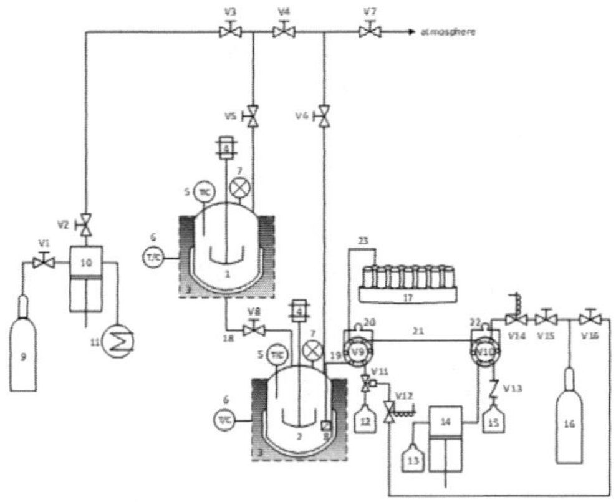

Fig. 1 Schematic of the experimental system. 1, Vessel 1; 2, Vessel 2; 3, heater jacket; 4, stirring assembly; 5, 6, thermocouple; 7, pressure sensor; 8, filter; 9, CO_2 cylinder; 10, syringe pump; 11, chiller; 12, 15, waste bottle; 13, solvent storage bottle; 14, syringe pump for solvent; 16, N_2 cylinder; 17, fraction collector; 18, transfer tubing; 19, connecting tubing between vessel 2 and sampling system; 20, sample loop; 21, connecting tubing between two sampling valves; 22, solvent loop; 23 connecting tubing between fraction collector and sampling valves. V1 ~ V8, V15, V16, manual valves; V9, V10, sampling valves; V11, on/off valve; V12, V14, solenoid valves; V13, check valve.

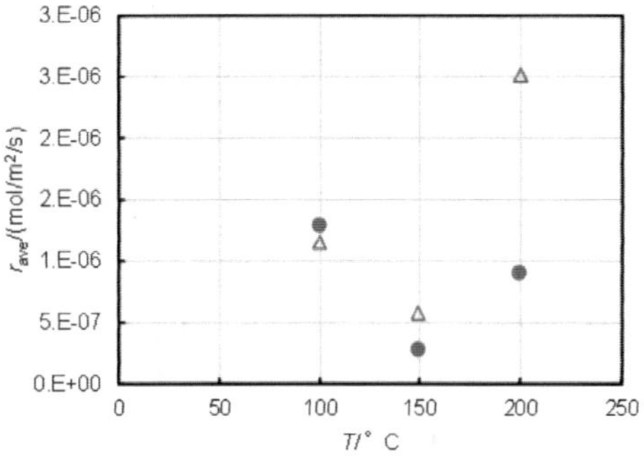

Fig.2 Olivine dissolution rates against temperature. △, 76 bar; ●, 155 bar.

352g A Quantitative Approach to Assessing CO_2 Permanence for California Air Resources Board (CARB) CCS Protocol

Zahra Finnigan, Stacy Kouba, Travis C. McGrath, Mark C. Ciardelli, Nicholas Azzolina

The Energy & Environmental Research Center (EERC) and Foth Infrastructure & Environment, LLC (Foth) developed a general methodology for assessing storage permanence as part of a site-based risk assessment conducted to address California Air Resources Board (CARB) Carbon Capture and Sequestration (CCS) Protocol (). This presentation will summarize a general methodology for assessing storage permanence under the CARB CCS Protocol, 2.2 Risk Assessment (f), which states: "Only sites in which the fraction of CO_2 retained in the storage complex is very likely (greater than 90% probability of occurrence) to exceed 99% over 100 years post-injection will be eligible to receive Permanence Certification."

The CARB CCS Protocol requirement for demonstrating storage permanence is quantitative because it requires quantifying probabilities of the occurrence of CO_2 loss events and quantifying uncertainty in the magnitude of CO_2 loss.

Quantitative risk assessment can be conducted in a variety of ways, depending on the quantity being evaluated (in this case, the probability of retaining at least 99% of injected CO_2), the desired level of robustness and defensibility, and the resources (time, expertise, and money) available. Corresponding methodologies can range from direct "lumped assessment" of the desired quantity to highly sophisticated probabilistic, dynamic (time-stepping), disruptive-event-based numerical multiphase flow simulations, with various potential approaches in between.

In between these two ends of the approach spectrum are methods that balance adequate defensibility with resource cost. These methods involve analytical (nonnumerical) characterization of the probability of CO_2 retention. These approaches describe how the system components could potentially fail and release CO_2 from the storage complex.

One such approach identifies potential system or component "failure modes" that could occur and, if they occur, lead to "failure." Each failure mode is described in terms of two key elements:

- The probability of the failure mode occurring: Potential "trigger events" that could occur in various combinations and lead to the loss of CO_2 are characterized in terms of "fault trees." The probability of each trigger event is assessed using available data (from site characterization, testing, or modeling) and the judgment of subject matter experts (SMEs). These probabilities can then be combined through a probabilistic expression for the probability of the failure mode's occurrence, defined by the fault tree, which in turn reflects the nature of the trigger event combinations leading to failure (e.g., Trigger Event A and Trigger Event B must occur to lead to failure).

- The amount of CO_2 lost from the storage complex if the failure mode occurs: This potential loss can be characterized by a single (deterministic) value or an uncertain amount lost in the form of a probability distribution. These assessments also leverage available data and SMEs related to the movement of CO_2 under the failure conditions or analogs.

The occurrence of each failure mode can then be simulated numerically, and if it occurs, the amount of CO_2 lost can also be simulated numerically. The amount of CO_2 lost across all the failure modes over the life of the project can be combined (added) to determine the amount of CO_2 lost for one potential project future. Thousands of potential project futures can be simulated to infer a probability distribution for CO_2 loss and, by extension, the probability of CO_2 permanence.

This approach for evaluating the probability of CO_2 permanence was adopted because the general methodology (i.e., failure modes and fault trees) has a historical usage pedigree for complex, uncertain problems because it provides adequate defensibility for reasonable effort (e.g., National Aeronautics and Space Administration, 2002; U.S. Nuclear Regulatory Commission, 1999; Vick, 2002).

References

California Air Resources Board, 2018, Carbon capture and sequestration protocol under the Low Carbon Fuel Standard [Online]: https: https://www.carb.ca.gov/sites/default/files/2020-03/CCS_Protocol_Under_LCFS_8-13-18_ada.pdf (accessed March 2024).

National Aeronautics and Space Administration, 2002, Fault tree handbook with aerospace applications: Washington, D.C., Office of Safety and Mission Assurance.

U.S. Nuclear Regulatory Commission, 1999, Framework for risk-informed regulation in the Office of Nuclear Material Safety and Regulation: SECY-99-100, Washington, D.C.

Vick, S.G., 2002, Degrees of belief—subjective probability and engineering judgment: Reston, Virginia, American Society of Civil Engineers.

352h Impact of Impurities on CO_2 Transport in Saline Aquifers

Prince Otabir, Aaditya Khanal

Large scale sequestration of CO_2 in subsurface porous media is one of the most promising methods to limit the concentration of CO_2 in the atmosphere. The transport of the injected CO_2 in subsurface formations depends on various factors including the properties of the reservoir, aquifer brine, and operational conditions. The flow behavior of CO_2 also depends on the presence of various impurities that may be present in the captured gas. This work investigates the impact of H_2S, a common impurity, on different trapping mechanisms during a long-term sequestration of CO_2 in subsurface formations.

For this purpose, we use a compositional reservoir simulator to model the injection of CO_2 with different fraction of the impurities for 10 years of active injection followed by 15 years of post-injection phase. The impact of capillary pressure, and relative permeability hysteresis effect on various CO_2 trapping mechanisms is also considered . We also investigate the impact of well trajectory on near-wellbore phenomena of halite precipitation and permeability alteration for pure CO_2 injection case and the case with impurities. Furthermore, the impact of brine with different ionic concentrations in CO_2 dissolution is assessed using different salts, commonly present in saline aquifers with varying ionic concentrations.

The results from the study showed that the presence of H_2S had a significant impact on dissolution trapping. Compared to the case with no impurity, significant enhancement of CO2 dissolution was observed throughout the reservoir. Furthermore, the results also showed that the dissolution was increased for a well with horizontal perforations compared to the vertical perforations. The halite precipitation was also significantly lower in case with horizontal perforations which resulted in only a moderate reduction in permeability compared to the vertical case. However, higher halite precipitation was observed for the case with impurities. The results were used to provide qualitative and quantitative insights which could be used to evaluate and optimize CO_2 sequestration projects efficiently.

This study represents a comprehensive reactive transport model to investigate the CO_2 sequestration in subsurface formations. The results will provide critical insights that can be used to improve upon the existing knowledge on the effect impurities on CO_2 trapping by different mechanisms.

432a Managing a Gigaton Carbon Sequestration Future: A Framework for Prediction of Fault Activation Caused By Basin-Scale CO_2 Storage

Jens Birkholzer, Stanislav Glubokovskikh, Yves Guglielmi, Abdullah Cihan, Jonny Rutqvist, Meng Cao, Preston Jordan, Matthew Reagan

Introduction

After decades of research on geologic carbon storage, the world is finally moving from pilot tests and demonstration experiments to industrial-scale implementation. Many large-scale CO_2 storage hubs are currently in development across the world, notably in the United States, Europe and China. Even further expansion of carbon capture and sequestration to perhaps a gigaton scale is needed to achieve a net zero clean energy future. Sedimentary basins with interconnected reservoirs will likely host multiple CO_2 storage sites between which pressure interference can be expected. The cumulative effect of multiple CO_2 projects in a given region may limit its overall storage capacity, may enlarge the project area that needs to be characterized and monitored, and may cause unwanted geomechanical effects such as generating seismic events and caprock failure per reactivation of critically stressed faults. Such risks need to be carefully managed.

Here, we focus on regional-scale CO_2 storage management with particular emphasis on geomechanical risks and discuss our ongoing development of a versatile framework for simulation-driven storage optimization at a regional scale. First, we use coarse poroelastic flow simulations across an entire basin (with multiple CO_2 storage projects) to identify potentially critically-stressed faults. Second, we simulate rupture on these faults using coupled hydromechanical simulations of fault slip at the scale of individual CO_2 storage projects. Finally, we use results from the project-scale simulations to provide parameters for sophisticated simulations of dynamic rupture propagation using rate and state friction and Cam-Clay models.

The state-of-the-art for assessment of induced seismicity and caprock failure risks relies on overpressure as the main control for the fault stability via the Coulomb failure criterion. Such a simplified view disregards several key mechanisms. First, poroelastic stresses triggered by waste water injection proved to be an important factor driving the increase of seismicity in Oklahoma. Second, a few recent field tests, including the controlled injection tests at Mont Terri (Switzerland), showed a clear relationship between the seismic/aseismic fault slip and permeability increase in the caprock faults. Third, multiphase effects that accompany CO_2 filtration through reservoir faults may boost their susceptibility to slip. Our simulation-driven framework aims to incorporate these physical mechanisms explicitly across a broad range of scales. In the following, we present a hypothetical basin model along with several injection scenarios, which highlight the importance of project-to-project interaction as well as accurate upscaling of the fault physics.

Simplified Poroelastic Basin-scale Simulations

The key challenge of regional-scale optimization arises from the fact that pressure/stress changes in a given basin occur over several years/decades and have characteristic dimensions of tens of kilometers, while fault reactivation is rapid and local, and thus requires frequent time steps as well as accurate representation of the fine-scale fault geometry and rheological properties. We implement a number of practical strategies to bridge the gap between the basin-scale computations of pressure and stress changes and sophisticated coupled hydromechanical simulations of localized fault rupture. The simplest approach locates the segments of faults with sufficiently high Coulomb stress on the faults in the coarse-scale poroelastic simulations. These segments are then passed through a suite of project-scale coupled hydromechanical simulations with the boundary and initial conditions extracted from the coarse-scale simulations.

We illustrate the coarse-scale poroelastic simulations using a synthetic benchmark model shown in Figure 1a-c. The model is based on the local geology in the San Francisco Bay Area and some scaling relationships between the fault parameters, such as throw, length, spacing. This purely hypothetical study allows us to test optimizing the computational parameters as well as quantifying the impact of various physical effects, such as 2-phase flow versus single-phase injections, boundary conditions, hydraulic connectivity to the basement, and the injection schedules for different projects. We find for example that only one fault becomes critically stressed for a scenario with constant injection volumes and simultaneous injection start of five CO_2 storage projects (Figure 1d). We analyze the potential for the failure of this fault in the next section.

Advanced Simulations of a High-Resolution Rupture Patch for Complex Faults

On the next level of detail, we use the TOUGH-FLAC simulator to capture the reactivation of the fault with complex geometry highlighted in Figure 1d. Figure 2a shows the computational mesh for TOUGH-FLAC simulations with a complex fault geometry. The size of the mesh is 5 km x 5 km x 1.4 km. Three layers (caprock, reservoir, and basement ordered from the top to bottom) are used. We consider injection of 0.5 Mt of supercritical CO_2 injection at 2,100 m depth into the hanging wall of a normal fault, which is offset from the injection by 140 m. After 2 years, the rupture nucleates at the boundary between the reservoir bottom and the underlying low-permeability rocks (Figure 2b), while red indicates past reactivation after 20 years of injection (Figure 2c). A majority of the slipped elements gets activated within the first 2 years of the injection. The rupture nucleates at the top of the basement, below the bottom of the storage reservoir. As more CO_2 is injected, the rupture propagated upwards a little bit, but it mainly spreads horizontally: about 1,400 m along the fault strike and about 140 m up-dip. Caprock integrity is thus not affected. Interestingly, in the case of an injection in the foot wall of the fault, the rupture occurs at the top of the reservoir and tends to extend horizontally and down-dip.

Estimating Rupture Acceleration to Seismicity

Thus far, we have considered a Coulomb-Mohr failure criterion on the faults, which precludes estimation of the magnitude of the triggered events. Here, we derive the range of seismic magnitudes using highly complex models for rupture acceleration to seismicity on the relatively small fault zone modeled in Figure 2. We compare two theoretical frameworks for the fault stability (Figure 3): (1) the Rate and State Friction model (RSF) and the Modified Cam-Clay model (MCC). In both models, the injection into the hanging wall of the fault creates a 5 MPa pore pressure increase in the fault zone after 2 years. Poroelastic stress changes induce a small 8 mm/year tangential displacement on the fault and 3 mm/year normal compression (Figure 3a). Preliminary results are: (1) the RSF and MCC model predictions provide similar estimates of fault slip acceleration and amplitude (Figure 3b-e: interval 3-4 of plastic instability, 15cm slip, 30-40cm/s); (2) the RSF model underestimates the preceding slow slip period which could have strong implications on fault permeability variation associated to dilatant rupture (Figure 3d-e: interval 2-3 of plastic stability, 30 cm slip with MCC and 8mm slip with RST); (3) the MCC is better suited to consider the effects of fault bulk plastic deformation on slip and leakage.

Conclusions

We present a workflow for simulation-driven assessment of risks of induced seismicity and fault leakage potentially caused by basin-scale CO_2 storage. The workflow starts with simplified multiphase poroelastic flow simulations to detect potentially critically-stressed faults, then zooms in on the detailed hydromechanical rupture behavior of these faults with advanced coupled hydromechanical simulations, and finally estimates seismic magnitudes of relevant fault slip scenarios with various constitutive laws for the dynamic rupture propagation. We propose this workflow as a practical strategy for multi-scale simulation of both slow and seismogenic slip on faults as well as estimation of the resultant magnitude.

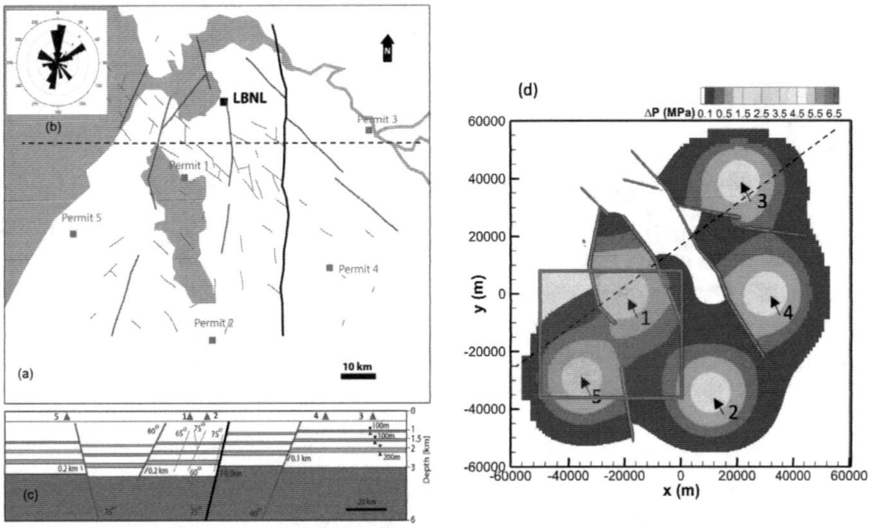

Figure 1. *Poroelastic large-scale simulations using a synthetic model, LBNL-basin. The in-plane view of the basin (a) shows five hypothetical storage projects along with three types of normal faults (black, red, and green solid lines) with varying orientations (b) and vertical offset (c). The pressure increase (d) critically stressed the lower segment of the most western fault (red rectangle).*

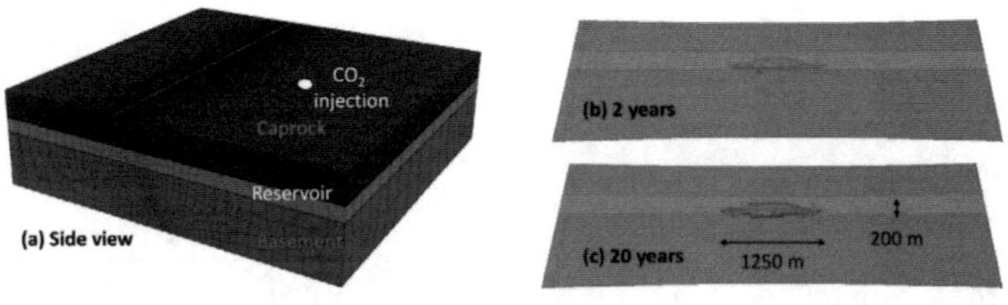

Figure 2. *Advanced hydromechanical simulations on the project-scale using TOUGH-FLAC3D. The 3D computational mesh represents a normal curved fault intersecting a 3-layer system. The rupture patches on the fault plane after 2 years (b) and 20 years (c) show that the fault slip extends laterally rather than vertically, so that the risk of leakage into the overburden is minimal.*

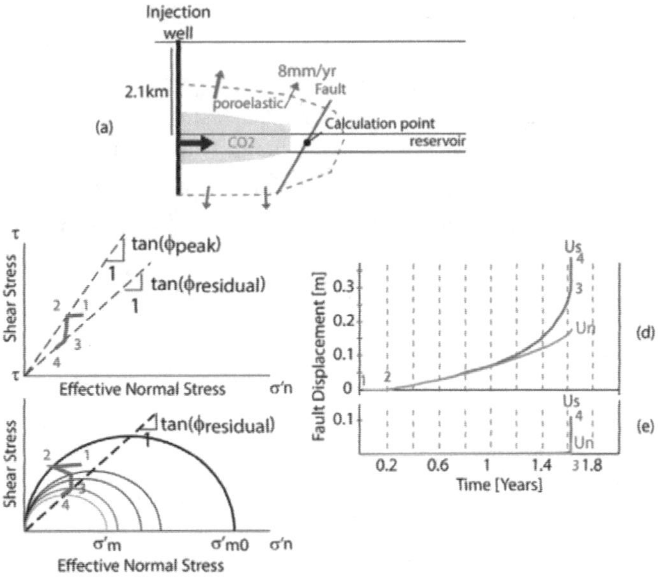

Figure 3. *1D analytical modelling of fault rupture dynamics. A diagram of the 1D simplified model (a) that corresponds to the computations in Figure 2. Stress-paths are comparable for the RSF slip-weakening fault (b) and the MCC model (c). Similarly, the time of occurrence and value of the accelerated slip is consistent between the MCC model (d) and RSF model (e). However, the slow slip is drastically underestimated by the latter one.*

432b Three-Dimensional Pore Networks in Miocene Stevens Sandstone, California: Implications for CO_2 Injection and Trapping

Liaosha Song

The Miocene Stevens sandstone in San Joaquin basin of California is a promising target for CO_2 geological sequestration. Among the four trapping mechanisms that have been identified so far, structural trapping which relies on the impermeable caprock, and residual trapping which relies on the capillary effect of non-wetting fluid are the most important during the injection stage, which occurs up to several decades. Significant heterogeneity in reservoir properties has been noticed in the Stevens sandstone due to the complicated geology in Miocene sedimentation in San Joaquin basin. To investigate the impact of heterogenous and complicated pore structures on fluid flow and trapping, core samples from Stevens sandstone were scanned by high-resolution X-ray micro-computed tomography (micro-CT), a non-destructive imaging method.

A three-dimensional characterization of the pore networks of these samples was conducted. Three rock phases were identified, namely mineral grains, pores, and cement. The cement phase includes cement, matrix, and microporosity that can hardly be resolved by micro-CT scanning. The three-dimensional pore systems of samples were segmented, and porosity values were measured. Absolute permeability simulation was then conducted using Lattice Boltzmann method. An idealized pore network model (PNM) was extracted from each sample, in which pores and pore throats are simplified as spheres and cylinders respectively. A two-phase fluid-flow simulation with CO_2 and water was then conducted at reservoir temperature and pressure conditions. The pore system model was validated through comparison with lab measured porosity, particle sizes, and permeability. While the relative permeability is consistent with published works on a number of sandstone reservoirs.

The tested samples demonstrated diverse reservoir properties, with porosity ranges from 15.3% to 19.3%, and permeability ranges from 0.55 md to 271 md. Several wettability scenarios were simulated, including strong water-wet, intermediate water-wet, mixed-wet with 25% CO_2-wet, and mixed-wet with 50% CO_2-wet. Initial and residual CO_2 saturations were simulated for each scenario.

Results indicate that microporosity is widely distributed in the Stevens sandstone, which can contribute up to 30% of the total porosity, yet it hardly contribute any to the permeability of a sample. Initial CO_2 saturation is largely controlled by the irreducible water saturation which is decided by the pore and pore throat sizes of pore system. In addition, Wettability exerts major influence on the residual trapping saturation. Although positively correlated with initial CO_2 saturation during primary drainage, a mixed-wet system provides the highest residual trapping capacity. Lastly, for heterogeneous sandstone reservoirs, injected CO_2 will likely form layered flow that first saturates the high-permeability beds.

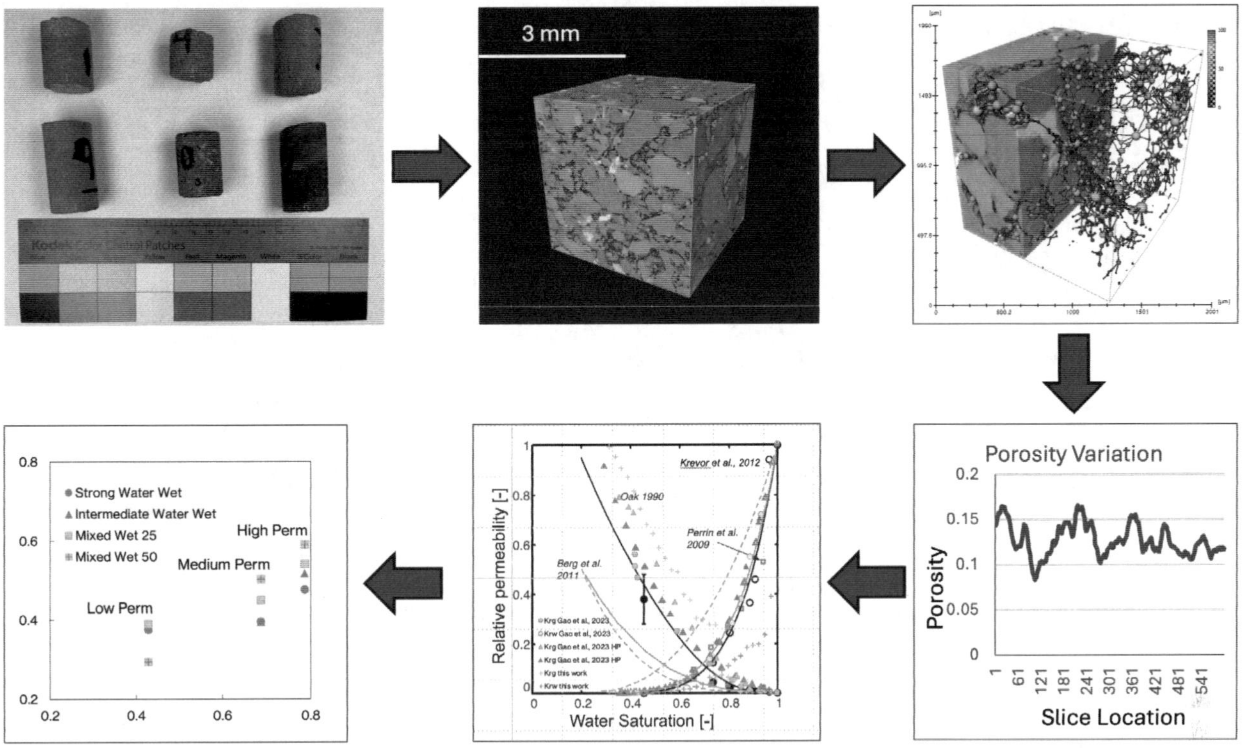
3 mm
Porosity Variation
Porosity
Slice Location
Relative permeability [-]
Water Saturation [-]
Krevor et al., 2012
Oak 1990
Perrin et al. 2009
Berg et al. 2011
Strong Water Wet
Intermediate Water Wet
Mixed Wet 25
Mixed Wet 50
High Perm
Medium Perm
Low Perm

432c Progress and Challenges in Developing CCS Clusters in the Midwest and Northeast Quadrant of United States

Neeraj Gupta, Mark Kelley, Joel Sminchak, Jared Hawkins, Joy Frank-Collins, Matthew Young, Christopher Korose, Dawn Deel

The midwestern and northeastern quadrant of the United States, spanning about 20 states, is experiencing a major spurt of growth in the development and deployment of Carbon Capture and Storage (CCS). The region, subject to CCS research and pilot-scale demonstrations since mid-1990s is fast transitioning into multiple CCS project clusters. Several factors make this diverse region a microcosm for large-scale development of CCS and a precursor to how CCS will evolve globally over next several decades, beyond the initial "low-hanging fruit". This presentation will summarize the past and ongoing efforts in the region, with a focus on opportunities fostering the ongoing commercial scale deployment and pertinent challenges, which could derail the momentum. Specific topics include:

1. Research and pilot projects implemented since 1990s and their impact on CCS development.
2. Current commercial interest in the area and the key drivers.
3. Cross-cutting opportunities, including blue hydrogen, direct air capture with CCS (DACCS), geothermal, bioenergy with CCS, and oilfield brine disposal, leading the need for integrated subsurface resources management.
4. Regional regulatory, business, and community benefit factors that will enable or hinder the foreseen deployment.

The region has long been the industrial heartland of the country accounting for over 900 million tonnes of CO_2 annually in recent years. Many of these sources, which include industry, manufacturing, and electric power plants, are hard to decarbonize and need additional solutions in a carbon-constrained future. The region features several storage prospects, including onshore saline in three major basins, depleted oil and gas fields, unconventional shale resources, basalt rift basins, and offshore saline reservoirs. The progress in this region is spurred in large part from the US federal investments under the Inflation Reduction Act (IRA), which expanded the 45Q tax credits, and the Bipartisan Infrastructure Law (BIL), which includes more than $20 billion to support CCS deployment.

Foundation for CCS in the Region – CCS research in the region started as early as mid-1990s, with initial techno-economic analyses aimed at CO_2 storage in the Mt. Simon Sandstone. Beginning in the early 2000s, there was a major growth based on the US Department of Energy (DOE) funded MIDCARB project, Midwest Regional Carbon Sequestration Partnership (MRCSP) and Midwest Geological Storage Consortium (MGSC). These Partnerships conducted mapping of CO_2 resources, assessed stakeholder outreach strategies, and evaluated infrastructure scenarios. Most importantly, they conducted several small and large-scale pilots in the Illinois, Michigan, and Appalachian Basins, in diverse geologic settings ranging up to a million tonnes (at ADB ethanol plant in Illinois). The region also saw successful pilot testing of CO_2 capture and storage at the AEP Mountaineer Plant in West Virginia and completion of several Class V injection well permits and Class VI permits for ADM (with industrial-scale injection) and FutureGen Sites. Starting in 2020, the Midwest Regional Carbon Initiative (MRCI), formed with the combination of MRCSP and MGSC, has been fostering CCS deployment in the region. The MRCI aims to advance and derisk CCS by addressing key technical challenges, facilitating regional infrastructure planning, performing regional technology transfer, and engaging stakeholders.

All these research and field programs, conducted with safe outcomes, boosted stakeholder confidence and industrial acceptance of technical viability of CO_2 storage for emission reduction, even if the business cases were not yet conducive to deployment. The rich technical foundation also identified uncertainties and disparate distribution of storage resources beyond the established Mt. Simon Sandstone corridor, resulting in the need for characterization of deeper mature sedimentary basins, dominated by dense carbonates and shales. The Mt. Simon Sandstone in the Illinois Basin is a world class reservoir with proven capacity; however, additional reservoirs like the onshore Cambro-Ordovician carbonates and saline reservoirs of the Mid-Atlantic offshore need more characterization to evolve beyond prospective storage resources. In a similar vein, integrated subsurface management is required to effectively utilize the pore space in these reservoirs while planning and mitigating potential project interference issues.

Emerging Commercial Deployment – Numerous ethanol plants with high purity CO_2 are the first movers for CCS deployment in the region with or without the DOE funding due to investor interest driven by 45Q tax credits. Most initial projects have focused on sub-million tonne scale storage at/near the source or in the centralized hubs supplied by pipeline networks. However, plans for regional pipelines have run into significant challenges due to insufficient regulation and local public opposition, resulting in cancellation of at least one major regional gathering and injection project. Despite the challenges, the region is home to about 10 projects across seven states under the DOE funded Carbon Storage Assurance Facility Enterprise (CarbonSAFE) Initiative that began in 2016 with the goal of addressing the key gaps on the critical path towards CCS deployment. With more than $300M in combined budget for feasibility studies and detailed characterization leading up permitting of the facilities and financial decisions, these projects represent CO_2 sourced from ethanol, industrial (steel, cement, refining, hydrogen), and power generation sources. Finally, the region has been selected for multiple DACCS projects and three major clean hydrogen hubs, including ARCH2 in Appalachian Basin and MachH2 in Illinois Basin, announced in 2023 that will lead to more than 40 million tonnes in captured CO_2, with commensurate need for CO_2 transport and storage network within the next ten years. Two of the hubs named the ARCH2 in the Appalachian Basin and MachH2 in Illinois Basin include blue hydrogen production combined CCS. These full value chain projects will create tens of thousands in new jobs, facilitate energy transition and offer significant community benefits, which will also help gain stakeholder acceptance.

Regulatory Progress underpinning Deployment –As of January 2024, there are 27 Class VI injection well permits under review across 11 sites in the region, with only one recent permit in Indiana approved for construction so far. This number is likely to grow substantially in the coming year, based on foreseen activity. In addition to permitting delays, progress has been hampered by lack of clear regulations for interstate pipelines, local county level requirements, and emerging framework for long-term liability, pore space ownership, eminent domain, and need to address local community benefits. States that are proactively developing a clear business and community framework to address these challenges are seeing enhanced project development activity, making them early movers in regional decarbonization. Overall, the opportunities, challenges, and growth path seen in the northeast/Midwest region of the US is seen as a key case study on the CCS deployment that will evolve in many parts of the world.

Keywords: Regional CCS deployment; geologic storage clusters and hubs; regulatory development; CCS challenges; Midwest USA; CCS Infrastructure, MRCI

432d Applicable Insights of Micp-Derived Capillary Pressure on CO_2 Injection for Wyoming Carbonsafe Project: When Does Capillary Pressure Affect the Pressure Front the Most?

Ying Yu, Tao Bai, John Jiao, Jonathan 'Fred' McLaughlin, Scott Quillinan

Mercury injection capillary pressure (MICP) needs conversion with the IFT and contact angle at both laboratory and reservoir conditions to be applied to the reservoir simulation to better estimate storage capacity and area of review, where the data are often unavailable. In this work, these data were collected and studied for their sensitivity and the effect of MICP variance on storage capacity and pressure front, adopting different injection strategies for the Wyoming CarbonSAFE project. First, reservoir IFT and contact angle data of storage formations (Lower Cretaceous Lakota Sandstone, JurassicHulett Sandstone, and Pennsylvanian Minnelusa Formation) were collected and analyzed for their sensitivities as the conversion factor for MICP on the storage capacity and pressure front under bottom hole pressure (BHP) control. Then, optimal conversion factors were fixed to study the effect of the MICP variance of storage formations on the pressure front adopting rate control mode.

Brown (1951) discussed the MICP-saturation curve's applicability to the water-gas capillary pressure-saturation curve using Purcell's conversion factor (1950). The conversion factors depend on the contact angle of water-CO_2-rock and IFT between water and CO_2 at the reservoir conditions, where contact angle and IFT synthetically contribute to the reservoir capillary pressure derived from MICP and less water-wet (larger contact angle) rock surface and lower IFT contribute to higher storage capacity. However, while under BHP control in CO_2 injection simulation, the difference of the capillary pressure caused by the variance in conversion factors is considered insignificant (0.17 – 1.7 psi at the end points) compared to the pressure difference between BHP and initial reservoir pressure and thus results in minimum fluctuation towards actual injection rate and an almost identical CO_2 injection amount (**Fig. 1a**). This step suggests the negligible effect of the conversion factors based on comparable contact angle and IFT results on CO_2 storage capacity under BHP control.

Then, small conversion factors were selected and applied to MICPs. The heterogeneity of the samples regarding different MICPs was studied under the rate control mode. Rates were set up as 50 and 90 percent of the equilibrating rates under the BHP control mode. We expected that the influence of capillary pressure difference on the pressure fronts (with fixed USDW-based critical pressure) formed by 50% rates could be better visualized than 90% rates as the order of the magnitude of the pressures is closer.

There are two major findings. First, for CO_2 to enter the first grid cell, the wellbore pressure must conquer capillary pressure. Once CO_2 enters the cell and saturation accumulates, the cell (pore) pressure continues to build up in every direction; meanwhile, the gravity of CO_2 mass from the first cell lowers the needed entering pressure for the adjacent vertical cell. Thus, it is easier for the injected CO_2 to enter the adjacent vertical cell than the lateral cells. For the low capillary pressure case, increased pore pressure could effortlessly surpass the capillary pressure threshold. Then, CO_2 expands horizontally with a subtle delay (**Fig. 1b**) rather than in the high capillary pressure case, where CO_2 invades first vertically and thoroughly before moving horizontally (**Fig. 1b**). As a result, for both cases, although the injection amounts are the same due to same injection rates, the horizontal expansion of CO_2 in the low capillary pressure case is further and vertically shallower, which leads to higher pore pressure increments at the top layer cells. As the pressure front is delineated by thresholding the pressure increment based on the same critical pressure, the low capillary pressure case has a further pressure front compared to the high capillary pressure case, which is more obvious for the 50% injection rate scenario around a single well (i.e., well at pad 5 in **Fig.1c**). Second, for 90% injection rate case, it shows a minor difference between pressure fronts under high and low capillary pressures (**Fig. 1d**) as the influence of capillary pressure difference is considered insignificant under high pore pressures caused by high injection rates. Note that the pressure interference among the clustered wells (i.e., wells at pads 1 to 4 in **Fig.1d**) can overwhelm the influence of capillary pressure difference on the high injection rate scenario.

In this study, we have learned that from the injection simulation perspective, when only considering capillary pressure, the conversion factors (i.e., contact angle and IFT) to obtain water-CO_2 capillary pressure from MICP are much less influential than the rock samples' petrophysical properties. Thus, statistically selecting the representative samples from the petrophysical perspective for MICP experiments is the key to better estimating the pressure front and achieving a thoughtful injection simulation for carbon storage projects.

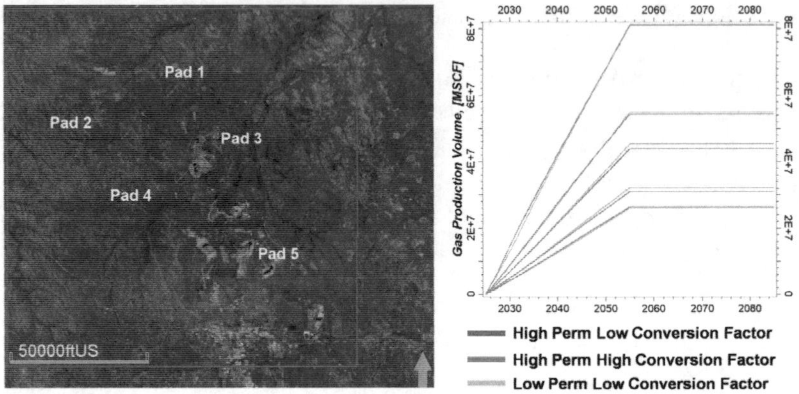

Figure 1a: (Left) well pad locations at the study area for Wyoming CarbonSAFE project and (right) cumulative gas injection amount for each well at the Lakota Sandstone.

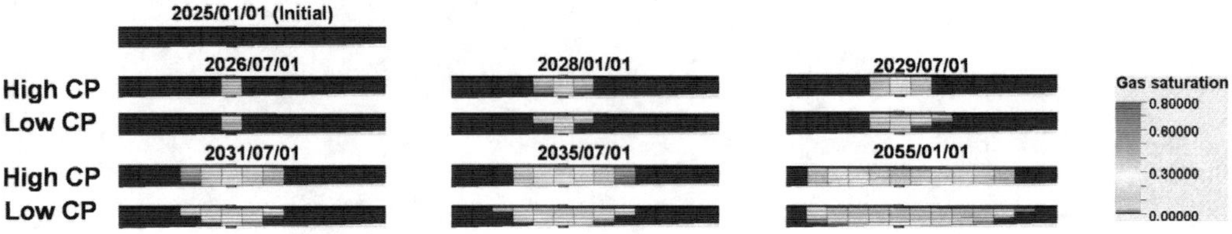

Figure 1b: Vertical slices of the Lakota Sandstone showing the movement of CO_2 plume at well pad 4 for the scenarios of (top) low capillary pressure and (bottom) high capillary pressure at some selected time steps.

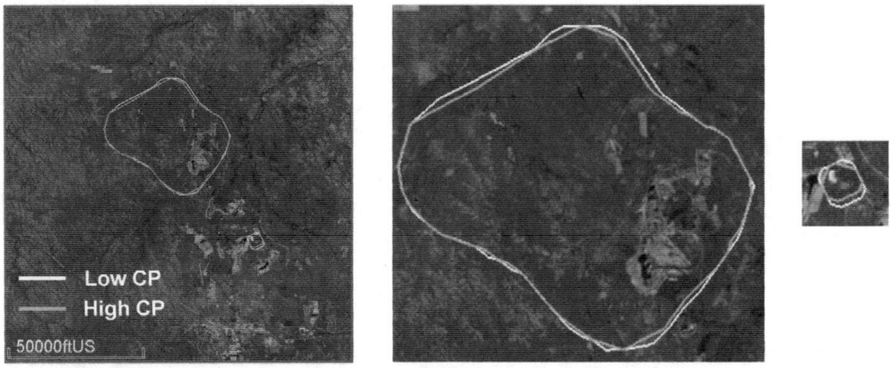

Figure 1c: Pressure front at the Lakota Sandstone for the case of 50% injection rate considering the scenarios of low capillary pressure (yellow) and high capillary pressure (green). The right most two figures show an exaggerated view of the pressure front (same exaggeration ratio).

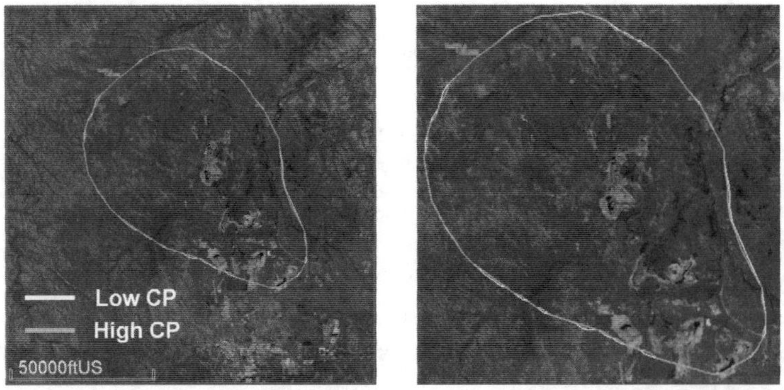

Figure 1d: Pressure front at the Lakota Sandstone for the case of 90% injection rate considering the scenarios of low capillary pressure (yellow) and high capillary pressure (green). The right figure shows an exaggerated view of the pressure front.

432e Leveraging 3D Seismic to Understand Induced Seismicity Risk Associated with CO_2 Injection: An Example from the One Earth Energy Sequestration Project in Gibson City, Illinois

Kendall Taft, Sherilyn Williams-Stroud

One of the key risks in a CO2 sequestration project is induced seismicity caused by the reactivation of pre-existing faults in the subsurface. During the injection of CO2 into saline aquifers, the existing reservoir pressure will increase in response to the injection. This increase in pressure may cause faults to be reactivated and slip along the fault plane, potentially causing seismic events. Induced seismic events that are large enough to be felt by landowners above or surrounding the injection site, particularly if there is property damage, could result in a pause or an entire shutdown of the CO2 sequestration project. Understanding the risk and the potential for felt seismicity is a key component of the overall risk profile of a CO2 sequestration project. Three dimensional (3D) seismic surveys can be a key tool to help characterize the potential induced seismicity risk.

The One Earth Energy site in Gibson City, Illinois, is one of the two sites that make up the Illinois Storage Corridor project, a U. S. Department of Energy supported Carbon Storage Assurance Facility Enterprise (CarbonSAFE) project. The subsurface characterization of this site consisted of a 2164 meter characterization well drilled in 2021, four lines of 2D seismic totaling 43 linear miles acquired during 2019 and 2021, and a 7 square mile 3D seismic survey acquired in 2022. The 3D survey covers the southern portion of the predicted CO2 plume that results from simulating the total planned injection of 4.5 million tonnes of CO2 per year via three injection wells into the lower Mt. Simon Sandstone storage reservoir.

The 3D seismic survey revealed the presence of at least 11 small faults, varying in length from 150 to 1,200 meters, that emanate from the Precambrian basement and die out in the overlying Argenta sandstone or Lower Mt. Simon Sandstone. Both reverse and normal faults were identified with the 3D seismic, with varying orientations. No faults were identified on the 3D survey that completely transect the Mt. Simon Sandstone storage reservoir, and hence no faults penetrate the overlying Eau Claire Formation confining zone.

The previously acquired 2D seismic data suggested that within the area of the 3D survey, between 1 and 3 faults would be identified. However, after the 3D survey was acquired and interpreted, at least 11 faults were identified. This higher density of faulting increases the risk for potential induced seismicity at the One Earth Energy site. However, the relatively small size of the faults, if reactivated, may not lead to felt seismicity. For slip to occur, the fault strike must have an orientation favorable for slip with respect to the maximum horizontal stress direction of 68° in this area of the Illinois Basin. At least three faults have an orientation that is favorable for slip. The magnitude of an earthquake is limited by the area of the slipped patch on the fault, which in turn is generally smaller than the area of the fault surface. Slip along a patch larger than 1 km in diameter could lead to felt seismicity, but only one of the faults, which is also not favorably oriented for slip, is interpreted to have a map length of more than 1 km.

From previous CCS work in the Illinois Basin the risk for induced seismicity was observed to be greatly mitigated by the presence of the Argenta sandstone between the basement and the injection interval in the lower Mt. Simon Sandstone. At locations where the Argenta is thin or missing, more induced seismicity occurred. In addition, thin, very low permeability baffles within the Mt. Simon inhibit vertical pressure migration and leads to primarily horizontal migration, further reducing the likelihood of activating faults in the basement. Because the Argenta appears to be uniformly present in the One Earth Energy injection area with a thickness averaging 125 meters, basement fault reactivation risk could be reduced despite the presence of multiple faults.

432f Assessment of CO_2 Storage and Mineralization in Hawaiian Basalt: Subsurface Characterization and Modeling Results

Donald DePaolo, Nicole Lautze, Donald Thomas, Bhavna Arora, Ziqiu Xue, Saeko Mito, Shuo Zhang, Pramod Bhuvankar, Xunfeng Lu, John Christensen

Most studies of subsurface CO_2 storage have focused on sedimentary formations. However, there is increasing interest in using basalt formations for CO_2 disposal because of the potential for relatively rapid transformation of CO_2 to carbonate minerals, which could decrease long-term leakage risk and monitoring requirements. Basalt formations are widespread in both continental and oceanic regions, but their suitability for CO_2 disposal is uncertain. We have been focused on the Hawaiian Islands, which are composed of basalt lavas and clastic rocks piled up on the ocean floor to thicknesses greater than 10 km (DePaolo et al., *IJGGC*, 2021). The likely target formations are below sea level but accessible with onshore wells and consist of subaerial and submarine lava flows interlayered with low-permeability glassy clastic material. Although the volume of basalt is sufficient for gigatons of CO_2 disposal on each island, there is uncertainty about the scale of interconnected porosity and the permeability, both horizontal and vertical. We are working to close the knowledge gaps, first with indirect characterization methods and modeling and then with downhole tests carried out in an existing 3500 meter deep well drilled and cored in 1999 – 2006 in Hilo, HI (Stolper et al., *Scientific Drilling*, 2009). Geochemical analyses of archived fluids collected from this well, combined with temperature logs, provide information on porosity and fluid residence time. Inversion of water level data with Earth and ocean tides provides information on large-scale permeability below the well casing (3000 meters). Experiments and modeling provide information on likely mineralization timescales. Available data indicate that permeability in the basalt formations varies from several Darcy at 500 to 1000 meters depth (from geochemistry, temperature profiles and hydrological modeling) to less than about 1mD at depth >3000 meters (water level – tidal forcing inversion). Fluid residence time varies from 1000's of years at shallow depths to >40,000 years below 2000 meters. Reactive transport simulation of an injection of 50 million tons of supercritical CO_2 over 100 years at 2500 meters depth suggests all CO_2 will be dissolved or mineralized within 400 years, and none leaks to the surface, but this result depends on details of permeability and mineral-fluid reaction kinetics. Most of the horizontal permeability is in relatively narrow zones spaced 10's of meters apart, an inference based on Sr and O isotopes and widely spaced water entries evident on temperature logs. Experiments aimed at measuring mineralization rates indicate that rates are likely to be high initially but may slow markedly with time, an issue that is applicable to all mineralization strategies. Our indirect approaches will be extended with additional downhole fluid sampling and pump tests combined with temperature logs. More direct measures of permeability are expected to come from downhole injections. Overall assessment is that the basalt formations under the northeast portion of the Island of Hawaii could be suitable for gigaton scale CO_2 disposal if the porosity-permeability characteristics prove favorable.

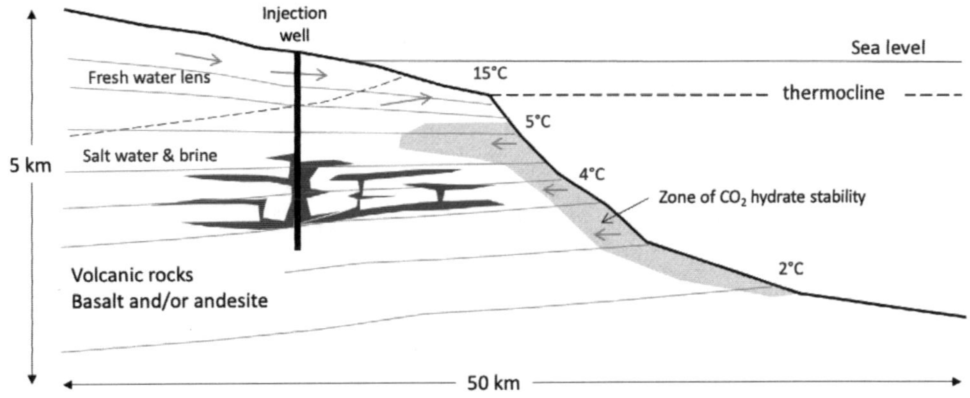

432g Water Thin Films on Kaolinite Gibbsite and Edge Surfaces and Their Effects on Surface Wettability in Relation to Geological Carbon Sequestration

Minjunshi Xie, Zhehui Jin

CO_2, as a major source of greenhouse gases (GHG), has increasingly drawn public attention in the recent decade. Geological CO_2 sequestration (GCS) is a viable option to mitigate this problem. GCS relies on various mechanisms, among which structural trapping plays an important role. Within the framework of structural trapping, the storage capacity is strongly dependent on CO_2-water-rock contact angle. Kaolinite is one of the common rocks in the caprock where structural trapping takes place. Even though CO_2-water-rock contact angle on kaolinite basal surfaces have been studied, the knowledge about kaolinite edge surface wettability remains limited.

In this work, we use molecular dynamics (MD) simulations to study CO_2-water-kaolinite contact angle on kaolinite gibbsite and edge surfaces under a typical GCS condition (330 K and 200 bar). The common belief is that surface wettability in the presence of water is dependent on the surface hydroxyl (-OH) group density. However, we find that surface -OH group density is not the only pivotal factor determining surface wettability. While the surface -OH group density of kaolinite edge surface is much lower than that of gibbsite surface, both kaolinite gibbsite and edge surfaces are strongly water-wet. Edge surface is better hydrated than gibbsite surface as both silanol and aluminol groups can form hydrogen bonding with water molecules thanks to large effective water accessible volume around them, and pocket water can further anchor water film. Our study provides important insights for CO_2 structural trapping as it illustrates that surface heterogeneity is crucial to wettability.

432h Key Challenges of CO_2 Transport Via Pipelines

Isaac Gamwo, Deepak Tapriyal, Ward A. Burgess

In the carbon capture and storage infrastructures, pipelines ensure continuous transport of CO2 from the source facilities operating in a continuous manner, such as fossil fuel power plants, to the storage site. The CO2 streams ready for sequestration typically contain impurities which cause additional challenges.
This presentation will review key challenges of CO2 transport via pipelines that are related to its integrity, flow assurance, and health, safety and environmental factors.

228a Optimal Design of Hydrogen-Blended Natural Gas Pipeline Network Considering Separation System

Shiya Gu, Yachao Dong, Jian Du

Important energy challenges that the world is facing now include the security of supply, environmental pollution, greenhouse gas emissions and the consequent issue of climate change. As a relatively clean fuel that can be produced from a range of sustainable sources and gasification of coal, hydrogen is expected to become an alternative energy carrier capable of addressing these challenges, as well as improving the overall energy supply security. To promote hydrogen utilization, establishing economically feasible delivery infrastructure represents a key issue that must be addressed. Rather than constructing specialized pure hydrogen pipeline, blending hydrogen into existing natural gas pipeline is considered the best option for long-distance and large-scale hydrogen transportation for the early market of hydrogen economy.

In this presentation, we propose a mixed-integer nonlinear programming model of injecting hydrogen from multiple sources into natural gas pipelines, and obtain commercial-grade hydrogen gas using separation and purification technologies downstream to satisfy the purity requirement and demand. The model mainly includes two parts, respectively for pipeline network and separation system. (1) In the pipeline network part, several adjustments are applied to the original natural gas pipeline to ensure that our design scheme can satisfy the stability and safety requirements of gas transportation, such as introducing new compressor stations that are compatible with hydrogen environment and reconsidering operation pressure of pipeline to avoid the adverse effects of hydrogen-induced material failure. (2) For the separation part, we develop a universal H_2/CH_4 separation system based on a stage-wise superstructure that accounts for products with a variety of concentration requirements. In the superstructure, membranes and pressure swing adsorption (PSA) are considered as potential separation units, with the possibility of recovering retentate gas of membranes and off gas of PSA. Based on information such as the pressure, composition, and flow rate of the feed streams, as well as the purity and capacity requirements of the products, the superstructure helps to select appropriate separation units and optimizes the operational parameters of each unit, thereby generating the optimal process scheme. All these separation requirements will be considered together with the pipeline network design to ensure that the important decisions, such as the hydrogen blending ratio, are optimized.

This integrated design model for hydrogen-blended natural gas pipeline network and separation system is a large-scale MINLP model. To address the computational complexity resulting from the large-scale and nonlinear nature of practical design problems, a decomposition algorithm is tailored to the proposed model, with two sub-models determining separation superstructure and pipeline configuration, respectively. Several case studies, including one based on the pipeline of West-East natural gas transmission project in China, are conducted; the results indicate that the proposed model and decomposition algorithm can provide good solutions, solving the issue that the original model is computationally intractable. Hydrogen supply from each source, mixing locations of hydrogen in the pipeline network, layout of compressor stations, and separation schemes for each sink can be obtained by this model. We will also show the quantitative estimation and analysis of the hydrogen contents that are most cost-effective in the transported gas passing through the existing infrastructures of the natural gas transmission. Therefore, this work can provide a decision-support tool for the optimal design of hydrogen-blended natural gas pipeline network and downstream separation processes adaptable to various product demand.

228b Operational, Economic, and Environmental Analysis of Hybrid Renewable Energy Systems

Ali Khosravani, Elaheh Safaei, Matthew DeHaan, Blake Billings, Michael Reynolds, Kerry Kelly, Kody Powell

The global push to achieve 100% renewable future energy system by 2050 is a key focus for sustainable development. To achieve this goal, previous studies have extensively investigated the potential transition to a renewable energy system by equating the electricity demand of a region with its available renewable energy resources. However, these studies often overlook the dynamic nature of energy systems and the intermittent characteristics of renewable sources. Furthermore, many studies predominantly focused just on the possibility of transition to 100% renewable energy system and lack a thorough economic analysis of such a transition. Notably, a 100% renewable energy system cannot be achieved without the elimination of natural gas consumption and using electricity instead, which puts extra load on the electric grid due to heating purposes during colder months. Our study conducts a detailed techno-economic analysis on an hourly basis to assess the cost implications of transitioning to 100% renewable energy. Additionally, we examine the impact of building electrification in residential and commercial sectors (which account for 62% of total electricity consumption in the U.S.) on the entire energy system. There have been large efforts to address the challenges of moving towards a 100% renewable energy system, yet challenges remain, particularly regarding transmission losses, environmental considerations, land use, and societal acceptance, which warrant further investigation.

In this study, using HOMER software tool, we performed a comprehensive techno-economic analysis of hybrid renewable energy systems (HRESs) to evaluate the technical feasibility and economic viability of integrating renewable energy sources into existing energy systems. In an HRES, as shown in Figure 1, Renewables are integrated with a battery system and there is a natural gas power plant as a back-up to help the energy system to meet the electric load when renewables are not available and battery is emply. Additionally, we investigated the impact of building electrification in residential and commercial sectors on the entire energy system. To simulate this electrification scenario, illustrated in Figure 2, we employed building simulations using EnergyPlus and OpenStudio, analyzing the electrified load of an entire city. Notably, in our simulations, we replaced conventional fossil fuel-powered heating, ventilation, and air conditioning (HVAC) systems with heat pumps, which operate solely on electricity. Our analysis yielded insightful results regarding the optimal renewable fraction necessary to minimize the levelized cost of electricity (LCOE) across different regional contexts. We found that the value of the optimum renewable fraction is heavily influenced by the regional availability of renewables, varying substantially between different case studies. Moreover, as we approached a 100% renewable fraction, we observed a remarkable exponential increase in the LCOE. This trend was mirrored in the sizing of renewable components such as photovoltaics (PVs) and wind turbines, as well as battery systems, all of which increased exponentially to meet peak electric demand. Additionally, our sensitivity analysis, examining the impact of carbon taxes, fuel prices, and renewable component costs, revealed significant shifts in the optimal renewable fraction. For instance, increasing the carbon tax from \$0 to \$120 per tonne of CO_2 led to an increase in the optimum renewable fraction from 5% to 26%. Similarly, an increase in natural gas prices from \$0.26/m^3 to \$0.7/m^3 and considering estimated renewables costs for 2050 resulted in notable rise in the optimal renewable fraction from 5% to 60% and 56%, respectively. A summary of sensitivity analysis is shown in Figure 3. Overall, our study underscores the importance of a cost-aware approach, urging policymakers to strike a balance between mitigating carbon emissions and ensuring economic feasibility when adopting renewable energy solutions.

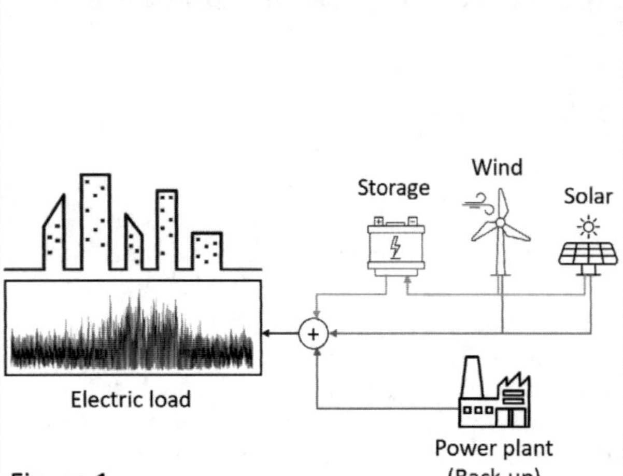

Figure 1

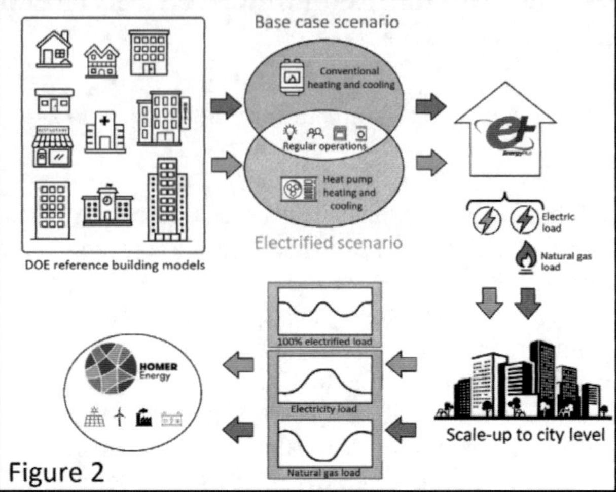

Figure 2

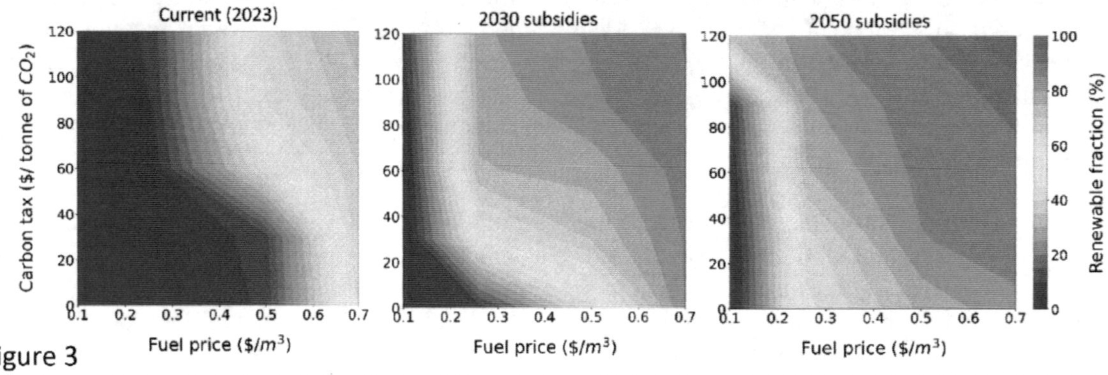

Figure 3

228c The Influence of Technical, Economic, and Site-Specific Factors on Optimal Ammonia Production from Renewables

Alexander Klimek, Tibor Svitnic, Caroline Ganzer, Kai Sundmacher

The chemical industry is still heavily dependent on fossil raw materials and consequently responsible for large quantities of anthropogenic emissions. Ammonia is one of the most commonly synthesized bulk chemicals globally, with reactant production primarily relying on steam methane reforming and coal gasification. Sustainable ammonia production using only water, air, and electrical energy from renewables is a way to reduce greenhouse gas emissions.

Large-scale chemical conversion processes, such as ammonia synthesis, are typically operated continuously throughout the year. Wind and solar resources, however, are inherently intermittent in their availability. Coupling fluctuating energy supply with the operation of chemical plants presents a significant challenge for sustainable systems. To investigate this issue, we formulate a large-scale superstructure-based mathematical mixed-integer linear program within this work. While the majority of existing studies consider either an optimal design and scheduling [1–3] or a supply chain problem [4], classical facility location constraints are added herein to solve a combined design, scheduling, and facility location problem, as shown in Figure 1. Thus, in addition to the optimal process route and the optimal operation over time, the best allocation of five potential production locations with respect to a demand center within an exemplary German scenario is determined. In addition to the availability of renewable resources, other location factors such as the land potential for the construction of wind power and photovoltaic plants, taxes, and transportation costs are also taken into account.

According to the optimization results, the construction of a single large plant is economically more favorable compared to several distributed small plants. In particular, decreasing marginal costs for increasing plant capacities due to economies of scale for chemical conversion processes are the reason for this. It is shown that neglecting economies of scale and assuming linear relationships between plant capacity and costs when solving such superstructure optimization problems can provide misleading results. However, the construction of one central ammonia plant requires the maximum installable capacity of the electricity-generating processes at one location to be sufficient to satisfy the specified ammonia demand. In particular, high wind speeds throughout the year and available space for the installation of wind turbines are identified as decisive location factors for the optimal site.

Although the optimal process design and scheduling suggests the technical feasibility of green ammonia production, negative after-tax net-present values show that it is currently not economically viable. Sensitivity analyses demonstrate that the main decisive factors for the selection of the production site in Germany are the installable wind turbine capacity and the prevailing wind speeds throughout the entire year. Especially neglecting land availability can be misleading and result in an overestimation of a location's potential. Solar irradiance, corporate income taxes and transport costs (at least over short and medium distances within Germany), on the other hand, are less crucial.

Our work underscores the location dependency of the production of chemical compounds utilizing renewable energy sources. For this reason, it is necessary to determine the process design and scheduling coupled with the optimal site allocation. A superstructure-based MILP formulation combined with real-world weather data and site-specific parameters is proven to be suitable for analyzing the coupling of wind and solar energy with chemical conversion processes.

References:

[1] Ganzer, C. and Mac Dowell, N. 2020. A comparative assessment framework for sustainable production of fuels and chemicals explicitly accounting for intermittency. *Sustainable Energy Fuels* 4, 8, 3888–3903.

[2] Palys, M. J., Mitrai, I., and Daoutidis, P. 2021. Renewable hydrogen and ammonia for combined heat

and power systems in remote locations: Optimal design and scheduling. *Optim Control Appl Meth*, 1–20.

[3] Wang, H., Daoutidis, P., and Zhang, Q. 2021. Harnessing the Wind Power of the Ocean with Green Offshore Ammonia. *ACS Sustainable Chem. Eng.* 9, 43, 14605–14617.

[4] Wang, H., Daoutidis, P., and Zhang, Q. 2023. Ammonia-based green corridors for sustainable maritime transportation. *Digital Chemical Engineering* 6, 100082.

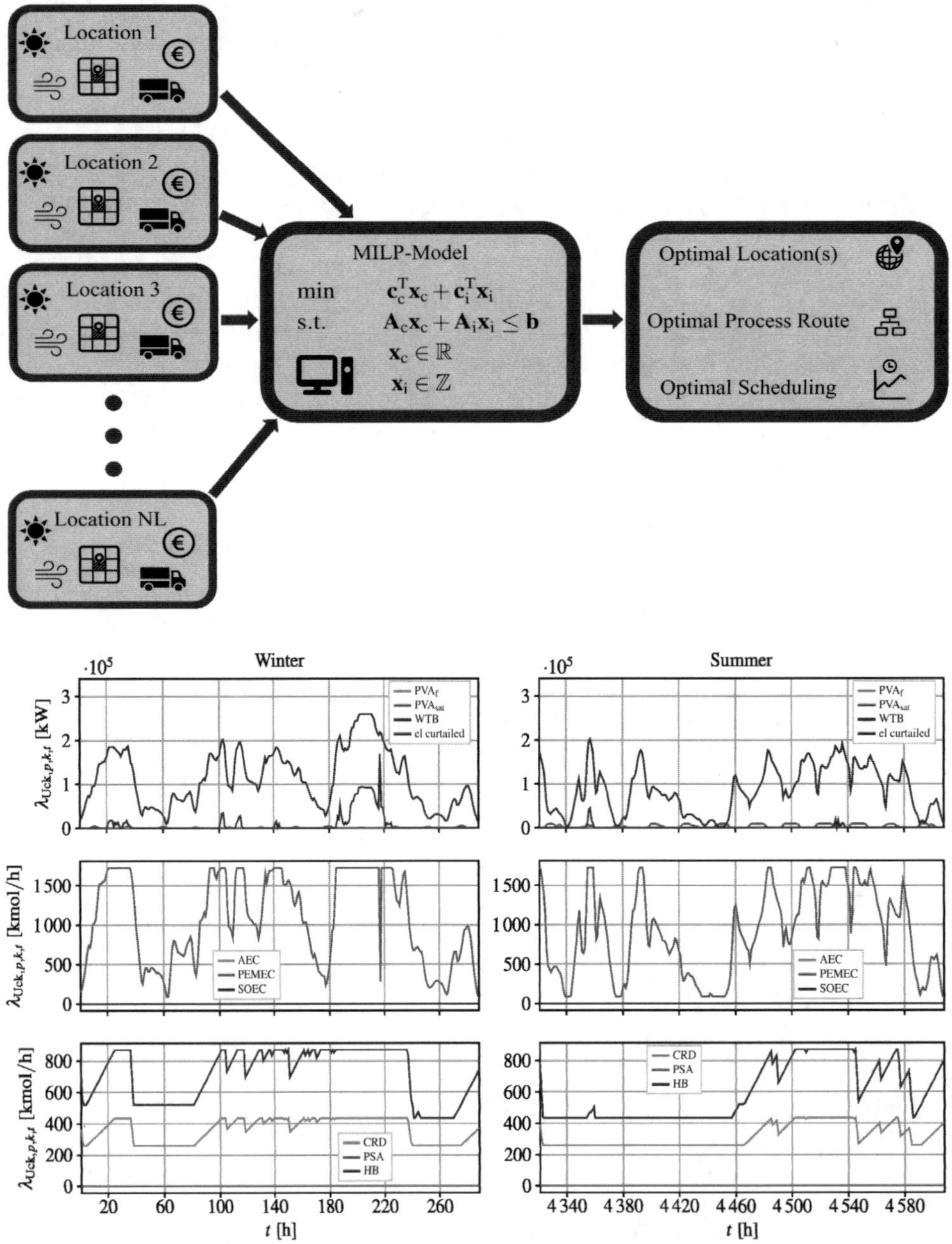

Figure 1: Top: General idea of the combined design, scheduling, and facility location problem, incorporating location factors such as wind speeds, solar irradiance, land availability, taxes, and transportation routes. The designation NL is representative for the general number of locations. Bottom: Scheduling results for typical sequence in winter and in summer of the combined design, scheduling, and facility location problem (AEC: alkaline electrolyzer; CRD: cryogenic distillation; el: electricity; HB: Haber-Bosch; PEMEC: proton exchange membrane electrolyzer; PSA: pressure swing adsorption; PVA$_f$: fixed photovoltaic array; PVA$_{sat}$: single-axis tracking photovoltaic array; SOEC: solid oxide electrolyzer; WTB: wind turbine).

228d Drivers of Blue and Green Hydrogen Deployment in Net-Zero Emissions Scenarios

Jun Wen Law, Dharik Mallapragada

As the emphasis of decarbonization shifts from the electric power sector to other energy sectors, there is growing interest in hydrogen (H_2) and enabling technologies for decarbonizing end-use sectors in which direct electrification may be difficult or costly. The versatility of H_2, with multiple technologies for production, storage, transportation, and end-use, is one factor driving its growing appeal. However, this versatility also leads to multiple potential pathways and cross-sectoral interactions that complicate the assessment of H_2 infrastructure and its role in deeply decarbonized energy systems. For example, H_2 produced from electrolysis could be powered by co-located variable renewable energy (VRE) or by electricity sourced largely from VRE generators connected to the power grid. While the former is an islanded system, the latter directly impacts the electric power sector both in terms of increasing aggregate electricity demand and by potentially changing the time when peak demand occurs, thus affecting the integration of variable renewable energy (VRE).

In this study, we use an integrated energy systems approach to identify the impact of H_2 infrastructure flexibility and negative emission technologies (NETs) on the deployment of "green" H_2 (referring to electrolyzers) and "blue" H_2, which uses natural gas with carbon capture and sequestration (CCS). Specifically, we use an open-source, multi-sector capacity expansion model (MACRO) to investigate the drivers of green and blue H_2 deployment in a cost-optimized, net-zero energy system. The model represents the coupling between the electric power grid, H_2 infrastructure (production, storage, and transmission), and spatially resolved CO_2 storage and biomass resources. In addition, temporal variability in energy supply (e.g., VRE) and demand are captured through modeling hourly operations of the energy system over representative weeks, which enables us to capture the value of temporally-coupled resources such as battery and H_2 energy storage. We use the MACRO model to study the impact of H_2 infrastructure flexibility and availability of NETs on the cost-optimal deployment of H_2 production technologies in a 2050 net-zero case study of the contiguous United States. H_2 infrastructure flexibility is represented via scenarios with and without H_2 storage and pipelines, as well as alternative assumptions about the operational flexibility of blue H_2. Availability of NETs is varied by considering cases allowing direct air capture (DAC) alone, or by allowing both DAC and bioenergy with carbon capture and sequestration (BECCS).

We find that the mix of H_2 production is primarily affected by H_2 infrastructure flexibility such as H_2 storage and pipelines, and less so by NETs availability. When H_2 storage and pipelines are unconstrained, green H_2 accounts for the largest share of H_2 production in a net-zero system. Under these conditions, green H_2 exploits the flexibility of H_2 infrastructure to maximize electrolytic H_2 production at locations and times of low electricity prices and vice versa. On the other hand, blue H_2, which requires additional investments in NETs and CO_2 transport and storage, accounts for the largest share of H_2 production when the flexibility of green H_2 production is limited, such as when H_2 storage and pipeline deployment is constrained. Counterintuitively, increasing blue H_2 flexibility with limited H_2 storage and pipeline deployment increases green H_2 deployment, as it also enables operational flexibility of green H_2. Across all scenarios, we find that H_2production is generally co-located with demand, implying a smaller role for H_2 transmission relative to electricity transmission.

Our analysis also sheds light on the relative value of grid-connected vs. islanded H_2 systems under cost-optimized net-zero emissions constraints. We find that grid-connected systems meet a greater share of exogeneous H_2 demand using electrolyzers compared to islanded systems across different scenarios of H_2 infrastructure flexibility. This finding highlights potential synergies of co-optimizing electricity and H_2 production in decarbonized energy systems.

228e A System-Level Decision-Making Analysis of Liquid Organic Hydrogen Carrier (LOHC) Processes: A Case Study of Korea-Australia

Byeongchan Ahn, Wangyun Won

The transition from a fossil fuel-based economy to a hydrogen economy is imperative, primarily due to the sustainability of hydrogen. Hydrogen only produces water after its utilization with greater gravimetric energy density compared to conventional fossil fuels. Especially, hydrogen can be utilized as an energy carrier of intermittent renewable energy. hydrogen produced using electricity generated from renewable energy sources can reduce the reliance on fossil fuels without CO_2 emission during its production, thereby contributing to achieving carbon neutrality. However, the transition to a hydrogen economy is challenged by its low volumetric energy density, which leads to an increase in transportation costs under ambient conditions, necessitating the appropriate hydrogen transportation method for long-distance.

Hydrogen storage may be majorly classified into two main categories: physical and chemical storage. Physical storage, such as high-pressure tanks and liquefaction, presents a lower complexity in processes and a higher technology readiness level (TRL), making it a more immediately viable option. However, the requirement for extreme conditions, such as cryogenic temperatures or high pressures, leads to continuous energy expenditure and hydrogen loss during transportation, posing a significant risk. On the other hand, chemical storage methods, which involve the hydrogenation of compounds such as liquid organic hydrogen carriers (LOHCs) or ammonia, offer a promising alternative. Although these methods currently exhibit a relatively lower TRL and require additional separation processes during dehydrogenation, their ability to transport hydrogen in a liquid state at ambient conditions with lower energy consumption makes them particularly suitable for long-distance transportation. Especially, LOHC is particularly advantageous due to its compatibility with established transportation systems for petroleum and chemical products, as well as its recyclable characteristics.

However, the vast array of LOHCs available complicates the selection of an optimal LOHC, as comparing all types is constrained by practical limitations. Therefore, in this study, we conducted decision-making of the process that uses the following LOHC systems: toluene/methylcyclohexane, dibenzyltoluene/perhydro-dibenzyltoluene, N-ethylcarbazole/perhydro-N-ethylcarbazole, and CO_2/methanol, which are named strategy 1 through 4. As decision-making criteria, the exergy efficiency, profitability, and global warming potential of the process are considered. To investigate those factors, a comprehensive analysis including exergy and techno-economic analyses, and life-cycle assessment is conducted based on a scenario transporting hydrogen from Australia to Korea as a representative.

Strategy 1 exhibits the least exergy efficacy (74.0%) because of its substantial energy consumption. However, despite employing a sophisticated separation process due to the volatile nature of TOL, the lowest minimum selling price of hydrogen (MHSP; \$5.9/kg) was achieved. Conversely, strategy 2 exhibits the highest exergy efficiency and second lowest in terms of the MHSP, even with the most substantial global warming potential (GWP) primarily attributable to the substantial influence of dibenzyltoluene. In comparison to strategy 2, strategy 3 exhibits the most favorable exergy efficiency (77.4%) due to its minimal H_2 loss and reduced energy consumption. However, despite the effective hydrogen transportation capability of N-ethylcarbazole, strategy 3 incurred the highest MHSP (\$9.6/kg) due to its high cost of N-ethylcarbazole. Lastly, since strategy 4 utilizes CO_2 captured from flue gas as LOHC, Strategy 4 was determined to be the most environmentally favorable alternative.

Based on the results obtained from the analyses, decision-making was conducted. Strategy 1 was found to be overwhelmingly beneficial, particularly when profitability takes precedence, while strategies 2 and 4 were selected at the vertices of exergy efficiency and GWP, respectively. In summary, this research provides valuable insights for policy-making by explicating the intricate dynamics of critical factors that are essential for devising successful strategies to transport hydrogen and achieve global carbon neutrality.

228f Impact of Technology Development and Climate Characteristics on the Joint Expansion Planning of Green Hydrogen and Renewable Energy

Mingyu Kim, Yechan Choi, Seongmin Heo

ABSTRACT

For the sustainable transition to the low carbon society, electricity from renewable energy is a must, whose generation is heavily dependent on the weather conditions, which exhibit high intermittency [1]. One option to alleviate this intermittency is the utilization of a hydrogen system including production, storage, and transportation facilities. Especially, green hydrogen produced from the electrolysis of water is attracting a significant attention as it produces high-purity hydrogen without carbon emissions [2]. Despite a high synergistic potential between the renewable electricity and green hydrogen productions, extensive study is rather scarce, where expansion planning problems for such productions are jointly considered. Currently, joint expansion planning for hydrogen and electricity is typically addressed in the connection with the existing infrastructure (e.g. fossil fuel-based electricity, grey hydrogen) [3].

To this end, we propose a joint planning model considering generation, storage, transmission and transportation of renewable electricity and green hydrogen. In the proposed model, the planning is formulated as a mixed-integer linear programming (MILP) problem, where a global optimum is guaranteed [4]. The objective of problem is the minimization of the overall cost (for both electricity and hydrogen systems), while satisfying both electricity and hydrogen demands. The major decision variables are the specifications for facility capacities (for different types), and hourly operational decisions of each facility. We initially consider three different options for renewable electricity generation (i.e. on-shore and off-shore wind turbines, and photovoltaic panel) for the joint expansion planning in South Korea to illustrate the application of the proposed model.

Then, we conduct several case studies to provide deeper insights into the joint development of renewable electricity and green hydrogen infrastructures. Firs, we add an alternative option for the generation part, namely bio gas steam reforming, to evaluate its impact. Also, we consider additional regions with vastly different climatic and regional characteristics (compared to South Korea) to analyze their influences. Finally, a range of policy and technology scenarios are considered to provide some future outlooks.

REFERENCES

[1] Reza Artis, Mojtaba Shivaie, Philip D. Weinsier (2023). A flexibility-based multi-objective model for contingency-constrained transmission expansion planning incorporating large-scale hydrogen/compressed-air energy storage systems and wind/solar farms. *Journal of Energy Storage*, 70, 108086.

[2] N.S. Hassan, A.A. Jalil, S. Rajendran, N.F. Khusnun, M.B. Bahari, A. Johari, M.J. Kamaruddin, M. Ismail (2024). Recent review and evaluation of green hydrogen production via water electrolysis for a sustainable and clean energy society. *International Journal of Hydrogen Energy*, 52, Part B, Pages 420-441.

[3] T. Klatzer, U. Bachhiesl, S. Wogrin (2022). State-of-the-art expansion planning of integrated power, natural gas, and hydrogen systems. *International Journal of Hydrogen Energy*, 47(47), 20585-

20603.

[4] Luca Urbanucci (2018). Limits and potentials of Mixed Integer Linear Programming methods for optimization of polygeneration energy systems. *Energy Procedia*, 148, 1199-1205.

228g Ammonia As an Energy Carrier: Cost and Emission in the Global Ammonia Supply Chain

Woojae Shin, Bosong Lin, Haoxiang Lai, Daniel R Cohn, Leslie Bromberg, Guiyan Zang

Ammonia is expected to play an important role as an effective energy carrier for the global decarbonization agenda. Its inherent molecular properties, existing infrastructure for fertilizer production, and correlated high technology readiness levels position ammonia as a promising candidate for medium- to long-distance energy transport. However, transitioning existing ammonia supply chains to low-carbon alternatives presents multifaceted challenges. The challenges include sustainable system scaling, large-scale production, cost-effective storage and distribution networks, regional differences in fossil and renewable energy availability, and regional specific hydrogen demand. To avoid pitfalls like 'emissions lock-in' and ensure a seamless shift away from conventional fossil-based supply chains, this work designs a future ammonia supply chain that considers 1) current and potential ammonia supply and demand volumes specified by regions and countries; 2) levelized costs of ammonia production and distribution, varied by technologies, regions, and supply chain routes; 3) life cycle greenhouse gas (GHG) emissions of the total supply chain; and 4) current and forthcoming global ammonia supply chain with government projections. These factors lead to different supply chain optimization conclusions, which are also influenced by each region's environmental-economic policy options (e.g. carbon constraints and clean hydrogen support schemes).

The objective of this research is to conduct a comprehensive techno-economic analysis (TEA) and life cycle analysis (LCA) on potential global ammonia supply chain configurations and to present a supply chain optimization model that can reflect the above factors. The optimization model aims to optimize the total cost and carbon emissions inside the supply chain system. Furthermore, appropriate constraints including regional primary energy availability, hydrogen demand, and carbon policies are included. Based on this modeling methodology, this work determines the most feasible supply chains for each supplying and demanding region, to advise strategic decision-making by both industry stakeholders and governments, and to provide insights into the implications of decarbonization efforts on a global scale. Our analysis provides a groundwork for supply chain optimization.

228h Assessing the Key Factors in the Transition to a Hydrogen-Based Economy through Agent-Based Simulation

Venkataramanan Swaminathan, Rajagopalan Srinivasan

Emerging concerns regarding global warming and climate emergency have induced many countries to consider environmentally-benign cleaner fuels that help in decarbonizing the national energy basket. It is imperative to consider and evaluate viable technological alternatives that can serve both economically as well as environmentally in the transition to a low-carbon economy. Among the different greener fuels of interest, hydrogen has recently drawn global attention and is increasingly considered a cleaner alternative that can act as a bridge fuel in making a swift and just transition. With regards to supply, though H_2 comes with a variety of recipes based on the nature of production, the shift towards green H_2 production has the potential to accelerate renewable energy deployment in the near future. Towards this, decarbonizing the power sector is the priority as renewable-based electrolysis production of H_2 will be useful only if the grid is deeply decarbonized[1]. IRENA estimates hydrogen to cover up to 12% of global energy use by 2050 with about two-thirds of the demand being met through green hydrogen production, while hard-to-abate sectors like Iron and Steel, refineries, chemicals, and cement are expected to drive the demand for green hydrogen with supportive and proactive policies in place[2,3]. To enable the transition and find uptakers, critical unlocks like reducing the production and distribution cost must be addressed rather quickly[4]. The success of a transformation to an H_2 based economy is contingent on the establishment of suitable distribution networks that would ensure cost-effective access to different end-users. **This paper seeks to develop a systematic framework to assess the key factors influencing the transition to a green hydrogen economy. Specifically, our focus is on countries like India with large geographies, characterized by spatially-distributed, geographically-disparate, and heterogeneous end-users of energy.**

The green H_2 supply chain (GHSC) comprises a green hydrogen production hub, a logistics and infrastructure distribution network, and the end-users who use either liquid or gaseous H_2 as fuel or feedstock. The production hubs are typically established in the vicinity of sources of renewable electricity production. Hence, locations with well-established solar and wind farms serve as potential supply sites. With respect to logistics and distribution infrastructure, constructing dedicated H_2 pipeline grids might be capitally intensive, but utilizing the existing NG pipeline network to transport blended H_2 in sufficient quantities is found to be both economically and operationally feasible. Additionally, pipeline transportation of gaseous H_2 can be supplemented by transport by liquid H_2 carriers or compressed tube trailers (especially for shorter distances) via either road or rail. The list of end-users includes energy-intensive industries such as iron and steel, fertilizer, refineries, chemicals, and cement, as well as transport or distribution fuel consumers. Existing literature lacks quantitative approaches that can **help identify the key factors (such as economics, logistics, or regulatory factors) that will influence the transition. We seek to address this gap in our work.**

We consider a market-based economy governed by laws of supply and demand. We assume that all the entities in the GHSC are rational and seek to maximize their economic profit. Hence, the decision of an end-user to switch to H_2 would be a consequence of either regulatory action (environmental regulations) or an economic choice depending on the competitiveness of utilizing H_2. Further, estimating the demand for H_2 would necessitate estimating its landed price for each end-user. Transport of H_2 through pipelines is economically viable only for transport in large quantities.

Similarly, road and rail-based transport have higher unit costs of delivery but are economically attractive for shorter distances. This unique interplay between demand, the economics of supply, and distribution makes planning the GHSC a challenging problem. **We, therefore, propose an agent-based simulation optimization model in this work.**

The proposed methodology comprises a **Demand Estimation Model** that is developed using an agent-based modeling framework to assess the spatially distributed demand for H_2. Here, each entity in the supply chain is modeled separately as an agent. Green H_2 supplier agents (production hubs) and end-user agents are distributed spatially and modeled as rational, self-interested decision-makers that seek to maximize their own objectives subject to government regulations. Each end-user agent may wish to switch to H_2 at a future time, subject to availability, affordability, and regulations. The switch to H_2 further creates opportunities for rail/road transport operator agents to enable transport of the necessary quantity from the supplier to the end-user. With established GIS routing maps and tools that can be utilized to model road/rail-based transport of H_2 to end-users, we have considered both the aforementioned transportation modes as the H_2 distribution infrastructure in the logistic segment of the supply chain, while transportation through pipelines will be addressed as part of future avenues of research.

The agent-based demand estimation model has been implemented using AnyLogic modeling and simulation platform. In this paper, we present the application of the developed framework to a case study focusing on the green H_2 demand potential for a period from 2020 to 2040 in Tamil Nadu, a state in India. The model comprises ~60 spatially distributed end-users and two potential production hubs. Different scenarios involving H_2 economics (based on predicted prices), carbon taxations, and environmental regulations have been considered, and the Spatio-temporal H_2 demands arising from end-users decisions over the 20 years horizon are estimated. In this paper, we will report **a spectrum of insights relating to the key role played by H_2 prices, national-level environmental regulations, and carbon taxations in influencing the transition to a hydrogen-based economy.Additionally, our results reveal granular details on the extent of fossil fuels that can be displaced by hydrogen and the preferred transportation modes categorized by various industrial end-use sectors.**

References

1. Mac Dowell, N. *et al.* The hydrogen economy: A pragmatic path forward. *Joule* **5**, 2524–2529 (2021).
2. Green Hydrogen for Industry: A Guide to Policy Making. */publications/2022/Mar/Green-Hydrogen-for-Industry* https://irena.org/publications/2022/Mar/Green-Hydrogen-for-Industry.
3. International Energy Agency. *Global Hydrogen Review 2021.* (OECD, 2021). doi:10.1787/39351842-en.
4. Hydrogen Insights 2021 - Hydrogen Council. *https://hydrogencouncil.com/en/* https://hydrogencouncil.com/en/hydrogen-insights-2021/.

162a How Carbon Capture and Storage Can Fail and Why It Won't

D. Nathan Meehan

This paper discusses challenges that may hinder or delay widespread adoption of Carbon Capture and Storage (CCS). Focusing on technical, economic, social, and regulatory aspects, it aims to pinpoint specific obstacles, such as the financial burden on energy production, uncertainties in long-term storage impacts, and the intricate regulatory framework governing CCS projects and pipelines. Reaching net zero emissions will be virtually impossible without widespread CCS; however, challenges and risks remain widespread.

The following challenges are addressed for capture, transportation and storage.

- Technological Challenges: These include energy use, capture efficiency, storage integrity, transportation, and infrastructure, and limited "use" applications.
- Economics and Financial Risks: Reducing capital and operating costs, addressing long-term liabilities, uncertainty of revenue streams, and carbon pricing/taxation.
- Regulatory and Policy Risks: Potential for shifting regulatory support, absence or inadequacy of carbon pricing, tax incentives or subsidies, authorization of permits, requirements for monitoring and actions, and long-term liabilities.
- Environmental and Safety Risks: Geological risks including assessing and mitigating geological risks associated with CO_2 storage, such as potential leakage from various routes, induced seismicity, and subsurface interactions. Minimizing ecological impacts of CCS projects on terrestrial and aquatic ecosystems, including effects on biodiversity, water quality, and land use.
- Public Perception and Social Risk: The perception that CCS exists primarily to extend the commercial life of oil and gas activity, the NIMBY and BANANA problem, public skepticism, or opposition due to concerns about safety, environmental impacts, and reliance on fossil fuels. "Not in my back yard" or NIMBY refers to opposition to projects near ones residence while "build absolutely nothing near anybody" or BANANA refers to people who oppose almost any new project.
- Challenges with community engagement to ensure transparency, address concerns, and secure social license for CCS projects, particularly regarding site selection and environmental impact assessments. Social equity concerns also remain.
- Startup and Scaleup Risks: Costs and performance of first-of-a-kind (FOAK) facilities have often been problematic and challenges for implementing n_{th}-of-a-kind (NOAK) facilities will continue to the degree that each facility needs to have bespoke design characteristics. Massive cost reductions in wind and solar power generation costs and batteries have occurred with widespread commercial adoption, manufacturing efficiencies and rapid technological improvements. The pace of cost reductions in the scaleup of CCS is likely to be much less pronounced.

More than a dozen projects are discussed in terms of challenges associated with startup and operations along with lessons learned.

162b Tuning Process Parameters for Enhanced Low CO_2 concentration Capture at a 0.7 MWe Post-Combustion Pilot Plant

Reynolds Frimpong, Heather Nikolic, Kunlei Liu

A post-combustion CO_2 capture technology at the 0.7 MWe scale implemented at the E.W. Brown Generating Station at Harrodsburg, Kentucky, a PPL Corporation facility, primarily funded by the U.S. Department of Energy, National Energy Technology Laboratory (DOE NETL) employs multi-pronged technological approaches developed over several years of experience to continue to lower the cost of carbon capture. The University of Kentucky (UK) CO_2 capture technology is applicable to any second-generation advanced solvent and can be adapted for a wide array of CO_2 concentrations such as for coal-fired or natural gas plants and has been demonstrated for several solvent campaigns where tailored conditions are used to maximize solvent carbon loading at absorber outlet to ultimately reduce the associated energy penalty of the process.

In this work, process insights will be shared on how CO_2 absorption is enhanced in a solvent campaign with a low inlet CO_2 concentration of about 4%, indicative conditions for capture from natural-gas plants. To obtain the desired low inlet concentration, a modified process draws in a stream of ambient air to dilute the flue gas generated at the plant from coal which typically have CO_2 concentrations at ~12-14%. While the reduced driving force for CO_2 transfer from the gas phase into the solvent is a limitation that results in associated high energy of the process from rigorous solvent regeneration conditions, our experimental work has shown that with appropriate systematic tuning of process parameters such as ensuring adequate wetting of packing and maintaining appropriate temperature profile in the absorber, the absorption of CO_2 can be significantly improved to reduce the energy penalty. With reduced liquid-gas (L/G) circulation rates for the low CO_2 concentration, it is necessary process conditions are tuned for increased wetted area. Results for impacts of different solvent circulation rates, steam flow rates with particular emphasis on absorber bulge temperatures in improving capture by ~20% will be discussed. Additionally, the contribution of the effect of reducing the absorber bottom temperature from a cooled flue gas feed to the bottom of the absorber to an overall reduced energy of regeneration for the process will also be highlighted.

162c Electricity Planning for Ecosystems and Society: A Spatial-Temporal Approach Considering Cost, Societal Benefit and Environmental Justice

Xinyu Zhen, Bhavik Bakshi

Recent years have seen an increase in both the frequency and severity of extreme weather events such as heatwaves, floods, and freezing conditions[1], with climate change and greenhouse gas (GHG) emissions playing a significant role. Emissions Gap Report 2023 by UN Environmental Programme (UNEP) emphasizes the urgent need for global actions to meet the Paris Agreement goals: by 2030 the predicted GHG emissions must decrease by 28% for the 2 °C pathway[2]. Furthermore, according to data from World Health Organization (WHO), hazardous air pollutants are contributing to high mortality rates of around 7 million per year, and almost 99% of people in the world live in places that do not meet the WHO air quality guidelines levels[3]. These facts highlight the pressing importance of implementing environmental health and safety measures. Among all the sources of GHG emissions and air pollutants, fossil fuel electricity power plants are one of the top contributors[4,5]. Studies show that some minority groups such as people of color, people with lower income, and places with high unemployment rate, are more likely to be vulnerable and exposed to worse air conditions, which leads to environmental injustice[6–8]. Therefore, our work is motivated by these facts and focuses on finding solutions to improve air quality and consider the needs of people. Previous work related to sustainable electricity planning[9,10] only focused on minimizing economic costs, and they rely on imposing environmental constraints and traditional equipment as ways to control the emissions. Our work is inspired by nature's ability of removing pollutants and we include ecosystems as unit operations to work in synergy with engineering technologies to improve air quality, and this method is called Techno-Ecological Synergy (TES)[11]. Studies in this area design ecosystems and quantify their capacity in purifying air or water quality to achieve harmony between engineering systems and nature[12,13], however, TES has not been applied to electricity generation planning. Quantifying societal costs and environmental justice have also been mostly ignored. In this work, we develop a spatial-temporal model that integrates hourly electricity generation planning and strategic design of reforestation locations over time. Moreover, this model will consider societal cost and environmental justice as important factors in decision-making.

We employ the Techno-Ecological Synergy (TES) framework that integrates traditional emission control equipment with natural ecosystem processes to remove carbon dioxide and hazardous air pollutants emitted by power plants. In this work, four traditional emission control technologies are considered: selective catalytic reducer, flue gas desulfurization, baghouse filter, and aqueous monoethanolamine (MEA) solution for removal of NO_2, SO_2, PM_{10}, and CO_2, respectively. Trees are considered as an ecological unit operation in this work. The capacity of trees is quantified by dry deposition flux which is related to vegetation parameters. Air pollution modeling performed by CALPUFF version 7, an advanced non-steady-state meteorological and air quality modeling system, was utilized to obtain the spatial concentration and dry deposition maps for NO_2, SO_2, and PM_{10} before and after land use change from original to tree covers. Besides, we used i-Tree Canopy to calculate carbon sequestration by trees, and the growth dynamics of trees were calculated by Forest Vegetation Simulator (FVS). To quantify societal costs, we obtain spatial social impact maps from BenMAP-CE, an open-source software tool designed to estimate the number of fatalities and illnesses attributable to air pollution. For incorporating environmental justice considerations, the block group level demographic index (calculated based on the average of five socioeconomic indicators; low-income, unemployment, limited English, less than high school education, and low life expectancy) was retrieved from EJScreen, an Environmental Protection Agency (EPA)'s tool for mapping and screening environmental justice concerns. This data was then integrated with our case study's specific geographic units using QGIS, ensuring a comprehensive analysis that aligns with our defined parameters. We conducted a case study over the region of Louisville, KY with nine power plants that utilize a mix of energy sources including coal, natural gas, and hydroelectric power. The optimization problem was formulated as a multi-objective mixed-integer linear program (MILP) and was solved in Julia by Gurobi solver to minimize engineering costs, societal costs, and environmental injustice. We explored three different scenario groups including conventional, techno-centric, and TES to find the best solution to balance the three objectives above and emission goals.

This work incorporates hourly electricity planning, emission control technologies, and spatial-temporal landscape design, in which the engineering systems work with nature in synergy to achieve both air quality and people-positive targets. The findings reveal that Techno-Ecological Synergy (TES) scenario groups demonstrate superior performance over time with lower overall costs. As trees become mature, their capacity to remove carbon dioxide and air pollutants will be higher than the demand of power plants, achieving greater environmental benefits by taking up additional carbon and emissions emitted by other sources. Furthermore, the cost of planting trees will be much less than traditional technologies. With the presence of more tree cover, electricity power plants will have more flexibility in running without the burden of worsening air quality during high-demand days. We also find out by setting additional objectives of social cost and environmental justice index, the results of time and location for tree planting can be very different: these objectives will prioritize reforestation at locations with high population densities and high percentages of minority groups and in the meantime start to plant trees at earlier years. From the Pareto front, we notice there are trade-offs between (1) engineering costs & social costs; (2) engineering costs & environmental justice index. The reason could be these locations have a relatively low dry deposition flux after land-use change according to air pollution modeling, so more trees and technologies are needed which makes the ecological and technological costs higher. However, people-positive is essential and our results show the necessity to consider it in decision-making. In summary, this work optimizes hourly electricity schedules, emission control technologies usage, time and locations for reforestation. It highlights the benefits of including trees as unit operations to improve air quality and emphasizes the significance of considering societal and minority groups in design strategies. Our aim is to foster a sustainable, nature-positive, and people-positive world, demonstrating how environmental solutions can be beneficial to both nature and diverse human communities .

Reference

1. Climate change fuelled extreme weather in 2023; expect more records in 2024. *World Weather Attribution* (2023).
2. United Nations Environment Programme. *Emissions Gap Report 2023: Broken Record – Temperatures Hit New Highs, yet World Fails to Cut Emissions (Again)*. (United Nations Environment Programme, 2023). doi:10.59117/20.500.11822/43922.
3. WHO fact sheets-Ambient(outdoor) air pollution. https://www.who.int/news-room/fact-sheets/detail/ambient-(outdoor)-air-quality-and-health.
4. United States Environmental Protection Agency. Sources of Greenhouse Gas Emissions. https://www.epa.gov/ghgemissions/sources-greenhouse-gas-emissions.
5. United States Environmental Protection Agency. *Our Nation's Air 2022*.
6. Wang, Y. *et al.* Location-specific strategies for eliminating US national racial-ethnic PM2.5 exposure inequality. *Proc. Natl. Acad. Sci. U.S.A.* **119**, e2205548119 (2022).
7. Pope, R., Wu, J. & Boone, C. Spatial patterns of air pollutants and social groups: a distributive environmental justice study in the phoenix metropolitan region of USA. *Environmental Management* **58**, 753–766 (2016).

8. Banzhaf, H. S., Ma, L. & Timmins, C. Environmental Justice: Establishing Causal Relationships. *Annu. Rev. Resour. Econ.* **11**, 377–398 (2019).
9. Nazari, M. E., Ardehali, M. M. & Jafari, S. Pumped-storage unit commitment with considerations for energy demand, economics, and environmental constraints. *Energy* **35**, 4092–4101 (2010).
10. Shuai Lu *et al.* Unit commitment considering generation flexibility and environmental constraints. in *IEEE PES General Meeting* 1–11 (IEEE, Minneapolis, MN, 2010). doi:10.1109/PES.2010.5589501.
11. Bakshi, B. R., Ziv, G. & Lepech, M. D. Techno-Ecological Synergy: A Framework for Sustainable Engineering. *Environ. Sci. Technol.* **49**, 1752–1760 (2015).
12. Shah, U. & Bakshi, B. R. Toward Nature-Positive Manufacturing by Adapting Industrial Processes to Pollution Uptake by Vegetation. *ACS Sustainable Chem. Eng.* **9**, 16709–16718 (2021).
13. Charles, M. & Bakshi, B. R. Designing industrial landscapes for mitigating air pollution with spatially-explicit techno-ecological synergy. *AIChE J* **67**, (2021).

162d Carbon Accounting for Sustainable Supply Chains.

Nadin Moustafa, Nilay Shah

In the pursuit of global sustainability and the ambitious commitment to achieve net-zero emissions, the significance of robust carbon accounting methodologies becomes paramount. Net-zero emissions entail achieving a balance between the total amount of greenhouse gases emitted and the amount removed from the atmosphere, often through carbon offsetting or removal technologies. Various countries, including the United Kingdom, Canada, India and Japan, have established net-zero targets, aiming to achieve this balance by a specified date. In fact, approximately 89% of the world's greenhouse gas emissions are now under some form of net-zero target, whether in policy documents, international agreements, or corporate pledges[1]. Carbon accounting is essential for organizations and industries aiming to quantify and mitigate their carbon footprints. As nations, businesses, and communities increasingly commit to net-zero goals†, accurate and transparent carbon accounting is necessary for tracking progress, identifying areas for improvement, and implementing effective strategies to reduce greenhouse gas emissions. The ability to measure and account for carbon emissions not only facilitates compliance with regulatory frameworks but also provides a foundation for informed decision-making, resource allocation, and the formulation of targeted interventions.

The need for standardized carbon accounting was highlighted by the World Economic Forum Net-Zero Industry Tracker in 2023 as a key takeaway[2], as challenges persist, including reporting standards, supply chain instability, and transparency gaps. The complexities of carbon accounting extend beyond the immediate operational boundaries of organizations, encompassing what are known as Scopes 1, 2, and 3 emissions. Scopes 1 and 2 cover direct emissions from sources owned or controlled by an organization (Scope 1) and indirect emissions from purchased electricity, heat, or steam (Scope 2). These scopes are relatively straightforward to address, as they involve emissions that organizations have direct control over or are closely linked to their operations. Consequently, many countries have already incorporated reporting requirements for Scopes 1 and 2 emissions into their regulatory frameworks, reflecting their importance in understanding and mitigating carbon footprints.

However, the challenge intensifies with Scope 3 emissions, which encompass all other indirect emissions along the value chain, including those from upstream and downstream activities such as supply chains, transportation, and product use. Scope 3 emissions often represent the largest portion of an organization's carbon footprint and are significantly more challenging to measure and manage due to their complex and extensive nature. According to the CDP, there are 11.4 times more emissions in a company's supply chain compared to its direct operations[3]. While reporting on Scope 3 emissions is not yet as widespread as for Scopes 1 and 2, there is a growing recognition of their importance in comprehensive carbon accounting. Countries like France have already begun to include Scope 3 reporting requirements in their policies, signalling the need for organizations to be prepared to incorporate these emissions into their accounting methodologies[4]. Addressing Scope 3 emissions effectively will require enhanced collaboration across supply chains, transparency in data sharing, and innovative approaches to measurement and reporting. As such, the inclusion of Scope 3 emissions in regulatory frameworks underscores the urgency for organizations to develop robust carbon accounting practices that capture the full extent of their emissions footprint and enable informed decision-making towards achieving net-zero objectives.

This research intricately examines the landscape of carbon accounting with a specialized focus on the decarbonization of the logistics industry incorporating maritime shipping, aviation, and road transport. The logistics industry plays a pivotal role in global supply chains, serving as a crucial link connecting producers with consumers across the globe. However, it is also a significant contributor to carbon emissions with the international trade-related freight transport accounting for around 12% of global greenhouse gas (GHG) emissions and energy and with transportation accounting for more than 75% of the GHG emissions embedded in international trade[5]. Unlike some industries that may have more localized emissions, logistics involves extensive transportation networks, making it challenging yet critical to develop effective strategies for carbon reduction. By focusing on the logistics sector, this research aims to address the unique challenges and opportunities associated with carbon accounting in an industry that spans various modes of transportation and supply chain complexities.

The significance of examining carbon accounting within the logistics industry is further underscored by the emergence of carbon border adjustment mechanisms and the increasing recognition of consumption-based emissions. With the implementation of carbon border adjustment mechanisms on the horizon, there is a growing imperative for organizations to accurately measure and report their carbon emissions throughout the entire supply chain. Additionally, the focus on consumption-based emissions highlights the need to account for emissions associated with the production and transportation of goods consumed in one region but manufactured in another. This necessitates a comprehensive understanding of the carbon footprint embedded in international trade, particularly within the logistics sector, which is essential in connecting global producers and consumers. By addressing the unique challenges and opportunities inherent in carbon accounting for logistics, this research aims to provide insights that contribute to the development of effective strategies for emissions reduction and compliance with evolving regulatory frameworks.

Building upon our critical review and assessment of CO_2 footprint methods and analysis, our research sheds light on the strengths, weaknesses, and gaps within existing carbon footprint methodologies. Through a thorough examination of factors within the CO_2 accounting and decarbonization ecosystem, we identify areas where current methods may fall short and propose innovative solutions to address these shortcomings. By scoring different factors within the CO_2 accounting framework, we provide a comprehensive assessment of its effectiveness in guiding emissions reduction strategies.

Furthermore, our research leverages diverse datasets to advance our understanding of carbon accounting methodologies. Collaborating with our industrial partners, primarily including Hitachi Ltd., we explore how different policies and regulations can be integrated into existing frameworks. Through modelling techniques, we investigate future interventions aimed at significantly reducing emissions from logistics and built environment functions, while also assessing how the benefits of such interventions can be effectively monitored and measured.

In addition to regulatory considerations, our summary analysis and prioritization of challenges faced by organizations in achieving Net Zero highlight common obstacles such as regulatory complexity, data quality issues, and the complexities of Scope 3 emissions. Through a structured analysis, we identify the most pressing challenges and propose strategies for overcoming them. Prioritizing challenges enables organizations to better allocate resources and develop targeted interventions to accelerate progress towards Net Zero goals.

In conclusion, our research bridges the gap between theoretical frameworks and practical applications in the realm of carbon accounting and emissions reduction. Through collaborative efforts leveraging diverse datasets, we identify intervention measures tailored to the unique needs of the logistics and built environment sectors. By delving into existing methodologies, regulatory landscapes, and industry challenges, this work equips organizations with the knowledge and tools necessary to navigate the complex ecosystem of carbon accounting. The study aims to propose innovative methodologies for quantifying the environmental impact of carbon accounting practices within a regulatory context. This collaborative exploration anticipates not only advancing carbon accounting practices but also providing tailored recommendations for the integration of these activities into regulatory frameworks on a global scale. The research aims to bridge knowledge gaps between regions, fostering global knowledge sharing and advancing transparent and credible accounting practices for carbon emissions associated with various industries.

**Corresponding author e-mail address: nm1217@imperial.ac.uk*

[1] Net Zero Tracker (2023). https://zerotracker.net/insights/net-zero-targets-among-worlds-largest-companies-double-but-credibility-gaps-undermine-progress

[2] World Economic Forum (2023). https://www3.weforum.org/docs/WEF_Net_Zero_Tracker_2023_REPORT.pdf

[3] Global Supply Chain Report (2020). https://www.cdp.net/en/research/global-reports/transparency-to-transformation

[4] S&P Global (2024). https://www.spglobal.com/marketintelligence/en/news-insights/blog/investment-banking-essentials-april-3

[5] World Trade Organization (2021). https://www.wto.org/english/news_e/news21_e/clim_03nov21-4_e.pdf

162e A Sequential Decision-Making Framework for Long-Term Energy Transition Planning Under Uncertainty

Molly McDonald, Christos Maravelias

Many countries around the world have set net zero emission targets within the next 30 years in an effort to mitigate the impacts of climate change. Achieving this ambitious goal will require large-scale investments into renewable and low-carbon technologies that will need to be efficiently integrated into our current energy system in order to meet net zero goals at the lowest cost. Energy system models can show how we can expand and operate a given set of technologies in order to meet specific goals. In most previous studies, the developed models have assumed to have perfect information about the future and that a decision maker will only make decisions at the beginning of the time horizon. But in reality, it is uncertain how parameters will change over time, such as the demand for liquid fuels. Also, decision makers are able to re-evaluate their previous decisions and can update their choices based on new information.

Uncertainty in energy system models is commonly studied through sensitivity analysis, allowing a decision maker to see how decisions change as the level of the input parameters change, as was done in the Net Zero America study [1]. While this method is beneficial for observing the effect of varying parameter values on the system, it does not allow for making unique decisions today that address the different parameter levels. Another method to study uncertainty is stochastic programming, which accounts for uncertainty through scenarios. In stochastic programming, there are first stage decisions that are made before any uncertainty is realized and are the same across all scenarios, and there are decisions made in the following stages as uncertainty is realized. In multistage stochastic programming (MSSP), uncertainty is realized successively through multiple stages [2]. Through the use of MSSP, a decision maker is able to make unique decisions today that address the different scenarios.

A common way energy system models are solved is with a forward-looking approach, where a model is solved once over the full horizon, meaning the energy transition is predicted at the beginning of the planning horizon. In reality, an energy transition over a long horizon will involve decisions made in a sequential manner which means that a rolling horizon approach can be applied, where a decision maker's ability to consider feedback and observe the development of parameters is better represented. With this approach, a decision maker will select which technologies to invest over a given planning horizon, for example over the next 30 years. After a few years, the decision maker is able to re-evaluate their choices and can make new investments considering how parameters have changed as well as new information about the future that they did not know previously.

We develop a linear mixed-integer MSSP model for capacity expansion to study the energy sector for the United States using real data [4]. The system includes 10 liquid fuel producing technologies, 9 electricity generating technologies and 2 intermediate technologies over a multiperiod horizon divided into 5-year time periods. The model is solved to meet liquid fuel and power demand over the horizon at the lowest cost and considers uncertainty in the demand, capital cost of select renewable technologies and policy related to carbon emission reduction. In addition to the MSSP model, we develop a deterministic model (meaning we assume perfect foresight) for the same system. We solve the models separately with a forward-looking and with a rolling horizon approach. This allows us to evaluate the benefit of accounting for (1) uncertainty with MSSP under both solution approaches and (2) a decision makers ability to re-evaluate decisions. We also present additional ways to simulate challenging real-world characteristics such as time delays, lagged capacity installation, or the impact of social acceptance with a rolling horizon approach.

When the models are solved with the forward-looking approach, it is seen that the solution of the stochastic model has larger initial investments into technologies producing electricity and liquid fuel compared to the solution of the deterministic model. This is because the stochastic model makes a single decision today that considers multiple scenarios with varying demand for liquid fuels and electricity. When the rolling horizon approach is used, the stochastic model has a wider variety of technologies installed throughout the time horizon. This is again due to the uncertainty considered in each iteration with the stochastic model, leading to more diverse investments.

We also consider the effect of incorrect parameter predictions for both the forward-looking and the rolling horizon approach. When the investment decisions of the solution obtained using the forward-looking approach are subject to different observed parameter values, the average optimal objective value of the stochastic model is lower and has a smaller standard deviation than the average optimal objective value of the deterministic model. A similar trend is observed when the parameter values are varied with the rolling horizon approach. It is also seen that the average optimal objective value of both the deterministic and stochastic models are lower when paired with the rolling horizon approach compared to the forward-looking approach, which demonstrates that there is a benefit to reoptimizing investment decisions as new information is observed. This trend is seen because there are more diverse investments made when a decision maker is able to reconsider their past choices and make new investments as they observe how parameter values develop compared to their original predictions.

We close with a short but broader discussion of energy system models. The proposed models allow us to study how real-world constraints impact the energy transition in the United States. For example, we study how investment decisions change to focus more on renewable and low-carbon technologies when we enforce net zero emissions by 2050. When we pair our model with a rolling horizon approach to better represent real-world decision making, we observe the benefit of being able to re-evaluate decisions, particularly when predictions about parameter values over the horizon are incorrect.

References

1. Larson, E., Greig, C., Jenkins, J., et al. (2021). Net-Zero America: Potential Pathways, Infrastructure, and Impacts, Final Report Summary. *Princeton University.*
2. Birge, J.R., Louveaux, F. (1997). Uncertainty and modeling issues. *Introduction to Stochastic Programming*, 54-61.
3. Rawlings, J.B., Mayne, D.Q. (2009). Model Predictive Control. *Model Predictive Control: Theory and Design*, 91-107.
4. "U.S. Energy Information Administration - EIA - Independent Statistics and Analysis," https://www.eia.gov/state/search/#?5=126&r=false.

162f CO_2 Conversion to Syngas Coupled with a Membrane Separation: Multi-Scale Modeling and Simulation.

Seckin Karagoz

Fossil fuels continue to become the major energy sources around the world. About 6,870 million metric tons (MMT) of greenhouse gas are emitted into the atmosphere each year because of the fossil fuels-based energy production actions, over 80% of which is in the form of CO_2. As a result of this process, the world is facing the serious environmental problems, and, thus, reducing the carbon emissions has become an urgent issue. The reverse water gas shift reaction (RWGS) has garnered significant interest as a possible method for the broad usage of CO_2 by generating synthesis gas. The reverse water gas shift (RWGS) reaction is a promising method for converting CO_2 and H_2 into CO and water. The resulting CO can then be utilized in downstream Fischer-Tropsch (FT) or MeOH synthesis processes. Nevertheless, because the RWGS reaction is endothermic, it necessitates elevated temperatures in order to attain equilibrium CO_2 conversions.

Integration of multiple operations (e.g., reaction and separation) in a single unit is a robust tool to improve the existing process' efficiency, reduce energy consumption, and unwanted outputs/by-products. Such intensification has the potential to enhance process efficiency/economics by offering drastic process technologies improvements, regarding a variety of metrics to accomplish such intensification, PI encompasses several techniques such as adsorption, absorption and membranes. Especially, membrane reactor (MR) receives considerable attention to removal, and recovery of the produced gases from fuel gas. The reaction products is selectively removed by the membrane, so the equilibrium is shifted to the product side, enabling higher conversions and the produced gas capture processes to be combined in one single step. Operating at lower reaction temperatures, reducing material costs, increasing operation safety, eliminating the need for excess steam in the reaction, minimizing the need for downstream purification and reducing the amount of catalyst for the desired conversion level are the other remarkable features of these technologies.

The objective of this work is to demonstrate the integration of membrane separation with RWGS for CO_2 conversion to syngas. The applicability of various (alternative to the conventional process) novel and efficient reactor configurations that include co- or counter-current membrane reactors in sequence is investigated. Innovative designs for the proposed processes are determined based on comprehensive multi-scale modeling and design. A comprehensive, multi-scale, multiphase, steady-state/dynamic, computational fluid dynamics (CFD)-based process model is developed that quantifies the many underlying complex physicochemical phenomena occurring at the pellet and reactor scales.

559a Inverse Design of Perovskite Materials through the Crossfit CEF and Bayesian Data Selection Algorithms for Solar Thermochemical Hydrogen Production

Steven Wilson, Matthew Park, Ellen B. Stechel, Christopher Muhich

The thermodynamic properties of metal oxides as a function of off-stoichiometry are crucial materials design attributes in metal oxide-based oxygen-exchange chemical processes. It has been shown that the compound energy formalism (CEF) is a powerful framework to describe these thermodynamic properties. Recently, we developed the CrossFit CEF method for fitting the CEF model that overcomes the current fitting challenges associated with non-unique fits in part, due to the thousands of possible liner combinations of the parameters involved resulting in dynamic behavior between enthalpy and entropy trends. The key innovations are: 1) the combination of density functional calculations with experimental data that delineates the enthalpic/entropic contributions to the Gibbs free energy; 2) a systematic determination of the important CEF model terms, removing thermodynamic predetermining human intervention; 3) a self-consistent solution of the starting oxygen off-stoichiometry (δ_0) of thermogravimetric measurements. With the advent of this state-of-the-art algorithm our method enables the reliable extraction of off-stoichiometric metal oxide thermodynamic properties and facilitates rapid materials compositional screening, and reliable process design of systems dependent on off-stoichiometric redox-active metal oxides. We build on this concept by examining the use of a Bayesian approach to selecting data points which should be gathered next. To date, this active data selection technique has not been applied or refined for thermodynamic characterization of metal oxide reduction/re-oxidation cycle materials. We focus our study on the investigation of perovskite ($ABO_{3-\delta}$) metal oxides for solar thermochemical water separation (STCH). The combination of the CrossFit CEF and Bayesian methods provides more reliable thermodynamic characterization and identifies, a priori, critical points to examine in the temperature, O_2 partial pressure, compositional landscape, thus minimizing the number of points which must be examined. Specifically, we show that a model identical to a ground truth model can be realized with half the data from careful selection of data points through the Bayesian approach. Furthermore, we show that the Bayesian informed approach does better than pure random sampling of data. We then couple the robust thermodynamic models to a STCH system level elucidating high efficiency materials at various operating conditions. Likewise, in an inverse design process, hypothetical materials can be inferred from the STCH system level model and in a high throughput process with the CrossFit CEF and Bayesian data selection algorithms a material can be identified to match the STCH system modeled material. Overall, implementation of our method will save significant time in data collection, allowing for more materials to be investigated and lower research costs. Further, it opens the possibility of a hands-off high-throughput process for metal oxide material selection and design and while not yet directly interfaced with TGA or other data acquisition software, these algorithms could easily be integrated for seamless informed data collection by TGA where a thermodynamic model is built in real time as the TGA collects more data.

559b Mixed Metal Oxides for Enhanced Solar Fuel Production Via Thermochemical Redox Cycling

Asim Riaz, Christopher I. Bodger, Jingjing Chen, Adrian Lowe, Takuya Tsuzuki, Wojciech Lipiński

The area of solar thermochemistry aims at direct thermochemical production of chemical fuels and advanced thermochemical processing of materials. Cheap and efficient solar production of synthesis gas (syngas), the precursor to synthetic drop-in hydrocarbon fuels such as petrol, diesel and kerosene, is an intriguing approach to transform today's fossil-based to tomorrow's renewable-based transportation sector. Solar-driven thermochemical redox cycles have been demonstrated as a promising approach to water and carbon dioxide splitting. While direct on-sun high-temperature reduction has been demonstrated with various metal oxides, cerium dioxide (ceria) remains the reference metal oxide. Main challenges are related to long-term cyclic chemical and mechanical stability of metal oxide structures at high temperatures required for the direct reduction step. The process conditions become less challenging when solar-driven carbothermal reduction is employed. We have investigated ceria-based and ceria-doped mixed metal oxide partial methane oxidation coupled to redox cycling for enhanced syngas production, materials stability and system scalability. Intriguing interplay effects were found to improve structural and catalytic properties in ceria-containing compositions, of which some resulted in promising fuel production rates using reticulate porous ceramic structures.

559c An Efficient and Scalable Approach to Solar Thermochemical Fuel Production Via an Iron Aluminate-Based, Pressure-Swing Redox Cycle

Kent J. Warren, Liam Taylor, Dragan Mejic, Robert Anderson, Justin Tran, Deepak Dileepkumar, Alan W. Weimer

Technologies like solar thermochemical hydrogen (STCH) production, which leverage the oxygen-exchange properties of metal oxides to split water (and/or carbon dioxide), are attractive because they can harness the entire solar spectrum [1], whereas photovoltaic-driven alternatives (e.g., water electrolysis) must rely only on specific wavelength ranges. Generally, this approach involves alternating between two steps namely, (1) the endothermic removal of oxygen from an oxide and (2) the exothermic oxidation of the reduced oxide with water (and/or carbon dioxide) to produce hydrogen (and/or carbon monoxide). The endo/exothermic nature of each reaction implies that opposite changes in temperature are required to drive the cyclic conversion of, for example, water into hydrogen. However, in practice, implementing a temperature swing between reduction-oxidation (redox) regimes imposes significant irreversibilities that arise from sensible heating of the solid, thus resulting in inefficient operation unless the heat rejected during cooling is recovered via ultra-high temperature thermal energy storage – the feasibility of which continues to remain uncertain. Therefore, in this work, a scalable fluidized bed reactor – situated within an extremely well-insulated cavity receiver – was designed and operated such that the oxidation (or fuel producing) step was intentionally initiated at temperatures near where reduction is favorable, thereby avoiding the complications associated with solid-solid heat recuperation. To mitigate the undesired thermodynamic consequence of performing the exothermic oxidation reaction at such higher temperatures, iron aluminates, a low-cost class of materials with an exceptional isothermal (and near-isothermal) capacity for fuel production [2], were considered for this prototype demonstration. To further boost performance, the oxidation step was additionally performed at elevated pressures, as doing so has recently been shown to improve both the equilibrium extent and rate of the reaction [3], despite it being equimolar. Our state-of-the-art experimental results, obtained under concentrated radiation from a 45 kW_e high-flux solar simulator, provide compelling motivation for the commercial viability of solar thermochemical routes for the production of fuel.

[1] Warren, K. J., & Weimer, A. W. (2022). Solar thermochemical fuels: Present status and future prospects. *Solar Compass*, *1*, 100010. https://doi.org/10.1016/j.solcom.2022.100010

[2] Warren, K. J., Tran, J. T., & Weimer, A. W. (2022). A thermochemical study of iron aluminate-based materials: A preferred class for isothermal water splitting. *Energy & Environmental Science*, *15*(2), 806-821. https://doi.org/10.1039/D1EE02679H

[3] Tran, J. T., Warren, K. J., Mejic, D., Anderson, R. L., Jones, L., Hauschulz, D. S., Wilson, C., & Weimer, A. W. (2023). Pressure-enhanced performance of metal oxides for thermochemical water and carbon dioxide splitting. *Joule*, *7*(8), 1759-1768. https://doi.org/10.1016/j.joule.2023.07.016

559d Electric Field Enhanced Thermochemical CO_2 Splitting

Jayni Hashimoto, Jordan Monroe, Alicia Bayon, Alonzo Mendez, Olivia Tamburro, Christopher L. Muhich

Thermochemical gas splitting has been identified as a promising method of harnessing solar energy to produce solar fuels. In thermochemical CO_2 splitting metal oxides, MO_x are thermally reduced (T > 1200 °C) to $MO_{x-\delta}$ using concentrated solar energy, followed by an oxidation step with CO_2 to regenerate MO_x and form CO, which can be mixed with H_2 to form liquid fuels. Ceria has been identified as the state-of-the-art material due to high stability and fast oxidation kinetics. However, solar thermochemical processes are plagued by high solar field costs due to the extraordinarily high temperatures required. Methods of lowering the reduction temperature of ceria have been well-studied, but can have detrimental effects on the conversion of CO_2 to CO. Previous studies have demonstrated the ability to reduce ceria under only the application of a strong electric field in a reversible manner at 25 °C. The E-field induced reduction of ceria changes the chemical potential of lattice oxygen, making the formation of O vacancies more favorable. As changes to the formation of O vacancies is tunable with the applied potential, removal of the potential from the system, allows ceria to re-establish its original thermodynamics and oxidation favorability. One major challenge to the usage of an electric field in thermochemical systems is the method by which a sufficiently large field is applied on a large-scale. In this work we use molten salts to generate an electronic double layer (EDL) at the electrolyte-ceria interface. The EDL acts as a parallel plate capacitor, with nanometer scale plate spacing, allowing for large specific capacitances to amplify low applied voltages into large electric fields to drive the reduction of ceria. We demonstrate an electrically enhanced two step thermochemical CO_2 splitting process, where a low potential is applied to the ionic liquid to generate an electric field via EDL in order to reduce ceria, then the field is removed and ceria is oxidized using CO_2, producing CO. Specifically, we show molten salt compositions compatible with generating the electronic double layer with CeO_2 and their ability to generate high electric fields. We also demonstrate the ability of these EDL electric fields to reduce ceria at temperatures below 1000 °C and perform isothermal CO_2 splitting cycles. By allowing for significant reductions in the needed temperatures, the E-field enhanced CO2 splitting process significantly drives down the CO production costs as the number of heliostats needed decreases, the radiative losses are lowered, and the constraints on materials of construction are loosened.

559e Kinetics and Techno-Economic Analysis of the Boudouard Reaction System Applied to Long Distance Thermal Energy Transport

Riley Choquette, Benjamin Phillips, Abigail Joseph, Annette von Jouanne, Alexandre Yokochi

The delivery of high-quality process heat derived from clean sources with high exergetic efficiency to distant locations (e.g., tens of kilometers) would be an enabling technology to decarbonizing the industrial sector. Our lab has been exploring the use of the Boudouard family of reactions including the simple Boudouard reaction [1,2]:

$2CO \leftrightarrow CO_2 + C$ $\Delta H = -171$ kJ

And the Sorption Assisted Boudouard Reaction [3]:

$2CO + MO \leftrightarrow MCO_3 + C$ $\Delta H = -350$ kJ ~ -440 kJ

where M represents a Group 2A element.

This is practical because a) the large positive entropy term causes the reaction to favour the left hand side of the equations at high temperature, yielding a strongly endothermic reaction that enables heat capture; b) in the presence of an adequate catalyst at suitable temperatures, and partial pressures of the reactant and product gases the reaction will favour the right hand term; and c) since the CO is kinetically stable at ambient temperatures, thermal energy can be stored indefinitely, transported, and released again. The modeled roundtrip exergetic efficiency of the process may be as high as 80% at the ideal heat release temperature, or achieve slightly lower efficiency at temperatures over 1000°C.

In order to enable design of the heat capture and heat regeneration reactors, the reaction kinetics of the forward and reverse reactions must be known. In this presentation, these kinetics will be presented, and an initial technoeconomic model describing the integration of the technology into existing industrial applications will be shared.

References:

[1] Boudouard, O. Influence De La Vapeur D'eau Sur La Réduction De L'anhydride Carbonique Par Le Charbon *C. R. Hebd. Acad. Sci.* **1905**, *141*, 252– 253.

[2] See, for example, a) Yokochi, A. et al. Thermal Energy Storage Using Sorption-Assisted Boudouard Processes *2021 AIChE Annual*, **Nov 7-19 2021**; b) 737e. Alexandre Yokochi, Riley Choquette, Benjamin Phillips and Annette von Jouanne, School of Engineering and Computer Science, Baylor University, Waco, TX *2022 AIChE Annual*, **Nov 11-22 2022**; 264c Thermal Energy Transport Using the Sorption-Assisted Boudouard Reaction; c) Alexandre Yokochi, Riley Choquette, Benjamin Phillips and Annette von Jouanne, School of Engineering and Computer Science, Baylor University, Waco, TX *2023 AIChE Annual*, **Nov 4-10 2023**;

[3] Rout, K.R. et al. Highly Selective CO Removal by Sorption Enhanced Boudouard Reaction for Hydrogen Production *Catal. Sci. Technol.* **2019**, *9*, 4100–4107.

559f Ammonia-Based Thermochemical Energy Storage: Design and Optimization of Heat Recovery Cycle Using Conductive Ammonia-Selective Absorbents

Mahdi Malmali, Tanveer Mahtab, Sufian Humayun, Shahriar Rahman, Amir Safikhani, Alon McCormick

Concentrating solar power (CSP) offers large-scale, dispatchable solar electricity that is highly valued for the increased penetration of intermittent renewable energy. However, the levelized cost of electricity for CSP is still higher than the cost targets set by DOE (5¢ per kWh) required for large-scale adoption. Recent empirically-observed learning rates for CSP with thermal energy storage (TES) indicate that there is a significant opportunity for future cost reductions through the development of advanced storage subsystems with reduced cost. To date, three TES pathways have been proposed for the CSP power cycle: sensible heat, latent heat of fusion/vaporization, and thermochemical energy storage (TCES). Each pathway has at least one significant techno-economic or reliability risk. There is ongoing research to design economically viable storage technologies that can be integrated with CSP to offer increased grid reliability and resilience by providing solar dispatchability and grid-support functions.

Ammonia-based TCES (Fig.2) has received attention; a fixed inventory of ammonia-nitrogen-hydrogen mixture undergoes alternating charging (dissociation) and discharging (synthesis) cycles. Ammonia cycling is a promising choice for storage in CSP because:

- it is the most studied process in the chemical industry, with large-scale production optimized over a century. Thus, one can take advantage of existing technology and off-the-shelf catalysts for future Gen3 integrated TCES-CSP.
- there is **no** side reaction for synthesizing/dissociating ammonia.
- in conventional technology, ammonia can be separated from reactants with condensation.
- reactants are abundant and can be cost-effectively stored using salt caverns or shaft drilling.

Our team is advancing the design of the heat recovery cycle in ammonia-based TCES. This will be achieved by devising advanced separations, derisking the reaction and reactor design, and optimizing the storage using advance process system engineering approaches. In this presentation, I will report on the design and progress in developing a 1 kWh_{th} heat recovery cycle via reaction-absorption process.

559g Hydrogenation and Hydrodeoxygenation of Oxygenates and Nitroarenes at Ambient Conditions Via a Two-Step Redox Cycle

Chongyan Ruan, Kunran Yang, Ryota Akutsu, Junchen Liu, H. Henry Lamb, William Martin, Caitlin Beckett, Benjamin Lessin, Fanxing Li

Introduction

Hydrogenation is a critical reaction in the chemical industry. The chemoselective hydrogenation of biomass-derived oxygenates[1] and aromatic nitro compounds[2, 3] holds significant scientific and technological value for producing crucial intermediates of a variety of products, including renewable fuel additives, pharmaceuticals, agrochemicals, dyes, pigments, and others. However, conventional hydrogenation typically requires pressurized hydrogen, high temperatures and pay severe energetic debt. Therefore, cost-effective and sustainable catalytic processes for efficient upgrading is urgently needed.

Herein we report a two-step solar thermochemical hydrogenation process, sourcing hydrogen directly from water and concentrated solar radiation for bio-oils upgrading and nitroarenes hydrogenation. A metal or reduced metal oxide, obtainable from solar thermal or electrochemical reduction, provides the active sites for reactants adsorption and H_2O dissociation. Fundamentally different from conventional hydrogenation, the in situ generated reactive H atoms hydrogenate biomass-derived oxygenates and nitroarenes, eliminating the barriers for H_2 dissolution and mass transfer in the reaction medium and the subsequent dissociation at the catalyst surface.

Materials and Methods

Zn, Fe, Sn, Al, Al alloy, Mg, Mn were purchased from Sigma-Aldrich and used without further treatment. CeO_2 was purchased from Noah Technologies Corporation. Prior to the experiment, the as-received CeO_2 was reduced in 10% H_2/Ar (100 mL/min) at 800 °C for 2 h. The supported catalysts loaded on CeO_2 with nominal Ru loadings of 1 wt% was prepared via the incipient wetness impregnation method.

Results and Discussion

We proposed and validated a two-step solar thermochemical/electrochemical hydrogenation (ST/ECH) process, sourcing hydrogen directly from water and concentrated solar radiation for biomass-derived oxygenates and nitroarenes upgrading. Fig. 1a illustrates simplified schematic of the two-step ST/ECH process. During the upgrading step, a metal or oxygen deprived metal oxide provides the active sites for furan and aromatic rings adsorption as well as a bulk source of reducing potential to dissociate H_2O for in situ hydrogen generation. The reactive H atoms are highly effective for upgrading biomass-derived oxygenates and nitroarenes to hydrogenated products. During the regeneration step, the lattice oxygen is abstracted from the reoxidized metal oxide at elevated temperatures with concentrated solar energy or through well-established electrochemical reduction using renewable electricity. The oxophilic zero-valent metals including Zn, Sn, Fe, Al, Mg and Mn as well as a non-stoichiometric $CeO_{2-\delta}$ redox catalyst were employed to investigate the feasibility. Notably, efficient upgrading of furfural and nitrobenzene were achieved at room temperature and atmospheric pressure, with H_2 utilization efficiency 1-2 orders of magnitude higher than molecular-hydrogen-based HDO processes (Fig. 1c, d). Using suitable metal/reduced metal oxide catalysts with different oxophilicity, the reducing potential for H_2O dissociation and the subsequent hydrogenation can be manipulated. Mechanistic investigations indicate that zero-valent metals or surface oxygen vacancies are active redox centers for in-situ generation of H atoms from H_2O, via a reverse Mars–van Krevelen mechanism (Fig. 1e, f). The efficient STCH process was also extended to converting other biomass derivatives and substituted nitroarenes with similar functional groups, exhibiting superior catalytic performance.

Significance

This work provides an efficient and versatile strategy for bio-oils and nitroarenes upgrading and a promising pathway for renewable energy storage. The general approach can be extended to catalytic upgrading of other platform compounds with similar functional groups.

References

1. M. Saidi, F. Samimi, D. Karimipourfard, T. Nimmanwudipong, B. C. Gates and M. R. Rahimpour, *Energy Environ. Sci.*, 2014, **7**, 103-129.
2. A. Corma and P. Serna, *Science*, 2006, **313**, 332-334.
3. F. A. Westerhaus, R. V. Jagadeesh, G. Wienhofer, M. M. Pohl, J. Radnik, A. E. Surkus, J. Rabeah, K. Junge, H. Junge, M. Nielsen, A. Bruckner and M. Beller, *Nat. Chem.*, 2013, **5**, 537-543.

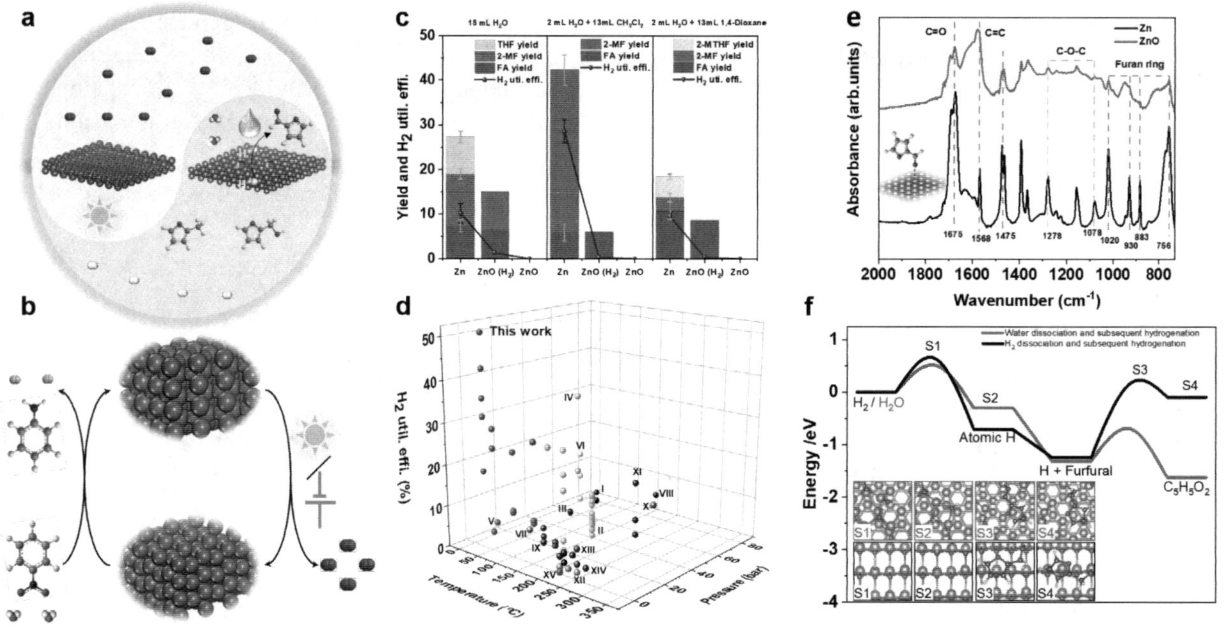

729a >90% Purity Syngas Generation for Direct Reduction of Iron Ore By Integration of Biomass-Based Chemical Looping with Carbon Dioxide Capture

Ishani Karki Kudva, Shekhar Shinde, Tanay Jawdekar, Sudeshna Gun, Rushikesh Joshi, Ashin Sunny, Sonu Kumar, Liang-Shih Fan

Decarbonizing the iron and steel-making industry is critical in achieving the net zero carbon emission goal by 2050. The iron-making sector is responsible for 7% of global greenhouse gas emissions, making it a significant contributor to global warming. As such, these industries must strive to achieve a net zero carbon emission goal. The state-of-the-art technology for producing iron is the MIDREX process, which utilizes syngas (primarily contains carbon monoxide (CO) and hydrogen (H_2)) produced from natural gas reforming for direct reduction of iron (DRI). As part of this process, the off-gas from the shaft furnace is burned to generate heat for the reformer, leading to carbon dioxide (CO_2) emissions. The MIDREX process utilizes acid gas remover to capture CO_2, which is highly energy intensive. Therefore, processes that can facilitate efficient CO_2 capture while also utilizing renewable and carbon-neutral fuels such as biomass for syngas generation would lead to fossil-free iron and steel making.

The Ohio State University's (OSU) moving bed Chemical looping technology leverages biomass to produce syngas. This technology can be coupled with DRI as it offers several advantages. It eliminates the need for costly investments in a tar reformer and air separation unit. Additionally, the moving bed design and catalytic properties of oxygen carrier particles ensure full conversion of biomass char. The integration of OSU's moving bed chemical looping technology and the DRI process will be helpful in making the DRI process carbon-negative. The integration of these technologies involves redirecting off-gases from the shaft furnace into a chemical looping system with the aim of achieving carbon negativity.

The proposed process scheme couples off gases from the DRI plant into the chemical looping system to generate a capture-ready CO_2 stream and also generate high-purity syngas from carbon-neutral biomass, which can be used as a reducing gas in the DRI process. The proposed scheme consists of three reactors. The off-gas stream from the DRI plant is fed into moving bed reactor 1, where the off-gas reacts with iron-based oxygen carrier (OC) particles, generating sequestration-ready carbon dioxide in moving bed reactor 1. These oxygen carrier particles are partially reduced in reactor 1 and then moved to moving bed reactor 2. These particles get further reduced, donating their lattice oxygen for the gasification of lignocellulosic biomass to generate high-purity syngas in moving bed reactor 2. The reduced oxygen carrier particles leaving reactor 2 get regenerated using air in fluidized bed reactor 3, called a combustor. These regenerated oxygen carrier particles are sent back to reactor 1 via the riser, completing the loop.

This process scheme was simulated in Aspen Plus, and the preliminary experimental studies were conducted in a 2.5 kW thermal bench scale setup. The process simulations have shown that the off-gas from the DRI plant in reactor 1 generate sequestration-ready CO_2 while reducing oxygen carrier particles. The dry purity of carbon dioxide from reducer 1 is approximately greater than 95%. In reactor 2, these oxygen carriers react with biomass at high temperatures, producing high-purity syngas. In the proposed method, the syngas has a purity greater than 90%. Moreover, the quality of syngas required for the DRI process is measured by a parameter called reducing potential. For effective DRI, the reducing potential should be greater than 9. Presently, the process produces syngas with a reducing potential of 10, surpassing the required threshold. The experimental results are coherent with the process simulations.

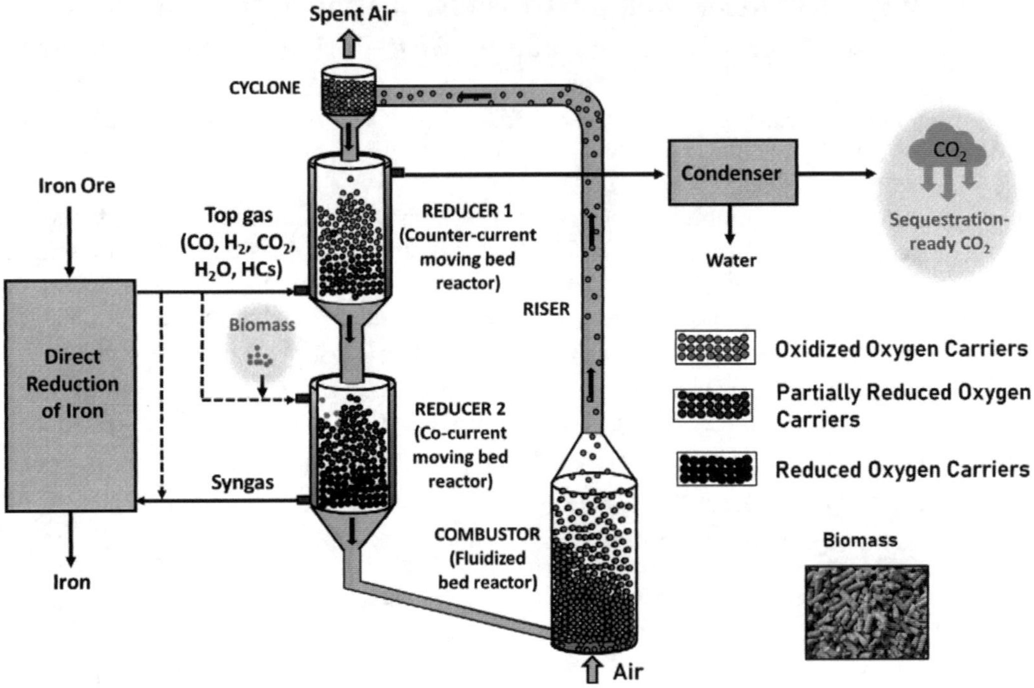
Spent Air
CYCLONE
Iron Ore
Top gas
(CO, H2, CO2,
H2O, HCs)
REDUCER 1
(Counter-current
moving bed
reactor)
Condenser
Water
CO2
Sequestration-
ready CO2
Direct
Reduction
of Iron
Biomass
RISER
Oxidized Oxygen Carriers
Partially Reduced Oxygen
Carriers
Reduced Oxygen Carriers
REDUCER 2
(Co-current
moving bed
reactor)
Syngas
COMBUSTOR
(Fluidized
bed reactor)
Biomass
Iron
Air

729b Advancing Low-Carbon Hydrogen Production: Innovations and Challenges

Teerawit Prasomsri, Junchen Liu, Ashten Molley, Brooke Ames

Climate change is urging us to cut down on CO_2 emissions, particularly from the industrial sector, which contributes 30% globally [1]. Hydrogen emerges as a transformative solution in the energy transition, offering a clean and reliable energy source as it does not emit CO_2 when burned. However, the process currently used to produce the bulk of the hydrogen, called steam methane reforming (SMR), does produce CO_2 and consumes a lot of energy.

Decarbonizing hydrogen production can be achieved through reconfigured reformer designs and improved catalysts, positioning SMR as a scalable and industry-ready source of low-carbon hydrogen. Collaborative efforts between Clariant and Technip Energies have introduced the Enhanced Annular Reforming Tube for Hydrogen (EARTH®), an innovative heat-recuperative reformer concept aimed at enhancing SMR efficiency, capacity, and sustainability. Commercial implementation of EARTH has yielded significant benefits, including in one case a nearly 40% reduction in fuel consumption, improved radiant efficiency, and a 20% decrease in CO_2 emissions while in another case boosting capacity by up to 20% compared to conventional SMR [2].

To further reduce the carbon footprint, minimizing steam consumption is essential. However, this may increase coke formation on catalysts due to a reduced steam-to-carbon ratio (S/C). Therefore, understanding the relationship between S/C and coke formation under industry-relevant conditions is critical for catalyst development. Recent advancements in research have enabled the study of coke formation under high-temperature and high-pressure conditions, mirroring realistic operating conditions. This study investigates the interplay between coke formation, S/C ratio, temperature, and pressure using Ni-Al-O containing catalysts and explores the impact of dopants on coke formation dynamics. The insights will allow the advancement of low-carbon hydrogen production processes.

References:

1. DOE, Industrial Decarbonization Roadmap, 2022, https://www.energy.gov/eere/doe-industrial-decarbonization-roadmap
2. Clariant's media release: https://www.pressreleasefinder.com/prdocs/2021/Clariant_Media_Release_EARTH_Tech_Launch_20211111_EN.pdf

729c Life Cycle Analysis of NO_x Emissions from Combustion of Hydrogen Blends in Natural Gas

Xiaoyue Pi, Vincenzo Cappello, Pradeep Vyawahare, Pingping Sun, Amgad Elgowainy

Hydrogen (H_2) has no tailpipe emissions (end-use emissions) and is considered a clean energy carrier when produced with renewable or low-carbon energy sources. It also has the potential to reduce carbon emissions by replacing fossil fuels in various end-use applications. For example, H_2 can be combusted to energize commercial, industrial, and utility applications (with appropriate equipment modification) where natural gas (NG) dominates today. Due to the absence of widespread H_2 infrastructure, enabling clean H_2 production at scale requires a corresponding market demand. Blending clean H_2 with NG in existing NG pipelines can provide a market liftoff for clean H_2 production while decarbonizing NG supply to various end-use applications. However, this approach comes with challenges, one of which is the potential increase in nitrogen oxides (NO_x) generated from the combustion of H_2 in the blend with NG. Therefore, careful evaluation of NO_x emissions from the combustion of H_2-NG gas blends in various end-use applications is needed before practical implementation.

This study focuses on analyzing NO_x emissions using available data sources and covering all stages of the gas life cycle, from fuel feedstock recovery to end-use combustion. Our life cycle analysis includes different technologies for clean H_2 productions (e.g., SMR-CCS, solar/wind power, etc.), two blended gas delivery scenarios (constant volumetric flow rate and constant energy throughput), and various end-use applications (e.g., boiler, engine, etc.). For the considered cases, the life cycle NO_x emissions at different blending ratios of H_2 with NG are reported and discussed with respect to the feasibility of using gas blends in various end-use applications. Results show that NO_x emissions vary at different H_2 blending ratios, and they may not be significantly increased by the H_2 addition in most cases. In addition, emissions control technologies may be employed to mitigate any increase in NO_x emissions.

729d Real-Time Emission-Driven Modeling and Optimization Framework Under Uncertainty for Sustainable Hydrogen Production

Marco De Sousa, Saatvi Suresh, Betsie Sara Monserrat Montaño Flores, Rahul Kakodkar, Harsh B. Shah, Dustin Kenefake, Iosif Pappas, Xiao Fu, C. Doga Demirhan, Mete Mutlu, Shayan Niknezhad, Efstratios Pistikopoulos

The U.S. hydrogen economy produces 10 million metric tons of hydrogen annually, which emits 41 million metric tons of carbon dioxide equivalent emissions [1]. As the demand for hydrogen increases and new emission regulations take effect, there is an emerging need to investigate the integration and evolution of both conventional and innovative hydrogen production systems [2]. To facilitate a smooth transition during the expansion of the hydrogen production infrastructure, these hybrid systems must be meticulously designed and operated with due consideration for future uncertainties. To this end, we present a real-time emission-driven optimization framework, which is implemented through a mixed-integer linear programming (MILP) formulation to determine optimal design configurations and operation schedules while simultaneously mitigating emissions by utilizing electricity price forecasts, sporadic weather data, and supply and demand variability [3]. Life cycle assessment (LCA) criteria are directly incorporated in the optimization model to evaluate different environmental concerns that are segmented into on-site emissions (Scope 1), indirect emissions (Scope 2), and upstream and downstream emissions (Scope 3) [4]. The methodology avoids assuming perfect information by implementing a rolling-horizon optimization approach that allows decision-makers to investigate the sensitivity of capacity expansion plans concerning different look-ahead periods. The framework, illustrated by a detailed hydrogen production case study, results in Pareto optimal solutions that capture the trade-offs between environmentally friendly and economically competitive designs and operational profiles.

References

[1] Z. Guiyan, G. Edward J., M. Darik, H2 production through natural gas reforming and carbon capture: A techno- economic and life cycle analysis comparison, International journal of hydrogen energy (2023) 1288–1303.
[2] M. Wappler, D. Unguder, X. Lu, H. Ohlmeyer, H. Teschke, W. Lueke, Building the green hydrogen market - current state and outlook on green hydrogen demand and electrolyzer manufacturing, International Journal of Hydrogen Energy 47 (79) (2022) 33551–33570.
[3] R. Kakodkar, G. He, C. Demirhan, M. Arbabzadeh, S. Baratsas, S. Avraamidou, D. Mallapragada, I. Miller, R. Allen, E. Gençer, E. Pistikopoulos, A review of analytical and optimization methodologies for transitions in multi-scale energy systems, Renewable and Sustainable Energy Reviews 160 (2022) 112277.
[4] A. Hugo, E. Pistikopoulos, Environmentally conscious long-range planning and design of supply chain networks, Journal of Cleaner Production 13 (15) (2005) 1471–1491.

729e Screening Cobalt-Based Single Atom Alloy Catalysts for Low-Temperature Fischer-Tropsch Synthesis

John El Berch, Rashad Ahmadov, Marc Porosoff, Giannis Mpourmpakis

To reach net-zero CO_2 emissions by 2050, the installed carbon capture, utilization, and sequestration capacity needs to increase ~15 times by 2030[1]. Among the CO_2 utilization applications, the production of synthetic fuel (also known as synfuel) has the highest CO_2 use capacity due to the vast worldwide energy demands[2]. Moreover, synfuels can supply low-carbon alternatives to sectors not yet ready for electrification, such as aviation and maritime[3]. One of the main CO_2-synfuel production routes consists of sequentially pairing the Reverse Water-Gas Shift (RWGS, $CO_2 + H_2 \leftrightarrow CO + H_2O$) with Fischer-Tropsch Synthesis (FTS, $CO + 2H_2 \rightarrow$ -(CH_2)- + H_2O)[4]. One drawback of this pathway is the separation of the CO_2 effluent from the RWGS step, as typical FTS catalysts are also active for the Sabatier reaction ($CO_2 + 4H_2 \rightarrow CH_4 + 2H_2O$)[5]. Considering CO_2 activation requires high temperatures, usually above 300 °C [6], low-temperature (LT) FTS (~200 °C) can reactively separate the CO/CO_2 inlet.

Cobalt is the most popular LT-FTS catalyst, often enhanced by doping (or promoting) with precious metals[7]. In our previous work, Ru-doped Co Single Atom Alloys (SAAs), in which isolated Ru atoms are dispersed on the Co surface, were investigated for LT-FTS. The Ru-Co SAA catalyst showed enhanced reducibility and improved efficiency in the use of Ru (Co:Ru atomic ratio decreased ~20 fold compared to co-impregnated catalysts)[8]. In this work, we search for cost-effective alternatives for Ru that result in active LT-FTS catalysts by using density functional theory to screen various Co-based SAAs. The overall catalytic performance is evaluated by comparing the energetics of three FTS elementary steps: 1) CH_2 coupling ($2CH_2 \rightarrow C_2H_4$) to probe the formation of long-chain hydrocarbons; 2) CH_3 hydrogenation ($CH_3 + H \rightarrow CH_4$) to evaluate the formation of undesired methane; and 3) CH_3 dissociation ($CH_3 \rightarrow CH_2 + H$) to gauge the likelihood of finding low-saturated carbon chains. Among the tested dopants, Re is identified as a cost-efficient alternative to Ru, with similar reaction energetics, whereas, V is found to have the most favorable energetics for long-chain hydrocarbons (unfavorable CH_4 formation and low CH_2 coupling barriers). The selectivity predictions on these two SAAs are experimentally validated by temperature-programmed reactor studies. Finally, fitted linear scaling relationships using common (de)hydrogenation descriptors show deviations with varying adsorption configurations, suggesting the importance of site-sensitive descriptors. Overall, this work introduces a framework for the effective screening of SAA candidates in complex multi-step reactions.

References

1. International Energy Agency, Capacity of current and planned large-scale CO2 capture projects vs. the Net Zero Scenario, 2020-2030. Paris https://www.iea.org/data-and-statistics/charts/capacity-of-current-and-planned-large-scale-co2-capture-projects-vs-the-net-zero-scenario-2020-2030, 2023.

2. International Energy Agency *Putting CO2 to Use*; 2019.

3. Anika, O. C.; Nnabuife, S. G.; Bello, A.; Okoroafor, E. R.; Kuang, B.; Villa, R., Prospects of low and zero-carbon renewable fuels in 1.5-degree net zero emission actualisation by 2050: A critical review. *Carbon Capture Science & Technology* **2022,** *5*, 100072.

4. Porosoff, M. D.; Yan, B.; Chen, J. G., Catalytic reduction of CO2 by H2 for synthesis of CO, methanol and hydrocarbons: challenges and opportunities. *Energy & Environmental Science* **2016,** *9* (1), 62-73.

5. Visconti, C. G.; Martinelli, M.; Falbo, L.; Fratalocchi, L.; Lietti, L., CO2 hydrogenation to hydrocarbons over Co and Fe-based Fischer-Tropsch catalysts. *Catalysis Today* **2016,** *277*, 161-170.

6. Spennati, E.; Garbarino, G.; Savio, L.; Vattuone, L.; Riani, P.; Busca, G., CO2 methanation vs reverse WGS activity on Co/γ-Al2O3 catalysts at atmospheric pressure: effect of cobalt loading and silica addition on selectivity and stability. *Catalysis Today* **2023,** *420*, 114164.

7. Khodakov, A. Y.; Chu, W.; Fongarland, P., Advances in the Development of Novel Cobalt Fischer–Tropsch Catalysts for Synthesis of Long-Chain Hydrocarbons and Clean Fuels. *Chemical Reviews* **2007,** *107* (5), 1692-1744.

8. Liu, R.; El Berch, J. N.; House, S.; Meil, S. W.; Mpourmpakis, G.; Porosoff, M. D., Reactive Separations of CO/CO2 mixtures over Ru–Co Single Atom Alloys. *ACS Catalysis* **2023,** *13* (4), 2449-2461.

433a Life Cycle Analysis and Techno-Economic Analysis Tools to Evaluate Carbon Conversion Technologies

Sheikh Moni, Michelle Krynock, Samuel Henry, Gregory Hackett

Conversion of captured carbon dioxide (CO_2) emitted from power and industrial sources into environmentally responsible and economically valuable products can reduce greenhouse gas (GHG) emissions and enable the achievement of a decarbonized economy by 2050. Rigorous and structured technology analysis, such as life cycle- and technoeconomic analysis, are crucial to determining these technologies' potential role in decarbonization. Life cycle assessment/analysis (LCA) as a tool and framework can help to determine the potential environmental impacts and carbon footprint of conversion technologies and evaluate policy options from a consistent and unbiased viewpoint. The U.S. Department of Energy (DOE) and the federal government are increasingly requiring LCA as a condition for funding primary research, grant programs such as Utilization Procurement Grants (UPGrants) for eligible entities to procure and use carbon conversion products, and tax incentives such as "45Q" for carbon oxide utilization projects (Treasury Regulations Section 1.45Q-4 (26 C.F.R. § 1.45Q-4), 2021).

The DOE National Energy Technology Laboratory (NETL) has developed a CO2U LCA Toolkit to support the creation of useful and consistent LCAs of carbon conversion projects funded by the DOE Carbon Dioxide Conversion Program and specific addenda to support the development of LCAs for UPGrants and 45Q. The toolkit has also been used to guide a wider audience outside the Carbon Conversion Program in conducting LCAs on carbon conversion systems and advance the global discussion on carbon conversion LCA. The Toolkit includes LCA guidance, life cycle inventory data, and other relevant tools to perform LCAs of both emerging and existing carbon conversion technologies and enable users to compare different products/pathways in terms of environmental performance. The guidance document should be considered a companion to the requirements established in ISO 14040/14044, providing specific guidance on methodological issues for scoping and completing the LCA. The guidance, data, and tools are publicly available and free.

Techno-economic analysis (TEA) can supplement LCA by assessing the technical and economic performance of conversion technologies as compared to established pathways. Recently, NETL has developed a CO2U TEA Toolkit to enable users to create useful and consistent TEAs of carbon conversion technologies. The toolkit includes TEA guidance for carbon conversion studies, example TEAs of established carbon conversion pathways, training videos to support TEA development, a spreadsheet tool to compare results from completed TEAs, a spreadsheet tool to support capital cost scaling, and other ancillary tools and templates. The guidance, data, and tools will be publicly available and free.

This presentation will provide an overview of both the LCA and TEA Toolkit, the utility of these tools to the LCA and TEA community, and the value of the toolkits in assessing emerging and existing carbon conversion technologies.

References:

Treasury Regulations Section 1.45Q-4 (26 C.F.R. § 1.45Q-4) (2021). Retrieved from https://www.irs.gov/pub/irs-drop/td-9944.pdf on November 18, 2021

Disclaimer: This project was funded by the Department of Energy, National Energy Technology

433b Analysis of a Novel Porous Bi-Functional Polymer Catalyst for CO_2 Conversion to Formic Acid: Techno-Economic Analysis and Life Cycle Assessment

Ikenna Okeke, Canan Karakaya, Michelle Kidder, Aye Meyer

To decarbonize the U.S. economy, technological innovations are required in the manufacturing sector to reduce the CO_2 emissions associated with the current practices. To this purpose, a novel polymer-based ruthenium catalyst is developed for the capture and conversion of CO_2 in industrial flue gas. This work presents the design and systems-level performance analysis for the thermochemical conversion of the captured CO_2 to formic acid. Based on the experiment, a power law kinetic data was developed which was employed in a flow reactor in computing CO_2 conversion and product distribution. Two scenarios that consider different sources of hydrogen for formic acid production were evaluated: conventional steam methane reformer and renewable-powered electrolysis hydrogen. The designs were modeled in Aspen Plus v14 for the estimation of materials and energy flows. Utilizing the converged flowsheet data, a techno-economic analysis and life cycle assessment were conducted to ascertain the financial feasibility and environmental impact of the proposed CO_2-based formic acid. Performance metrics such as energy and CO_2 conversion efficiencies, minimum formic acid selling price, and cost of CO_2 avoided were computed which were compared to the conventional and electrochemical routes of formic acid production. Furthermore, a sensitivity analysis was conducted to assess the impacts of a set of design and economic parameters used in the study.

433d High-Temperature CO_2 Electroreduction into Carbon Nanostructures in Molten Salt Electrolyte for Carbon-Negative MSCO2RR

Andrew Wong

The electrochemical CO_2 reduction reaction (CO_2RR) represents a critical approach toward the carbon-neutral or carbon-negative production of value-added products. In this presentation, we will present design rules to control the electrochemical parameters and microenvironment towards CO_2 conversion to carbon nanotubes and other nanostructured carbons with high Faradaic efficiency (>80%) in molten salt electrolyte. These carbon nanostructures are particularly attractive to produce fixed forms of carbon for carbon-negative processes that do not lead to the re-emission of CO_2.

To this end, we will describe our efforts in molten salt CO_2 reduction reaction ($MSCO_2RR$) to valuable carbon nanostructures. Overall, carbon nanostructures also represent useful value-added products with market value for CNTs on the order of $100 USD/kg. In contrast to aqueous CO_2 electroreduction, molten salt electrochemistry offers a microenvironment that can convert CO2 to cathodic solid carbon via an electrochemical CO_2 reduction reaction. Through systematic experiments, we elucidate the effects of changing electrolyte, catalyst composition, and temperature, amongst other factors, to control carbon nanostructure growth, as will be described in this presentation. Finally, informed by the design principles that we have discovered, we will share our findings on the technoeconomic analysis of $MSCO_2RR$, which suggest that $MSCO_2RR$ to CNTs can be a profitable CO_2-negative technology at a commercial scale, suggesting a viable route to commercialization of this CO_2-negative technology.

433e Two-Dimensional Steady State Numerical Simulation of Electrochemical CO_2 Conversion Using a Zero-Gap Electrolyzer

Ahmad Ijaz, SeyedSepehr Mostafayi, Mohammad Asadi, Javad Abbasian, Hamid Arastoopour

Zero-gap membrane electrode assembly (MEA) is emerging as a potential electrochemical reactor for the CO_2 conversion into value added products. MEA requires optimization of parameters such as local CO_2 concentration, local pH in the catalyst zone, and local water flow to the catalyst area. To obtain a better understanding of the local phenomena, a 2-D multiphysics numerical model is developed for MEA. Ideally wetted catalyst (assuming 64% saturation), and partially wetted catalyst using Van-Genuchten equation for capillary pressure have been modeled to mimic potential flooding scenarios in MEA. Kinetic parameters used in the MEA model are obtained from the data in the literature. The MEA model is validated with experimental data by comparing products current density and selectivity. A parametric study is conducted to investigate design and operating parameters of MEA including gas-diffusion layer thickness, catalyst layer thickness, catalyst layer porosity, membrane thickness, use of different electrolyte, and strength of an electrolyte.

433f Effect of Sorbent Incorporation to a $ZnZrO_2$ Catalyst in Its CO_2 Capture Capacity and Catalytic Activity

Laura Proano, Katlo Galefete, Guanhe Rim, Christopher Jones

Recently process intensification through reactive carbon capture (RCC) using catalytic sorbents (CS) or dual function materials (DFM) has been explored as a potential process in the transition to a CO_2 circular economy.[1] So far, most of the CS reported in literature for CH_4, syngas and MeOH synthesis consist of metal-based supported catalysts that are impregnated with an alkali/alkaline salt or physical contact (mixing or stacking) of the catalyst with an alkali-promoted MgO-based sorbent. Wirner et al. studied RCC to MeOH using stacked beds of NaO/Al_2O_3 adsorbent and $Cu/ZnO/Al_2O_3$ catalyst and reported a MeOH productivity of 12 μmol/g.[2] On the contrary, Jeong-Potter et al. reported higher productivity (59 μmol/g) when using a 5 wt% K_2O impregnated $Cu/Zn/Al_2O_3$ CS.[1] However, it is not clear how the different configurations impact the two processes occurring during RCC: CO_2 adsorption and catalytic hydrogenation of CO_2. Thus, in this work we evaluated the effect of sorbent incorporation into a $ZnZrO_2$ catalyst on the CO_2 uptake and CO_2 hydrogenation catalytic activity of the material. Two different configurations were evaluated, (i) a physical mixture with a Mg_3Al hydrotalcite derived oxide (Mg_3AlO) with a weight ratio of (6:4) and (ii) sorbent and catalyst impregnation with 10 wt.% of $NaNO_3$.

Figure 1a presents TGA uptake curves for the different materials. It can be observed that $ZnZrO_2$ without sorbent modification presents the lowest CO_2 uptake (0.05 mmol/g). The physical mixture of $ZnZrO_2$ with Mg_3AlO sorbent in a ratio of 6:4 results in an increased CO_2 uptake of 0.15 mmol/g. The highest CO_2 uptake capacity was obtained for the 10% $NaNO_3$ impregnated materials. Figure 1b presents CO_2 conversion vs. MeOH selectivity at 150 psi and 300 °C for all the materials. The physical mixture of $ZnZrO_2$ with Mg_3AlO did not significantly affect the catalytic activity. On the contrary, the impregnation with $NaNO_3$ decreased the CO_2 conversion and MeOH selectivity, especially when done directly on the catalyst domain. This could be related to the titration of the O defects in ZrO_2, which have been reported to be the active sites for CO_2 activation.[3] This negative impact was reduced when the impregnation was done on the Mg_3AlO domain. CO_2-TPD experiments revealed that $ZnZrO_2$ basic sites are mostly weak and moderate with desorption peaks between 100-200 °C, which are associated to the catalyst high activity. On the other hand, Mg_3AlO has mostly strong CO_2 adsorption (desorption at 500 °C). Impregnation with $NaNO_3$ resulted in an increase in all basic sites. Thus, while the physical mixture of $ZnZrO_2$ with Mg_3AlO maintains the high selectivity of the catalyst with a moderate increase in total CO_2 uptake, it is possible that strongly bound CO_2 species are not reactive during RCC. On the other hand, $NaNO_3$ impregnation enhanced the total CO_2 uptake, along with weak and moderate basic site, and the poisoning effect on catalytic activity can be moderated by depositing $NaNO_3$ on the Mg_3Al sorbent. This understanding of CS configuration in the separated process provides insights for future rational design of CS that require high CO_2 uptakes and catalytic activity.

References

1 Jeong-Potter, C. *et al. EES Catalysis* **2** (2024)

2 Wirner, L. C. *et al. Chemical Engineering Journal* **470** (2023).

3 Araújo, T. P. *et al. Advanced Energy Materials* **13** (2023).

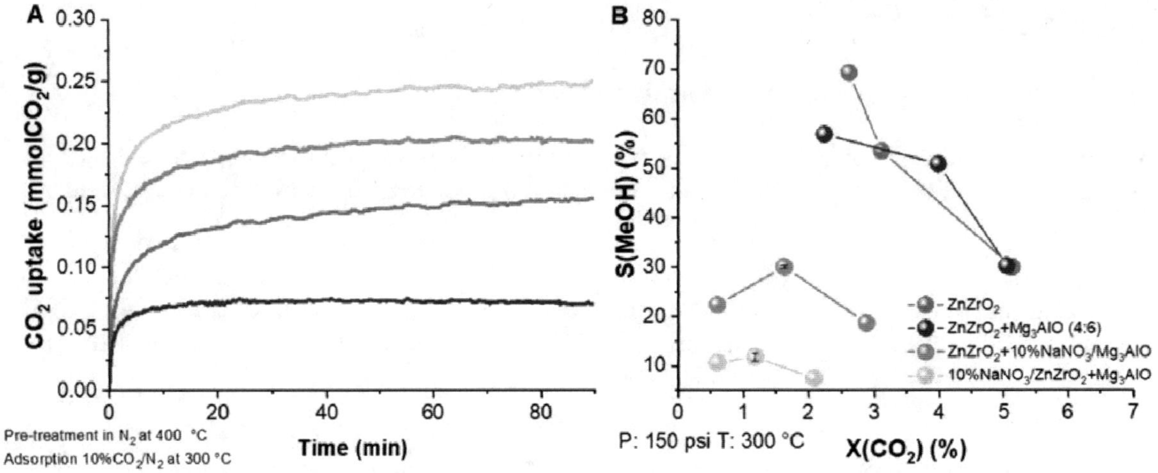

A
B
CO2 uptake (mmolCO2/g)
Time (min)
Pre-treatment in N2 at 400 °C
Adsorption 10%CO2/N2 at 300 °C
S(MeOH) (%)
X(CO2) (%)
P: 150 psi T: 300 °C
ZnZrO2
ZnZrO2+Mg3AlO (4:6)
ZnZrO2+10%NaNO3/Mg3AlO
10%NaNO3/ZnZrO2+Mg3AlO

433g Oxide-Encapsulated Electrocataysts for Selective and Stable CO_2 Upgrading from Reactive Carbon Capture Solutions

Zhexi Lin, Nathaniel Blake, Zhirui He, Xueqi Pang, Reza Mirshekari, Oyinkansola Romiluyi, Yoon Jun Son, Suryansh Kabra, Daniel Esposito

The direct upgrading of CO_2 from bicarbonate-laden reactive carbon capture solutions is a promising approach for lowering energy consumption and enhancing CO_2 utilization of the process.[1] However, the high concentration of bicarbonate ions in reactive carbon capture solutions promote the bicarbonate-mediated hydrogen evolution (HER) reaction, thus lowering the selectivity towards CO.[2] Additionally, the trace amounts of metal impurities present in the electrolyte can deposit on the cathode and further shift the selectivity towards HER.[3] In this work, we demonstrate a simple yet effective strategy to overcome these challenges by encapsulating Ag nanoparticle electrocatalysts with a semi-permeable TiO_2 nano-membrane to block bicarbonate ions from accessing the buried Ag active sites and therefore mitigate the bicarbonate-mediated HER. The TiO_2-encapsulted Ag (Ag@TiO_2) electrocatalyst showed a 50% lower HER partial current density at potentials more negative than -0.7 V vs. RHE, resulting in a 15% increase in CO faradaic efficiency. In situ surface-enhanced Raman spectroscopy (SERS) further revealed the absence of bicarbonate ions and the presence of CO_2 reduction intermediates near the buried interface, confirming the bicarbonate-blocking ability of the TiO_2 overlayer. During electrode stability evaluation conducted in the presence of 5 ppm Fe and 0.02 M ethylene diamine tetra-acetic acid (EDTA), Ag@TiO_2 demonstrated 15% higher steady-state faradaic efficiency than bare Ag. X-ray fluorescence (XRF) characterization results on spent electrodes indicated that the Fe deposition rate on Ag@TiO_2 was significantly lower than that on bare Ag, showing the positive effect of the TiO_2 overlayer in blocking Fe impurities. This work demonstrates the combination of oxide encapsulation and impurity scavenging with EDTA as an effective strategy for selective and stable syngas production in reactive carbon capture solutions.

References

1. Lee, G.; Rasouli, A. S.; Lee, B.-H.; Zhang, J.; Won, D. H.; Xiao, Y. C.; Edwards, J. P.; Lee, M. G.; Jung, E. D.; Arabyarmohammadi, F.; Liu, H.; Grigioni, I.; Abed, J.; Alkayyali, T.; Liu, S.; Xie, K.; Miao, R. K.; Park, S.; Dorakhan, R.; Zhao, Y.; O'Brien, C. P.; Chen, Z.; Sinton, D.; Sargent, E., CO_2 electroreduction to multicarbon products from carbonate capture liquid. *Joule* **2023,** *7* (6), 1277-1288.

2. Marcandalli, G.; Goyal, A.; Koper, M. T. M., Electrolyte Effects on the Faradaic Efficiency of CO_2 Reduction to CO on a Gold Electrode. *ACS Catalysis* **2021,** *11* (9), 4936-4945.

3. Hori, Y.; Konishi, H.; Futamura, T.; Murata, A.; Koga, O.; Sakurai, H.; Oguma, K., "Deactivation of copper electrode" in electrochemical reduction of CO_2. *Electrochimica Acta* **2005,** *50* (27), 5354-5369.

500a Synergistic Effect between Plasma and Catalyst on Dry Reforming of Methane

Dong-Wook Kim, Jinwoo Kim, Jaehee Chang, Jun Kyu Kim, Jong-Min Lee, Hyuk Jae Kwon

Dry reforming of methane (DRM) was performed by plasma-catalyst hybrid method. Plasma provides plentiful active species such as electrons, ions, radicals, excited molecules. While plasma can induce dry reforming reaction ($CO2+CH4 \rightarrow 2CO+2H2$) by itself, the combination of plasma and catalyst has been known to enhance the reaction. The active species generated by plasma interact with the catalyst, so that the reaction is promoted by the synergistic effect. The purpose of this study is to investigate the synergistic effect between plasma and catalyst on dry reforming of methane. In this study, two types of plasmas; a dielectric barrier discharge (DBD) and a rotating arc plasma (RAP), respectively, were combined with the catalyst reactor to confirm which type of plasma shows higher synergistic effect with the catalyst. DBD provides relatively low gas temperature (~120°C) and high electron temperature (~10eV), whereas RAP provides relatively high gas temperature (~1,500°C) and low electron temperature (~3eV). The different type of hybrid reactor was introduced according to the type of plasma. For DBD, the catalyst bed was located inside the discharge region, so that the plasma is generated on the surface of catalyst. On the other, for RAP, the catalyst bed was located at the afterglow region to avoid thermal damage by high gas temperature of RAP. The experiments were performed using each hybrid reactor with/without the catalyst to investigate the synergistic effect. All experiments were performed under atmospheric pressure without any external heat source. The total conversion and the energy efficiency were evaluated at each experiment. A total conversion is an average of conversion of CO2 and CH4 considering mole fraction. An energy efficiency is a ratio of the required energy for 100% conversion to an enthalpy of dry reforming reaction. In the case of the DBD hybrid reactor, the total conversion and the energy efficiency increased from 4.3% and 11.2% to 36.4% and 35.2% by combining with the catalyst. In the case of the RAP hybrid reactor, the total conversion and the energy efficiency increased from 62.1% and 76% to 86.8% to 99.6% by applying the catalyst. The results shows that the DBD could create the higher synergistic effect, but the absolute values of the total conversion and the energy efficiency are lower in comparison to the RAP. Such a high synergy is considered to be the result of active interaction between the reactive species and the catalyst inside the discharge region. However, it seems that the low gas temperature of DBD didn't allow to obtain the higher conversion by overcoming thermodynamic limit because dry reforming reaction is largely endothermic. Meanwhile, the use of RAP led to the higher conversion and the higher energy efficiency regardless of the presence or absence of the catalyst, although the increasing ratio of the conversion and the energy efficiency was not high as that of the DBD. Herein, the effect of the various experimental parameters including the types of catalysts, the ratio of CO2/CH4 and the flow rate of diluting gas on the synergy will be also discussed.

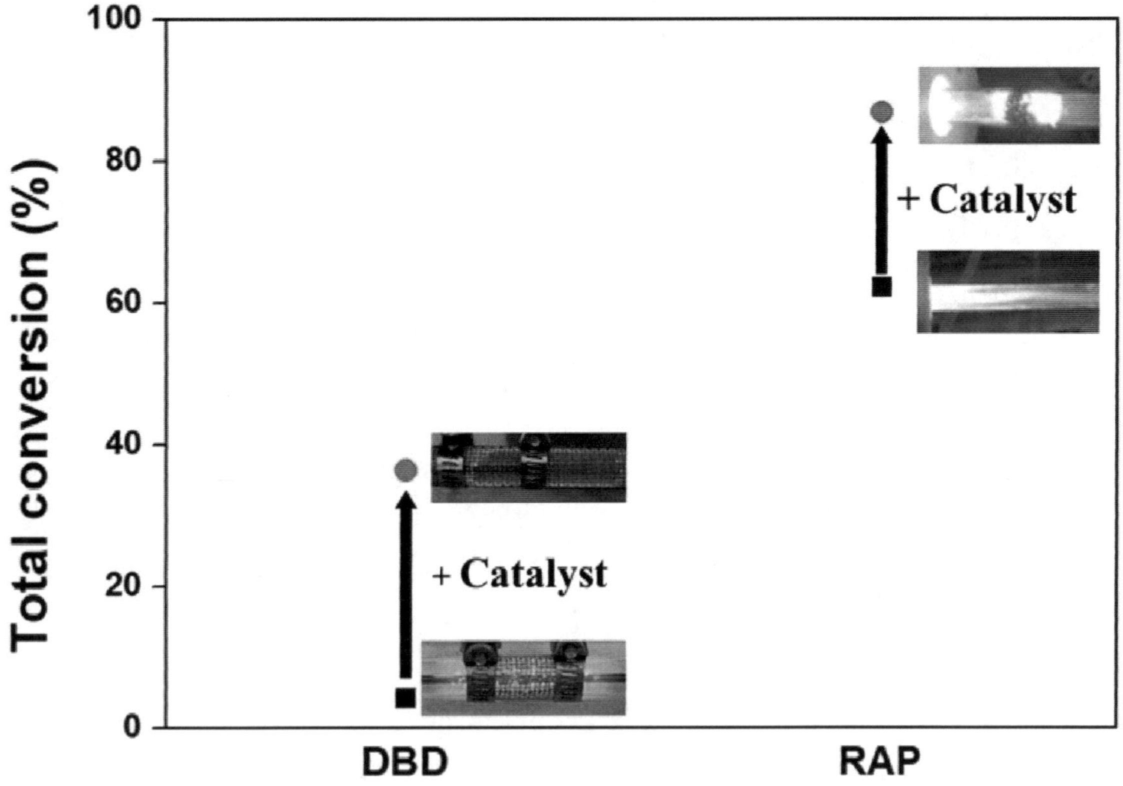

Total conversion (%)
100
80
60
40
20
0
+ Catalyst
+ Catalyst
DBD
RAP

500b Autothermal Sabatier Reactor for Biogas Conversion to Renewable Natural Gas: Design, Experimental Proof-of-Concept, and Parametric Study

Yichen Zhuang, David Simakov

PROBLEM STATEMENT

Commercially available biogas upgrade technologies utilize adsorption, absorption, cryogenic, and membrane separation to upgrade biogas to high quality biomethane based on physical or chemical separation of CO_2 followed by removal of additional impurities including water vapor, H_2S, and siloxanes. Depending on the composition of biogas, significant portion of N_2 and O_2 can be removed via pressure swing adsorption (PSA) on molecular sieves or membrane separation. Despite the decent performance, high operation costs and large capital investment associated with the abovementioned technologies limit the economic viability of the biogas utilization on an industrial scale. New upgrade technologies such as biological methods are still under development. One of the biggest challenges of biological methanation is that it's hard for water and gas phase to have sufficient contact due to low solubility of hydrogen in water.

An alternative way is direct methanation of biogas, which involves thermocatalytic conversion of CO_2. Producing synthetic CH_4 via the thermocatalytic conversion of CO_2 involves Sabatier reaction, which is accompanied by reverse water gas shift and CO methanation. When using renewable (or low-carbon footprint) electricity to generate H_2 via water electrolysis, the resulted synthetic CH_4 can be considered as renewable natural gas (RNG). Thermocatalytic conversion of CO_2 offers the benefits of fast reaction rates and high conversion efficiencies allowing for compact, high throughput operation. Also, direct CO_2 methanation via the Sabatier reaction can eliminate the huge capital and operating cost of CO_2 separation.

SUGGESTED SOLUTION

The suggested process of biogas direct upgrading is conceptually shown in the attached figure. H_2 required for CO_2 methanation can be obtained from water electrolysis powered by renewable or (low carbon footprint) surplus electricity, since the electrolysis consumes most of the electricity among all processes. A number of different gas cleanings solutions has been suggested and tested in recent years, such as adsorption on active carbon and other solid sorbents under ambient temperature to remove impurities such as H_2S, VOCs and siloxanes. After the cleaning step, pretreated biogas is fed to the Sabatier reactor, e.g., packed bed. A pipeline grade RNG is obtained from the reactor outlet after water condensation and further product upgrading (removal of N_2 and unreacted CO_2 and H_2), if required. Electrolysis, as well as all cleaning, condensation and upgrading stages are commercially available and the economic viability of the concept presented was recently assessed in our group, leading to positive conclusions. However, reactor design and catalysis still present significant technological challenges, mainly due to the significant heat of reaction from the reactor, which leads to sintering of catalyst as well as conversion limitation due to the thermodynamic equilibrium.

METHODOLOGY

In the study reported herein, a simple tube-and-shell design was adopted in order to investigate the possibility of the autothermal operation of a Sabatier reactor, i.e., standalone operation without any feed preheating or reactor heating. The reactor was made from stainless steel tubing using standard (Swagelok) connectors and a commercial Ni/Al_2O_3 catalyst. Nitrogen was used for cooling. Three prototypes were investigated with different sizes and configurations. It was demonstrated that, after the initial feed preheating for ignition, the reactor can be continuously operated as a standalone autothermal unit, achieving 91% CO_2 conversion and 100% selectivity to CH_4 generation using a synthetic biogas feed.

MAIN RESULTS

An air-cooled, autothermal Sabatier reactor was experimentally investigated using simulated biogas composition, analyzing the effects of feed temperature, space velocity, pressure, and cooling rate on the reactor performance in terms of CO_2 conversion and selectivity to CH_4 generation. The stainless steel, tube-and-shell reactor comprised a commercial Ni/Al_2O_3 catalyst in the shell compartment, with the inner tube being used for active cooling using compressed air. The performance of the commercial Ni/Al_2O_3 catalyst was first assessed with real raw biogas. Air-balanced synthetic biogas was found to be a good model system for reactor testing as a substitute for treated raw biogas (after H_2S and VOC removal). All reactor prototypes were demonstrated to be able to operate autothermally using the air-balanced synthetic biogas as a feed. The final prototype achieved 91% CO_2 conversion, while maintaining complete selectivity to CH_4 for over 100 h at 3,100 L/(kg h) with stable reactor temperatures.

INNOVATION & SIGNIFICANCE

Our findings provide important general insights into operation of lab-scale Sabatier reactors. Due to the simplicity of assembly, such laboratory reactors are very useful for investigating effects of various operating parameters on reactor performance. Our study also provides a novel perspective on direct biogas upgrade, an approach that can potentially reduce RNG generation costs due to the elimination of CO_2 separation, promoting biogas utilization. The experimental data collected in this study can be also used to calibrate a reactor model, which can be used for further scale-up (this work is underway and will be reported elsewhere).

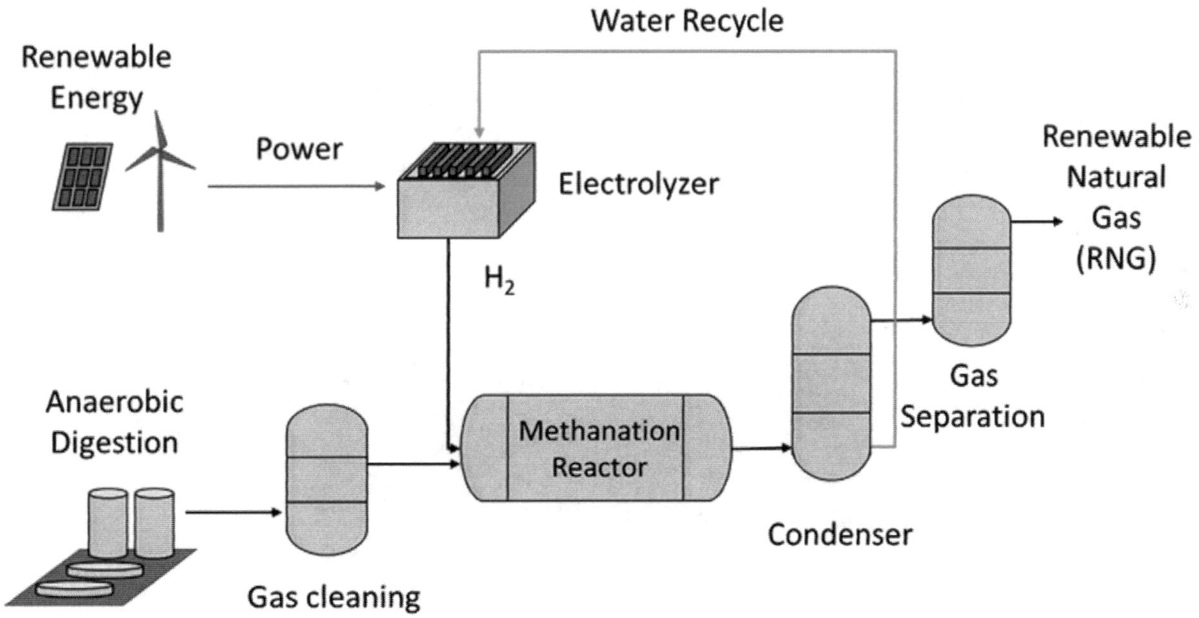

500c Towards Self-Regenerative Sorption-Enhanced Catalysts for Integrated CO_2 Capture and Dry Reforming of Methane

Kandis Leslie Gilliard-Abdulaziz, Seongbin Jo

Improving the thermal stability of sorption-enhanced catalysts (CO_2 sorbent + embedded heterogeneous catalyst nanoparticles)[1] is one of the most difficult and crucial parameters to control for integrated carbon capture and conversion processes. For example, sintering of the catalyst nanoparticle and CO_2 sorbent is associated with a loss of surface area and lower CO_2 capture capacity and activity. At industrial processes that operate at high temperatures, such as dry reforming of methane, the degree of deactivation of the sorbent and catalyst is particularly amplified.

This talk will introduce the use of self-regenerative Ni-doped $CaTiO_3$/CaO nanocomposites as multifunctional materials (MFMs) for the integrated CO_2 capture and subsequent dry reforming of methane (ICCDRM), where Ni-doped $CaTiO_3$ is a catalytically active material, and CaO is a sorbent for CO_2 capture. The incorporation of Ni into $CaTiO_3$ and in-situ exsolution of Ni from $CaNi_xTi_{1-x}O_3$ double perovskite were investigated. In-situ exsolved Ni nanoparticles, which interact strongly with the host $CaTiO_3$ perovskite, were evenly distributed throughout $CaTiO_3$ under H_2 or CH_4 (reductive conditions) – ***Fig 1a***. The exsolved Ni nanoparticles were re-dispersed back into the bulk of $CaTiO_3$ under CO_2, indicating a self-regenerative process. Ni-doped $CaTiO_3$/CaO MFMs showed relatively stable CO_2 capture capacity and syngas productivity during 30 cycles of ICCDRM – ***Fig 1b***. Strong interaction between Ni and $CaTiO_3$ and self-regeneration prevented Ni nanoparticles from sintering and decreased the deactivation of Ni due to both sintering and coke accumulation. Lastly, we summarize our findings to discuss rational design concepts for using MFMs for in situ carbon capture and conversion.

1. Jo, L. Cruz, S. Shah, S. Wasantwisut, A. Phan and K. L. Gilliard-AbdulAziz, *ACS Catalysis*, 2022, **12**, 7486-7510.

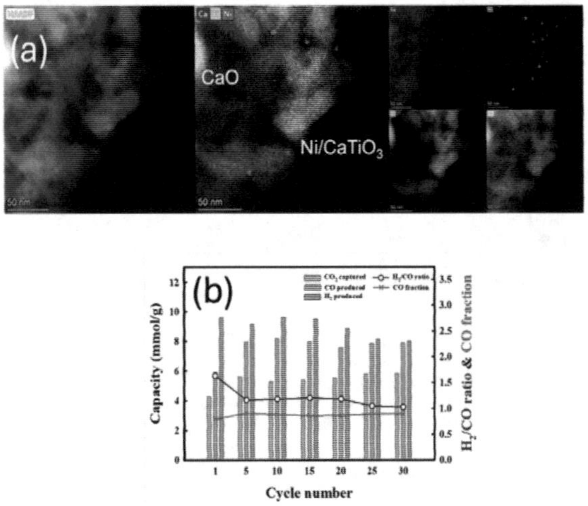

Figure 1. *Representative (a) image of exsolved Ni-CaTiO$_3$/CaO catalysts and (b) 30 cyclic capture and conversion of CO_2 to syngas.*

500d Material and Process Optimization of Reactive Carbon Capture to Methanol Using Cu-ZnO-Al_2O_3 Dual Function Materials

Chae Jeong-Potter, Martha A. Arellano-Treviño, Wilson McNeary, Alexander Hill, Daniel Ruddy, Anh To

Reactive carbon capture (RCC), an integrated CO_2 capture and conversion process that does not require generating a purified CO_2 stream, is an attractive carbon management strategy that can reduce costs and energy requirements associated with traditionally separate capture and conversion processes. Dual function materials (DFMs) comprised of co-supported sorbent and catalytic sites have emerged as a promising material design to enable RCC. DFMs have been extensively studied for methane production, but the noncompetitive economics of renewable methane necessitate the development of DFMs that can target more valuable, useful, and versatile products, such as methanol (MeOH).

Herein, we report the development and evaluation of modified Cu-Zn-Al (CZA) mixed oxide DFMs for the combined capture and conversion of CO_2 to MeOH. Previously, we demonstrated that a CZA DFM doped with a potassium sorbent (K/CZA) had promising performance with high conversion of adsorbed CO_2 (94.4%) and methanol selectivity (46%), leading to higher MeOH production (59.0 $\mu mol/g_{DFM}$) than the parent CZA or Ca/CZA DFM. In this study, we present a continuation of our efforts on the improvement of the DFM composition (alkali identity and loading) and process conditions (adsorption temperature, hydrogenation parameters) through a full factorial design of experiments, targeting higher MeOH production.

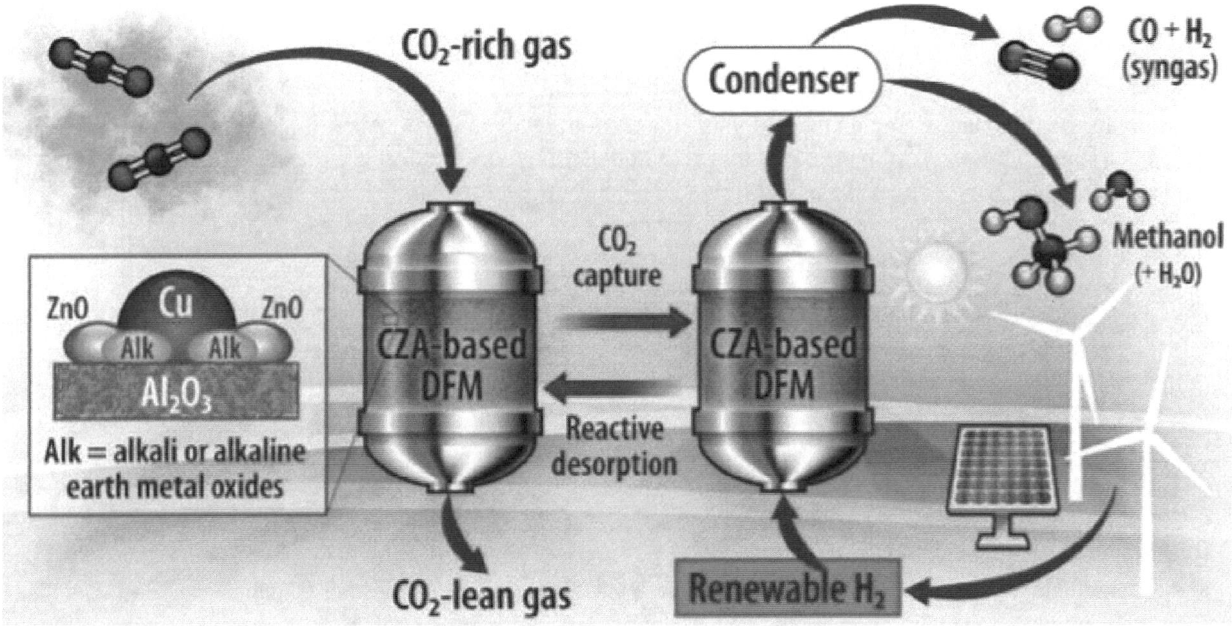

500e Hydrophobic h-BN Supported Cu/ZnO Catalysts for CO_2 Hydrogenation to Methanol

Jui-Chi Hu, Jeffrey Wu, Hsi-Yen Hsu, Yen-Chih Chen

Methanol (CH_3OH) is widely used as a common chemical feedstock for industrial purposes. Because of its high hydrogen-to-carbon ratio and being liquid under ambient conditions, methanol is widely recommended as hydrogen energy carrier in literature. CO_2 hydrogenation to methanol not only stores hydrogen as methanol but also achieves the purpose of CO_2 recycling. Industrially, $Cu/ZnO/Al_2O_3$ catalyst is used in CO_2 hydrogenation to methanol, which needs to be operated under very high pressure (5-10 MPa) from the thermodynamic point of view. In addition, water, an inevitable by-product during the reaction, not only decreases the methanol selectivity but also promotes the oxidation of Cu and the aggregation of ZnO, resulting in the deactivation of the catalyst.

In this study, we used hydrophobic material, hexagonal boron nitride (h-BN), as the support of Cu/ZnO catalyst for CO_2 hydrogenation to methanol. h-BN was pretreated by sonication or modified Hummers method to increase the adhesion of metal oxide. Cu/Zn metals were loaded on h-BN by the deposition-precipitation method. We designed different procedures and loading parameters to synthesize catalysts, including h-BN pretreatment method, catalyst synthesis method, metal loading, and promoter. The reaction was carried out at 230 °C and 1 MPa. The result indicated that all catalysts using h-BN as support showed higher methanol selectivity compared with those by Al_2O_3 support. We found that the catalyst using 40 wt% $Cu/ZnO/La_2O_3$ with 0.3 mol% Mg loaded on h-BN pretreated with modified Hummers' method showed the highest activity in this study. The space-time yield of methanol reached 105 $mg_{MeOH}/g_{cat}/hr$ at 1.0 MPa and 200 °C. The instruments XRD, SEM, EDS, H_2-TPR, CO_2-TPD, XPS, and contact angle were applied to evaluate the characteristics of catalysts and explain the activity test results. Chemical kinetic analysis was conducted based on the results obtained under different reaction conditions. Furthermore, a high-pressure reaction was carried out at the Industrial Technology Research Institute (ITRI) to confirm the excellent catalyst performance under industrial high-pressure operation, thus worth the further development of h-BN-supported catalysts.

Hydrophobic h-BN Supported Cu/ZnO Catalysts for CO_2 Hydrogenation to Methanol

Jui-Chi Hu[1], Jeffrey C. S. Wu[1], Hsi-Yen Hsu[2], Yen-Chih Chen[2]

[1]Department of Chemical Engineering, National Taiwan University, Taipei 10617, Taiwan
[2]Material and Chemical Research Laboratories, Industrial Technology Research Institute, Hsinchu 310401, Taiwan

Methanol (CH_3OH) is widely used as a common chemical feedstock for industrial purposes. Because of its high hydrogen-to-carbon ratio and being liquid under ambient conditions, methanol is widely recommended as hydrogen energy carrier in literature. CO_2 hydrogenation to methanol not only stores hydrogen as methanol but also achieves the purpose of CO_2 recycling. Industrially, $Cu/ZnO/Al_2O_3$ catalyst is used in CO_2 hydrogenation to methanol, which needs to be operated under very high pressure (5-10 MPa) from the thermodynamic point of view. In addition, water, an inevitable by-product during the reaction, not only decreases the methanol selectivity but also promotes the oxidation of Cu and the aggregation of ZnO, resulting in the deactivation of the catalyst.

In this study, we used hydrophobic material, hexagonal boron nitride (h-BN), as the support of Cu/ZnO catalyst for CO_2 hydrogenation to methanol. h-BN was pretreated by sonication or modified Hummers method to increase the adhesion of metal oxide. Cu/Zn metals were loaded on h-BN by the deposition-precipitation method. We designed different procedures and loading parameters to synthesize catalysts, including h-BN pretreatment method, catalyst synthesis method, metal loading, and promoter. The reaction was carried out at 230 °C and 1 MPa. The result indicated that all catalysts using h-BN as support showed higher methanol selectivity compared with those by Al_2O_3 support. We found that the catalyst using 40 wt% Cu/ZnO/La_2O_3 with 0.3 mol% Mg loaded on h-BN pretreated with modified Hummers' method showed the highest activity in this study. The space-time yield of methanol reached 105 $mg_{MeOH}/g_{cat}/hr$ at 1.0 MPa and 200 °C. The instruments XRD, SEM, EDS, H_2-TPR, CO_2-TPD, XPS, and contact angle were applied to evaluate the characteristics of catalysts and explain the activity test results. Chemical kinetic analysis was conducted based on the results obtained under different reaction conditions. Furthermore, a high-pressure reaction was carried out at the Industrial Technology Research Institute (ITRI) to confirm the excellent catalyst performance under industrial high-pressure operation, thus worth the further development of h-BN-supported catalysts.

Keywords: CO_2 hydrogenation, Methanol, Boron Nitride, Hydrophobic, Cu/ZnO catalyst

500f Process Modeling and Design for Methanol Synthesis Using Membrane Reactor and Vacuum Pressure Swing Adsorption

Akihide Wakita, Tomoyuki Yajima, Yuta Nishikawa, Kohei Yoshikawa, Nobuyuki Shigaki, Yoshiaki Kawajiri

Various Carbon dioxide Capture and Utilization (CCU) approaches have been proposed to reduce CO_2 emissions. Conversion of CO_2 to methanol is one of CCU approaches. It is a promising CCU approach because methanol is a raw material for many chemicals, including chemical fibers, pharmaceuticals, and plastics. Methanol, which is currently produced from coal and natural gas, can be synthesized from CO_2, which would reduce dependence on fossil fuels and mitigate CO_2 emissions simultaneously. However, methanol production by CCU is more expensive than the current production method. This is due to the high cost of CO_2 separation from the exhaust gas and the severe thermodynamic constraint in the methanol synthesis from CO_2.

To reduce the cost of CO_2 separation, chemical absorption, and physical adsorption methods have been studied. Among them, Vacuum Pressure Swing Adsorption (VPSA), one of the physical adsorption methods, allows flexible operations to reduce costs when high-purity CO_2 is not required. There have been many studies in the past on purifying CO_2 from various emission sources [1,2].

Furthermore, a solution to the thermodynamic constraints in methanol synthesis is to use a membrane reactor. This concept is a multifunctional reactor that integrates a membrane and a reactor into a single unit. By continuously removing a reaction product by the membrane, the reaction equilibrium can be shifted in the direction where the conversion is higher, exceeding the thermodynamic limit.

This study performs multi-objective optimization to analyze the cost reduction of a methanol synthesis process using blast furnace gas from a steelwork with a membrane reactor and VPSA. The CCU system is illustrated in Fig.1. which includes a water gas shift (WGS) reactor, VPSA unit, and membrane reactor. These units are modeled mathematically by taking material and energy balances. In particular, the VPSA model requires time-consuming calculations because it is complex and must be calculated to the cycle steady state (CSS). To overcome this challenge, we employed a surrogate model to approximate the input-output relationship to reduce the computational cost. A surrogate model was created using training data from the detailed VPSA model simulated under various operating parameters. The entire flowsheet including these units is optimized to minimize the energy consumption and maximize productivity, where the economic trade-offs are revealed as Pareto fronts.

References

[1] Z. Liu, C. A. Grande, P. Li, J. Yu, and A. E. Rodrigues, "Multi-bed Vacuum Pressure Swing Adsorption for carbon dioxide capture from flue gas," *Separation and Purification Technology*, vol. 81, no. 3, pp. 307–317, Oct. 2011.

[2] N. Shigaki, Y. Mogi, H. Kijima, T. Kakiuchi, T. Yajima, and Y. Kawajiri, "Performance evaluation of gas fraction vacuum pressure swing adsorption for CO2 capture and utilization process," *International Journal of Greenhouse Gas Control*, vol. 120, p. 103763, Oct. 2022.

Acknowledgment

This study was funded by New Energy and Industrial Technology Development Organization (Grant Number JPNP 16002).

Fig. 1. Process flow of CCU-methanol system

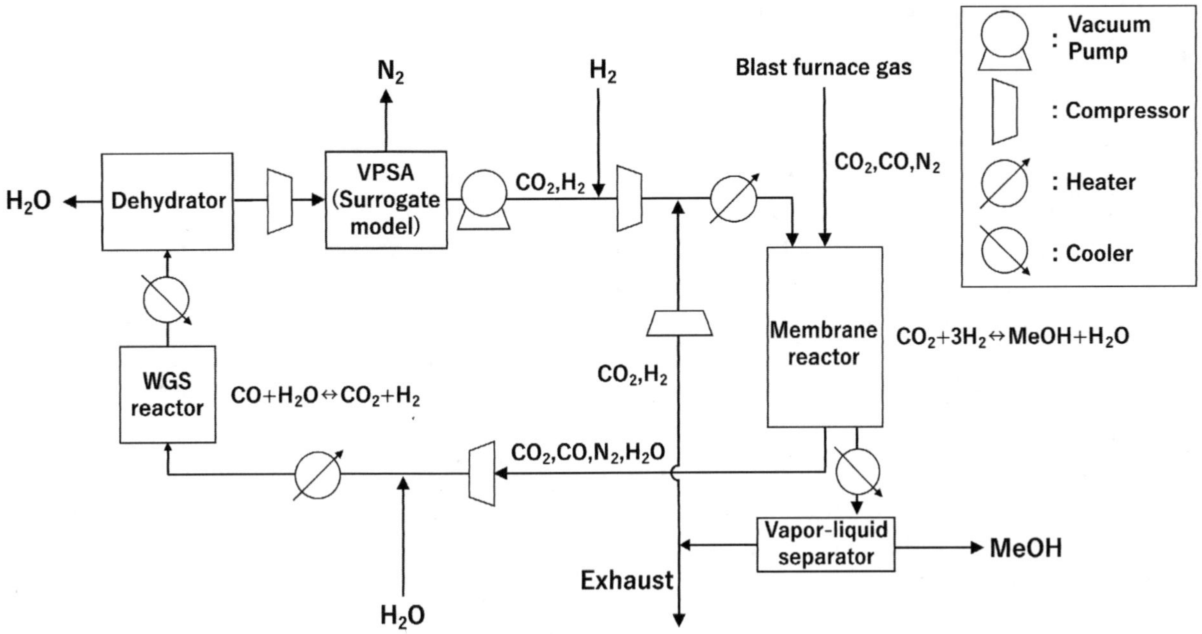
Vacuum Pump
: Compressor
: Heater
: Cooler
N_2
H_2
Blast furnace gas
H_2O
Dehydrator
VPSA (Surrogate model)
CO_2,H_2
CO_2,CO,N_2
Membrane reactor
$CO_2+3H_2 \leftrightarrow MeOH+H_2O$
WGS reactor
$CO+H_2O \leftrightarrow CO_2+H_2$
CO_2,H_2
CO_2,CO,N_2,H_2O
Vapor-liquid separator
MeOH
Exhaust
H_2O

500g Dehydration Membrane Reactor for the Production of Valuable Chemicals from CO_2 and H_2

Weiwei Xu

Production of liquid fuels or chemicals from captured and removed CO_2 and renewable H_2 presents a new approach to producing clean, low-carbon fuels domestically. While significant progress has been made in renewable electricity generation from solar and wind, many processes for producing liquid fuels/chemicals from renewable electricity are constrained by thermodynamic limitations, making them prohibitively expensive and impractical. We are overcoming these limitations and developing catalytic membrane reactor processes with high yields and low energy penalties. We developed a technology for the production of renewable liquefied petroleum gas (rLPG) from CO_2 and H_2 using a novel catalytic membrane reactor. Due to its environmentally benign characteristics, LPG is being used as a clean fuel in heating appliances, cooking equipment, and vehicles. Global LPG production was approximately 330 million tonnes in 2022. By producing rLPG through the catalytic conversion of captured CO_2 and renewable H_2, this process will produce renewable liquid transportation and heating fuels and a means of large-scale utilization of captured CO_2.

In our LPG synthesis process, CO_2 and H_2 are fed to a hollow-fiber, catalytic membrane reactor, at 20-35 bar and 220-320°C. The reactor contains a bi-functional catalyst that promotes two reactions, methanol (MeOH or CH_3OH) synthesis ($CO_2 + 3H_2 \leftrightharpoons CH_3OH + H_2O$) and LPG synthesis (MeOH $\leftrightharpoons$ hydrocarbon pool $\leftrightharpoons$ LPG + H_2O), into a one-step process to produce LPG. A $Cu/ZnO/ZrO_2/Al_2O_3$ (CZZA) catalyst is used for methanol synthesis and is coupled with a Pd-β zeolite catalyst for LPG production. However, combining these two reactions results in increased water production which inhibits catalytic activity. Here, our recently developed Na^+-gated, water-transport membrane (*Science, vol. 367, pp. 667, 2020*) removes water in situ, shifting the thermodynamic equilibrium towards product formation while decreasing kinetic inhibition from water adsorption onto the catalyst surface. Our membrane showed H_2O/CO_2, H_2O/H_2, H_2O/CO, and H_2O/MeOH selectivities of 560, 190, 170, and 80, respectively. In laboratory-scale testing using this membrane reactor, the one-pass LPG yield of 60.5% with a CO_2 conversion of 90.2% was achieved, which significantly exceeds the literature results of traditional packed-bed, catalytic reactors.

560a Experimental and Computational Efficiency Analysis of Perovskites-Driven Solar Thermochemical CO2 Conversion Cycle

Rahul Bhosale, Sayma Akhter

Syngas serves as a crucial input for the Fischer Tropsch process, which is responsible for generating liquid transportation fuels such as gasoline, diesel, and jet fuel. The syngas that is supplied as feedstock to the Fischer Tropsch process is typically produced through the gasification of fossil fuels such as coal, natural gas, or petroleum coke. The disproportionate and unchecked consumption of fossil fuels has been a major cause of concern in recent times, as it not only depletes their finite resources but also contributes significantly to the alarming rise in global environmental issues. A alternate technique to generate syngas incorporates a perovskite-based thermochemical redox cycle that splits water (H_2O) and carbon dioxide (CO_2). The perovskite based thermochemical redox cycle involves a two-step procedure. In the first step, perovskites are thermally reduced. This is followed by the second step in which the reduced perovskites are re-oxidized through the splitting of H_2O/CO_2. This results in the production of fuel such as hydrogen or syngas. In the present research, we have successfully synthesized $A_{0.5}Sr_{0.5}MnO_{3-\delta}$ perovskites (where A represents lanthanides) employing the combustion method. We have further evaluated their efficacy for the thermochemical splitting of H_2O/CO_2. The $A_{0.5}Sr_{0.5}MnO_{3-\delta}$ perovskites were subjected to a thorough characterization process using various analytical techniques, including Powder X-ray diffraction (PXRD), Brunauer-Emmett-Teller (BET) surface area analysis, Scanning Electron Microscopy (SEM), Transmission Electron Microscopy (TEM), and Inductively Coupled Plasma (ICP) analysis. The perovskites were then subjected to thermochemical redox reactions using a high-temperature Thermogravimetric Analysis (TGA) setup. The resulting experimental data obtained was used to estimate the solar-to-fuel energy conversion efficiency of the $A_{0.5}Sr_{0.5}MnO_{3-\delta}$ perovskites-driven thermochemical H_2O/CO_2 splitting cycle. This estimation was made possible by performing computational thermodynamic analysis on the experimental results.

560b Integrated Technology for Cost-Effective CO_2 Capture and Formic Acid Production: Modeling, Optimization, and Economic Analysis

Maojian Wang, Jinliang Ma, Miguel A. Zamarripa, Benjamin P. Omell

Due to the significant costs associated with conventional carbon capture and storage (CCS) technologies, there is a growing emphasis on exploring alternative approaches to mitigate CO_2 emissions. Converting CO_2 into chemicals and energy products shows promise in not only reducing emissions but also creating economic value due to high market potential. Revenue generated from utilizing CO_2 can also offset a portion of the CO_2 capture cost. Integrating CO_2 capture and conversion can eliminate the need for costly CO_2 transport and storage in conventional CCS applications, reducing costs by up to $45 per ton [1].

The National Energy and Technology Laboratory (NETL) has been exploring alternative carbon capture and utilization methods that offer high efficiency and competitive costs. A novel reactive technology is being investigated that electrochemically converts CO_2 into valuable chemicals, particularly formic acid. This work focuses on identifying the optimal design and operation of an integrated membrane-based CO_2 capture unit with the electrochemical conversion process. In this setup, the CO_2 in the flue gas permeates through a CO_2-selective membrane and enters an electrolyzer to produce formic acid, creating an integrated reaction module. To refine the chemical product, gas products from the electrolyzer are directed to a pressure swing adsorption unit, while the liquid product undergoes refinement to achieve commercial-grade formic acid using reactive distillation.

A membrane CO_2 capture model and an electrochemical conversion model have been developed using the IDAES Integrated Platform (Institute for the Design of Advanced Energy System) [2], facilitating rigorous flowsheet modeling and process design and optimization. The membrane model demonstrates good performance in CO_2 capture across a variety of conditions, ranging from 4% to 12% CO_2 concentration in volume. A rigorous electrochemical conversion model based on first principles has also been developed, with partial differential equations discretized in two dimensions to describe the heat, mass, and charge transfers along multiple layers of the electrolyzer cell, as well as the kinetics of catalytic electrochemical reactions on both cathode and anode sides. In this work, the integrated analysis shows the optimal process design and operation and provides valuable insights into competitive low- and no-carbon formic acid production routes.

[1] Smith E, Morris J, Kheshgi H, Teletzke G, Herzog H, Paltsev S. The cost of CO2 transport and storage in global integrated assessment modeling. International Journal of Greenhouse Gas Control. 2021 Jul 1;109:103367.

[2] Lee A, Ghouse JH, Eslick JC, Laird CD, Siirola JD, Zamarripa MA, Gunter D, Shinn JH, Dowling AW, Bhattacharyya D, Biegler LT. The IDAES process modeling framework and model library—Flexibility for process simulation and optimization. Journal of advanced manufacturing and processing. 2021 Jul;3(3):e10095.

Disclaimer:

This project was funded by the Department of Energy, National Energy Technology Laboratory an agency of the United States Government, through a support contract. Neither the United States

560c Tri-Reforming of CH_4 over Non-Noble (Cu, Ni) Metal Catalyst Supported on Basic Oxide (MgO, ZrO_2) for the Production of Syn-Gas with the H_2/CO Ratio of 1.5-2

Akansha Pandey, Prakash Biswas

CO_2 and CH_4 are the two most problematic greenhouse gases that create adverse effects on the environment. The conversion of these two anthropogenic gases to syn-gas and hydrogen through tri-reforming process is a promising route. The syn-gas with the H_2/CO ratio of 1.5-2 is highly desirable as a feedstock in the Fischer-Tropsch process for the synthesis of various important chemicals, including dimethyl ether (DME), higher alkenes, methanol, and substitute natural gas (SNG), etc. In this study, various non-noble metal (Ni, Cu) catalysts supported on basic oxide (MgO, ZrO_2) were synthesized by precipitation followed by the wetness impregnation technique. The physico-chemical properties of these catalysts were characterized by several N_2 adsorption-desorption, X-ray diffraction (XRD), temperature programmed reduction (TPR), CO_2-temperature programmed desorption (CO_2-TPD), etc. The tri-reforming activity of these catalysts were evaluated in a continuous downflow tubular packed bed reactor at different temperatures (650-800 °C), feed ratios, and at atmospheric pressure. At 800 °C, and at the optimum feed (CH_4: CO_2: H_2O: O_2: N_2) ratio of (1: 0.5: 0.0125: 0.1: 1.0), Cu-Zn-MgO catalyst demonstrated the CH_4 and CO_2 conversion of 40 %, and ~20%, respectively. The catalyst activity was significantly improved for 5 wt.% Ni/ZrO_2 catalyst. At lower temperature of 700°C, higher CH_4 (86%) and CO_2 conversion (~28%) was. The obtained H_2/CO ration over the 5 wt.% Ni/ZrO_2 catalyst was ~1.6, and it was almost constant for a longer period with negligible coke deposition. The activation energy with respect to CH_4 and CO_2 was calculated as 47 kJ/mol and 191 kJ/mol, respectively.

560d Low Intensity Pathways for Carbon Mineralization: Utilizing Mafic Resources in Arizona

Adam Sjolund, Thomas Sasser, Olivia Andrea Wrenn, Lisa A. Thompson, Klaus S. Lackner, Jennifer Wade

Natural mineral resources in abundant quantities, geologic stability, and relatively quick reaction rates make mineral carbonation a promising pathway for long-term carbon sequestration (Lackner, 1995). High pressure and temperature carbonation of mafic and ultramafic mineral in CO2 saturated environments is a known and demonstrated method of stable carbon sequestration. (Gadikota, 2014) Ex-situ applications of undisturbed materials have largely been deemed cost-prohibitive due to the economics of crushing and transporting large-scale quantities of hard rock, particularly compared to the lower costs of in-situ methods. (National Acadamies Press, 2018) The performance of mafic resources under comparably low intensity conditions remains relatively unstudied, and highly glassy and porous scoria from Arizona shows promise for ex-situ processes due to inherently high surface area and lower crushing costs, arising from the naturally porous and glassy structure formed by the rapid cooling and young age of the associated volcanic events. (Valentine, 2021)

This work explores the reaction rates of local mafic and ultramafic resources in an ambient temperature, CO2 saturated aqueous solution; materials show considerable carbonation extent over monthslong timespans. Benchmarked using an ultramafic San Carlos (Arizona) olivine/pyroxene specimen samples demonstrate a reaction extent exceeding 10 wt% CO2 (100g CO2/1kg rock) over 128+ day experiments. Consistent magnesite, siderite, and dolomite carbonate phases are observed under thermogravimetric analysis (TGA) which are not seen under in-situ conditions. The combination of mild conditions and preferable rock characteristics indicate greater economic feasibility and potential for new carbon mineralization pathways.

References

Klaus S. Lackner, Christopher H. Wendt, Darryl P. Butt, Edward L. Joyce, David H. Sharp, Carbon dioxide disposal in carbonate minerals, Energy, Volume 20, Issue 11, 1995, Pages 1153-1170, ISSN 0360-5442, https://doi.org/10.1016/0360-5442(95)00071-N. (https://www.sciencedirect.com/science/article/pii/036054429500071N)

Gadikota G, Matter J, Kelemen P, Park AH. Chemical and morphological changes during olivine carbonation for CO2 storage in the presence of NaCl and NaHCO3. Phys Chem Chem Phys. 2014 Mar 14;16(10):4679-93. doi: 10.1039/c3cp54903h. PMID: 24469156.

National Academies of Sciences, Engineering, and Medicine; Division on Earth and Life Studies; Ocean Studies Board; Board on Chemical Sciences and Technology; Board on Earth Sciences and Resources; Board on Agriculture and Natural Resources; Board on Energy and Environmental Systems; Board on Atmospheric Sciences and Climate; Committee on Developing a Research Agenda for Carbon Dioxide Removal and Reliable Sequestration. Negative Emissions Technologies and Reliable Sequestration: A Research Agenda. Washington (DC): National Academies Press (US); 2018 Oct 24. PMID: 31120708.

Greg A. Valentine, Michael H. Ort, Joaquín A. Cortés; Quaternary basaltic volcanic fields of the American Southwest. *Geosphere* 2021;; 17 (6): 2144–2171. doi: https://doi.org/10.1130/GES02405.1

560e Transition Metal-Free Dual Function Materials for Highly Selective CO_2 Capture and Conversion

Anh To, Alexander J. Hill, Wilson McNeary, Chae Jeong-Potter, Martha A. Arellano-Treviño, Daniel Ruddy

Scaling conventional carbon capture and utilization methods can be limited by high costs and underdeveloped infrastructure for transporting captured CO_2 to upgrading sites. Reactive carbon capture (RCC) can eliminate these concerns by unifying the CO_2capture and conversion steps in a single unit operation. In RCC, a single sorbent-catalyst material (known as a dual function material, or DFM) is used to capture CO_2 from the atmosphere or a point source, most often via interactions with basic surface sites. The bound CO_2 is then converted to products on adjacent catalyst sites by changing the reaction conditions to promote reactive desorption alongside a reactive agent (i.e., H_2).

The process intensification approach of RCC implies minimal cleanup of the input gas stream for CO_2 capture and conversion. It is therefore important to design DFMs that tolerate the poisons common to point source and atmospheric emissions. Oxygen is one such poison, which readily oxidizes the transition metals used in conventional CO_2 hydrogenation such as nickel and copper. These metals must be reduced at elevated temperature prior to CO_2 conversion in RCC applications. This increases H_2 consumption, cycle time, and decreases product yield through the desorption of the captured CO_2.

Here, we report transition metal-free DFMs to address the challenges imposed by the oxidation of conventional CO_2 hydrogenation catalysts. Mixed metal oxide catalysts comprised of zinc and aluminum were doped with several alkali species to facilitate CO_2capture. These zinc aluminate spinels ($ZnAlO_X$) were prepared at various Zn:Al ratios to understand the mechanisms that govern CO_2 and H_2 activation. The $ZnAlO_X$ DFMs were studied in model RCC experiments and demonstrate over 97% selectivity to CO under atmospheric pressure and temperature swing operation. Subsequent RCC cycling with oxygen probes the stability of these transition metal-free DFMs in more realistic CO_2 capture streams.

560g Direct Carbonylation of Glycerol with CO_2 to Organic Carbonates Catalyzed By Cerium Oxide/Zinc Oxide

Wei-Hong Hong, Bing-Hung Chen

Direct carbonylation of glycerol with CO_2 with cerium oxide/zinc oxide catalysts and 2-cyanopyridine (2-CP) as a dehydrating agent was studied under a CO_2 pressure (P_{CO2}) from 30 bar to 50 bar and a reaction temperature from 120°C to 200 °C in a Parr 4563 reactor. The incipient wetness impregnation was employed to load cerium oxide, using $Ce(NO_3)_3 \cdot 6H_2O$ as precursor, onto the surface of zinc oxide, forming cerium oxide/zinc oxide catalysts. The properties of the catalysts were identified using equipment and methods such as XRD, SEM, EDS, ICP-OES, XPS, BET porosimetry and CO_2-TPD. The experimental results showed that cerium oxide/zinc oxide synthesized at a weight ratio of 30:70 could achieve generally the highest yield of glycerol carbonate. Notably, increasing the reaction pressure demonstrated a dual benefit in enhancing both yield and selectivity of glycerol carbonate. For example, a yield of glycerol carbonate from glycerol exceeding 40% could be attained from a reaction at 180 °C. In general, an insufficient amount of catalyst present in the reaction system could result in a lower yield of glycerol carbonate, probably owing to fewer active sites available, while catalyst present in excess led to a slightly inferior catalysis and a lower yield of glycerol carbonate as well. Thus, a catalyst dosage comprising 3 wt% of glycerol would be the optimal, based on empirical results available.

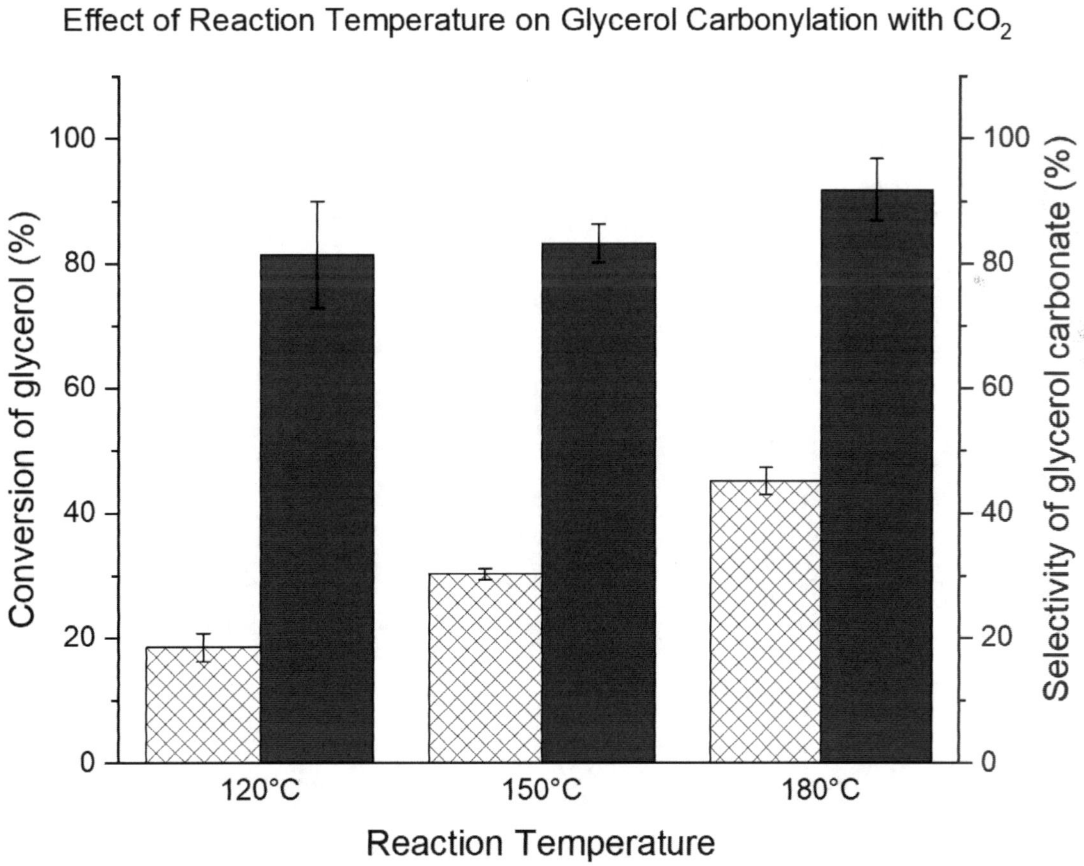

558b Electrification As a Pathway to Green Living: Economic and Environmental Outcomes in U.S. Multifamily Housing

Matthew Williams, Chris Parker, Amanda Dillon, Blake Billings, Kody Powell

As the urgency to address climate change intensifies, the electrification of multifamily buildings emerges as a crucial strategy for reducing carbon emissions and enhancing energy efficiency. This study delves into the environmental and economic impacts of transitioning multifamily residences from mixed-fuel to fully electrified systems across the United States. Utilizing calibrated building energy models (BEMs), this study assesses the performance of electrified buildings in terms of energy consumption, carbon dioxide emissions, and economic viability. Our findings reveal a notable decrease in energy usage and carbon emissions in nearly all electrified models, with 22 states showing enhanced or comparable performance to traditional systems. A sensitivity analysis further illustrates the economic feasibility of electrification, highlighting a potential shift towards affordability for tenants amidst evolving energy pricing and grid decarbonization efforts. Through case studies and policy analysis, this research underscores the importance of incorporating multifamily buildings into the broader electrification narrative, proposing electrification as an optimistic solution for achieving sustainability and affordability in the residential sector.

558c Direct Methane Capture Via Aerobic Methanotrophs

Waaseyaaban Landgren, Robert Handler, David Shonnard, Mary E. Lidstrom

According to the IPCC's Sixth Assessment Report (AR6), as of 2019, 18.6% of greenhouse gas emissions were methane, and as reported by the Environmental Protection Agency, in 2021 the three largest sources of methane emissions were natural gas and oil systems (32%), enteric fermentation (27%), and municipal solid waste landfills (14%). The quantity of emissions is important, as methane possesses a warming impact 86 times greater than CO2 on a 20-year timescale, and breaking this methane down into CO2 would provide near-term climate benefits. An example of these benefits, also assessed in AR6, is that reducing methane emissions would limit both peak warming levels and overshoot warming that would occur once peak global temperatures are reached. To combat methane emissions, a novel bioreactor is being developed as a potential short-term mitigation strategy. This bioreactor would be driven by the metabolic processes of methanotroph bacteria; a class of bacterium which survives solely off methane as its carbon and energy source. Microbes are being isolated which can thrive at methane concentrations of ~500 ppm, which is a common ambient methane concentration at sources like oil and gas wells, dairy farms, and landfills. Methane-rich air will be directed into this reactor, and reduced into CO2 and H2O by the bacteria, whose biomass can also be harvested as a potential value-added product. In order to ensure that this technology is adopted widely and confers a large environmental benefit, it is important to evaluate the potential environmental and economic benefits of bioreactor operations under several potential operating scenarios. In this presentation, we will present our current work to model the methane bioreactor system using spreadsheet tools and process simulation software, and evaluate its performance using life cycle assessment (LCA) and techno-economic assessment (TEA) methods. We will illustrate a range of potential system configurations that offer the best combination of environmental performance and economic benefit, and highlight key process assumptions that will warrant further research for the rest of our project.

558d A Novel “Dry” Biofilm Reactor for Atmospheric Methane Removal

Reza Khorasani, Q. Peter He, Jin Wang

Because of CH_4’s high global warming potential and relatively short half-time in the atmosphere, CH_4 removal can contribute substantially to near-term climate mitigation. It has been reported that there are hundreds of thousands of sites where the methane in the overlying air is estimated to be in the 200-2000 ppm range, but no proven scalable technology currently exists to capture CH_4 in this concentration range.

Bacteria (aerobic methanotrophs) in the soil and water provide the second-largest natural sink for atmospheric CH_4, accounting for ~7% of the total CH_4 sink. We believe engineering innovations can significantly increase this capability.

Biological atmospheric CH_4 removal has two major advantages: (1) it occurs at ambient temperature and pressure without releasing toxic byproducts (Lawton & Rosenzweig, 2016), providing a safe pathway with known impacts. (2) only about 50 mol % of CH_4 is converted to CO_2, while the other half is assimilated to produce microbial biomass (a co-benefit), which reduces the amount of CO_2 released back into the atmosphere.

However, the very low atmospheric CH_4 concentrations (2ppm overall, and 200 – 2000 ppm for areas with elevated CH_4 concentrations) pose significant challenges for biological CH_4 removal. Specifically, CH_4's low concentration and its very small solubility in water results in low throughput and significant energy consumption to overcome the mass transfer resistance of the gas substrate. This is why available CH_4 bioconversion technologies all target point-source methane with high concentrations (>5%). As such, existing biological CH_4 conversion approaches (all targeting point sources with >5% concentrations) are not suitable for CH_4 removal from air (La et al., 2018; Smith et al., 2001).

Recently, the Lidstrom lab at the University of Washington showed that *Methylotuvimicrobium buryatense* 5GB1C can efficiently remove CH_4 at 500 ppm (He et al., 2023). However, even with this efficient biocatalyst, to remove CH_4 from air at scale, we need a bioreactor that can effectively address the above-mentioned mass transfer-related challenges. To fill this gap, we introduce a new concept of “dry” CH_4 bioconversion and a novel scalable bioreactor design that has the potential for atmospheric CH_4 removal. In the proposed bioreactor, biocatalyst forms biofilm on a substratum tray placed horizontally. “Dry” means there is no bulk liquid phase and biofilm is directly exposed to gas phase. By eliminating the bulk liquid phase, we hypothesize that the “dry” biofilm reactor will not only drastically enhance CH_4 mass transfer, but also significantly reduce energy and water use.

In this talk, we will present our bioreactor prototype to demonstrate the new concept. In addition, preliminary experimental results are presented to support our hypothesis.

He, L., Groom, J. D., Wilson, E. H., Fernandez, J., Konopka, M. C., Beck, D. A. C., & Lidstrom, M. E. (2023). A methanotrophic bacterium to enable methane removal for climate mitigation. *Proceedings of the National Academy of Sciences*, *120*(35), e2310046120.

La, H., Hettiaratchi, J. P. A., Achari, G., & Dunfield, P. F. (2018). Biofiltration of methane. *Bioresource Technology*, *268*, 759–772.

Lawton, T. J., & Rosenzweig, A. C. (2016). Methane-oxidizing enzymes: an upstream problem in biological gas-to-liquids conversion. *Journal of the American Chemical Society*, *138*(30), 9327–9340.

Smith, P., Goulding, K. W., Smith, K. A., Powlson, D. S., Smith, J. U., Falloon, P., & Coleman, K. (2001). Enhancing the carbon sink in European agricultural soils: including trace gas fluxes in estimates of carbon mitigation potential. *Nutrient Cycling in Agroecosystems*, *60*, 237–252.

558e Eco-Efficient Processing Clusters: Blockchain-Enabled Peer-to-Peer Resource and Emission Trading Framework

Manar Oqbi, Dhabia Al-Mohannadi

Achieving sustainable development goals requires seamless integration of materials and energy across industrial processes, alongside effective implementation of an Emission Trading System (ETS). The core challenge is to foster transparency and accountability, while optimizing the transformation of raw materials and energy into valuable products and maintaining data privacy and traceability within the ETS framework. This study proposes a novel framework to develop optimal processing clusters and a dual blockchain-based trading system for resources and emissions, addressing these concerns. The framework utilizes a stepwise approach to synthesize integrated processing clusters and ETS, featuring two interactive systems: a Blockchain-based system for trading resources and another for emissions. The resource trading system enhances process transparency, allowing for precise tracking of materials and energy. Additionally, we introduce a Mixed-Integer Nonlinear Program (MINLP) integrated with Ethereum blockchain's smart contracts, enhancing data privacy and traceability and facilitating fair cost distribution among the processes to achieve environmental goals. This model calculates emission trade volumes and investments in carbon capture and storage (CCS), promoting system efficiency and reliability. By equipping decision-makers and regulatory bodies with a robust tool, this framework not only enhances understanding of resource and emission trading interactions but also supports the development of new policies and regulatory compliance through blockchain technology. This advancement signifies a pivotal step towards realizing sustainable industrial practices and achieving environmental targets.

558f A Systems Approach for Renewable Fuels and Innovations for the Shipping Industry

Nefeli Roufogali, Konstantinos Pyrgakis, Lazaros Karaoglanoglou, Antonios Kokossis

Maritime transport, although one of the most energy-efficient modes of transport, is also a large and growing source of greenhouse gas emissions. Hence, the EU has established a new regulation for reducing GHG emissions, which came into effect from January 2024, to achieve the reduction targets. Consequently, the shipping industry faces escalating pressure to decarbonize and alleviate its environmental footprint. Meeting the ambitious sustainability goals and regulations set for the industry necessitates a sophisticated portfolio of tailored technological solutions, meticulously developed and implemented.

Alternative fuel options considered in the study include LNG, methanol, ammonia, and hydrogen, alongside biocrude oil as a direct substitute for marine diesel. Additionally, bio-based production routes, according to the available state-of the art technologies, are also examined for all alternatives to assess their impact on the ship's environmental performance. The advantages and disadvantages of each option are summarized in Table 1.

Table 1 Pros and cons of different fuel options

The paper outlines a systematic and holistic approach that combines the LCA-based environmental impact and techno-economic feasibility assessment for the use of conventional and alternative marine fuel options, along with energy efficiency, integration, and on-board CO_2 capture technologies.

Methodologically a case study approach for specific vessel type and route configurations was employed for the simulation and flowsheeting of the available options. The process and energy integration, encompassing heat recovery, cogeneration, and cryogenic integration of low boiling fuels, was conducted to assess their potential application onboard. Additionally, an environmental impact assessment based on Life Cycle Assessment (LCA) using the ReCiPe 2016 Hierarchist (End point) method, along with the Global Warming assessment method, was utilized to evaluate the comparative impact of various ship operation scenarios on its environmental performance. Finally, the technoeconomic constraints of applying the available options to the existing ships, through retrofitting, was investigated.

***Energy integration**:* Indicative shifted composite curves for each of the five alternative fuel options applied in the engine simulation model are presented in Figure 1. The major energy needs, in the case of marine diesel use are for heating, and the pinch point value in this case is quite low, therefore heat exchange is expected to be beneficial only above the pinch temperature, where the heat exchange potential is larger; when LNG is considered high fuel preheating requirements arise, due to low storage temperatures, therefore the cooling needs are fully covered for any of the examined operation modes. It is noteworthy that despite the reduced comparative need for carbon dioxide separation from the other components of the flue gas, due to the lower emissions for the same energy production by LNG in comparison with marine diesel, the heating needs cannot be fully satisfied by the integration. This is due to the corresponding reduced heat recovery from the flue gas and the additional heating requirements for fuel preheating; the heating requirements for methanol are particularly high, as it can be observed in the respective diagram of Figure 1, and only under energy integration assumptions it can be considered feasible. The scenario without carbon dioxide capture seems to be more realistic; however, the feasibility of its implementation will depend on the environmental impacts which should be studied further; the thermal needs of the system for the use of ammonia can be fully satisfied with the implementation of a network of heat exchangers based on energy integration. The maximum possible heat recovery from the generated flue gases is significantly greater than the required heat supply. Consequently, it is considered that full heat recovery from the flue gases will not be achieved; the thermal requirements for hydrogen have been fully matched as well, with the difference lying in the ratio of actual to maximum heat recovery. Overall, the degree of integration is satisfactory, and minimum heat supply requirements are negligible.

Figure 1: Energy integration – Shifted composite curves for multiple fuel options, MCR 75% and Carbon Capture (except NH_3 and H_2)

Table 2 summarizes the examined scenarios for the operation of the main engine under different fuel, MCR and Carbon capture assumptions, as well as the expected impact of energy integration on the overall efficiency of ship energy consumption. The energy integration benefits are in the range of 8-20% in the case of marine diesel oil; 17-18% for the use of Methanol; 9-12% for LNG; 12-14% for ammonia and 10-12% for Hydrogen; depending on the selected operation mode for each fuel type.

Table 2: Fuel consumption before and after energy integration, for multiple fuel, MCR and Carbon Capture combinations (MDO: Marine Diesel Oil, MET: Methanol, AMM: Ammonia, HYD: Hydrogen)

LCA based environmental impact assessment

The environmental impact of the most promising operation scenarios for the specific ship and route considered in the case study, is presented in Figure 2. As it can be easily observed the fuel with the lowest impact is renewable hydrogen where up to 90% of the base case emissions (i.e. MDO without CC) can be achieved. For fuels where carbon dioxide is produced during combustion, biofuel options dominate within each category. More specifically the use of bio-oil results in 33%; biomethane 70%; bio-methanol 75% reduction of GHG emissions. The use of NH_3 can also produce benefits of up to 60% reduction of GHG emissions; but only under the condition its production follows a biological pathway.

It should be noted that the use of carbon capture technologies has a very significant impact on the GHG emission potentials of examined ship operation scenarios; i.e. 55% reduction can be achieved in the case of MDO, and 50% for the use of LNG, if assessed on a Well-to-Wake basis. Given the limited resource availability for the production of significant amounts of bio-based fuels to meet the demand of marine sector, the paper pointed to the CO_2 capture on board as the most realistic technology option to meet the targets set for the GHG emissions reduction.

Figure 2. Indicative LCA based, cradle-to-grave environmental impact of different fuel origins, use and carbon capture technologies for the case study ship and route scenarios

Table 1 Pros and cons of different fuel options

Fuel	Pros	Cons
LNG	***Lower Emissions***, not only lower SOx and NOx emissions but also lower CO_2 ***Abundant and established*** supply systems ***Improved Efficiency,*** more efficient than traditional marine engines ***Compliance*** with regulations for SOx emissions ***Biomethane,*** a relatively mature technology (Biogas upgrading route)	***Infrastructure Investment needed***, since the bunkering infrastructure is limited ***GHG emissions***, despite the reduction which can be achieved, still fossil fuel ***Safety challenges,*** due to storing and handling conditions
Methanol	***Lower Emissions***, not only lower SOx and NOx emissions but also lower CO_2 ***Flexibility in its production,*** fossil, bio-based and renewable (as a utilization option of captured carbon) ***Infrastructure,*** exist for production and handling ***Dual fuel technology,*** already exist, to	***GHG emissions*** related benefits can be really achieved only for bio-based or renewable methanol ***Energy density*** lower than diesel, space related challenges ***Bunkering infrastructures*** not yet in place

	be used in combination with diesel	
Ammonia	***Zero Carbon Emissions***, in its end use ***High Energy Density***, limited needs of storage space ***Existing infrastructure***, being a chemical produced in bulk amounts and available safety protocols ***Reduced air pollution*** not only for CO_2 emissions but also for SO_x, NO_x and particles.	***Safety Concerns***; accidents can cause severe impacts ***Infrastructure*** not available yet for its use as a fuel ***Efficiency***, lower than conventional marine engines ***Positive impact*** on CO_2 only if produced from renewable sources ***NOx emissions***, if not properly used
Hydrogen	***Renewable*** production options are available ***Fuel cells or internal combustion*** engines possible with ***zero emissions*** during its end-use ***Higher efficiency*** of fuel cells in the conversion of fuel energy to work	***High cost***, for renewable hydrogen production with existing technology ***Infrastructures***, not available due to technical difficulties and cost ***Low energy density per volume***, needs for advanced storage technologies
Bio oil	***Compatibility*** with existing engines and supply chain infrastructures ***Circular economy*** targets can be achieved if waste biomass used ***Significant CO_2 emissions*** reduction on a well to wake approach ***Emissions reduction*** for SOx as well.	***Cost*** of fuel is much higher and can cause competition with food sector ***Limited availability***, and its intensive use on ships will put further pressure on its price ***Limited CO_2 emissions*** benefits from its end-use, combustion, on ship ***It can be used only in blends***, according to its current quality and existing engine specifications

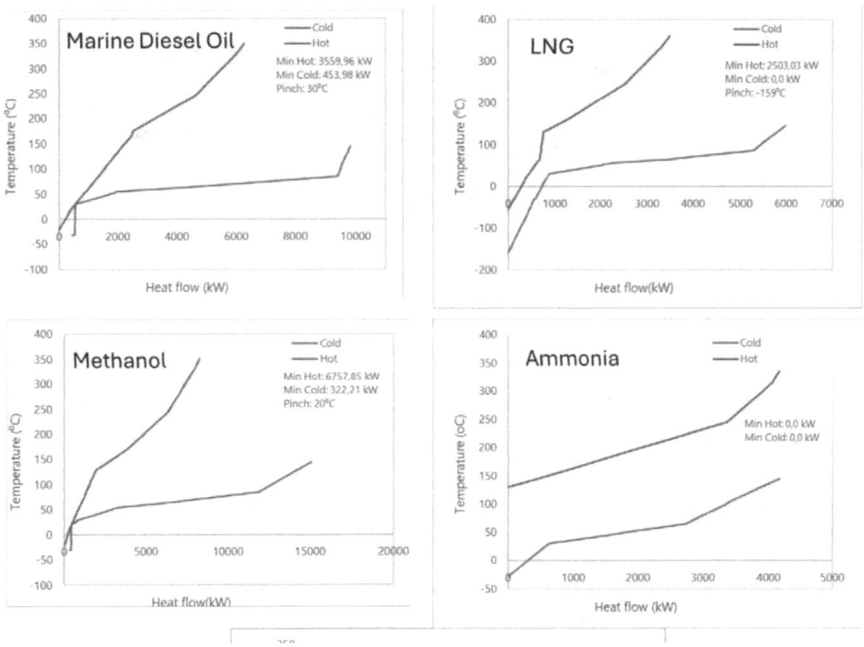

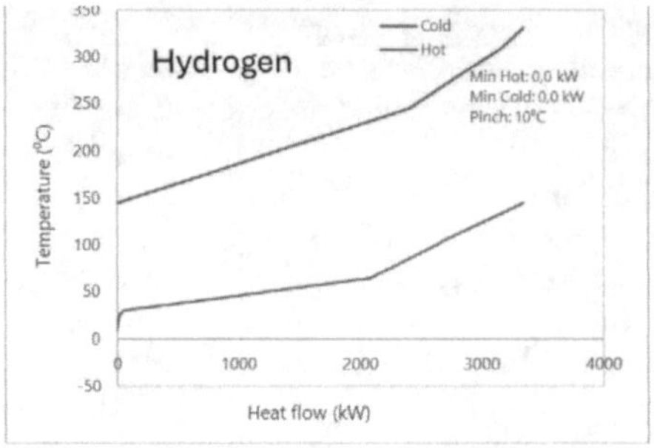

Figure 1: Energy integration – Shifted composite curves for multiple fuel options, MCR 75% and Carbon Capture (except NH_3 and H_2)

Table 2: Fuel consumption before and after energy integration, for multiple fuel, MCR and Carbon Capture combinations (MDO: Marine Diesel Oil, MET: Methanol, AMM: Ammonia, HYD: Hydrogen)

	Overall without Energy Integration (MW)	Overall with Energy Integration (MW)	Energy saved due to Energy Integration (MW)	% Energy saved due to energy Integration
MDO 60%+CCS	32,44	26,69	5,75	17,7%
MDO 75%+CCS	39,97	32,18	7,78	19,5%
MDO 90%+CCS	45,31	37,12	8,19	18,1%
MDO 75%	29,29	26,90	2,39	8,2%
LNG 60%+CCS	28,80	25,50	3,30	11,5%
LNG 75%+CCS	34,58	30,54	4,04	11,7%
LNG 90%+CCS	40,36	35,70	4,66	11,5%
LNG 75%	30,42	27,75	2,68	8,8%
MET 60%+CCS	37,58	30,96	6,62	17,6%
MET 75%+CCS	45,27	37,20	8,07	17,8%
MET 90%+CCS	52,96	43,40	9,55	18,0%
MET 75%	34,66	28,86	5,80	16,7%
AMM 60%	25,78	22,10	3,69	14,3%
AMM 75%	30,91	26,90	4,01	13,0%
AMM 90%	36,03	31,70	4,33	12,0%

AMM 90%	36,05	31,70	4,35	12,0%
HYD 60%	25,11	22,10	3,01	12,0%
HYD 75%	30,09	26,90	3,19	10,6%
HYD 90%	35,07	31,70	3,37	9,6%

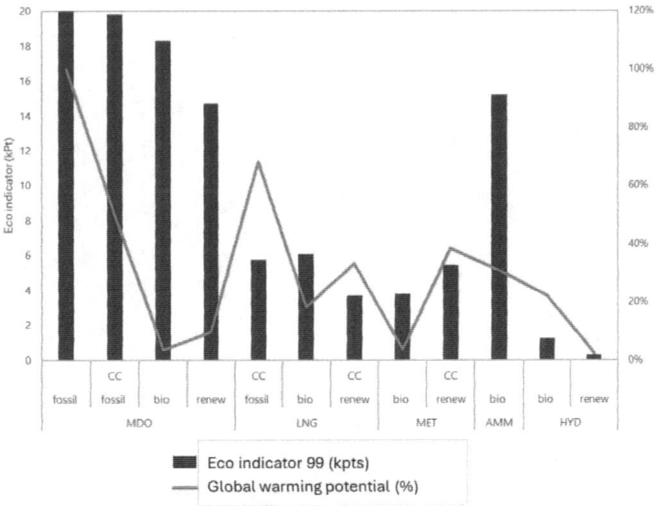

Figure 2. Indicative LCA based, cradle-to-grave environmental impact of different fuel origins, use and carbon capture technologies for the case study ship and route scenarios.

558g A 1-2-3 of Performing Industrial Life Cycle Assessments

Seshasayee Mahadevan, Deboleena Chakraborty, Mahesh Sawant, Chad English, Paul Doll

The purpose of this talk is to increase awareness of Dow's sustainability strategies for Path to Zero. The author will give an overview of how Life Cycle Assessment is currently performed within the company, what gaps exist and a call to action for what needs to be done to bridge that gap. Industry-Academia or collaborations between companies would be necessary to innovate in the technology space to help Dow achieve the steep sustainability targets.

Industrial Life Cycle Assessments are performed to understand the cradle to gate or cradle to grave impacts of processes, products sold, or application of products. The assessments are performed using primary (process) data from the company and secondary data from publicly available databases for emission factors. The initial assessments are accompanied by sensitivity analysis, impact reduction opportunities and data quality assessments. These in accordance with ISO standards are then validated with external entities and publicly disclosed afterwards.

These analytical quantifications typically happen internally within corporations with limited external visibility to the workflows and external validation process. This presentation aims to bridge this gap in visibility and knowledge. The authors aim to provide case studies on life cycle assessments developed internally and externally validated. These span broad businesses and use cases to provide a flavor for the types of life cycle assessments performed in industry.

Industry-academia collaboration is vital for achieving zero emission goals, as it combines academia's innovative research capabilities with industry's practical applications and resources. This partnership accelerates the development and implementation of sustainable technologies, ensuring that academic breakthroughs transition efficiently into market-ready solutions. By working together, both sectors can leverage their strengths to address climate change challenges more effectively and rapidly.

Novel Fruit Hydrochars: Eco-Friendly Solutions for Waste Management in Electronics, Environment, and Agriculture

Erick Mendizábal[1], María Fernanda Lecaro[1], Camily Rivadeneira[1], Yuliana Pullas Gallardo[1], Mathilde Reffet[2], Juan Fernando Posso[3], Pierina Mirabá[3], Victoria Alomía[4], Marco León[5], Darío Niebieskikwiat[6], Luis M. Prócel[3] and Andrea C. Landazuri[1,5,7,8]*

[1] *Universidad San Francisco de Quito USFQ, Colegio de Ciencias e Ingenierías, Applied Circular Engineering & Simulation Group (GICAS), Chemical Engineering Department, Diego de Robles y Vía Interoceánica, P.O. Box 17-0901, Quito, Ecuador*

[2] *Chemical Engineering Department, École Polytechnique Fédérale de Lausanne, Lausanne, Switzerland.*

[3] *Universidad San Francisco de Quito USFQ, Instituto de Micro y Nanoelectrónica, Diego de Robles y Vía Interoceánica, P.O. Box 17-0901, Quito, Ecuador*

[4] *Universidad San Francisco de Quito USFQ, Ingeniería en Agronomía, Diego de Robles y Vía Interoceánica, P.O. Box 17-0901, Quito, Ecuador*

[5] *Universidad San Francisco de Quito USFQ, Institute for Energy and Materials, Diego de Robles y Vía Interoceánica, P.O. Box 17-0901, Quito, Ecuador*

[6] *Universidad San Francisco de Quito USFQ, Physics Department, Diego de Robles y Vía Interoceánica, P.O. Box 17-0901, Quito, Ecuador*

[7] *Universidad San Francisco de Quito USFQ, Instituto IBioMed, Diego de Robles y Vía Interoceánica, P.O. Box 17-0901, Quito, Ecuador*

[8] *Universidad San Francisco de Quito USFQ, Instituto de Investigaciones Biológicas y Ambientales (Biósfera), Diego de Robles y Vía Interoceánica, P.O. Box 17-0901, Quito, Ecuador*

*Corresponding author: alandazuri@usfq.edu.ec

Abstract

This research explores the potential of hydrochars obtained from diverse exotic fruit residues such as mango endocarps, mangosteen rinds, African oil palm residues, moringa husks and blackberry bushes, as materials suitable for insulation, dielectric applications and for potential agricultural purposes. The electrical properties of hydrochars derived from these residues reveal that these materials exhibit elevated dielectric constants and minimal electrical conductivities, rendering them well-suited for application as insulating materials. Furthermore, the study highlights the influence of fruit type, hydrothermal carbonization conditions, and post-treatment processes on the electrical characteristics of hydrochars. These research outcomes hold potential for advancing sustainable waste management practices promoting circular bioeconomy and circular engineering practices.

Introduction

Waste management, particularly of organic materials, presents a significant environmental challenge in Ecuador. Agricultural residues and fruit byproducts often end up in landfills, contributing to soil and groundwater contamination (Ulloa, 2006). To address this issue, innovative approaches in the agricultural and electronic sectors can be implemented.

In the agricultural domain, Hydrothermal Carbonization (HTC) emerges as a promising solution. This process can convert organic waste into hydrochar, a valuable product for agricultural applications (Pauline & Joseph, 2020). Hydrochar derived from agricultural residues can be used as a biofertilizer, improving soil quality and crop yields (Wang et al., 2022). Additionally, a second-generation biorefinery approach can extract valuable compounds from fruit byproducts such as mango endocarps, mangosteen rinds, and moringa husks, which can be utilized in agricultural applications, reducing waste and promoting sustainability (Popa, 2018). Smart composting systems using Internet of Things (IoT) sensors can also optimize the composting process, ensuring efficient breakdown of organic waste (Chakraborty, 2021).

On the electronic front, recycling electronic waste can recover valuable metals and materials for reuse (Zeng et al., 2018). IoT-enabled waste bins and blockchain technology can improve waste management efficiency and transparency (Sosunova & Porras, 2022). Additionally, hydrochar derived from agricultural waste could serve as a sustainable alternative in electronic applications, such as dielectric materials or supercapacitors.

This study aims to demonstrate the potential of utilizing byproducts from mango endocarps, African oil palm residues, moringa husks, and other plant materials from the Galapagos Islands through a second-generation biorefinery process. By extracting valuable phytochemicals and bioactive compounds, this approach not only supports agricultural and electronic sectors but also explores additional applications, including biofuel production. Integrating these agricultural and electronic solutions can address Ecuador's waste management challenges while promoting sustainability and circular economy practices.

Methodology

African palm rachis, mango endocarps, mangosteen fruit rinds, moringa husk and blackberry bushes were cleaned, dried, and ground for further processing. Proximate composition analysis involved determining moisture, ash, total fat, proteins, and carbohydrates in the biomasses through specific methods like Soxhlet extraction and Kjeldahl method. Thermogravimetric analysis (TGA) was conducted on raw materials to measure weight variations during heating, providing insights into composition and thermal stability. In the Hydrothermal Carbonization (HTC) process, biomass samples were processed in a high-pressure reactor with varying temperatures and times, followed by vacuum filtration to separate hydrochar from liquors. Chemical analyses, including Fourier Transform Infrared (FTIR) spectroscopy, were performed to identify functional groups and structural changes in the materials. X-Ray Diffraction analyses were performed to identify the relevant crystalline compounds present in the samples. To further complement these results, samples were pressed into pellets and subjected to frequency sweeps using a Semiconductor Device Parameter Analyzer to measure voltage retention capacity at different frequencies.

A statistical multicategorical method was selected for optimization through the p-value criterion, in which a threshold of 0.05 was used. The optimization method consisted on minimizing both conductivity and loss tangent where the "desirability" parameter is obtained from this process, where values close to 1 indicate a better optimization.

Results and Discussion

Cumbayá blackberry bushes (CBH) electrical analysis showed that generally all the samples treated with HTC (despite the time) optimized the loss tangent, keeping this parameter stable and below the raw material in a wide range of frequencies (Fig. 1.). Galapagos blackberry bushes (GBH) only show this trend in high temperature hydrochars and short times (about an hour). This suggests that at some point during the carbonization, the chemical structure of GBH HTC could be severely changed resulting in compounds that maximize the thermal diffusion of electric energy, while in CBH HTC the structure keeps the resistive effect without significant thermal loses. Also, this behavior leads to suggest that both specimens have different entities in their native matrix, possibly due to the difference in soil nutrients, temperatures and altitudes where each species grows. As expected, conductivity trends (Fig. 1.) showed that native materials were more conductive than their HTC samples, this mainly due to the presence of inner water (absorbed during the metabolic cycle of the plant) between the crystals of their matrixes. Both species HTC samples show similarities of conductivity in the range of frequencies, only native materials presented a significant change which supports the chemical difference between both just like in other studies is mentioned (Nizamuddin et al., 2016).

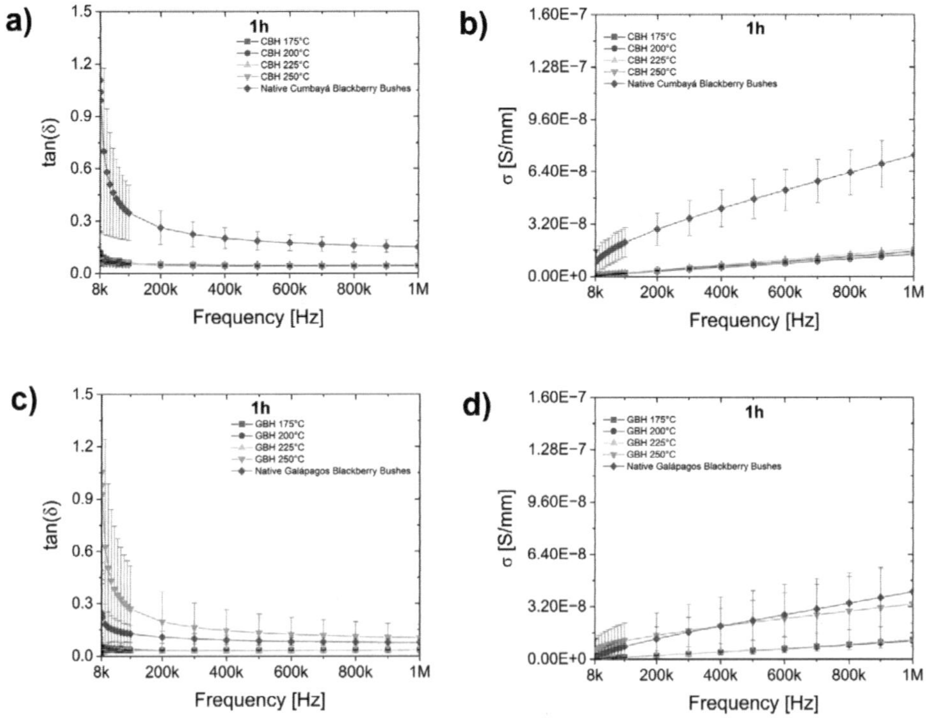

Fig. 1. Loss tangent of materials for (a) native Galápagos blackberry bushes and their hydrochars (GBH); and (b) their conductivity (c) native Cumbayá blackberry Bushes and their hydrochars (CBH); and (d) their conductivity, treated for 1 hour as a function of frequency (range 8 kHz to 1 MHz).

In addition, it was found that in Galapagos Islands samples, the thermal treatment reduced significantly the concentration of some metallic ions (specially the most common oligoelements) but incremented their percentage in the total oligoelements composition (Fig. 2). This is probably due to some substitution reactions that trap the ions onto the liquid matrix (remaining liquor), and in consequence leading to the formation of a "more organic" structure (Table 2). Also, these differences may be due to specific soil characteristics and further exploration is needed.

Looking into some physical-chemical characterizations, both blackberry samples showed that despite the hydrothermal treatments, the peaks of several functional groups keep being stable but with a minor intensity while the temperatures increase (Fig. 3.). This suggests that generally the hydrolysis performed in the reactor is capable to reduce considerably the original organic matrix and transform it into a functionalized one (Sagar et al., 2023). It is observed the presence of vibrations of alkanes (2900-2850 cm^{-1}), some carbonylic bindings, probably corresponding to carboxylic acids and ketones (1750-1700 cm^{-1}), and finally some alcoholic-ether bounds near 1700 cm^{-1}.

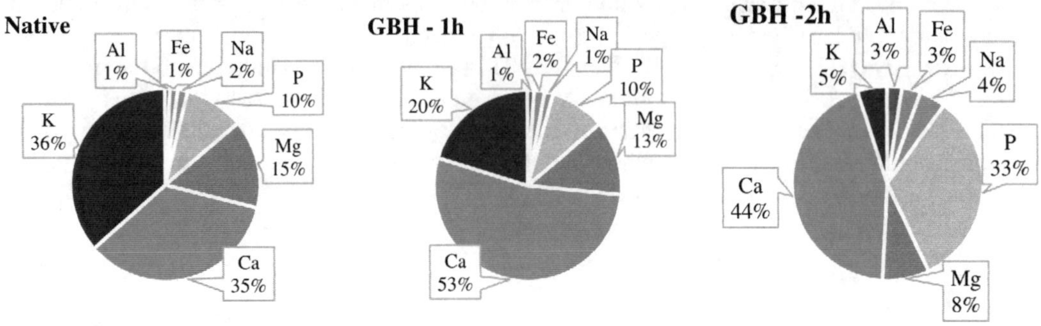

Fig 2. Percentage in concentration in ppm of different types of oligoelements found in the samples of native Galápagos blackberry and its hydrochars.

Table 2. Concentrations of common oligoelements in the samples.

Element	**Native**	**GBH 1h-250°C**	**GBH 2h-250°C**
Al	240.176 ± 7.460	132.968 ± 2.687	332.552 ± 8.098
Fe	357.565 ±10.310	198.761 ± 2.317	414.667 ± 10.439
Na	452.592 ± 8.006	164.945 ± 29.038	579.303 ± 10.668
P	2705.099 ± 24.181	1067.162 ± 10.172	4312.468 ± 18.061
Mg	3974.360 ±177.100	1421.318 ± 16.004	1033.473 ± 56.664
Ca	9326.974 ± 160.813	5984.410 ± 61.817	5798.771 ± 213.633
K	9740.476 ± 33.895	2286.543 ± 21.918	653.653 ± 37.882

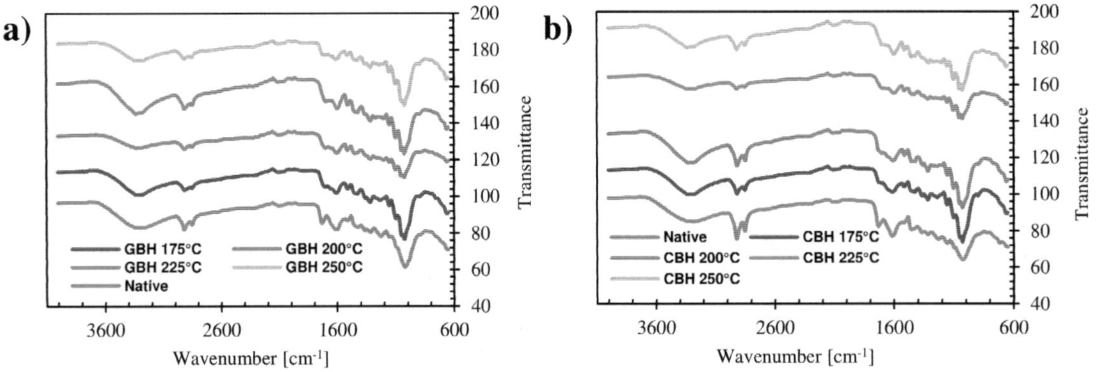

Fig. 3. FTIR spectra for a) Galápagos blackberry, and b) Cumbayá blackberry and their hydrochars treated for 1 hour, at different temperatures.

Searching for new methods to remove contaminants in water is a modern problem, so optimizing these new options for water treatment is a priority. Some parameters to consider in hydrothermal carbonization (HTC) are content of cellulose, conductivity, and particle size (Abdelhamid & Mathew, 2021). The cellulose present in these materials has the capacity to adsorb and eliminate contaminants such as dyes, pharmaceutical products, and heavy metals in water treatment. By using hydrothermal carbonization, the available surface area increases, but we must consider the temperature at which the cellulose degrades in each material, as the performance of the hydrochar decreases. If the size of the hydrochar is smaller, the better the productivity of the removal will be (Sevilla & Fuertes, 2009). The ability of a material to remove organic contaminants in water and the conductivity have an inverse relationship. Because the concentration of organic carbon in the samples increases, the conductivity decreases.

Referring to Mango Endocarp, has been found that its chemical structure has an important percentage of crystalline cellulose (reflected in its crystallinity index approx. 35%) (Fig. 4.). This contributes to form a highly porous surface (Fig. 5.), ideal for removing molecules via physisorption (Diakité et al., 2013). Even the raw material keeps this proportion opening for several cleaning and removal applications (Reza et al., 2015).

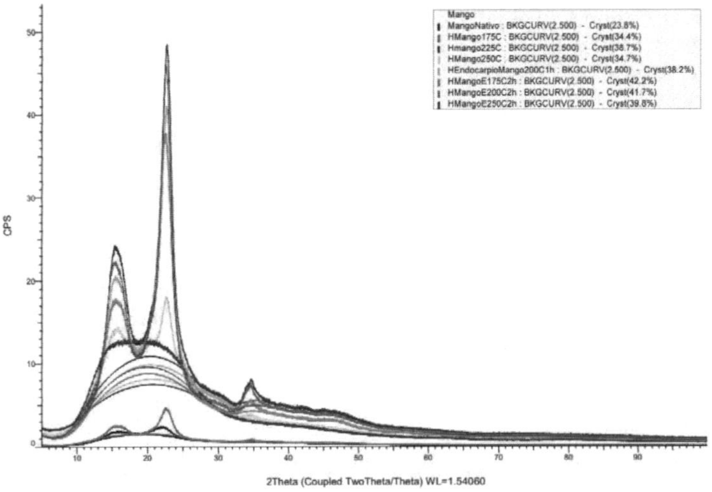

Fig. 4. X-Ray diffractogram analyses for several samples of Mango endocarp and its hydrochars showing their crystallinity indexes.

a) **b)**

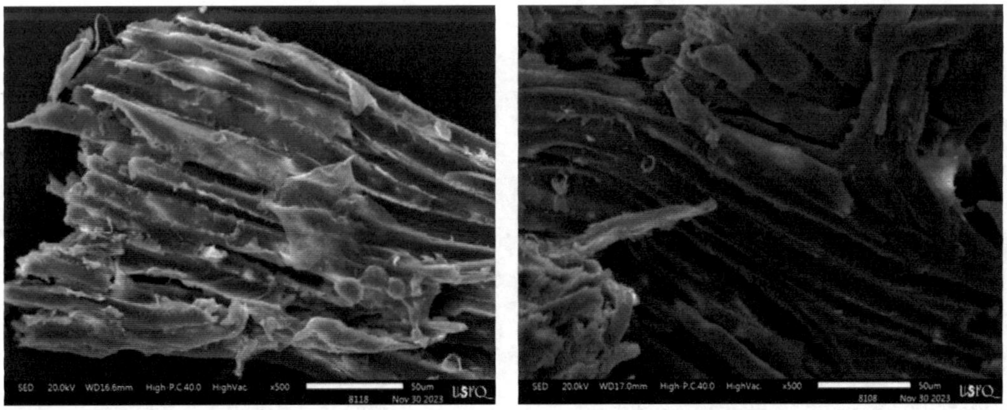

Fig. 5. SEM analysis for two samples of native mango endocarp with different particle sizes a) 125-250 and b) 355-500 μm.

The desirability trends observed in the frequency range for both Mango Endocarps Hydrochars (MEH) and Mangosteen Rinds Hydrochars (MRH) exhibit similarities (Fig. 6.), which could indicate a correlation with their compositions. Soil electrical conductivity (EC) is crucial in agriculture, indicating soil's ability to conduct electricity, affected by factors like moisture, minerals, and temperature. Higher

EC signifies more salt in soil, influencing nutrient availability and salinity. Examination of results show that both native mango and mangosteen materials share similar quantities of ash and proteins, and they are in the lowest range of the four materials analyzed. This suggests a plausible association between the properties of the hydrochar and these identified variables (Ding et al., 2013), providing a basis for further exploration and analysis.

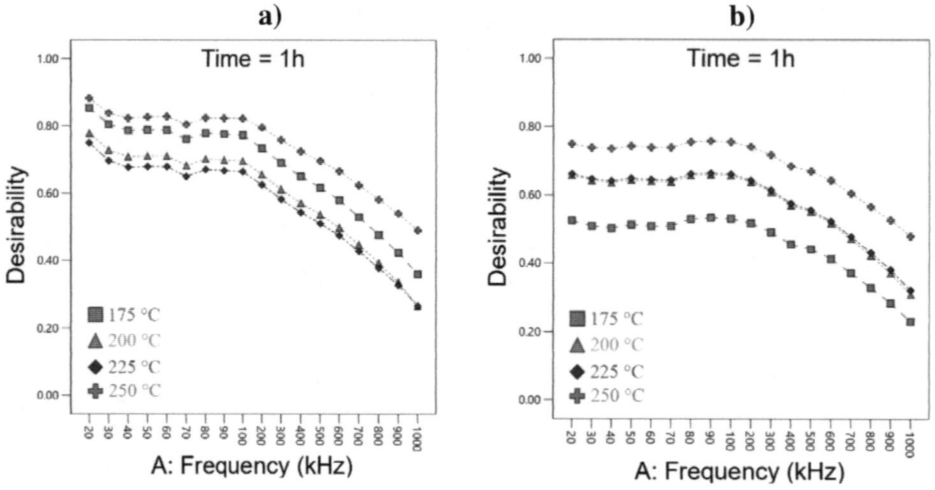

Fig. 6. Desirability trends for a) Mango Endocarps Hydrochars (MEH) and b) Mangosteen Rinds Hydrochars (MRH) treated for one hour.

Microscopic analysis of African palm rachis samples revealed a highly porous surface structure, particularly prominent in the native specimen (Fig. 7). This morphological characteristic suggests a potentially high capacity for molecular and nanoparticle adsorption. Subsequent experimental investigations, utilizing emerging contaminants as model adsorbates, have demonstrated the efficacy of African palm rachis as an adsorptive medium. The material's ability to capture these molecules through physisorption mechanisms is comparable to that observed in the previously discussed mango endocarp samples. These findings indicate that African palm rachis could serve as an effective substrate for the sequestration of various molecular species from aqueous environments.

Fig. 7. Sample and SEM analysis for African palm rachis with a particle size between 355-500 μm.

TGA analyses (Fig. 8) were performed on several exotic fruit samples to explore their physical characteristics. It was found that Moringa maintained relatively stable behavior in a temperature range between 100-200°C, while Palm and Mango samples exhibited fluctuations and a rapid transition to a physical change (possibly due to oxidation or evaporation of structural compounds). Beyond 400°C, all samples maintained a certain degree of stability. These results suggest that Moringa husks could potentially preserve their porosity and physical structure for longer periods under high-temperature conditions, making them ideal for applications that require elevated temperature inputs.

Furthermore, the analysis revealed that mangosteen rinds exhibited the least mass loss among the samples tested. This observation suggests that mangosteen rind particles are less susceptible to changes in moderate temperature ranges (Oumabady et al., 2024), and consequently, their structure tends to remain relatively stable under these conditions. Such thermal stability could be advantageous for applications requiring materials with consistent or predictable thermal behavior.

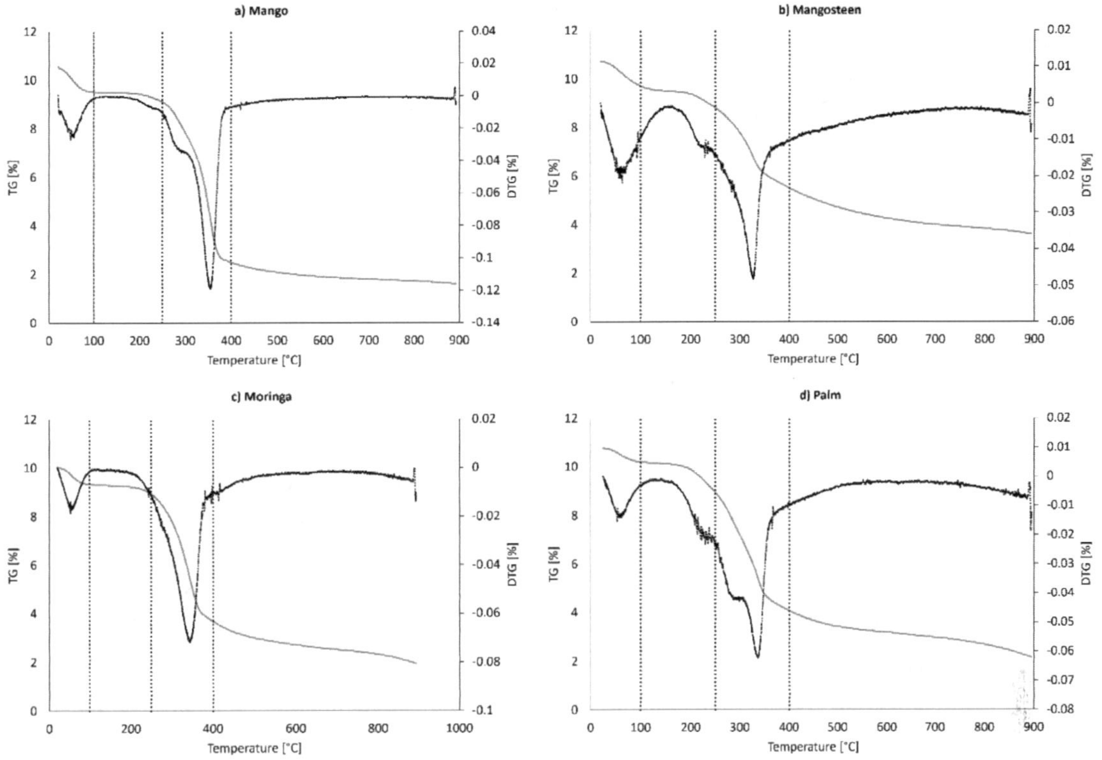

Fig. 8. TGA analyses for a) Mango, b) Mangosteen, c) Moringa, and d) Palm, native materials.

Conclusions

This study demonstrates the potential of hydrochars derived from various exotic fruit residues as eco-friendly materials for electronic and agricultural applications. Hydrochars from mango endocarps and mangosteen rinds exhibited high dielectric constants and low loss tangents, making them suitable for insulation applications. Blackberry bushes seem to be good candidates for dielectric applications, thanks to their stability across a wide range of frequencies, low conductivity values, and low loss tangent. Additionally, Galapagos blackberry can be further investigated for agricultural applications such as fertilizers, due to its specific content of important oligoelements such as potassium and calcium.

The study highlights how fruit type, hydrothermal carbonization conditions, and post-treatment processes affect the electrical and physical characteristics of hydrochars. Generally, the hydrothermal carbonization (HTC) treatment tends to create porous surfaces and a more stable organic structure. FTIR analyses of blackberry bushes revealed that hydrothermal treatment alters the chemical structure of these materials, potentially enhancing their functional properties and resulting in new matrices capable of several applications.

XRD analyses showed that the porous structure of some hydrochars, particularly those from mango endocarps, indicates potential use in water treatment for removing contaminants. This is mainly due to the high percentage of crystalline cellulose in their composition, which is supported by SEM images. The same applies to African palm rachis. TGA analyses showed varying thermal behaviors among the different fruit residues, indicating potential applications in temperature-sensitive environments (for construction materials and thermal insulators) in samples such as mangosteen and moringa.

These findings open new avenues for sustainable waste management and circular bioeconomy practices. By repurposing agricultural waste into valuable materials for electronics and agriculture, this research contributes to addressing environmental challenges while promoting resource efficiency. Future research directions could include exploring additional applications in areas such as energy storage or environmental remediation, and conducting long-term studies on the performance and environmental impact of these materials in real-world settings.

In conclusion, this study demonstrates the potential of fruit-derived hydrochars as sustainable alternatives in various applications, contributing to waste reduction and the development of eco-friendly technologies in the researchers' country and worldwide.

References

[1] Abdelhamid, H. N., & Mathew, A. P. (2021). Cellulose-Based Materials for Water Remediation: Adsorption, Catalysis, and Antifouling. *Frontiers in Chemical Engineering*, *3*. https://doi.org/10.3389/fceng.2021.790314

[2] Chakraborty, M. , et al. (2021). IoT-based composting system: A sustainable approach for organic waste management. *Sustainability*, *13*(5), 2867.

[3] Diakité, M., Paul, A., Jäger, C., Pielert, J., & Mumme, J. (2013). Chemical and morphological changes in hydrochars derived from microcrystalline cellulose and investigated by chromatographic, spectroscopic and adsorption techniques. *Bioresource Technology*, *150*, 98–105. https://doi.org/10.1016/j.biortech.2013.09.129

[4] Ding, L., Zou, B., Li, Y., Liu, H., Wang, Z., Zhao, C., Su, Y., & Guo, Y. (2013). The production of hydrochar-based hierarchical porous carbons for use as electrochemical supercapacitor electrode materials. *Colloids and Surfaces A: Physicochemical and Engineering Aspects*, *423*, 104–111. https://doi.org/10.1016/j.colsurfa.2013.02.003

[5] Nizamuddin, S., Mubarak, N. M., Tiripathi, M., Jayakumar, N. S., Sahu, J. N., & Ganesan, P. (2016). Chemical, dielectric and structural characterization of optimized hydrochar produced from hydrothermal carbonization of palm shell. *Fuel*, *163*, 88–97. https://doi.org/10.1016/j.fuel.2015.08.057

[6] Oumabady, S., Bhardwaj, S. K., Ramasamy, S. P., Dandare, S. U., Sakrabani, R., Doherty, R., Nanukuttan, S., & Kumaresan, D. (2024). Thermochemical characterisation of hydrochar from agricultural waste and its efficiency as a supplement with solid fuel. *Biomass and Bioenergy*, *187*, 107299. https://doi.org/10.1016/j.biombioe.2024.107299

[7] Pauline, A. L., & Joseph, K. (2020). Hydrothermal carbonization of organic wastes to carbonaceous solid fuel – A review of mechanisms and process parameters. *Fuel*, *279*(April), 118472. https://doi.org/10.1016/j.fuel.2020.118472

[8] Popa, V. I. (2018). Biomass for Fuels and Biomaterials. *Biomass as Renewable Raw Material to Obtain Bioproducts of High-Tech Value*, 1–37. https://doi.org/10.1016/B978-0-444-63774-1.00001-6

[9] Reza, M. T., Rottler, E., Tölle, R., Werner, M., Ramm, P., & Mumme, J. (2015). Production, characterization, and biogas application of magnetic hydrochar from cellulose. *Bioresource Technology*, *186*, 34–43. https://doi.org/10.1016/j.biortech.2015.03.044

[10] Sagar, V., Hardin, M., Kumar, N., & Lynam, J. G. (2023). Characterization and Energy Densification of Mayhaw Jelly Production Wastes Using Hydrothermal Carbonization. *Food Technology and Biotechnology*, *61*(1), 118–126. https://doi.org/10.17113/ftb.61.01.23.7783

[11] Sevilla, M., & Fuertes, A. B. (2009). The production of carbon materials by hydrothermal carbonization of cellulose. *Carbon*, *47*(9), 2281–2289. https://doi.org/10.1016/j.carbon.2009.04.026

[12] Sosunova, I., & Porras, J. (2022). IoT-Enabled Smart Waste Management Systems for Smart Cities: A Systematic Review. *IEEE Access*, *10*, 73326–73363. https://doi.org/10.1109/ACCESS.2022.3188308

[13] Ulloa, J. (2006). LOS RELLENOS SANITARIOS. *Revista de Ciencias de La Vida*, *4*, 2–17.

[14] Wang, L., Chi, Y., Du, K., Zhou, Z., Wang, F., & Huang, Q. (2022). Hydrothermal treatment of food waste for bio-fertilizer production: Formation and regulation of humus substances in hydrochar. *Science of the Total Environment*, *838*(1).

[15] Zeng, X., Mathews, J., & Li, J. (2018). Urban mining of e-waste is becoming more cost-effective than virgin mining. *Environmental Science & Technology*, *55*(8), 5313–5322.

434d Enhancing Dispersion Stability of Alpha-1,3-Glucan Using Adipic Acid Derivatization: A Novel Approach for Industrial Applications

Nigel Twi-Yeboah, Alan Allgeier, Khanh Le Vinh, Christian Lenges, Jure Zlopasa, Natnael Behabtu

Polysaccharides play a crucial role in many industrial applications and are of renewed interest as sustainable material options. However, polysaccharides typically face challenges related to dispersion stability and processing characteristics. Derivatized polysaccharides, with charged surfaces, often have improved dispersion stability. These charged polysaccharides may include carboxymethylcellulose, (CMC) and carboxymethyl glucan (CMG). This work investigates the utilization of polysaccharide esters made from alpha-1,3-glucan and adipic acid, a dicarboxylic acid, to impart charge on the resulting polysaccharide surface. The method is designed to selectively derivatize surface sites of insoluble particles. Unlike CMC and CMG which are polysaccharide-ethers, these anionic glucan esters, exhibit improved thermal stability (decomposition temperature = ~240 °C versus CMG = ~210 °C) and maintain commendable dispersion stability (zeta potential as low as -13.00 mV). Solid-state NMR spectroscopy via the magic-angle-spinning, direct excitation method is promoted for quantitative analysis of degree of substitution. The study demonstrates colloidal particle dispersion with attributes amenable to practical applications of these new alpha-1,3-glucan esters in areas such as home and personal care applications.

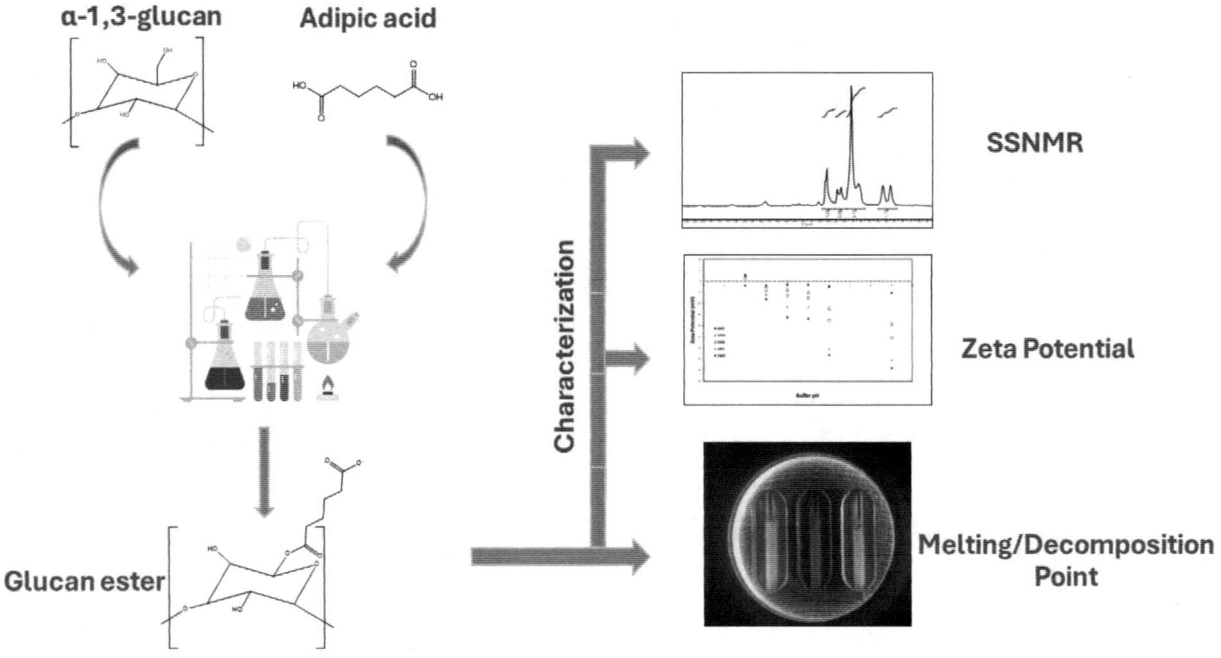

434e Acid-Catalysed Reflux-Free Silanisation at Room Temperature: Technical and Environmental Benefits

Deniz Etit, Tsaone Gosiamemang, Ethan Errington, Miao Guo, Jerry Heng

Silica nanoparticles are versatile materials, since their surface chemistry can be functionalised for specific applications by tuning the type and grafting density of surface modifying reagents [1]. A common technique to functionalise silica nanoparticles is silanisation, conventionally conducted by refluxing silica nanoparticles in organic solvents. However, reflux-based silanisation features disadvantages including prolonged reaction times, usage of organic solvents and energy need for refluxing, and excessive consumption of surface modifying reagents [2]. Hydrophobic and superhydrophobic silica nanoparticles have particular applications in catalysis, chromatography, and rubber industry [3], [4]. These surfaces are defined by their water contact angle (WCA), with the respective ranges of 90-150° for hydrophobicity and >150° for superhydrophobicity. Hydrophobicity is usually achieved by utilising alkyl-organosilanes as surface modifying reagents [5]. However, silane reagents with long alkyl side changes are associated with technical challenges, including finding suitable solvents and reaction inefficiencies driven by steric hindrances [1].

In attainment of hydrophobically-functionalised surface chemistry, this work presents the use of liquid acids to catalyse silanisation of silica nanoparticles at room temperature in toluene and cyclohexane, demonstrating reduced reaction times, energy usage, and silane consumption compared to the conventional refluxing method. By increasing acidic strength, acetic acid (AA) and formic acid (FA) were tested as weak acids, and dichloroacetic acid (DCA), trifluoroacetic acid (TFA), and methanesulfonic acid (MSA) were tested as strong acids to catalyse the reaction. Propyltrimethoxysilane (PTMS), octyltrimethoxysilane (OTMS), and octadecyltrimethoxysilane (ODTMS) were utilised as surface modifying reagents to attain hydrophobicity. Performance of the developed method was compared with the conventional refluxing-based method. Complementing the outcomes, environmental advantages of the developed method were analysed via process modelling and life cycle assessment (LCA).

Regarding the experimental methodology, initially, silica nanoparticles were obtained via preparing 200 ml mixtures comprised of 90% ethanol, 5% deionised water, and 5% tetraethyl orthosilicate. Then, 8ml ammonia is slowly added into the mixture, and the mixture was given 3h for reaction. After the reaction, produced silica nanoparticles were recovered via centrifugation, washed with deionised water, and dried under vacuum to finalise the process. Regarding the subsequent silanisation step, 0.5g of obtained silica nanoparticles were mixed with 50ml organic solvent (i.e., toluene or cyclohexane) and 0.9 mmol modifying reagent (i.e., PTMS, OTMS or ODTMS). Then, 0.8 mmol of acid catalyst (i.e., AA, FA, DCA, TFA or MSA) was added to the mixture, and the mixture was given 1h for reaction. After the reaction, produced functionalised silica nanoparticles were recovered via centrifugation, washed with v/10v ammonia/acetone solution, and dried under vacuum to finalise the process. To compare with the novel production method, silanisation was also conducted via conventional reflux-based method, involving 20h refluxing reaction of 0.5g produced silica nanoparticles with 1.9 mmol OTMS in 50ml solvent without the acid addition, followed by the same purification steps. To evaluate the performance of produced nanoparticles, thermogravimetric analysis (TGA) was used to quantify grafting densities (chains/nm^2), and WCA was measured to compare hydrophobicities.

Environmental impacts of the developed processes were quantified via LCA. Inventory for the LCA was estimated based on the amounts of silica material, respective acid catalyst, solvent usage, silanisation reagent, and energy use utilised in lab-scale experiments. Inventory data from Ecoinvent 3.6 [6] were referred to model background processes, and ReCiPE 2016 hierarchical midpoint method was used for environmental impact characterisation [7]. System boundaries were set from cradle to end of the silanisation reaction stage. LCA was conducted for the cases with OTMS as reagent; AA, FA, TFA and MSA as acid catalyst (since impact data for DCA was not found); and both toluene and cyclohexane as organic solvent. Environmental impacts of OTMS were estimated via designing an industrial process in Aspen software. Results regarding the climate change (CC) impact (kgCO$_2$eq/kg) were provided per reference flow of 1kg functionalised silica nanoparticles.

Silanisation performance with OTMS, by the used acid catalyst and organic solvent, is provided in Figure 1a. Stronger acids (DCA, TFA, MSA) yielded significantly higher grafting densities with >1.5 chains/nm^2, compared to the weaker acids (AA, FA) with <0.5 chains/nm^2. The enhanced silanisation with stronger acids was attributed to their increased release of [H^+] cations, translating into more effective silane hydrolysis. Contrasting the acidic strength, FA resulted in a lower grafting density than AA, and MSA resulted in a lower grafting density than DCA and TFA. That is reasoned by the higher affinity of FA and MSA with the silica surface, potentially competing with the reagents during the reaction. The WCA results reflected the grafting density outcomes, with higher hydrophobicities attained with the cases of highest grafting density. Cyclohexane and toluene resulted in a similar grafting density and WCA across all the selected acids, demonstrating that either solvent can be used to achieve the corresponding performance under the tested experimental conditions. Solvent choice resulted in a more sizable impact for the conventional refluxing method (Figure 1a), by respective grafting densities of 0.9 chains/nm^2 for cyclohexane and 1.3 chains/nm^2 for toluene. This difference is reasoned by reaction temperature, that is determined by the solvent's boiling point (BP) in the refluxing case. As the BP of toluene (111°C) is greater than cyclohexane (81°C), thermal hydrolysis of silanes is higher for the former, translating into higher silanisation performance. Performances of surface modifying reagents were compared for the case of TFA catalyst in toluene (Figure 1b). Attributed to increasing steric hinderance with the longer carbon chain length from PTMS (3) to OTMS (8) to ODTMS (18), the reagents resulted in respective grafting densities of 2.9, 2.6, and 2.1 chains/nm^2. Since longer carbon chain length increases hydrophobicity performance, the reagents ultimately demonstrated similar WCA results, being 145°, 159°, and 156°, respectively.

Results of the LCA are illustrated in Figure 2. The novel acid-catalysed route significantly reduced the overall CC impacts compared to the conventional refluxing process. Contributions of the acid catalyst, silane reagent, and process energy were minor for the novel route. Thus, the CC impacts were not sizably impacted by the catalyst choice. Toluene performed better than cyclohexane as organic solvent, reducing the CC impact for the novel route. With dominant process energy contributions, overall CC impacts were significantly higher for the conventional refluxing process for both solvents. Cyclohexane performed better than toluene for the conventional process, by requiring less process energy thanks to its lower boiling point. Overall, the novel process demonstrated considerable reductions for the CC impacts compared to the conventional process, across all catalyst types and both solvents. As a future direction, application-specific life cycle impacts (e.g., per adsorption capacity functional units) would describe the environmental benefits of individual catalysts and reagents for the novel process.

Briefly, a novel acid-catalysed silanisation process is presented in this work, that is operable at ambient temperature. The technique was validated by obtaining comparable grafting densities and WCA outcomes with nanoparticles functionalised through the conventional refluxing process. High acidic strength and low affinity with the original silica surface were detected to improve the catalytic performance. Longer alkyl chains of modifying reagents resulted in lower grafting densities, reasoned by steric hinderances. Toluene and cyclohexane demonstrated similar technical performance as organic solvents; however, toluene performed environmentally better, resulting in lower CC impact. Acid catalyst, silane reagent, and process energy contributed minorly to the overall CC impact for the novel process. The conventional process resulted in significantly higher CC impacts for both solvents, driven by high process energies. Altogether, the developed method is promising to mitigate the technical challenges and environmental impacts of the conventional process, by minimising reaction times, operating at ambient temperatures, removing refluxing energies, and reducing modifying reagents consumption.

References

[1] T. Gosiamemang and J. Y. Y. Heng, *Powder Technol*, vol. 417, p. 118237, Mar. 2023, doi: 10.1016/J.POWTEC.2023.118237.

[2] L. C. Mugica, B. Rodríguez-Molina, S. Ramos, and A. Kozina, *Colloids Surf A Physicochem Eng Asp*, vol. 500, pp. 79–87, Jul. 2016, doi: 10.1016/J.COLSURFA.2016.04.002.

[3] S. A. Kulkarni, S. B. Ogale, and K. P. Vijayamohanan, *J Colloid Interface Sci*, vol. 318, no. 2, pp. 372–379, Feb. 2008, doi: 10.1016/J.JCIS.2007.11.012.

[4] E. Moretto *et al.*, *ACS Omega*, vol. 7, no. 21, pp. 17692–17702, May 2022, doi: 10.1021/ACSOMEGA.2C00665/ASSET/IMAGES/LARGE/AO2C00665_0011.JPEG.

[5] N. García, E. Benito, J. Guzmán, and P. Tiemblo, *J Am Chem Soc*, vol. 129, no. 16, pp. 5052–5060, Apr. 2007, doi: 10.1021/JA067987A/ASSET/IMAGES/LARGE/JA067987AC00003.JPEG.

[6] G. Wernet, C. Bauer, B. Steubing, J. Reinhard, E. Moreno-Ruiz, and B. Weidema, *International Journal of Life Cycle Assessment*, vol. 21, no. 9, pp. 1218–1230, 2016, doi: 10.1007/s11367-016-1087-8.

[7] M. A. J. Huijbregts *et al.*, *International Journal of Life Cycle Assessment*, vol. 22, no. 2, pp. 138–147, 2017, doi: 10.1007/s11367-016-1246-y.

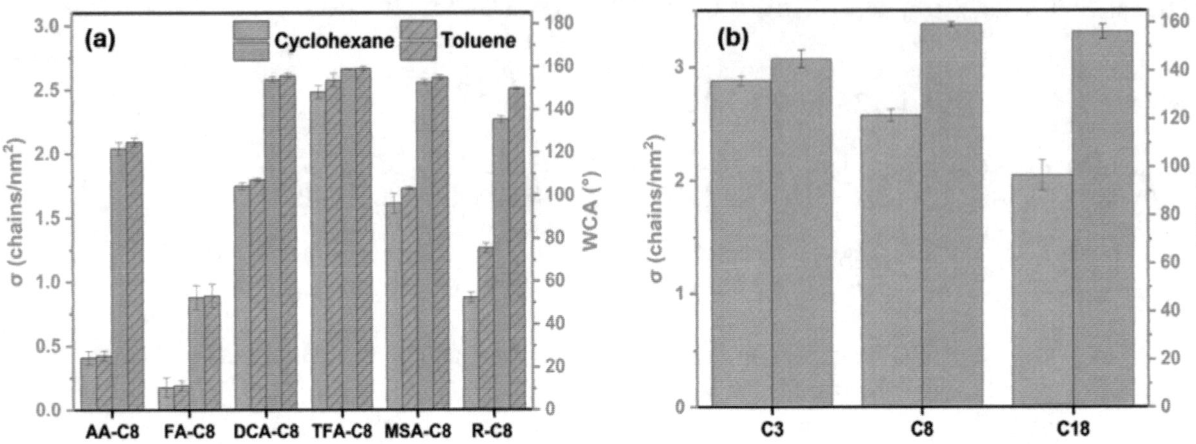

Figure 1. Grafting density (σ) and WCA by a) used acid catalyst and organic solvent, and b) used surface modification reagent with TFA as acid catalyst and toluene as organic solvent. Acetic acid, formic acid, dichloroacetic acid, trifluoroacetic acid and methanesulfonic acid are respectively abbreviated by AA, FA, DCA, TFA and MSA. The refluxing process is abbreviated by R. Surface modification reagents propyltrimethoxysilane, octyltrimethoxysilane, and octadecyltrimethoxysilane are respectively abbreviated by C3, C8 and C18.

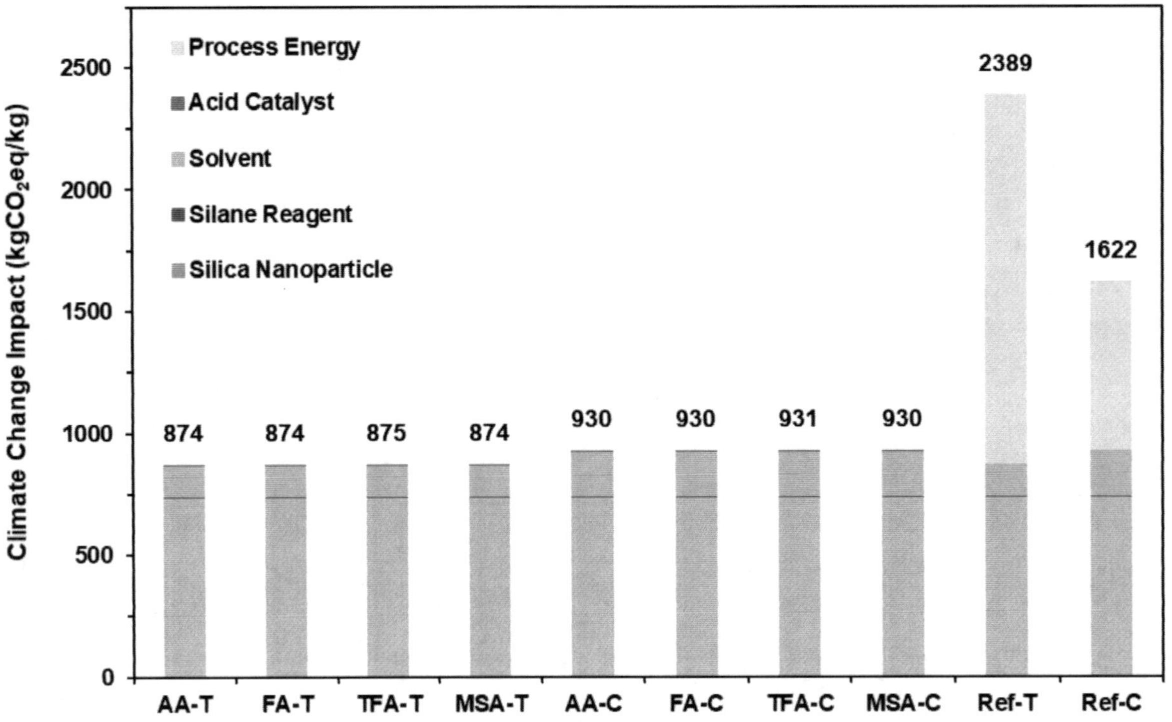

Figure 2. Estimated climate change impact ($kgCO_2eq/kg$) of functionalised silica nanoparticles by process route. Acetic acid, formic acid, trifluoroacetic acid and methanesulfonic acid are respectively abbreviated by AA, FA, TFA and MSA. The refluxing route is abbreviated by Ref. Toluene and cyclohexane solvents are respectively abbreviated by T and C.

434f Prospective Nipu and Biopu Models: A Push for Polyurethanes Sourced from Biomass

Jennifer B. Dunn, Ulises R. Gracida-Alvarez, Troy Hawkins, Yasheen Jadidi, Chloe Liang

Background: The use of fossil fuel and natural gas reserves depletes nonrenewable resources while releasing emissions that directly contribute to global warming. In 2016, 2.9 million tonnes of polyurethane (PU) were produced in the United States at the expense of 1.1 million tonnes of crude oil and 1.1 million tonnes of natural gas.[1] With a 5.5 % end-of-life recycling rate, PU is recycled at less than the average recycling rate of U.S. plastics in 2015.[1,2] To enhance PU sustainability, it is necessary to advance technologies for producing this material from bio-based feedstocks and to expand methods for recycling PU monomers. Another challenge with today's technology for manufacturing PU is the use of isocyanates, which pose a human health hazard. To alleviate issues with safety, the U.S. Environmental Protection Agency (EPA) has begun to phase out the use of isocyanates in the synthesis of PU.[3] Non-isocyanate polyurethane (NIPU) materials can be synthesized from biomass[4,5] and reprocessed into secondary (2°) NIPU materials.[6] NIPU materials eliminate the need for toxic and potentially harmful diisocyanates.[7]

Methods: Previously, our group performed a cradle-to-gate material flow analysis (MFA) of PUs in the United States[1] and a techno-economic analysis (TEA) on two NIPU materials.[8] The studies utilized various data sources including governmental resources, industry associations, and peer-reviewed literature. The MFA study involved estimating precursor quantities based on industrial surveys and market reports, calculating energy consumption, tracking waste generation and management, and incorporating historical production data for in-use phase analysis. These data and data from a baseline MFA on NIPU production push the dynamic systems modeling for introducing NIPUs into the market of PU materials. This model is primarily informed by key economic factors affecting the PU market over a temporal horizon of 25 years (2025 – 2050). These economic factors include feedstock costs, GDP growth, cost of PU, and cost of NIPU.[9]

Results: Our research efforts are focused on analyzing prospective scenarios for the entry of NIPUs into the market for PU materials. Initially, two baseline scenario is reported: (1) for the replacement of PU monomers with bio-based polyols and diisocyanates and (2) for the substitution of all PU materials manufactured within the United States with NIPU materials. We report optimistic and pessimistic potentials for augmenting biogenic NIPUs into the PU market through the year 2050. Furthermore, our prospective MFA provides valuable insights into the changing efficacy of recycling NIPU materials over time. Overall, this research contributes to the ongoing efforts to promote environmentally conscious practices and investigate isocyanate-free methods in the plastics industry.

[1] Liang, C., Gracida-Alvarez, U.R., Gallant, E.T., Gillis, P.A., Marques, Y.A., Abramo, G.P., Hawkins, T.R. and Dunn, J.B. 'Material flows of polyurethane in the United States.' *Environmental Science & Technology* 55.20 (2021): 14215-14224.

[2] Di, J., Reck, B.K., Miatto, A. and Graedel, T.E. 'United States plastics: Large flows, short lifetimes, and negligible recycling.' *Resources, Conservation and Recycling* 167 (2021): 105440.

[3] US Environmental Protection Agency. 'Methylene Diphenyl Diisocyanate (MDI) And Related Compounds Action Plan.' (2011).

[4] Bobbink, F. D.; van Muyden, A. P.; Dyson, P. J. En route to CO 2-containing renewable materials: catalytic synthesis of polycarbonates and non-isocyanate polyhydroxyurethanes derived from cyclic carbonates. *Chemical Communications* **2019**, 55(10), 1360-1373.

[5] Wu, S.; Luo, M.; Darensbourg, D. J.; Zeng, D.; Yao, Y.; Zuo, X.; Hu, X.; Tan, D. Non-isocyanate and catalyst-free synthesis of a recyclable polythiourethane with cyclic structure. *ACS Sustainable Chemistry & Engineering* **2020**, 8(14), 5693-5703.

[6] Zhao, L; Semetey, V. Recycling polyurethanes through transcarbamoylation. *ACS omega* **2021**, 6(6), 4175-4183.

[7] Khatoon, H.; Iqbal, S.; Irfan, M.; Darda, A.; Rawat, N. K. A review on the production, properties and applications of non-isocyanate polyurethane: A greener perspective. *Progress in Organic Coatings* **2021**, 154, 106124.

[8] Liang, C.; Jadidi, Y.; Chen, Y.; Gracida-Alvarez, U. R.; Torkelson, J. M.; Hawkins, T. R.; Dunn, J. B. Techno-economic Analysis and Life Cycle Assessment of Biomass-Derived Polyhydroxyurethane and Non-isocyanate Polythiourethane Production and Reprocessing. *Manuscript in preparation* **2024**.

[9] Döhler, N.; Wellenreuther, C.; Wolf, A. Market dynamics of biodegradable bio-based plastics: Projections and linkages to European policies. *EFB Bioeconomy Journal* **2022**, 2, 100028.

501b Comparative Study of Sustainable Aviation Fuel Produced from Wastewater Grown Algae Vs. Farmed Algae

Abhishek Kumar, Peter Valdez

Algae are a promising biomass source for low-carbon fuels owning to their higher growth rate, high CO_2 uptake high lipid content, comparatively higher area-specific yield, and negligible competitiveness with farmland and food supplies. Biocrude can be produced from algae using various techniques among which hydrothermal liquefaction (HTL) has shown the highest carbon efficiency leading to comparatively lower costs of biocrude generation. However, the yield of biocrude and the overall cost of biofuel is highly dependent on the source and cost of the algae. Algae cultivation can be used to treat wastewater (WW), removing nutrients such as nitrogen and phosphorus. Seeing the efficacy of algae for wastewater treatment in reducing nutrients from water resource recovery facilities (WRRFs) as well as recent stringent government policies on the levels of N, P, and chemical oxygen demand (COD) levels in the treated water discharge has made the use of algae treatment more favorable.[1]

We present a comparative study to produce sustainable aviation fuel (SAF) from WW-grown algae and farm-cultivated algae. While the fuel yield from algae from WRRFs is lower than the farmed algae, co-products such as struvite fertilizers may generate extra revenue. WW-grown algae could also be provided as a zero-cost feedstock because HTL reduces waste solids from the WRRF, Moreover, the CO_2 captured from air by algae during their growth brings in additional opportunities for low-carbon products and carbon credits. Assuming zero algal feedstock cost, the minimum selling fuel price (MFSP) of biofuel (which includes both gasoline and jet cuts) from WW-grown algae came out to be \$2.61 gasoline gallon equivalent (GGE) relative to farm grown algae that ranged from \$4.48-8.05 GGE.[2] However, the previous analysis assumed that algae cultivation wholly replaced the traditional wastewater treatment process, which is not feasible. Additional scenarios were considered to leverage existing WWT assets, while boosting the availability of HTL feedstock via algae cultivation. In the first scenario, wet-sludge is mixed with WW-grown algae for HTL leading to higher biocrude yield as well as lower biosolids processing and disposal cost. While in the second scenario, the wet sludge is processed via anaerobic digestion (AD) followed by their subsequent treatment of AD centrate via algae cultivation. We investigated other treatment streams at the WRRF for their viability for algae cultivation and blending algal and wastewater solids as combined feedstocks to maximize biocrude production and solids reduction.

References

(1) Clippinger, J.; Davis, R. *Techno-Economic Assessment for Opportunities to Integrate Algae Farming with Wastewater Treatment*; National Renewable Energy Lab.(NREL), Golden, CO (United States), 2021.

(2) Zhu, Y.; Schmidt, A. J.; Valdez, P. J.; Snowden-Swan, L. J.; Edmundson, S. J. *Hydrothermal Liquefaction and Upgrading of Wastewater-Grown Microalgae: 2021 State of Technology*; Pacific Northwest National Lab.(PNNL), Richland, WA (United States), 2022.

501c Environmental-Techno-Economic Analyses of Biodiesel Production from Animal Fat Waste Using Hydrothermal Process

Pyeong-Gon Jung, Semie Kim, Young-Il Lim

Producing fuel from human-generated waste presents a compelling path towards carbon neutrality. Hydrothermal conversion stands out as one of the most thermally efficient methods for biofuel production, significantly reducing the energy required for drying raw materials. This study employs hydrothermal processing to produce biodiesel, encompassing hydrothermal conversion, separation, and a utility area that includes wastewater treatment, heat and cooling sources. The bio-oil derived from this process is segregated into solids, liquids, and gases, where the liquid fraction further is distilled into gasoline, diesel, and heavy oil via a distillation column. The gaseous by-products are utilized as a heating source to enhance plant efficiency.

This study investigates the economic and environmental feasibility of hydrothermal pathway for biodiesel production from animal fat waste. Based on the process simulation using ASPEN Plus, the economic values such as return on investment (ROI), payback period (PBP) and minimum fuel selling prices (MFSPs) were calculated from total capital investment (TCI) and total production cost (TPC) to determine economic feasibility. Additionally, life cycle assessment (LCA) by using GREET was performed to quantify greenhouse gas emissions at the country level and to compare the environmental impact of conventional diesel and biodiesel. The findings from this study are expected to offer valuable insight into strategic decisions regarding biodiesel policy.

501d Techno-Economic Analysis with Uncertainties for Alternative Fuels Enabling Near and Mid-Term U.S. Rail Sector Decarbonization

Yuan Jiang, Shuyun Li

The Federal Highway Administration (FRA) expects a 47% increase in U.S. freight shipments and challenges the rail industry to achieve net-zero greenhouse gas (GHG) emissions by 2050 and meet the U.S. long-term decarbonization target (Hwang, 2022). While working towards complete electrification, finding alternative low-carbon fuels is crucial for near- and mid-term rail sector decarbonization, as they can be used as direct drop in or only require minor retrofitting of existing infrastructure. Various alternative fuel pathways have been explored in the open literature, but with different technology readiness level (TRL), model fidelities and economic assumptions, and therefore the results are not directly comparable for identifying top options. In addition, for emerging technologies, existing techno-economic analyses were often not based on the most up-to-date performance or optimized for the rail sector application.

To provide basis for screening alternative fuel options for rail sector decarbonizing, in this study, techno-economic analyses were conducted for 16 alternative fuel pathways with harmonized economic assumptions. The selected pathways include 7 fuel types of interest to U.S. rail sector (biodiesel, renewable diesel, bio-oils, ethanol, methanol, dimethyl ether, and green ammonia) and 4 feedstock types (1st generation, organic waste, cellulosic biomass, and renewable hydrogen). Technologies are either fully commercialized or at least demonstrated in bench scale. For emerging technologies, process data were generated from Aspen Plus or ChemCAD models, while for commercial technologies, subscribed databases (i.e., PEP Yearbook) were used. Monte Carlo simulation was also conducted to capture the uncertainties from technology maturity, knowledge gaps, model fidelities, and economic assumptions. Results were reported in minimum fuel selling price with uncertainty range and detailed cost breakdown.

References

Hwang H, Lim H, Chin S, Uddin M, Biehl A, Xie F, Hargrove S, Liu Y, Wang C. 2022. 'Freight analysis framework version 5 (FAF5) base year 2017 data development technical report.' Oak Ridge National Laboratory, Oak Ridge, Tennessee. [Data available online, accessed 6/30/2022: https://faf.ornl.gov/faf5/]

501e Implications of Integrating Carbon Capture and Storage Technology into Sustainable Aviation Fuel (SAF) Production

Alberto Ruiz, Mariano Martin, Katie Chong, Röder Mirjam, Thornley Patricia

From the early 20th century until today, air traffic has shown a relentless growth in passengers and goods mobilised, only disrupted by the mobility limitations imposed during the COVID-19 pandemic. At present, air travel accounts for 2-3% of global carbon emissions (ca. 800 Mt CO_2 per year). If no action is taken, international aviation is forecasted to become the second-largest emitter by 2050.[2] The International Air Transport Association (IATA) committed to cap the CO_2 emissions generated by international aviation to 2020 values and halve that output by 2050 by means of sustainable aviation fuels (SAFs). However, their high cost can jeopardize the economic performance of airlines limiting their use.[3] To meet emissions mitigation targets and endorse technology development for SAF utilization, global market-based measures like CORSIA have been implemented. Under the CORSIA framework, specific lifecycle emissions values are assigned to a list of certified SAF production routes, allowing airlines to claim for emissions reduction bonuses. Only fossil carbon was evaluated to calculate those values, so potential emission reductions originated from carbon capture and storage technologies (CCS) within any biomass-to-SAF routes are excluded.[4]

The objective of this research was to investigate a CORSIA-eligible fuel process, that is suitable to incorporate carbon capture and storage technology, and compare the life cycle emissions score computed in this study with the one provided by CORSIA. The Fischer-Tropsch Hydroprocessed Isoparaffinic Kerosene (FT-IPK) pathway, an ASTM-certified route to produce jet fuel from waste wood, was the selected process to be evaluated. This study additionally contributed to resolve the lack of comprehensive assessment in determining the carbon balance linked to the production of aviation fuel from biomass.[5] A complete and detailed model was developed in a validated and robust process simulation software (Aspen Plus) for the FT-SPK route for SAF production from wooden residue, which included feedstock pretreatment (biomass drying and torrefaction), biomass-conversion (entrained flow gasification, syngas conditioning and Fischer Tropsch island), carbon dioxide removal (Selexol process) and CO_2 compression, refining and upgrading stages (fractionation and isomerization), SAF yield maximization processes (oligomerization and wax cracking) and auxiliary unit operations (air separation unit and combined cycle gas turbine). The work examined the underpinning engineering process/steps and integrated a comprehensive lifecycle assessment (LCA) implemented in SimaPro software, with a well-to-wake approach (see Figure 1), were both fossil and biogenic carbon flows were accounted. The LCA enabled the tracking of all impact contributions to the final score of SAF, so the recommendations and conclusions derived from this study are based on rigorous scientific methodology and entail high confidence.

Results showed that, for the operational scale and process described, 145 t kg h^{-1} pine wood was required to generate 185.1 MW of liquid FT hydrocarbon fuels. The biomass-to-liquid fuel efficiency was 51 %, calculated on a dry basis (16.95 MJ kg^{-1}). The CDR unit removed 95% of the CO_2 content from the syngas, allowing to capture 72% of the CO_2 generated during the whole biomass conversion process. The total electricity and heat consumption were 33.4 MWe and 140.5 MWth, respectively. The heat demand could be internally satisfied after performing heat integration due to the highly exothermic nature of biomass gasification and FT synthesis. External power supply was required since just 85 % of the power requirements could be co-generated at the CCGT. Thus, the produced fuel reported an energy intensity of 0.01 MJ/MJ liquid fuel, measuring two orders of magnitude below e-fuel production pathways.[6,7]

The well to wake LCA reported 21.6 g of fossil CO_2e per MJ of FT-IPK produced, which indicates a 74% reduction in fossil carbon emissions (Jet A1 scores 81.8g CO_2e MJ^{-1}), a figure that rivals with the 50-90% reduction range reported by synthetic e-fuel pathways.[6,7] The value obtained is 49% higher than the

CORSIA default lifecycle emissions value indicated for this route (8.3 g CO_2e MJ^{-1}).[8] Differences on process configurations and computing assumptions where identified, which greatly affects the reported score. On this basis, transparency in the emissions estimation is pivotal when claiming for emissions reductions, to certainly achieve the emission mitigation targeted with this framework. Conversely, present jet turbine engines cannot run on this fuel due to the lack of aromatics content and low density. The ASTM certified a 50:50 mass FT-IPK and Jet A1 blend to operate a commercial aircraft, thus limiting the impact from using SAF, which emits 51.5 fossil CO_2e MJ-1 fuel used by the aircraft. When accounting fossil and biogenic carbon dynamics, i.e. the net CO_2 carbon balance of the system, using the ASTM-certified blend result in a positive CO_2 flow to the atmosphere estimated in 30.5 g CO_2e MJ^{-1}. Adapting the engines to operate with 100% SAF would, conversely, result in net-negative emissions. A negative carbon flow of -20.0 g CO_2e MJ^{-1} from the atmosphere to a geological could be generated. This finding suggests that a better strategy would be adapting the engines to operate this type of fuel, instead of attempting synthetic fuels to replicate the features of the conventional fuel to operate the existing engines.

References

[1] ICAO. Environmental Report Chapter Six - Climate Change Mitigation CORSIA 2019. **2019**.

[2] UK Governemnt. Department for Transport. Jet Zero Consultation. 2021.

[3] Watson MJ, Machado P, da Silva AV, Rivera Y, Ribeiro C, Nascimento C, Dowling AW. Sustainable aviation fuel technologies, costs, emissions, policies, and markets: A critical review. Journal of Cleaner Production **2024;** 141472. https://doi.org/10.1016/j.jclepro.2024.

[4] ICAO. CORSIA Supporting Document. COSIA Eligible Fuels - Life Cycle Assessment Methodology. **2019**.

[5] The Royal Society. Net zero aviation fuels: resource requirements and environmental impacts. **2023**.

[6] Liu CM, Sandhu NK, McCoy ST, Bergerson JA. A life cycle assessment of greenhouse gas emissions from direct air capture and Fischer-Tropsch fuel production. Sustain Energy Fuels **2020**;4:3129–42. https://doi.org/10.1039/C9SE00479C.

[7] Rojas-Michaga MF, Michailos S, Cardozo E, Akram M, Hughes KJ, Ingham D, et al. Sustainable aviation fuel (SAF) production through power-to-liquid (PtL): a combined techno-economic and life cycle assessment. Energy Convers Manag **2023**; 292:117427. https://doi.org/10.1016/j.enconman.2023.117427.

[8] ICAO. CORSIA Default Life Cycle Emissions Values for CORSIA Eligible Fuels. **2021**.

[9] Almena A, Siu R, Chong K, Thornley P, Röder M. Reducing the environmental impact of international aviation through sustainable aviation fuel with integrated carbon capture and storage. Energy Conversion and Management **2024**;303:118186

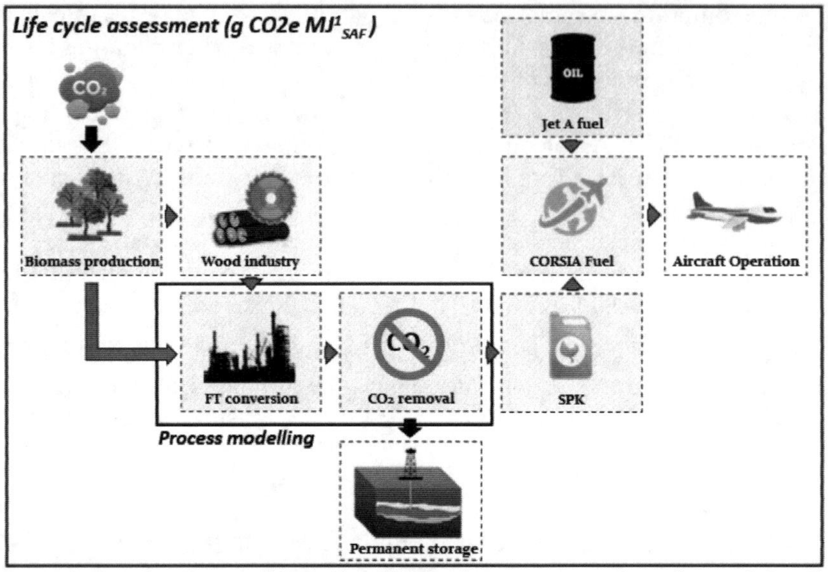

Figure 1. System boundaries for the well-to-wake life cycle assessment performed for the Fischer-Tropsch Hydroprocessed Synthesized Paraffinic Kerosene pathway including carbon capture and storage technology.[9]

501f Techno-Economic and Life Cycle Analysis of BioH_2 Production from Waste Streams through Process Integration

Arna Ganguly, Xinyu Liu, Hernan Delgado, Lili Sun, Pingping Sun, Amgad Elgowainy

Transportation sector and industrial sector are the two leading sectors in greenhouse gas emission. Low carbon hydrogen is of great interest to decarbonize both sectors by replacing fossil-based fuels and feedstocks. As hydrogen generated from electrolysis requires extensive electricity and fresh water, bioH2 generated from waste streams can provide particular advantage with lower electricity consumption. Moreover, contrary to electrolysis that consumes fresh water, bioH_2 production treats wastewater and generates fresh water. Relative to biomass, generating bioH_2 from waste streams can reduce the levelized production cost with revenues from wastewater treatment fee or solid waste tipping fee. This study explored the potential of bioH_2 production from industrial bio-waste streams namely, cheese whey (CW) and solid food waste (SFW) through dark fermentation (DF) and microbial electrolysis (MEC) integration. This integrated system also generates high purity CO_2 from both DF and MEC, respectively. Capture and storing the by-product CO_2 can lead to net CO_2 sequestration, resulting in a life cycle carbon intensity of around -67.2 g CO_{2e}/MJ H_2 produced, which makes the pathway potentially qualified under 45V tax credit. Techno-economic analysis results showed that CW can produce bioH_2 with cost ranging between \$3.5 to -\$2.2/ kg H_2 by considering 45V tax credit and revenues of wastewater treatment fee that varies with region. On the other hand, SFW as feedstock can lead to bioH2 cost of \$6.9/kg H_2, considering 45V credit and tipping fee.

232b Charting Development Pathways for Precision Fermentation through Agile System Design and Analyses

Sarang Bhagwat, Christopher V. Rao, Huimin Zhao, Vijay Singh, Jeremy Guest

To address climate change by mitigating greenhouse gas emissions, there is a pressing need to develop technologies that can facilitate replacing crude oil with alternative, renewable feedstocks to manufacture fuels and chemicals. Precision fermentation has emerged as promising for the sustainable manufacturing of biofuels and bioproducts. Developments in metabolic engineering of microbial candidates have enabled the biological production of commercially significant chemicals including succinic acid (e.g., by *Issatchenkia orientalis*), acrylic acid (via 3-hydroxypropionic acid, 3-HP; e.g., by *I. orientalis* or *Corynebacterium glutamicum*), and sorbic acid (via triacetic acid lactone, TAL; e.g., by *Yarrowia lipolytica*), among others. We leveraged BioSTEAM—an open-source platform—to design, simulate, and evaluate under uncertainty (via techno-economic analysis, TEA, and life cycle assessment, LCA) biorefineries producing these chemicals by fermentation of substrates obtained from renewable 1st-generation (e.g., sugarcane, sweet sorghum) and 2nd-generation feedstocks (e.g., corn stover, miscanthus). Further, we simulated and evaluated entire theoretical fermentation performance landscapes (e.g., formed by all potential combinations of fermentation yield, titer, and productivity under neutral and low-pH regimes) in each of these biorefineries, revealing fermentation development pathways to achieve system-wide sustainability targets. We showed recent fermentation developments could enable financially viable and environmentally beneficial production of bio-based succinic and acrylic acids, and identified targeted improvements in TAL production that could enable sustainable bio-based sorbic acid production. Finally, we showcase AutoSynthesis, a new open-source platform in Python to automate key aspects of process synthesis and design for biorefineries. We show how AutoSynthesis could be used to explore potentially synergistic developments across technologies and to enable end-to-end frameworks for computationally guided discovery of promising metabolic pathways.

For example, recent efforts to engineer *I. orientalis* strains to produce 3-HP at low pH can enable acrylic acid production via 3-HP with an estimated minimum product selling price (MPSP) of \$1.29–1.52$\cdot$kg^{-1} [5th-95th percentiles, hereafter in brackets], below the market price range of succinic acid in ~88% of Monte Carlo simulations. The biorefinery's estimated 100-year global warming potential (GWP_{100}) of 3.00 [2.53–3.38] kgCO_2-eq$\cdot$kg^{-1} and fossil energy consumption (FEC) of 39.9 [31.6–45.1] MJ$\cdot$kg^{-1} were both significantly below those of fossil-derived acrylic acid.

Further, *I. orientalis* strains have recently been engineered to produce succinic acid under low-pH conditions at pilot scale. We showed biorefineries leveraging these developments could produce succinic acid with an MPSP of 1.37$\cdot$kg^{-1} [\$1.23–1.54$\cdot$kg^{-1}], which was market-competitive in 100% of Monte Carlo simulations. The biorefinery's GWP_{100} of 1.67 kgCO_2-eq$\cdot$kg^{-1} (1.22–2.17 kgCO_2-eq$\cdot$kg^{-1}) and FEC of −0.21 MJ$\cdot$kg^{-1} (−7.08 to 6.47 MJ$\cdot$kg^{-1}) were also consistently below those of fossil-derived succinic acid.

To design biorefineries that could produce high-purity TAL using engineered *Y. lipolytica* strains, we experimentally characterized TAL solubility in water at various temperatures. We calibrated a solubility model to these experimental data and used it to design a process to separate TAL from fermentation broths through fractional crystallization. The biorefinery could produce TAL (>95.0 dry wt% purity) at an MPSP of \$6.13$\cdot$kg^{-1} [\$5.12–8.38$\cdot$kg^{-1}], comparable to the maximum viable TAL price range for sorbic acid production (approximately \$5.99–7.74·kg^{-1}), with net displacement of fossil energy consumption (FEC) in 87.7% of simulations by displacing marginal grid electricity. Advancements in fermentation TAL yield and titer could greatly enhance the biorefinery's financial viability (MPSP of 4.36·kg^{-1} [\$3.69–5.87·kg^{-1}], lower than the maximum viable price range in 96.1% of simulations) and environmental benefits (GWP_{100} of 3.02 [0.905–5.22] kg CO_2-eq·kg^{-1} and FEC of -21.8 [-51.7 to -3.26] MJ·kg^{-1}, with net displacement of FEC in 97.4% of simulations).

This work showcases how agile and robust system design and analyses can elucidate key drivers of system cost and environmental impacts across technological landscapes, chart roadmaps to navigate the opportunity space for precision fermentation, screen promising designs, navigate sustainability tradeoffs, and prioritize research, development, and deployment needs in biofuels and bioproducts development.

232d Evaluation of Alternative Processes for Grapeseed Oil Production from a Techno-Economic and Environmental Perspective

Rosa María Fonseca-Pérez, Alberto Ruiz, César Ramírez-Márquez, Adrián Bonilla-Petriciolet, Mariano Martin

Circular economy principles serve as the vehicle to establishing sustainable consumption and production patterns within the manufacturing sector. A more efficient use of the existing resources, including the revalorization of residues, facilitates the reduction of the environmental impact while fostering innovation and resilience in the process industry. As part of the food manufacturing sector, wine production is one of the largest contributors to solid waste generation, producing 0.2 kg solid residue per bottle of wine dispatched. This residue, comprising skins, seed, grape stalks, and wine lees, holds high intrinsic value and can be transformed into a wide range of high added value products.[1] The production of grapeseed oil for culinary and medicinal purposes using this residue has been investigated as a supplementary process for small and large wineries. Conventional oil extraction methods employing organic solvents like n-hexane offer simple recovery, low vaporization heat and high selectivity[2], but also involve negative human-health, safety, and environmental impacts.[3] To mitigate these issues, supercritical fluids, particularly CO_2, were suggested as safer and greener alternatives due to their recyclability, non-toxicity, and non-flammability features. Supercritical (SC) CO_2 extraction proved to prevent product degradation and yields highly pure oil with minimal solvent residue.[4] Furthermore, incorporation an organic co-solvent can improve the solubility and extraction efficiency of the targeted compounds. Ethanol, a green and safe compound widely used in food and drugs processing, has been proposed as a safe co-solvent to improve the yield of the SC-CO_2 extraction technology.[5]

This work systematically evaluated the performance of SC-CO_2 extraction, and combined SC-CO_2 and ethanol co-solvent extraction, which were compared with the conventional process using hexane for grapeseed oil recovery. Each solvent-oil system necessitated an appropriate prediction method to accurately replicate the thermodynamic equilibrium occurring during the extraction process, and thus estimate yield and optimal operating conditions. The assessment included conceptualization of the extraction process, process modeling and simulation, techno-economic analysis (TEA), and cradle-to-gate life cycle assessment (LCA). Aspen Plus modeled system performance and estimated costs, while SimaPro conducted LCA. These methods formed the basis for techno-economic and environmental assessment.

The three extraction systems were designed to process 15% of the seed residue generated in California, i.e. a production capacity of 900 kg/h. Results showed that hexane extraction technology can achieve the highest yield, recovering 130.6 g grapeseed oil per kg of grape seed entering the process. Using solely SC-CO_2 as extracting agent reduced by 12% the efficiency of the extraction (114.6 $g_{oil}\ kg_{seed}^{-1}$), while incorporating ethanol as co-solvent system in a 10% mass ratio could achieve a comparable yield to hexane extraction, reporting 129.8 $g_{oil}\ kg_{seed}^{-1}$. The advantage of the latter with respect to the conventional extraction is the absence of traces of toxic solvent. Oil produced via hexane extraction reported a residual hexane content of 500ppm, while CO_2 (45 ppm) and ethanol (0.3%) residual content is harmless to human health. The TEA performed to the hexane process estimated a total capital cost of $4.95 MM and a production cost of 6.31 USD per kg of extracted grape oil. The SC-CO_2+ethanol extraction system almost doubled capital cost ($9.72 MM) while the production cost was 36% superior (8.61 USD kg_{oil}^{-1}). However, the level of purity achieved with the SC-CO_2+ethanol system increases the selling price of the grapeseed oil,[6] so the profit of the former resulted approximately 3 to 4 times higher than the oil recovered using hexane. Finally, a cradle-to-gate LCA was performed to track both biogenic and fossil carbon dynamics. Different operating conditions of each extraction method identified resulted in a dissimilar environmental impact. Hexane extraction operated at atmospheric pressure and mild temperature conditions, thus reporting relatively low heat and power demands. Assuming biogenic carbon as climate neutral, hexane extraction generated 2.78 fossil $kgCO_2$-eq per kg oil produced. The use of supercritical CO_2 requires operating pressures above 150 bar, and elevated solvent to seed ratios. The greater solvent to grape seed mass ratio involved, together with the pressurisation and heating/refrigeration requirements caused the superior carbon emission score for the alternative processes. SC-CO_2 oil-extraction agent generated 7.35 kg CO_2-eq kg_{oil}^{-1}, while the CO_2-ethanol system reported the highest score estimated in 8.48 kg CO2-eq kg_{oil}^{-1}. However, when the electrification of utilities and renewable energy supply were assumed, the CO_2-ethanol scenario incurred the lowest GWP (i.e. 1.44 kg CO_2-eq kg_{oil}^{-1}) reducing by 25% the environmental impact of the conventional technology.

References

[1] Taifouris, M., El-Halwagi, M., Martín, M., **2023**. Evaluation of the economic, environmental, and social impact of the valorization of grape pomace from the wine industry. ACS Sustain. Chem. Eng. 11, 13718–13728. https://doi.org/10.1021/acssuschemeng.3c03615.

[2] Kumar, S.J., Prasad, S.R., Banerjee, R., Agarwal, D.K., Kulkarni, K.S., Ramesh, K.V., **2017**. Green solvents and technologies for oil extraction from oilseeds. Chem. Cent. J. 11, 1–7. https://doi.org/10.1186/s13065-017-0238-8.

[3] Lavenburg, V.M., Rosentrater, K.A., Jung, S., **2021**. Extraction methods of oils and phytochemicals from seeds and their environmental and economic impacts. Processes 9, 1839. https://doi.org/10.3390/pr9101839.

[4] Daud, N.M., Putra, N.R., Jamaludin, R., Norodin, N.S.M., Sarkawi, N.S., Hamzah, M.H.S., Nasir, H.M., Zaidel, D.N.A., Yunus, M.A.C., Salleh, L.M., **2022**. Valorisation of plant seed as natural bioactive compounds by various extraction methods: a review. Trends Food Sci. Technol. 119, 201–214. https://doi.org/10.1016/j.tifs.2021.12.010.

[5] Díaz-Reinoso, B., Moure, A., Domínguez, H., Parajó, J.C., **2006**. Supercritical CO2 extraction and purification of compounds with antioxidant activity. J. Agric. Food Chem. 54, 2441–2469. https://doi.org/10.1021/jf052858j.

[6] Duba, K., Fiori, L., **2019**. Supercritical CO2 extraction of grape seeds oil: scale-up and economic analysis. Int. J. Food Sci. Technol. 54, 1306–1312. https://doi.org/10.1111/ijfs.14104.

232e Integrated Techno-Economic and Life Cycle Analysis of Reductive Catalytic Fractionation and Base-Catalyzed Depolymerization Hybrid Processing for Lignin Valorization

Zahra Ebrahimpourboura, Aditya Ponukumati, Rhiannon Carr, Yu Gao, Zeyu Shang, Anirudh Krishnamurthy, Tae Seok Moon, Marcus Foston, Mark Mba Wright

The chemical depolymerization of lignin is characterized by its non-selective nature, yielding a diverse array of lignin breakdown products (LBPs). To tackle this challenge and focus on generating specific high-value chemical commodities, a novel approach has been developed. This study aims to validate the industrial relevance of reductive catalytic fractionation (RCF) and base-catalyzed depolymerization (BCD) hybrid processing. Utilizing a comprehensive approach, including techno-economic analysis (TEA) and life cycle assessment (LCA), the research evaluates the economic and environmental impacts of these processes. Employing the National Renewable Energy Laboratory (NREL) cellulosic ethanol production model, we assess the commercial viability of electrochemical lignin conversion through a TEA and LCA implemented in BioSTEAM and GREET® 2023, respectively, over a 20-year period. A Pd/C catalyst was used for reductive catalytic fractionation (RCF) of poplar biomass, and *Rhodococcus opacus* PD630 (R. opacus) was subsequently cultivated on the resulting LBPs. Results indicate that RCF decreases the ethanol selling price from $4.07 to $3.93/gallon, while BCD increases it to $4.18/gallon. Global warming potentials range from 29.21 to 30.26 for RCF and from 27.53 to 28.43 for BCD. Notably, the carbon-negative effects range from -0.38 to –0.85 grams of CO2eq/MJ of ethanol for RCF, compared to -0.02 to -0.3 for BCD. Increasing lipid production emerges as a key factor in enhancing the economics and environmental impact of the lignin-first biorefinery concept.

232f Techno-Economic and Life Cycle Analysis of Conversion of Natural Gas and Biomass to Hydrogen and Performance Carbons

Farhan Mashuk Haidary, Phuong Nguyen Thi, Eric Cochran, Jean-Philippe Tessonnier, Steven Crossley, Mark Mba Wright

The transition towards a sustainable energy future necessitates the development of technologies that can efficiently produce clean energy carriers and valuable by-products from renewable and abundant resources. Hydrogen presents a versatile alternative for heat and power generation in applications traditionally reliant on fossil hydrocarbons. Currently, steam methane reforming (SMR) dominates global hydrogen production, yielding 13.7 CO_2 (equiv.)/kg of net hydrogen, contributing to substantial greenhouse gas emissions [1]. SMR's dependence on finite, cost-effective natural gas and associated CO_2 release risks necessitate exploring alternatives. Methane pyrolysis emerges as a promising option, producing molecular hydrogen and solid carbon without CO_2 byproducts. Different catalysts such as nickel, molybdenum, and carbon have been investigated for pyrolysis at temperatures of 500-900°C. Still, the removal of carbon fibers from catalysts is challenging and, despite innovations in pyrolysis reactor designs, there has been little cost-effective management scheme identified [2]. While small-scale pyrolysis may yield additional revenue from carbon products, achieving significant global carbon emission reductions requires large-scale implementation with perpetual solid carbon storage. The cost of producing hydrogen can be reduced if the carbon produced from pyrolysis can be converted into high-performance carbon materials such as carbon nanotubes. These performance carbons possess unique properties, such as high strength, thermal and electrical conductivity, and adsorption capabilities, making them valuable for a wide range of applications such as in wastewater treatment and the fabrication of composite materials [3].

Another potential source of hydrogen production is biomass. Traditionally, biomass has been primarily viewed as a means to produce liquid biofuels, such as bioethanol and green diesel. However, the process of biomass gasification can produce a significant quantity of H_2, which then needs to be purified or separated from other gases produced during gasification [4]. Consequently, recent research has focused on assessing hydrogen production through co-feeding biomass with natural gas to identify technologies that show promise in terms of both yield and energy efficiency [5].

The objective of this study is to formulate a techno-economic analysis (TEA), comprising a process model, economic calculations, and sensitivity analysis, coupled with a Life Cycle Assessment (LCA) for the entire system. The design and research are based on research at Iowa State University and the University of Oklahoma. Utilizing a combination of data from experimental setups and commercial technology evaluations, we are developing a scalable process model using BioSTEAM that simulates the operational conditions optimal for maximizing hydrogen yield while ensuring the production of high-quality carbon nanomaterials. The design specifies a processing capacity of 2000 dry metric tons of corn stover and 1200 tonnes of natural gas per day and incorporates other sources for technologies that will ascertain the industrial-scale competitiveness of hydrogen against SMR. Subsequent analysis will assess the economic and environmental viability of incorporating catalytic pyrolysis into the system.

We conducted preliminary TEA using general models by Riley et al., 2021 [6] and Swanson et al., 2010 [7] for the pyrolysis of natural gas and gasification of biomass, respectively. We obtained a minimum fuel selling price of hydrogen at $1.63/kg through natural gas pyrolysis. The initial results suggest an immense potential for cost reduction, which motivates further research. Our findings aim to propel the dialogue on sustainable hydrogen production forward, emphasizing the dual benefits of energy carrier generation and value-added carbon material production as integral to achieving a carbon-neutral future.

Acknowledgements:

The authors gratefully acknowledge the funding for the project from the National Science Foundation (NSF) with the NSF Award number 2218070.

References:

- Abbas, H. F., Daud, W. W. (2010). 'Hydrogen production by methane decomposition: a review', *International Journal of Hydrogen Energy*, 35(3), 1160-1190.

- Parkinson, B. *et al.* (2017) 'Techno-economic analysis of methane pyrolysis in molten metals: Decarbonizing Natural Gas', *Chemical Engineering & Technology*, 40(6), pp. 1022–1030. doi:10.1002/ceat.201600414.

- Sánchez-Bastardo, N., Schlögl, R. and Ruland, H. (2021) 'Methane pyrolysis for zero-emission hydrogen production: A potential bridge technology from fossil fuels to a renewable and sustainable hydrogen economy', *Industrial & Engineering Chemistry Research*, 60(32), pp. 11855–11881. doi:10.1021/acs.iecr.1c01679.

- Tanksale, A., Beltramini, J.N., Lu, G.M. (2010) 'A review of catalytic hydrogen production processes from biomass', *Renewable and Sustainable Energy Reviews*, Elsevier, vol. 14(1), pages 166-182.

- Lalsare, A., Wang, Y., Li, Q., Sivri, A., Vukmanovich, R.J., Dumitrescu, C.E., Hu, J. (2019) 'Hydrogen-Rich Syngas Production through Synergistic Methane-Activated Catalytic Biomass Gasification', *ACS Sustainable Chemistry & Engineering*, 7(19), 16060-16071.

- Riley, J., Atallah, C., Siriwardane, R., Stevens, R. (2021) 'Technoeconomic analysis for hydrogen and carbon co-production via catalytic pyrolysis of methane', *International Journal of Hydrogen Energy*.

- Swanson, R. M., Platon, A., Satrio, J. A., & Brown, R. C. (2010). Techno-economic analysis of biomass-to-liquids production based on gasification. *Fuel*, *89*, S11-S19. https://doi.org/10.1016/j.fuel.2010.07.027

AUTHOR INDEX

Abarr, Miles ... 129
Abbasian, Javad ... 408
Abdel-Azeim, Safwat ... 310
Abdulla, Hussain ... 182
Abeyta, Dresean ... 147
Aboagye, Emmanuel ... 13, 21
Adaryan, Sarah ... 200
Adewuyi, Kasim ... 68
Agalliadou, Anna ... 165
Ahmad, Ali ... 14
Ahmadov, Rashad ... 402
Ahmed, Sheikh M. ... 148
Ahn, Byeongchan ... 269, 371
Aitbekova, Aisulu ... 146
Ajo-Franklin, Jonathan ... 342
Akhter, Sayma ... 423
Akulwar, Falguni ... 241
Akutsu, Ryota ... 395
Al Moinee, Abdullah ... 126
Alam, Muhammad N. ... 280
Albay, Alp ... 205
Al-Harbi, Alankaa ... 310
Ali, Mohammed A. M. ... 262
Aljoe, William ... 329
Allen, Joshua W. ... 28
Allgeier, Alan ... 450
Almarri, Mohammed J. ... 310
Al-Mohannadi, Dhabia ... 74, 434
Alomía, Victoria ... 441
Alvarez, Ulises G. ... 251
Alvarez-Pugliese, Christian E. ... 108
Ames, Brooke ... 399
Amini, Shahriar ... 7
Anderson, Robert ... 391
Anwar, Touhid B. ... 107
Arafat, Hassan ... 92
Arastoopour, Hamid ... 408
Arellano-Treviño, Martha A. ... 417, 428
Arora, Bhavna ... 361
Asadi, Mohammad ... 408
Askari, Zohreh ... 180
Atherton, Shelby ... 287
Atiyeh, Hasan ... 225
Attia, Fathy M. E. ... 130
Atwater, Harry ... 146
Aui, Alvina ... 215
AuYeung, Nick ... 139
Avendano, Marco ... 226
Avraamidou, Styliani ... 80
Azzolina, Nicholas ... 346
Babu, Lokesh V. ... 202
Badger, Nick ... 7
Badreldin, Ahmed ... 275
Baek, Kwang H. ... 178, 188, 190
Bahr, Mohamed ... 203
Bai, Baojun ... 31
Bai, Tao ... 356
Bai, Xianglan ... 265
Bakkaloglu, Semra ... 161, 249
Bakshi, Bhavik ... 17, 41, 49, 59, 61, 75, 78, 171, 380
Banat, Fawzi ... 103
Banerjee, Avishek ... 199
Barajas-Olalde, Cesar ... 185
Barbosa, Gabriel ... 104
Bardeen, Christopher J. ... 107
Barenthin, Torsten ... 9
Barkow, Milo ... 21
Barpaga, Dushyant ... 127
Barrett, William ... 106
Barton, Paul I. ... 207
Battistelli, Chiara L. ... 165
Baugh, Lisa S. ... 131
Bayon, Alicia ... 140, 302, 392
Beckett, Caitlin ... 395
Behabtu, Natnael ... 450
Benard, Andre ... 139
Benavides, Pahola T. ... 251
Benfenati, Emilio ... 165
Bennett, Quinn ... 304
Bergeron, Scott ... 9
Bever, Cordel ... 110
Bhagwat, Sarang ... 464
Bhartiya, Sharad ... 167
Bhat, Maanasa ... 191
Bhattacharjee, Tiasha ... 245
Bhosale, Rahul ... 423
Bhuvankar, Pramod ... 361
Biegler, Lorenz ... 132
Billings, Blake ... 203, 365, 430
Birkholzer, Jens ... 342, 349
Biswas, Prakash ... 426
Blake, Nathaniel ... 411
Blenner, Mark ... 224
Bobba, Pallavi ... 188
Bodger, Christopher I. ... 390
Bonilla-Petriciolet, Adrián ... 281, 466

Bonnesen, Peter V. 223
Bossa, Cecilia 165
Botte, Gerardine 108
Botterud, Audun 211
Bouman, Evert 165
Bourgé, Émilien 165
Bovy, Loic 280
Brambila, Carlos 71
Bresson, James 28
Brinker, Niklas 294
Briones, Paola A. M. 80
Brisson, Vanessa L. 315
Brito, Arleth G. 119
Bromberg, Leslie 375
Bronkema, Jason L. 28
Brooker, Paul 210
Brouwer-Milovanovic, Milena 165
Brown, Solomon F. 71
Brushett, Fikile 211
Bryan, Samuel 127
Buie, Cullen R. 315
Burgess, Ward A. 363
Bush, H. Evan 147
Cahill, John F. 223
Cai, Janny 28
Cai, Runxia 307
Calabro, David C. 131
Cameron, Caroline 111
Cano, Natalia 247
Cao, Meng 349
Cappa, Frederic 342
Cappello, Vincenzo 188, 190, 400
Carman, Carl 183
Carr, Rhiannon 468
Casey, John R. 315
Cavalcante, Celio 303
Cavaliere, Aurora 161, 249
Cetegen, Shaylin 207
Cha, Jaehyun 269
Chakraborty, Deboleena 440
Chan, Chin H. 216
Chang, Jaehee 412
Charlebois, Andrew 113
Chea, John D. 52, 106
Chen, Bing-Hung 429
Chen, Daniel 297
Chen, Jingjing 390
Chen, Qiaoyun 344
Chen, Yen-Chih 418
Chen, Yujuan 213
Chiang, Yadong 226
Chinn, Mari S. 225
Choe, Changgwon 312
Choi, Jounghwan 199
Choi, Soyoung 308
Choi, Yechan 373
Chong, Katie 460
Choquette, Riley 393
Chrisandina, Natasha 311
Christensen, John 361
Chung, Hoyong 114
Chung, Wonsuk 156
Ciardelli, Mark C. 346
Cihan, Abdullah 349
Cochran, Eric 83, 469
Coffin, Richard 182, 182
Cohn, Daniel R. 375
Conde, Laura R. 228
Conway, Matthew 21
Cook, Paul 342
Cordiner, Joan 148
Coronella, Chuck 110
Cortes-Pena, Yoel 316
Coventry, Joe 140
Cremaschi, Selen 254
Cross, Nicholas 125
Crossley, Steven 469
Da Silva, Aline V. 154
Damiani, Darin 329
Daniel, Stina 67
Darwish, Ahmad 103
Datta-Gupta, Akhil 216
Datye, Abhaya K. 147
De Haan, Matthew 365
De La Riva, Andrew T. 147
De Luna, Francisco M. T. 303
De Paolo, Donald 361
De Sousa, Marco 153, 401
Deel, Dawn 354
Delgado, Hernan 190, 463
Deliz-Quiñones, Katherine 105
Del-Mazo-Alvarado, Octavio 281
Demirhan, C. Doga 401
Deng, Lingyan 212
Deng, Sili 191
Diao, Jinjin 224
Dichtel, William 102
Dien, Bruce S. 316
Dileep, Dhananjay 83
Dileepkumar, Deepak 391
Dillon, Amanda 430
Ding, Ling 244
Dittmeyer, Roland 294
Diwekar, Urmila 27, 167

Dodd, Nathan 273
Doll, Paul 440
Domingos, Meire E. G. R. 319
Dominguez, Juan C. 257
Dong, Yachao 364
Donohoe, Bryon 244
Donohue, Timothy J. 292
Dos Santos, Lucas F. 162
Dosso, Cheick 132
Dou, Chang 244
Dowling, Alexander 154
Drochner, Alfons 285
Du, Jian 364
Dubey, Pallavi 267
Dunn, Jennifer 1, 23, 102, 455
Dyke, Kevin 260
Dziedzic, Owen J. 261
Ebrahimpourboura, Zahra 468
Eid, Sana 103
Eilts, Kristen 316
El Berch, John 402
Elfaki, Hesan 74
Elgowainy, Amgad 178, 188, 190, 400, 463
Elizalde-Solis, Octavio 231
Ellebracht, Nathan C. 215
Ellison, Christopher 70
English, Chad 440
Erguvan, Mustafa 7
Errington, Ethan 451
Esfahani, Milad 104
Esmaeili, Narjes 130
Esposito, Daniel 411
Essam, Ahmed 250
Etit, Deniz 451
Etzold, Bastian 285
Fan, Devin 260
Fan, Liang-Shih 241, 397
Fang, Jian 297
Fath, Brian D. 41
Fawcett, John 147
Featherman, Jarryd 261
Fee, Christopher 9
Fenton, James 210
Fereidoni, Sahar 291
Fereidooni, Leila 291
Fernandez, Antonio J. M. 53
Figueredo, Igor 303
Filho, Paulo R. C. F. R. 303
Finke, Cody 28
Finn, M. G. 131
Finnigan, Zahra 346
Firdaus, Nur H. W. 5
Flores, Betsie M. 153, 401
Fonseca-Pérez, Rosa M. 466
Foston, Marcus 267, 468
Fox, James 204
Franco-García, Laura 247
Frank-Collins, Joy 354
Fransen, Katharina 109
Freiberg, Lucas 139
Frimpong, Reynolds 379
Fu, Qiang 226
Fu, Xiao 401
Gager, Elizabeth 144
Galefete, Katlo 409
Gallardo, Yuliana P. 94, 441
Gamwo, Isaac 363
Gande, Vamsi V. 274
Ganguly, Arna 268, 463
Ganzer, Caroline 55, 367
Gao, Barnabas 13
Gao, Yu 468
Garcia, Maximiliano A. 292
García, Roberto C. 116
García-Morales, Ricardo 231
Ge, Congqin 323
Gershenfeld, Neil 9
Ghosh, Tapajyoti 3, 82
Gilliard-Abdulaziz, Kandis L. 416
Glass, Kay 246, 272
Glasser, David 204
Glubokovskikh, Stanislav 349
Gnangbe, Samirah 265
Godbille-Cardona, Gabrielle 109
Goldstein, Hannah M. 215
Gonzalez, Michael 52, 106
González-Arias, Sandro 231
Gosiamemang, Tsaone 451
Gotsis, George 9
Goyal, Amit 326
Goyal, Anjali 92
Gracida-Alvarez, Ulises R. 455
Gu, Shiya 364
Guerra, Inés H. 119
Guerrón-Orejuela, Edgar J. 119
Guest, Jeremy 316, 464
Guglielmi, Yves 342, 349
Guha, Sirshendu 193
Guillen-Gosalbez, Gonzalo 53, 162
Guleria, Richa 224
Gun, Sudeshna 241, 397
Gunukula, Sampath 242
Guo, Miao 451
Gupta, Neeraj 354

Gupta, Vijay 134
Gwinn, Rosa 102
Gypakis, Antonis 165
Hacioglu, Aycan 24
Hackett, Gregory 404
Haidary, Farhan M. 469
Haljasmaa, Igor 177
Halling, Maja 165
Hammack, Richard 175
Han, Fon Y. 263
Hancock, Jenny 201
Handler, Robert 431
Hangi, Morteza 140
Haq, Ahteshamul 38
Haque, Alamgir M. 108
Hart, Zachary 141
Harvey, Benjamin 147
Hasan, Shadi 92, 103
Hashimoto, Jayni 302, 392
Hawkins, Gary 48, 299
Hawkins, Jared 354
Hawkins, Troy 251, 455
Hayes, Michael 139
He, Q. Peter 46, 432
He, Zhirui 411
Heldebrant, David 127
Hellgardt, Klaus 227
Hellwig, Hubert 280
Heng, Jerry 451
Henry, Samuel 404
Heo, Seongmin 373
Hesketh, Robert 13, 21
Hicks, Andrea 44
Hildebrandt, Diane 204
Hill, Alexander 417, 428
Hill, Annabel 165
Hill, Caroline 144
Hillmyer, Marc A. 70
Hiromori, Kousuke 278
Holcomb, Paul 173
Holmes, Hannah 131
Hong, Wei-Hong 429
Hopkinson, David 132
Hopp, Chet 342
Hosseini, Erfan 254
Hosseini, Seyyed 341
Hsu, Hsi-Yen 418
Hu, Jianli 213
Hu, Jui-Chi 418
Huber, George W. 261
Humayun, Sufian 394
Hunt, John 185
Hussein, Ibnelwaleed 310
Iacovidou, Eleni 165
Iakovou, Eleftherios 187, 311
Iavicoli, Ivo 165
Ibrahim, Gasim 189, 191
Iglesias, Guillermo G. 116
Ijaz, Ahmad 408
Inoue, Shuhei 262
Intriago, Briggitte L. 119
Iseri, Funda 187, 311
Iseri, Halil 187, 311
Istrate, Ioan-Robert 53
Jadidi, Yasheen 455
Jaiswal, Priyank 338
Jakobs, Dion 162
Jaoude, Maguy A. 103
Jawdekar, Tanay 241, 397
Jeffrey, Paul 9
Jeong, Yeonseo 151
Jeong-Potter, Chae 417, 428
Jerpoth, Swapana S. 13
Jessop, Katy 134
Ji, Ruiyang 58
Jiang, Bojing 267
Jiang, Yuan 459
Jiang, Zhihua 254
Jiao, John 356
Jiménez, Araceli S. 165
Jin, Yong-Su 316
Jin, Zhehui 309, 362
Jo, Seongbin 416
Jones, Christopher 324, 409
Joo, Sangwook 276
Jordan, Preston 349
Joseph, Abigail 393
Joshi, Rushikesh 397
Jung, Pyeong-Gon 458
Juvekar, Vinay A. 167
Kabra, Suryansh 411
Kakodkar, Rahul 153, 401
Kanerva, Tomi 165
Kang, Xihui 244
Karagoz, Seckin 388
Karakaya, Canan 406
Karakitsios, Spyros 165
Karakoltzidis, Achilleas 165
Karaoglanoglou, Lazaros 435
Karimi, Iftekar 313
Karlen, Steven D. 292
Kärnman, Therese 165
Kasaeian, Alibakhsh 291
Kato, Makiko 278

Kawajiri, Yoshiaki 420
Kelley, Mark 354
Kelly, Kerry 365
Kenefake, Dustin 401
Kenney, David 113
Keum, Jong K. 223
Khan, Saad A. 307
Khanal, Aaditya 338, 348
Khanna, Vikas 38
Khatun, M. S. T. Sharifa 268
Khorasani, Reza 46, 432
Khosravani, Ali 209, 365
Khosravi, Mansour 180
Kida, Hajime 264
Kidder, Michelle 406
Kim, Beom-Sik 276
Kim, Bumkyu 292
Kim, Dong-Wook 412
Kim, Hyungseok 315
Kim, Jaesung 211
Kim, Jinsu 192
Kim, Jinwoo 412
Kim, Jun K. 412
Kim, Mingyu 373
Kim, Semie 458
Kim, Sunghoon 59
Kim, Yewon 293
Kim, Younghun 276
Kimbrel, Jeffrey A. 315
Klahn, Christoph 294
Klausner, James F. 139
Klimek, Alexander 367
Klinger, Jordan 245
Kodama, Mizuki 262
Kohnke, Colton 175
Koishybay, Aibolat 280
Kokossis, Antonios 435
Kolp, Andrew W. 297
Kong, Shik R. 322
Korose, Christopher 354
Kosari, Mohammadreza 307
Kouba, Stacy 346
Kovacevich, Justin 185
Kratzer, Madison 38
Kreider, Peter 140
Krishnamurthy, Anirudh 468
Krishnamurthy, Ramnarayan 24
Krynock, Michelle 404
Kuanyshev, Nurzhan 316
Kudva, Ishani K. 397
Kumar, Abhishek 127, 457
Kumar, Avan 17
Kumar, Sonu 241, 397
Kumawat, Piyush K. 203
Kurz, Bethany 185
Kusakabe, Katsuki 263
Kusuma, Victor A. 132
Kwon, Gihun 276
Kwon, Hyuk J. 412
Lackner, Klaus S. 427
Lagos, Andrés 94
Lai, Haoxiang 189, 212, 375
Lail, Marty 134
Lamb, H. Henry 395
Lamers, Patrick 3
Lan, Kai 253
Landázuri, Andrea C. 94, 119, 441
Landgren, Waaseyaaban 431
Lanyon, G. W. 342
Lao, Jianpei 226
Lastoskie, Christian 29
Lautze, Nicole 361
Lavallais, Chayse 23
Law, Jun W. 369
Le Vinh, Khanh 450
Leandri, Hugo 28
Leban, Martin 47
Lecaro, María F. 94, 441
Lee, Eunseo 224
Lee, Hakyung 224
Lee, Hung L. 280
Lee, Jiwoo 322
Lee, Jong-Min 412
Lee, Kyuha 188
Lee, Sanghun 293
Lee, Seungwon 269
Lee, Ung 156
Legako, Jerrad 5
Leggieri, Rosella T. 165
Lei, Fuqiong 139
Leisen, Johannes 131
Lemaoui, Tarek 92, 103
Lenges, Christian 450
León, Marco 94, 441
Lesnjak, Claire 261
Leso, Veruscka 165
Lessin, Benjamin 395
Lettieri, Paola 161, 249
Li, Fanxing 307, 395
Li, Lifeng 140
Li, Like 139
Li, Shuyun 459
Li, Wenqin 125, 215
Li, Wenzhen 283

Li, Yalin ... 14
Li, Yaoguo ... 175
Li, Ying ... 275
Liang, Chloe ... 455
Lidor, Alon ... 141
Lidstrom, Mary E. ... 431
Lim, Hankwon ... 312
Lim, Young-Il ... 458
Lin, Bosong ... 189, 212, 375
Lin, Haiqing ... 130
Lin, Sidney ... 84
Lin, Zhexi ... 411
Lindén, Jenny ... 165
Lines, Amanda ... 127
Lipiński, Wojciech ... 140, 390
Liu, Eric ... 28
Liu, Guoliang ... 58
Liu, Guoxiang ... 173
Liu, Junchen ... 395, 399
Liu, Kunlei ... 379
Liu, Xinyu ... 463
Liu, Y. ... 64
Lively, Ryan ... 131, 306
Lofstedt, Magnus ... 165
Longo, Jared ... 21
Lopez, Gary ... 243
Lou, Helen ... 297
Lowe, Adrian ... 390
Lu, Lu ... 306
Lu, Xunfeng ... 361
Lucci, Michael A. ... 260
Ma, Jinliang ... 424
Machado, Pedro G. ... 154
Mahadevan, Seshasayee ... 440
Mahtab, Tanveer ... 394
Mainka, Hendrik ... 48, 299
Maitra, Shraddha ... 316
Mali, Aniket ... 49, 171
Mallapragada, Dharik ... 369
Malmali, Mahdi ... 394
Mamrol, Natalie ... 109
Manjarekar, Siddhi ... 27
Manjur, S. Moumita ... 84
Maravelias, Christos ... 158, 327, 386
Maréchal, François ... 319
Martin, Mariano ... 116, 149, 281, 460, 466
Martin, William ... 395
Martinez, Jeimy J. B. ... 231
Martinez, Kenneth ... 39
Matamoros, Glenn ... 242
Materazzi, Massimiliano ... 161, 249
Matsumura, Yukihiko ... 262
Matthews, Leanne ... 140
Mayali, Xavier ... 315
McBride, Kyle ... 185
McCool, Ben ... 192
McCormick, Alon ... 394
McDonald, Molly ... 386
McGrath, Travis C. ... 346
McLaughlin, Jonathan ... 356
McNeary, Wilson ... 417, 428
McNeeley, Adam ... 64
McNelis, Michael ... 129
Meehan, D. Nathan ... 378
Mejic, Dragan ... 391
Mendez, Alonzo ... 302, 392
Mendizábal, Erick ... 441
Meramo, Samir ... 6
Mercangöz, Mehmet ... 205
Meyer, Aye ... 406
Meyer, David E. ... 106
Miller, James ... 261
Milton, Robert ... 71
Min, Kyoungseon ... 269
Minor, Ann-Joelle ... 55
Mirabá, Pierina ... 441
Mirjam, Röder ... 460
Mirshekari, Reza ... 411
Mito, Saeko ... 361
Mitsutake, Yuki ... 263
Miyai, Yuma ... 280
Mobley, Paul ... 134
Mokhtarnejad, Mahshid ... 48, 299
Molley, Ashten ... 399
Monbaliu, Jean-Christophe ... 280
Moni, Sheikh ... 404
Monroe, Jordan ... 302, 392
Moon, Tae S. ... 224, 468
Morales-Guio, Carlos ... 199
Mosalpuri, Manish ... 219, 283
Mostafayi, SeyedSepehr ... 408
Moustafa, Nadin ... 383
Mpourmpakis, Giannis ... 402
Muhich, Christopher ... 302, 389, 392
Munguía-López, Aurora D. C. ... 80
Munyaneza, Nuwayo E. ... 58
Mutlu, Mete ... 401
Myerson, Allan ... 280
Nabera, Abhinandan ... 53
Nagao, Masahiro ... 216
Nair, Sankar ... 226
Nandimandalam, Hariteja ... 272
Nascimento, Claudio A. O. D. ... 154
Nashef, Enas ... 92, 103

Nathan, Aditya 29
Nazemi, Farshid 41, 75
Nelson, Abraham 328
Neta, Marliete M. 303
Newman, Alex 11
Ng, Clarence 190
Nguyen, Huy 300
Nicholas, Christopher 112
Niebieskikwiat, Darío 94, 441
Nikiforou, Fotini 165
Niknezhad, Shayan 153, 401
Nikolic, Heather 379
Nino, Juan C. 144
Nishikawa, Yuta 420
Noguchi, Takashi 262
Noguera, Daniel R. 292
Noori, M. Tabish 227
Nowack, Bernd 165
Nussbaum, Christophe 342
Nwaogaraku, Nkechi M. 288
O'Connell, Margaret 102
Okeke, Ikenna 406
Okwen, Roland 180
Olsen, Bradley 109
Omell, Benjamin P. 424
O'Neill, Eric 158
Oqbi, Manar 434
Orejuela-Escobar, Lourdes M. 119
Orjuela, Alvaro 228
Orrego, Daniel F. 319
Ortiz-Ulloa, Juvenal 139
Otabir, Prince 348
Oyedeji, Oluwafemi 245
Ozalp, Nesrin 139
Pajntar, Anja 47
Pala, Jasneet 104
Panagakos, Grigorios 132
Pandey, Akansha 426
Pang, Simon H. 215
Pang, Xueqi 411
Panja, Palash 203
Papathanasiou, Maria M. 72
Pappas, Iosif 401
Parameswaran, Prathap 38
Park, Ah-Hyung A. 308, 325
Park, Hai W. 276
Park, Joungho 269
Park, Matthew 389
Park, Youngjune 322
Parker, Chris 430
Parks, James E. 245
Paroushi, Maryam S. 31
Passuto, Eleonora 6
Patel, Dhruv 260
Patricia, Thornley 460
Patwardhan, Siddharth 71, 201
Paulillo, Andrea 161, 249
Pazik, John 21
Pedrozo, Hector 132
Pérez-Ramírez, Javier 53
Petrasch, Joerg 139
Petrunia, Ievgenii 9
Pfau-Cloud, Michaela 70
Phillips, Benjamin 393
Pi, Xiaoyue 400
Pistikopoulos, Efstratios 153, 187, 311, 401
Ponnusamy, Ettigounder 51
Ponukumati, Aditya 468
Porosoff, Marc 402
Port, Jennifer 251
Posada, Carlos 58
Posso, Juan F. 441
Pottas, Roelof 140
Powell, Kody 203, 209, 365, 430
Prasomsri, Teerawit 399
Prather, Kristala 109
Prieto, Carlos 281
Proano, Laura 409
Prócel, Luis M. 94, 441
Puhar, Jan 47
Pulselli, Federico 75
Pyrgakis, Konstantinos 435
Qian, Xianghong 230
Qiao, Rui 58
Qualls, Anna 292
Quillinan, Scott 356
Radhakrishnan, Harish 265
Rahbar, Nima 113
Rahbari, Alireza 140
Rahman, Mohammad R. 227
Rahman, Shahriar 394
Rahmanian, Vahid 307
Rajan, Kalavathy 5
Ramírez, Cesar 281
Ramirez-Jimenez, Edgar 231
Ramírez-Márquez, César 466
Ramsey, Owen 139
Randhir, Kelvin 139
Ranjan, Durgesh 144
Rankin, Scott A. 261
Rao, Christopher V. 464
Razavi, Seyedamin 307
Reagan, Matthew 349
Realff, Matthew 192, 226

Redij, Mrinal M. ... 152
Reffet, Mathilde ... 441
Ren, Yi ... 131
Rentschler, Philipp ... 294
Resch, Susanne ... 165
Reynolds, Michael ... 365
Reza, Toufiq ... 118, 259
Rezaei, Fateme ... 126
Rhee, Chang H. ... 276
Riaz, Asim ... 390
Ribeiro, Celma O. ... 154
Richa, Kirti ... 251
Richards, Trevor ... 185
Rihko-Struckmann, Liisa ... 55
Riley, Christopher ... 147
Rim, Guanhe ... 409
Rivadeneira, Camily ... 441
Roberson, Luke ... 224
Robertson, Michelle ... 342
Robinson, Brandon ... 213
Roe, Benjamin ... 71
Roeb, Martin ... 136
Roldán-San Antonio, Jose E. ... 149
Rolison, Kristina A. ... 315
Romañach, Rodolfo ... 280
Romanov, Vyacheslav ... 177
Romiluyi, Oyinkansola ... 411
Rothman, Rachael ... 273
Roufogali, Nefeli ... 435
Roy, Ankana ... 131
Ruan, Chongyan ... 395
Ruddy, Daniel ... 417, 428
Ruiz, Alberto ... 460, 466
Ruiz-Mercado, Gerardo ... 106
Rukh, Mahe ... 307
Rutqvist, Jonny ... 349
Rydberg, Tomas ... 165
Saberi, Shadi ... 28
Safaei, Elaheh ... 365
Safikhani, Amir ... 394
Saha, Nepu ... 245
Saha, Swarna ... 118, 259
Saive, Rebecca ... 247
Sakai, Takuto ... 216
Salazar, Matthew ... 146
Sale, Kenneth ... 244
Saleh, Mohammed Ali ... 310
Sánchez, Antonio ... 281
Sarigiannis, Dimosthenis ... 165
Sarkis, Miriam ... 72
Sarmah, Anubhav ... 70
Sartape, Rohan ... 274
Sasser, Thomas ... 427
Sawant, Mahesh ... 440
Schefer, Senecio ... 342
Scheffe, Jonathan ... 144
Schimmels, Philipp ... 139
Scholz, Samuel ... 292
Schulte, Lisa ... 152, 257
Selvestrel, Gianluca ... 165
Sen, Amrita ... 17, 61, 75
Sener, Canan ... 292
Seo, Dongju ... 322
Shadoan, Tanner ... 342
Shah, Harsh B. ... 401
Shah, Nilay ... 72, 161, 249, 383
Shahnam, Mehrdad ... 245
Shamshina, Julia ... 5
Shang, Zeyu ... 468
Sharma, Anezka ... 165
Sharma, Niket ... 64
Sharma, Reena ... 49
Sharma, Vedanti ... 83
Sharma, Victor ... 223
Shi, Jian ... 243
Shi, Rui ... 8, 246, 260, 272, 298, 300
Shi, Xiaolei ... 261
Shibasaki-Kitakawa, Naomi ... 278
Shigaki, Nobuyuki ... 420
Shih, Chung ... 173
Shin, Woojae ... 375
Shinde, Shekhar ... 397
Shonnard, David ... 431
Siivola, Kirsi ... 165
Sim, Gyudae ... 322
Simakov, David ... 414
Simon, Terrence W. ... 140
Singh, Meenesh ... 274
Singh, Suraj P. ... 313
Singh, Vijay ... 316, 464
Siriwardane, Hema ... 173
Sjolund, Adam ... 427
Slater, C. Stewart ... 13
Smerigan, Adam ... 8, 298
Sminchak, Joel ... 354
Smith, Alexandra ... 224
Smith, Raymond ... 25, 106
Solarte, Steven ... 228
Son, Yoon J. ... 411
Song, Liaosha ... 352
Soom, Florian ... 342
Soong, Yee ... 177
Sosa, Ana G. A. ... 119
Soto, Lisciele ... 55

Spoerke, Erik ... 147
Sporchia, Fabio ... 75
Srinivasan, Rajagopalan ... 313, 376
Sripada, Sarada ... 292
Stacey, Neil ... 204
Stacey, Peter ... 204
Stamoulis, Petros ... 102
Stathatou, Patricia ... 9
Stechel, Ellen B. ... 389
Stelzer, Torsten ... 280
Stephanopoulos, George ... 61, 75
Stickel, Jonathan ... 26, 129
Stuart, Rhona K. ... 315
Su, Magel ... 146
Subrahmanyam, Nistala V. ... 167
Subramanian, Vrishali ... 165
Sukumara, Sumesh ... 6
Sun, Alex ... 173
Sun, Lili ... 463
Sun, Ning ... 244
Sun, Pingping ... 178, 188, 190, 400, 463
Sundmacher, Kai ... 55, 221, 367
Sunny, Ashin ... 397
Suresh, Saatvi ... 401
Svitnic, Tibor ... 221, 367
Swaminathan, Venkataramanan ... 376
Taft, Kendall ... 359
Taheri, Mahdiar ... 140
Tajipour, Melika ... 267
Tamburro, Olivia ... 302, 392
Tan, Eric ... 258
Tang, Yinjie ... 224
Tanner, Ralph S. ... 225
Tanoue, Kenichiro ... 264
Tanthana, Jak ... 134
Tanveer, Sheik ... 178
Tapriyal, Deepak ... 363
Taylor, Liam ... 391
Teixeira, Andrew R. ... 113
Tessonnier, Jean-Philippe ... 469
Thakker, Vyom ... 61
Thi, Phuong N. ... 469
Thomas, Donald ... 361
Thompson, Lisa A. ... 427
Thompson, Vicki ... 245
Thunuguntla, Rahul ... 225
Tian, Xu Y. ... 31
Tian, Yuxin ... 224
Timko, Michael ... 113
To, Anh ... 417, 428
Toor, Fatima ... 219
Tran, Justin ... 391
Tran, Phuc ... 158
Tren, Thien ... 132
Triantafyllou, Michael ... 9
Tribaldos, Verónica R. ... 342
Trinh, Thanh H. ... 263
Trost, Jenna ... 1
Trusler, Martin ... 344
Tse, Po K. ... 127
Tsezos, Marios ... 9
Tsuzuki, Takuya ... 390
Turner, Christoffer ... 104
Turner, Kevin ... 305
Twi-Yeboah, Nigel ... 450
Uemura, Yoshimitsu ... 263
Urgun-Demirtas, Meltem ... 67
Vaiana, Christopher A. ... 315
Valarezo, Alfredo ... 119
Valdez, Peter ... 457
Van Bodegraven, Martijn ... 165
Van Dijk, Joanke ... 165
Vasilyeva, Maria ... 182
Vedant, Shivam ... 311
Velazquez, Mileisha ... 105
Verduzco, Rafael ... 200
Vlaar, Cornelis ... 280
Von Jouanne, Annette ... 393
Vujanović, Annamaria ... 47
Vyawahare, Pradeep ... 400
Wade, Jennifer ... 427
Wadgama, Muhammad H. ... 246, 260
Waidmann, Guido ... 285
Wakita, Akihide ... 420
Wang, Bo ... 140
Wang, Chi-Hwa ... 140
Wang, Haotian ... 200
Wang, Jin ... 46, 432
Wang, Maojian ... 424
Wang, Sai ... 331
Wang, Shuai ... 113
Wang, Wenjia ... 261
Wang, Yangyang ... 223
Wang, Yifan ... 210
Wang, Zhichang ... 341
Wang, Zhiwu ... 268
Wangikar, Pramod ... 202
Warren, Kent J. ... 391
Watson, Madelynn ... 154
Weber, Peter K. ... 315
Weimer, Alan W. ... 391
Weston, Simon C. ... 131
Westra, Jaco ... 165
Wheeler, Clayton ... 242

Wickramasinghe, Ranil 230
Widjaja, Andreas 285
Wild, Katherine 257
Williams, C. Luke 242
Williams, Matthew 430
Williams-Stroud, Sherilyn 359
Wilson, Steven 389
Wittek, Marvin 285
Won, Wangyun 269, 371
Wong, Andrew 407
Wood, Todd 342
Wrenn, Olivia A. 427
Wright, Mark M. 83, 152, 219, 257, 265, 267, 268, 283, 288, 468, 469
Wu, Haoran 67
Wu, Jeffrey 418
Wu, Jiyue 331
Wu, Shunmei 323
Xia, Yidong 245
Xie, Minjunshi 309, 362
Xu, Weiwei 422
Xu, Yupeng 245
Xue, Ying 78
Xue, Ziqiu 361
Yajima, Tomoyuki 420
Yang, Kunran 395
Yao, Yuan 115, 253
Yeh, Daniel H. 224
Yenkie, Kirti 13, 21
Yokochi, Alexandre 393
Yokogawa, Yoshinobu 264
Yoshikawa, Kohei 420
Yossri, Widad 129
Young, Jeffrey 129
Young, Matthew 354
Yu, Ying 356
Yuan, Mengyao 125
Yuan, Yanhui 192
Yuan, Yue 223
Yuan, Zhihong 323
Yue, Chudi 199
Yue, Lindsey 140
Zamarripa, Miguel A. 424
Zang, Guiyan 189, 191, 212, 375
Zapata, José 140
Zappi, Alex 109
Zavala, Victor M. 80
Zeng, Yining 244
Zervoudakis, Aristotle 70
Zeszutek, Grant 260
Zhang, Bingquan 253
Zhang, Fuzhong 267
Zhang, Gengyi 130
Zhang, Mengli 262
Zhang, Qi 151
Zhang, Shuo 361
Zhao, Dongwei 211
Zhao, Huimin 464
Zhen, Xinyu 380
Zheng, Ziye 165
Zhu, Lingxiang 132
Zhu, Tianjie 280
Zhu, Zhexi 230
Zhuang, Yichen 414
Zlopasa, Jure 450